Suren N. Dwivedi
Alok K. Verma
John E. Sneckenberger
Editors

CAD/CAM Robotics and Factories of the Future '90

Volume 1: Concurrent Engineering

5th International Conference on CAD/CAM, Robotics, and Factories of the Future (CARS and FOF'90) Proceedings

International Society for Productivity Enhancement

With 192 Figures

Springer-Verlag
Berlin Heidelberg New York London
Paris Tokyo Hong Kong Barcelona

Suren N. Dwivedi
Department of Mechanical and
  Aerospace Engineering
West Virginia University
Morgantown, WV 26506-6101
USA

Alok K. Verma
Chairman
Department of Mechanical Engineering Technology
Old Dominion University
Norfolk, VA 23508
USA

John E. Sneckenberger
Concurrent Engineering Research Center
West Virginia University
Morgantown, WV 26506
USA

Library of Congress Cataloging-in-Publication Data
International Conference on CAD/CAM, Robotics, and Factories of the
  Future (5th: 1990: Norfolk, VA.)
    CAD/CAM, robotics, and factories of the future: 5th International
  Conference on CAD/CAM, Robotics, and Factories of the Future (CARS
  and FOF'90) proceedings /S.N. Dwivedi, Alok Verma, John
  Sneckenberger.
      p. cm.
    "International Society for Productivity Enhancement."
    Includes bibliographical references.
    Contents: v. 1. Concurrent engineering—v. 2. Flexible
  automation.
      ISBN 3-540-53398-2(set).—ISBN 0-387-53398-2(set).—ISBN
  3-540-53399-0 (Berlin Heidelberg New York:v. 1).—ISBN
  0-387-53399-0 (New York Berlin Heidelberg:v. 1).—ISBN
  3-540-53400-8 (Berlin Heidelberg New York:v. 2).—ISBN
  0-387-53400-8 (New York Berlin Heidelberg:v. 2)
      1. CAD/CAM systems—Congresses. 2. Flexible manufacturing
  systems—Congresses. 3. Robotics—Congresses. 4. Manufacturing
  processes—Automation—Congresses. I. Dwivedi, Suren N.
  II. Verma, Alok. III. Sneckenberger, John. IV. International
  Society for Productivity Enhancement. V. Title.
  TS155.6.I5818 1990
  670.42'7—dc20   90-19478

Printed on acid-free paper.

©1991 by Springer-Verlag Berlin Heidelberg, except for pp. 563–585—Copyright 1990 Ford Motor
Company.
This work is subject to copyright. All rights are reserved, whether the whole or part of the material is
concerned, specifically the rights of translation, reprinting, reuse of illustrations, recitation, broadcasting, reproduction on microfilms or in other ways, and storage in data banks. Duplication of this
publication or parts thereof is only permitted under the provisions of the German Copyright Law of
September 9, 1965, in its version of June 24, 1985, and a copyright fee must always be paid. Violations
fall under the prosecution act of the German Copyright Law.
The use of general descriptive names, trade names, trademarks, etc., in this publication, even if the
former are not especially identified, is not to be taken as a sign that such names, as understood by the
Trade Marks and Merchandise Act, may accordingly be used freely by anyone.

Camera-ready copy prepared by the editors.
Printed and bound by Edwards Brothers, Inc., Ann Arbor, MI.
Printed in the United States of America.

9 8 7 6 5 4 3 2 1

ISBN 3-540-53399-0 Springer-Verlag Berlin Heidelberg New York
ISBN 0-387-53399-0 Springer-Verlag New York Berlin Heidelberg
ISBN 3-540-53398-2 Two-volume set
ISBN 0-387-53398-2 Two-volume set

Conference Objective

The last decade has seen the emergence of a unified approach for product design which attempts to combine traditionally distinct tasks like design, management, marketing, analysis, manufacture and materials. Often called "Concurrent Engineering" or "Simultaneous Engineering", this new philosophy aims at improving cost competitiveness by reducing waste of time, money, and other resources inherent in the iterative traditional methods. In view of the importance of this new philosophy, Concurrent Engineering is selected as the theme for this conference.

The main objective of the conference is to bring together researchers and practitioners from government, industries and academia interested in the multi-disciplinary and inter-organizational productivity aspects of advanced manufacturing systems utilizing CAD/CAM, CAE, CIM, Parametric Technology, AI, Robotics, AGV Technology, etc.

Conference Organization

Sponsors

International Society for Productivity Enhancement (ISPE), USA

Center for Innovative Technology, Virginia, USA

Old Dominion University, Norfolk, Virginia, USA

Concurrent Engineering Research Center (CERC)
West Virginia University, West Virginia, USA

Department of Mechanical and Aerospace Engineering,
West Virginia University, West Virginia, USA

Committee Chairpersons

Conference General Chairperson:
Alok K. Verma, Old Dominion University, USA.

Conference co-chairs:
Suren N. Dwivedi, West Virginia University, USA.
Gary Crossman, Old Dominion University, USA.

Program Chairpersons:
John Sneckenberger, West Virginia University, USA.

Technical Chairperson:
Virendra Kumar, General Electric, USA.

International Chairperson:
Jean Marie Proth, INRIA, France.

Reception Chairperson:
John. M. Jeffords, Old Dominion University, USA.

Workshop Chairpersons:
Stewart Shen, Old Dominion University, USA.
Bharat Thacker, Universal Computer Services, USA.
Sumitra Reddy, CERC, West Virginia University, USA.
Hal Schall, Ford Motor Co., Dearborn, USA.

Printing and Publication Chairperson:
Sacharia Albin, Old Dominion University, USA.

Abstract and Paper Review Chairperson:
Jean Hou, Old Dominion University, USA.
Resit Unal, Old Dominion University, USA.

University Chairperson:
Robert Ash, Old Dominion University, USA.

Industrial Chairpersons:
Gary Crossman, Old Dominion University, USA.
Larry Richards, Old Dominion University, USA.

Plenary Chairperson:
Suren N. Dwivedi, West Virginia University, USA.

Exhibit Chairperson:
Thomas Houlihan, Jonathan Corporation, USA.

Student Chairpersons:
Drew Landman, Old Dominion University, USA.
Francis M. Williams, Old Dominion University, USA.

International Coordinators

H. Bera (U.K.)
M. Dominguez (Spain)
K. Ghosh (Canada)
V.M. Ponomaryov (USSR)
J.M. Proth (France)
R. Sagar (India)
T.-P. Wang (Taiwan)
T. Yamashita (Japan)

Committee Rosters

Abstract and Paper Review Committee

Dr. Ralph Wood
Dr. John Spears
Dr. Cheng Y. Lin
Dr. Duc Nguyen
Dr. Nageswara Rao

Program Committee

Donald W. Lyons
John Sneckenberger
Suren N. Dwivedi
Sumitra Reddy
Larry Banta
John Hackworth
Bob Creese
B. Gopalkrishnan
Waeik Iskander
Bruce Kang
Ken Means
Jacky Prucz
Nithi Sivaneri
Emil Steinhardt

Editorial Board and Publication Committee

Donald W. Lyons
Ralph Wood
John Spears
William Bentley
Zenon Kulpa
Michael Sobolewski
Sati Maharaj
Sisir Padhy
Bin Du
Prashanth Murthy
Deepak Kohli
Dandamudi Venugopal
Dhananjay Salunke

Industrial Committee

Thomas Houlihan
Moustafa R. Moustafa
Ed Wilson
Jim Fox
Larry Wilson

Reception Committee

Taj Mohieldin
Linda Vahala
Nancy Short

Professional Relations Committee

John Jurewicz
Donald W. Lyons
Ramana Reddy
Ralph Wood
Biren Prasad
Kumar Singh
John Spears

Conference Coordinators

William Bentley
Nancy Short
Georgette Ingram

Conference Staff

Sati Maharaj
Robin Johnson
Joette Claiborne
Marylin Host
Indira Dwivedi
Iva Dwivedi
Fern Wood
Vicki Grim
Pat Logar
Jean Shellito

Letter from the President, ISPE

The International Society for Productivity Enhancement (ISPE) is entering its seventh year. The Conference you are attending is our fifth of the international series on CAD/CAM, Robotics and Factories of the Future (CARS & FOF). The fourth conference was held at the Indian Institute of Technology, New Delhi, India in 1989. During the past seven years, we have expanded our activities significantly. The membership interest and international participation are also growing. During the past year alone, the Society has made tremendous progress in the following major frontiers:

JOURNAL: The Society now has its own journal entitled The International *Journal of Systems Automation Research and Applications* (SARA), an international, multidisciplinary research and applications-oriented journal to promote a better understanding of systems considerations in interdisciplinary automation using computers. The Journal contains important reading for design, engineering, and manufacturing persons as well as those with interest in research and development and applications of productivity tools, concepts and strategies to multidisciplinary systems environments. The Journal will only publish original, quality papers. To receive more information about this Journal, write to: Editor-in-Chief, ISPE, SARA Journal Department, P.O. Box 731, Bloomfield Hills, Michigan 48303-0731.

PROCEEDINGS: Starting this year (with the Fifth Conference), the Society is now making the Conference Proceedings available at the Conference. Selected papers from this Proceedings will also be considered for publication in *SARA*.

CONFERENCES: ISPE's annual conferences are now book until 1994. The Sixth International Conference will take place at South Bank Polytechnic, London from August 19-21, 1991. The Seventh and Eight International Conferences will be held in Leningrad, USSR and France, respectively.

COOPERATIVE PROGRAMS: In 1989, ISPE started a new cooperative program called the Indo-U.S. Forum for Cooperative Research and Technology Transfer (IFCRTT) in cooperation with West Virginia University and the National Science Foundation (NSF). The first joint meeting of the IFCRTT was held from December 17-18, 1989, in New Delhi, India. The meeting attracted a large body of scholars from industry, universities, and research institutions from both the United States and India. Similar cooperative programs are being arranged in the U.K., U.S.S.R. and France.

As you can see, we have made great strides, but significant changes are taking place in the manufacturing sectors due to global competitiveness and economic factors. Productivity enhancement needs are even larger than before, and such needs require us to be more dynamic and resourceful. ISPE is looking for a few good people to take leadership positions in its organization and committees for sponsored events. If you would like to help us build our technical program or if you would like to work on ideas of your own, please write to us. There are openings in the following areas:

* *SARA* Journal - Readers' Committee
* Productivity Directors
* Workshop and Tutorial Organizers
* CARS & FOF Conferences: University, Industry, International Representatives, Session Organizers, and Technical and Program Chairpersons.

We are still a very young organization and your leadership can play a significant role. Please do not hesitate to write us with your ideas and opinions.

Biren Prasad, Ph.D.
ISPE, P.O. Box 731, Bloomfield Hills, MI 48303-0731, USA.

Acknowledgments

The Fifth International Conference on CAD/CAM, Robotics, and Factories of the Future (Cars & FOF '90) was hosted by the College of Engineering and Technology at Old Dominion University and was endorsed by more than ten societies, associations, and international organizations. The conference was held in Norfolk, Virginia at the Omni International Hotel from December 2-5, 1990. Over 200 presentations organized into 40 specialty sessions, three plenary sessions, and eight workshops were conducted during the four days. Authors, plenary session speakers, and participants from 17 different countries around the world converged in Norfolk for this Conference. In view of the ever-increasing importance for integrating different facets of manufacturing with design process, the organizing committee selected "Concurrent Engineering" as the theme of the Conference.

I wish to acknowledge, with many thanks, the contributions of all the authors who presented their work at the Conference and submitted the manuscripts for publication. It is also my pleasure to acknowledge the role of banquet, luncheon, and plenary session speakers who shared their vision of the manufacturing industry and issues related to productivity. My sincere thanks to the session organizers, session chairs, and members of the Organizing Committee both at Old Dominion University and West Virginia University without whose cooperation this Conference would not be possible. Thanks are due to Ms. Georgette Ingram and other staff members in the MET Department for their patience and hard work. Financial support from the Center for Innovative Technology and industrial sponsors also made this Conference possible.

I acknowledge, with gratitude, the help and support received from Dr. James V. Kock, President, and Dr. Ernest J. Cross, Dean of the College of Engineering and Technology at Old Dominion University. From West Virginia University, I thank Dr. Donald W. Lyons, Chairman, MAE Department, for his support; Drs. Ralph Wood and John Spears for their help in reviewing conference papers and for allowing us to use the facilities of the Concurrent Engineering Research Center; and Ms. Sati Maharaj for her assistance in coordinating the conference. In addition, I extend my deepest gratitude to Dr. Suren N. Dwivedi for providing me with support and encouragement in organizing this conference. Furthermore, I express my sincere thanks to all my colleagues, friends, student volunteers, and family members who extended their help in organizing this conference.

I also acknowledge with great appreciation the excellent work done by Springer-Verlag in publishing both volumes of the proceedings.

Alok K. Verma
Conference General Chairperson

Preface

According to the Concurrent Engineering Research Center (CERC) at West Virginia University, "the concurrent engineering (CE) is a rapid simultaneous approach where research and development, design, manufacturing and support are carried out in parallel". The mission of concurrent engineering is to reduce time to market, improve total quality and lower cost for products or systems developed and supported by large organizations. The purpose of the concurrent design methodology is to let the designer know the consequences of his design decisions in the manufacturing and assembly stages as well as in subsequent operations. Design for manufacture and assembly, design for reliability and testability, CAD/CAM/CAE, knowledge based systems, cost analysis and advanced material technology are the major constituents of concurrent engineering. The need for concurrent engineering can be justified from the fact that in every production cycle, the design phase approximately takes 5 to 10% of the total cycle, but overall it influences 80% of the production cycle.

This volume contains articles from a wide spectrum dealing with concepts of concurrent engineering. The importance of the knowledge-based systems in the CE environment is significant as they provide the common platform to achieve the same level of expertise to the designers and manufacturers throughout the organization for the specific task. Their role in "*do it right the first time*" is very important in providing aid to the designers and manufacturers to optimize the design and manufacturing setups for a cost-effectiveness and reduced production time. The application of neural networks in various manufacturing areas has been presented. The papers on the feature based design, process simulation, design automation and quality control are discussed. A special section has been devoted to printed circuit boards, recognizing their importance in a CAD/CAM environment from both design and manufacture standpoint. This volume also presents articles describing the payoffs of concurrent engineering in advance materials development. The final section discusses the implementation of CE technology.

Suren N. Dwivedi

Contents, Volume 1

Conference Objective	v
Conference Organization	vi
Committee Rosters	viii
Letter from the President, ISPE	xi
Acknowledgments	xiii
Preface	xv

Chapter I: General Issues	1
Introduction	1
Concurrent Engineering: An Introduction SUREN N. DWIVEDI and MICHAEL SOBOLEWSKI	3
Quality Design Engineering: The Missing Link in U.S. Competitiveness H. BARRY BEBB	17
New Activities in the Manufacturing Domain to Support the Concurrent Engineering Process KEITH WALL and S.N. DWIVEDI	31

Chapter II: Intelligent Information Networks	37
Introduction	37
Artificial Intelligence in Concurrent Engineering A. KUSIAK and E. SZCZERBICKI	39
An Open Ended Network Architecture for Integrated Control and Manufacturing K.C.S. MURTY and RAMESH BABU	49
A Data Base Inconsistency Checker for EASIE K.H. JONES, S. OLARIU, L.F. ROWELL, J.L. SCHWING, and A. WILHITE	55

Database Exchange in the CAD/CAM/CIM Arena
BRUCE A. HARDING 61

Formation of Machine Cells: A Fuzzy Set Approach
CHUN ZHANG and HSU-PIN WANG 67

Design Assessment Tool
DANIEL M. NICHOLS and SUMITRA REDDY 73

Support of PCs in Concurrent Engineering
NARESH C. MAHESHWARI and BRADLEY S. BENNETT 80

Chapter III: Neural Network 87

Introduction 87

Artificial Neural Networks In Manufacturing
KENNETH R. CURRIE 89

Using Artificial Neural Networks For Flexible Manufacturing System Scheduling
LUIS CARLOS RABELO and SEMA ALPTEKIN 95

Machine-Part Family Formation Using Neural Networks
RAM HUGGAHALLI and CIHAN DAGLI 102

Neural Networks in Process Diagnostics
SOUNDAR R.T. KUMARA and NAJWA S. MERCHAWI 108

Chapter IV: Knowledge Based Engineering 115

Introduction 115

DICEtalk: An Object-Oriented Knowledge-Based Engineering Environment
M. SOBOLEWSKI 117

Data Models of Mechanical Systems for Concurrent Design
KIRK J. WU, FOOK CHOONG, and S. TWU 123

Manufacturing Knowledge Representation Using an Object Oriented Data Model
RASHPAL S. AHLUWALIA and PING JI 130

Knowledge-Based Evaluation of Manufacturability
SIPING LIU, VASILE R. MONTAN, and RAVI S. RAMAN 136

Knowledge-Based Graphic User Interface Management Methodology
STEWART N.T. SHEN and JIH-SHIH HSU 142

Knowledge Augmentation Via Interactive Learning in a Path Finder
  Q. ZHU, D. SHI, and S. TANG 148

Graphical User Interface with Object-Oriented Knowledge-Based
Engineering Environment
  Z. KULPA, M. SOBOLEWSKI, and S.N. DWIVEDI 154

Knowledge Automation: Unifying Learning Automation and
Knowledge Base
  A. CHANDRAMOULI and P.S. SATSANGI 160

Developing a Knowledge Based System for Progressive Die Design
  PRATYUSH KUMAR, P.N. RAO, and N.K. TEWARI 166

An Expert System Model for the Use in Some Aspects of
Manufacturing
  R.B. MISHRA and SUREN N. DWIVEDI 172

Chapter V: Feature Based Design and Manufacturing 179

Introduction 179

Using a Feature Algebra in Concurrent Engineering Design and
Manufacturing
  RAGHU KARINTHI and DANA NAU 181

Feature Recognition During Design Evolution
  HYOWON SUH and RASHPAL S. AHLUWALIA 187

Extraction of Manufacturing Features from an I-DEAS Universal File
  JONG-YUN JUNG and RASHPAL S. AHLUWALIA 193

Feature Based Design Assembly
  SISIR K. PADHY and SUREN N. DWIVEDI 199

Feature Based Machining Analysis and Cost Estimation for the
Manufacture of Complex Geometries in Concurrent Engineering
  B. GOPALAKRISHNAN and V. PANDIARAJAN 205

Use of Part Features for Process Planning
  S.K. GUPTA, P.N. RAO, and N.K. TEWARI 211

Chapter VI: CAD and FEM 217

Introduction 217

Model Based 3-D Curved Object Recognition Using Quadrics
  M. HANMANDLU, C. RANGAIAH, and K.K. BISWAS 219

Finite-Element Model for Modal Analysis of Pretwisted Unsymmetric Blades
  N.T. SIVANERI and Y.P. XIE 225

Computer Based Life Prediction Methodology for Structural Design
  T.L. NORMAN, T.S. CIVELEK, and J. PRUCZ 231

Chapter VII: Process Modeling and Control 237

Introduction 237

Processing of Superalloys in the 1990s
  F. ROBERT DAX 239

Application of the Finite Element Method in Metal Forming Process Design
  SHANKAR RACHAKONDA and SUREN N. DWIVEDI 253

Strategic Value of Concurrent Product and Process Engineering
  EDWIN R. BRAUN and JASON R. LEMON 259

The Design Process for Concurrent Engineering
  NICHOLAS J. YANNOULAKIS, SANJAY B. JOSHI,
  and RICHARD A. WYSK 265

Modeling Concurrent Manufacturing Systems Using Petri Nets
  KELWYN A. D'SOUZA 271

Production Planning and Control in the Factory of the Future
  W.H. ISKANDER and M. JARAIEDI 281

Expert Control of Turning Process
  P.S. SUBRAMANYA, V. LATINOVIC, and M.O.M. OSMAN ... 287

Expert System for Milling Process Selection
  B. GOPALAKRISHNAN and M.A. PATHAK 293

Forging Die Design with Artificial Intelligence
  S.K. PADHY, R. SHARAN, S.N. DWIVEDI, and D.W. LYONS .. 299

Chapter VIII: Process Simulation and Automation 307

Introduction 307

Simulation Modeling in CIM Systems Design
  COLIN O. BENJAMIN, MELINDA L. SMITH, and DEBRA A. HUNKE 309

EXSEMA-An EXpert System for SElecting Simulation Software for Manufacturing Applications
  COLIN O. BENJAMIN and OSSAMA A. HOSNY 315

Group Technology Analysis for Manufacturing Data
   ABDELLAH NADIF, RENE-PIERRE BALLOT, and BERNARD MUTEL 321

Dispatching Mobile Robots in Flexible Manufacturing Systems:
The Issues and Problems
   HIMANSHU BHATNAGAR and PATRICK D. KROLAK 327

Automation of Prototype General Aviation Aircraft Development
   GEORGE BENNETT... 334

Determining Organizational Readiness for Advanced Manufacturing
Technology: Development of a Knowledge-Based System to Aid
Implementation
   DONALD D. DAVIS, ANN MAJCHRZAK, LES GASSER,
   MURRAY SINCLAIR, and CARYS SIEMIENIUCH 340

Planning and Realization of Skill Based Flexible Automation for
Developing Countries
   S. KUMAR and A.K. JHA.. 346

Chapter IX: PCB Manufacturability and Assembly 353

Introduction 353

An Expert System Based Concurrent Engineering Approach to PCB
Assembly
   K. SRIHARI 355

Real Time Production Scheduling and Dynamic Parts Routing for
Flexible Assembly Lines
   J.P. BOURRIERES, O.K. SHIN, and F. LHOTE 361

A Knowledge-Based proach for Manufacturability of Printed Wiring
Boards
   SISIR K. PADHY and S.N. DWIVEDI 369

Design of an IGES Post Processor and Integration with a Robotic
Workcell
   R.H. WILLISON and G.M. PALMER.. 376

Discrete Optimum Assembly Methods for Automated Workcells
   KENNETH H. MEANS and JIE JIANG... 382

Trajectory Planning for Obstacle-Avoided Assembly of Planar Printed
Circuit Boards
   TAK-LAI LUK and JOHN E. SNECKENBERGER 388

Development of a Vision Assisted Optimal Part-To-Pad Placement
Technique for Printed Circuit Board Assembly
   S.H. CHERAGHI, E.A. LEHTIHET, and P.J. EGBELU............... 394

Chapter X: Quality Control Techniques 401

Introduction 401

Quality Function Deployment, a Technique of Design for Quality
CHIA-HAO CHANG 403

Quality Value Function and Consumer Quality Loss
FU QIANG YANG, MAJID JARAIEDI, and WAFIK ISKANDER 409

Implementation of a Computer Aided Quality System (CAQ) in CIM Environment: Advantages and Disadvantages
M. DOMINGUEZ, M.M. ESPINOSA, J.I. PEDRERO, and J.M. PEREZ 415

Computer Aided Quality Assurance Systems
V.K. GUPTA and R. SAGAR 421

Quality Consideration During DFA Analysis
SUDERSHAN L. CHHABRA and RASHPAL S. AHLUWALIA 436

Selection of Acceptance Sampling Plans Through Knowledge Based Approach
S.S.N. MURTY and D. CHANDRA REDDY 443

Chapter XI: Cost Analysis Concept 451

Introduction 451

Analysis of Quality Costs: A Critical Element in CIM
RESIT UNAL and EDWIN B. DEAN 453

A Databased Time and Cost Estimation Algorithm for Piece Part Design and Manufacturing
K.W.-N. LAU and M. RAMULU 458

Improvement Curves in Manufacturing
R.C. CREESE and MADHU SUDHAN 466

Chapter XII: Materials: Composite 473

Introduction 473

The Payoffs of Concurrent Engineering in Advanced Materials Development
JACKY C. PRUCZ 475

A Practical Engineering Approach for Predicting Interlaminar Stresses in Composites
JACKY PRUCZ and MARIOS LAMBI 487

Interactive Optimum Parametric Design of Laminated Composite Flange
B.S.-J. KANG, JACKY PRUCZ, and F.K. HSIEH 506

Computer Aided Dynamic Analysis of Laminated Composite Plates
ALEXANDER E. BOGDANOVICH, ENDEL V. IARVE, and SUREN N. DWIVEDI 519

Integration of Rigid-Plastic Simulation Engines into Engineering Database System for Advanced Forging
TATSUHIKO AIZAWA and JUNJI KIHARA 528

Chapter XIII: Implementation 535

Introduction 535

Transitioning CE Technology to Industry
S.N. DWIVEDI, RAVI PRASAD, and D.W. LYONS 537

Integrated Models of FMS in Concurrent Engineering Environment
A.A. LESKIN and S.N. DWIVEDI 553

Implementing QFD at the Ford Motor Company
HAROLD F. SCHAAL and WILLIAM R. SLABEY 563

Contents, Volume 2

Conference Objective	v
Conference Organization	vi
Committee Rosters	viii
Letter from the President, ISPE	xi
Acknowledgments	xiii
Preface	xv

Section A: Implementation of Intelligent Manufacturing

Chapter I: Factory Enhancements ... 1

Introduction ... 1

From the Existing Manufacturing System to CIM
D.S. LVOV, E.I. ZAK, and YU.M. ZYBAREV ... 3

Flexible Manufacturing System in Manufacture of Precision Engineering Components - Key Issues in Implementation
V.K. GUPTA and R. SAGAR ... 9

A Survey of CIM Strategic Planning in U.S. Industry
MARK D. PARDUE and FREDERICK J. MICHEL ... 19

Modelling and Optimization of a Flexible Manufacturing System
R.N. CHAVALI, S. KESWANI, and S.C. BOSE ... 25

Computer Based Safety System for the FMS - Management Logic
C.F. MARCIOLLI ... 31

CIM Repositories
H. T. GORANSON ... 45

The Selection and Prospect of CAD/CAM System for Diesel Engine Design and Manufacturing
GU ZE-TONG and HU GANG ... 56

A Model for the Factory of the Future for Industrialized Housing
AHMAD K. ELSHENNAWY, MICHAEL A. MULLENS, WILLIAM W. SWART, and SUBRATO CHANDRA ... 62

Enabling Automation Technologies for an Automated Mail Facility of the Future
JAY LEE and GARY HERRING ... 71

Some Optimization Problems of Scheduling in a Flexible
Manufacturing System
   TOMASZ AMBROZIAK 77

Some Methods of Modeling for Computer Integrated Workshop
   V.N. KALACHEV and YE.N. KHOBOTOV 83

Combined Procedures for Simulation of Manufacturing Systems
   SUREN N. DWIVEDI and YE.N. KHOBOTOV 88

Expert Systems in CIM
   V.M. PONOMARYOV, V.V. IVANISTCHEV, A.A. LESKIN,
   and N.N. LYASHENKO 94

Chapter II: Production Planning 100

Introduction 100

A Taxonomy on Event-Driven Production Systems
   S.S. IYENGAR, NITIN S. NAIK, and RAJENDRA SHRIVASTAVA 105

An Improved Lot Sizing Policy for Variable Demand
   M.D. SREEKUMAR, C. ESWARA REDDY,
   and O.V. KRISHNAIAH CHETTY 120

Simulation for Real-Time Control: Advantages, Potential Pitfalls,
Opportunities
   C.M. HARMONOSKY 126

Decomposition Approach for the Job-Shop Scheduling Problem
   H.D. LEMONIAS and Z. BINDER 132

Evaluation of the Impact of Plant and Production Management
Automation on Job-Shop Manufacturing Performances
   M. PESSINA, A. POZZETTI, and A. SIANESI 138

Role of Non-Productive Time in the Evaluation of Computer Generated
Process Plans
   N.K. MEHTA, P.C. PANDEY, and A.V.S.R.K. PRASAD 144

Chapter III: Process Technology 151

Introduction 151

Computer Managed Process Planning for Cylindrical Parts
   AMY J. KNUTILLA and BOBBY C. PARK 153

An Application of Non-Linear Goal Programming in Electrodischarge
Machining of Composite Material
   M. RAMULU, H.-W. SEE, and D.H. WANG 166

An Expert System for Metalforming
  K. HANS RAJ and V.M. KUMAR 173

Optimal Process Planning for Robotic Assembly Operations
  SHYANGLIN LEE and HSU-PIN WANG 179

Effect of Angular Errors in Part Registration for PC Board Assembly
  T. RADHAKRISHNAN 185

An Evaluation Framework for AGVS Within FMS
  P.F. RIEL and M.S. JONES 191

Computer Aided Machine Loading Technique
  USHIR SHAH, SAUMIL TRIVEDI, KETAN SHAH, and P.B. POPAT 197

An Optimal Parallel Algorithm for Channel-Assignment
  STEPHEN OLARIU, JAMES L. SCHWING, and JINGYUAN ZHANG..... 203

Chapter IV: Product Engineering 217

Introduction 217

Design Using Case-Based Reasoning
  COSTAS TSATSOULIS 219

An Interactive Programming System for Design of Mechanical Clutches
  B. SATYANARAYANA, K.V. MOHAN, and M. MALLIKHARJUNA RAO 225

An Expert System for the Design and Selection of Ball Bearing
Parameters
  M.A. PATHAK and R.S. AHLUWALIA.. 231

Computer-Aided Optimal Design of Gears
  HUNGLIN WANG and HSU-PIN WANG 237

CAD for Underground Structure
  GU HANLIANG 243

A Microcomputer Aided Design of Technical Systems
  W. PRZYBYLO 249

Solid Modeling With Tension
  DA-PAN CHEN 259

Integration of Design Optimization in Finite Element Analysis
  FRED BAREZ 265

Automatic Generation of Finite Element Modeling for Integrated CAD
and CAE
  TATSUHIKO AIZAWA 273

Three Dimensional Mesh Generation: A New Approach
  M.H. KADIVAR and H. SHARIFI 279

Effective Modeling of Elastic Mechanical System Through
Objective-Aimed Finite Element Strategies
  V.H. MUCINO, W.G. WANG, and J.E. SMITH 285

Design and Evaluation of Shock Isolation of Trailer Mounted Electronic
Equipments
  V. SUNDARARAMAN 297

Chapter V: Workcell Operations 303

Introduction 303

Group Technology: Cell Formation Using Simulated Annealing
  J.M. PROTH 305

Cost Considerations for Cell Design in Group Technology
  KENNETH R. CURRIE 311

Application of CAD/CAM in the Textile Industry
  P.B. JHALA 317

CAD/CAM of Cams for Use in Automatic Lathes
  P.C. PANDEY, N.K. MEHTA, and AATUL WADEGAONKAR 323

An Objective SIMTOOL in FMS
  C. ESWARA REDDY, O.V. KRISHNAIAH CHETTY,
  and D. CHAUDHURI 329

A Methodology for Automating the Redressing of the Grinding Wheel
  A.C.S. KUMAR and U.R.K. RAO 336

Experimental Investigations on Tool Vibrations in Turning for On-Line
Tool Wear Monitoring
  D.N. RAO, P.N. RAO, and U.R.K. RAO 342

μ_p-Based Industrial Grade Multi-Channel Temperature Controller For
Sugar and Allied Industries
  H. SINGH, S.M. SHARMA, and C.R.K. PRASAD 348

Use of Sensors for Safety of Personnel in Robotic Installations
  K. GHOSH, J.-J. PAQUES, and Y. BEAUCHAMP 355

Section B: Developments in Applied Robotics and Automation

Chapter VI: Industrial Applications 361

Introduction 361

Determining the Workspace Design of Robotized Cells in Pre-Determined Environments
  LOUISE CLEROUX 363

Judicious Selection of a Robot for an Industrial Task - An Expert System Approach
  SURENDER KUMAR and ALOK VARMA 369

Fixtureless Robotic Assembly Workcell
  LARRY BANTA and THOMAS BUBNICK 374

Design of a Wall-Scaling Robot for Inspection and Maintenance
  BEHNAM BAHR and SAMI MAARI 381

A Telemanipulator for Hazardous Mining Operations
  M.R. UDAYAGIRI, T.R. RANGNATH, K.C.S. MURTY, S. RAGHUNATH, and PRAVEEN DHYANI 388

Adoption of Robotic System for Inter-Station Handling Operations for Nagpur Milk Scheme, India
  J.P. MODAK, R.D. ASKHEDKAR, and A.V. PESHWE 394

Integration and Realtime Monitoring of Robotic Controllers
  SUDHAKAR R. PAIDY and MICHAEL SHEA 400

On the Applications of Part Image Reconstruction Systems in Automated Manufacturing
  SAEID MOTAVALLI and BOPAYA BIDANDA 406

Kalman Filter Application to Tridimensional Rigid Body Motion Parameter Estimation from a Sequence of Images
  R. VASQUEZ and J. MAYORA 412

Optimization Techniques for Mathematical Routines Available through High-Level Source Code
  S. ROY and A. CHAUDHURI 421

Chapter VII: Task Performance 427

Introduction 427

Sensing and Analysis of End-Effector Forces for Precision Assembly
  ANTHONY DE SAM LAZARO, ECHEMPATI RAGHU, and BERAT GUROCAK 429

Accuracy Test From Kinematic Parameter Errors in a Closed-Loop Robot
CHENG Y. LIN, ALOK K. VERMA, and LOUIS J. EVERETT 435

The Effect of Robot Kinematic Parameter Errors on Joint Torques
JING TIAN and JOHN E. SNECKENBERGER 441

Kinematic Error Budgeting to Obtain the Best Feasible Task Performance for a Specified SCARA Manipulator
TONY M. LAMB and JOHN E. SNECKENBERGER 448

Demonstrating Robot Calibration in a Manufacturing Environment
KEN PFEIFFER and LOUIS J. EVERETT 454

On-Line Robot Calibration
F. TUIJNMAN and G.R. MEIJER 460

Expert System for Robot Hand Design Using Graph Representation
M. CHEW, G.F. ISSA, and S.N.T. SHEN 466

PreGrasp Pose Estimation of Objects Using Local Sensors on Dexterous Hands
V.H. PINTO, L.J. EVERETT, and M. DRIELS 472

Chapter VIII: Motion Specification 479

Introduction 479

Approximate and Hierarchical Path Planning
NAGESWARA S.V. RAO, WENCHENG WU, and PAI-SHAN LEE 481

Bandlimited Trajectory Planning for Continuous Path Industrial Robots
J.T. HUANG 487

Trajectory Planning and Kinematic Control of a Stewart Platform-Based Manipulator
CHARLES C. NGUYEN, SAMI S. ANTRAZI, and ZHEN-LEI ZHOU 493

Planning and Execution of Polynomial Manipulator Trajectories
FRED BAREZ 500

Efficient Trajectory Planning Algorithm for Coordinately Operating Multiple Robots
YUNG-PING CHIEN and QING XUE 509

Path Planning for Coordinated Planar Robot Arms Moving in Unknown Environment
YUNG-PING CHIEN and QING XUE 515

A Theory of Collision Avoidance on Visual Guidance of Robot Motion in Dynamic Environment
Q. ZHU and J. LIU 521

A Collision Prediction System for a Robotic Environment
  NANCY SLIWA, WILLIAM BYNUM, and CHARLES WATLAND 527

Multivalue Coding: Application to the Autonomous Robots
  A. PRUSKI 533

Information Management for Off-Line Robot Programming
  H. AFSARMANESH, G.R. MEIJER, and F. TUIJNMAN 539

Chapter IX: Manipulator Mechanics 545

Introduction 545

Computer Aided Analysis of a Planar Robot
  SHAILESH SHAH and YOGESHWAR HARI 547

Computer Aided Analysis of the First Three Links in a Puma Robot
  DURAISWAMI PALANIVELU and YOGESHWAR HARI 553

Work-Space Calculation of a Robotic Arm Using the Articulated Total
Body Model
  XAVIER J.R. AVULA, INTS KALEPS, and LOUISE OBERGEFELL 559

Effect on Flexibility on Manipulator Dynamics
  H. ASHRAFIUON and C. NATARAJ 565

Dynamics of Flexible Manipulators With Application to Robotic
Assembly
  E. WEHRLI and P. COIFFET 576

Kineto-Elastodynamic Effect on the Design of Elastic Mechanisms
  ECHEMPATI RAGHU and A. BALASUBRAMONIAN 585

Scheme for Active Positional Correction of Robot Arms
  SAEED B. NIKU 590

Feedback Control of Robot End-Effector Probable Position Error
  Y.C. PAO and L.C. CHANG 594

An Intelligent Signal Recognition System
  T. HOU and L. LIN 600

State Estimation Under Unknown Noises - A Least-Squares Approach
  CHUNG-WEN CHEN and JEN-KUANG HUANG 606

Chapter X: Educational Endeavors 613

Introduction 613

Combined EE-ME Senior Capstone Projects In Robotics at
West Virginia University
  NIGEL T. MIDDLETON and LARRY BANTA 615

Automated Manufacturing at Western Kentucky University
   R.I. EVERSOLL, H.T. LEEPER, and L.T. ROSS 621

Manufacturing Systems Engineering Education at the University of Pittsburgh
   JOHN H. MANLEY 626

Human Factors Considerations in the Design of a Teach Pendant
   ALOK K. VERMA and CHENG Y. LIN 632

Chapter I
General Issues

Introduction

Concurrent or Simultaneous Engineering basically addresses and accounts for all the product life cycle issues viz. functionality, manufacturability, maintainability, reliability, cost, performance, etc., right at the product conception and design stages. The objective is to design it "right the first time" in as short a time as possible.

The first paper focuses on the need and the underlying concepts of concurrent engineering methodology for product development. It also makes an attempt to clarify the prevailing misconceptions about concurrent engineering. The realization of functional integration necessitates a versatile and flexible information management system within the organization. Thus, the information management system can be visualized as being composed of three basic blocks: data architecture, information framework, and software tools and services.

Moreover, CE is dependent, to a great extent, on an information network for managing the communications between various groups involved in product life cycle activity design. There is a need for an information framework capable of accomplishing the cooperation and coordination necessary for concurrent engineering. The CE information management system envisages to provide a whole range of software tools and services that will support an economical and an optimum product design. In addition to a multitude of CAD/CAE/CAM tools, there will also be a host of tools for project management, process planning, etc. Services primarily include friendly user interfaces, network monitoring, and rapid prototyping.

The second paper demonstrates that engineering design process capability is the "missing link" in efforts to achieve world class competitiveness in mechanically intense systems found in most homes, offices, military, aerospace, and factory products using the Competitive Benchmarking Process. The final paper takes a look at creative activities that can be applied to the engineering process for the generic development of new products which are viewed from the manufacturing domain and seek to explore nontraditional roles in the early and late stages of product development.

Concurrent Engineering: An Introduction

SUREN N. DWIVEDI
Mechanical and Aerospace Engineering
West Virginia Unversity, Morgantown, WV 26506
MICHAEL SOBOLEWSKI
Concurrent Engineering Research Center
West Virginia University, Morgantown WV 26506

Summary

The paper focuses on the need and the underlying concepts of concurrent engineering approach. It lays out the concurrent engineering methodology for product development. An attempt is made to clarify the prevailing misconceptions about concurrent engineering. Integration is the working foundation for computer-based approaches to concurrent engineering. An architecture of a concurrent engineering system is considered as four levels consisting of an object-oriented data base, an intelligent data base engine, a high-level interface and high-level tools. This architecture combines technologies such as object-oriented programming, expert systems, hypermedia, visual programming, data bases and information retrieval. Within this architecture the integration of processors, humans, tools and methods to form a concurrent engineering environment for the development of products is treated in the context of cooperating knowledge bases.

1. Introduction: Why Concurrent Engineering

There has been a tremendous decline in American industrial productivity over the last two decades, but the Japanese have steadily captured world dominance in industry after industry, relegating U.S. far behind. As a result, the methodology of product development needs to be changed. A strong, dependable product development methodology has to be developed which can be continuously upgraded and modified. Such a methodology should lead to a significant reduction in cost and development time without sacrifing any of the desired product specifications. Moreover, it should be simple to comprehend, easy to implement and easily adaptable to a diverse nature of product development activities [4,8]. Concurrent engineering is the approach which provides all the above capabilities, and it can prove to be the panacea of all the ills plaguing the American industry.

2. What Is Concurrent Engineering: Definition

Concurrent or simultaneous engineering, as the name suggests, is the approach of doing all the activities at the same time. It is the unision of all the facets of product life cycle to minimize modifications in the prototype i.e. to decrease the design iterations performed during product design [4]. A comprehensive definition of concurrent engineering is given in the IDA (Institute for Defense Analysis) report on concurrent engineering [9]:

"Concurrent engineering is the systematic approach to the integrated, concurrent design of products and related processes including manufacture and support. This approach is to cause the developers, from the outset, to consider all the elements of product life-cycle from conception through disposal including quality, cost, schedule and user requirement."

3. Methodology of Concurrent Engineering

Concurrent Engineering (henceforth CE) is characterized by a focus on customer's requirement. Moreover, it embodies the belief that quality is inbuilt in the product, and that it (quality) is a result of continuous improvement of a process. This concept is not new, in fact, the approach is quite similar to the "tiger team" approach characteristic of small organizations. The "tiger team" essentially comprises of a small group of people working closely for a common endeavor, which might be product development. The magnitude of the problem is usually small with few conflicting constraints. The approach works well for small organizations; however, in large organizations the technique needs to be modified and restructured. It is here that CE comes into picture. CE envisages to translate the "tiger team" concept to big organizations and such "tiger teams" will work with a unified product concept. Although the team members can be at geographically different, networked locations, this requires far-reaching changes in the work culture and ethical values of the organization [5].

The methodology of CE, to be discussed here, can be possible only if the traditional work culture (existing in most present day enterprises) paves the way for a more open, equable and trustful environment, where the hierarchical position of an individual is not a barrier in information exchange. The design group can effectively utilize the CE approach only when it works cooperatively and consistently. The group should have the capability to analyze individual viewpoints and reach a common decision and technological agreement.

The objectives that CE will meet are:
    a) elimination of waste,
    b) reduction in lead time for product development and product delivery,
    c) improvement of quality,
    d) reduction of cost,
    e) continuous improvement.

The goals are broad and ambitious, and attaining them requires fundamental changes not only in the functioning of the various groups involved in product development, but in the basic attitude of the group as well. The group must understand, appreciate and accept ALL the goals of the task it is working for. This is very important as many problems arise from incomplete understanding of the objectives of an exercise. In addition, the group should have a natural inclination for innovation and change. The organizational values should constantly encourage improvements wherever possible. The PHILOSOPHY of the design process in an enterprise is very critical for realizing the benefits of concurrent engineering.

Present day design activity involves sequential information transfer from "concept designers" to "design finalizers". As the activities are finished, the people involved in them get detached from the design chain. Thus, the people involved in earlier design phases do not interact with people in the later stages. A natural consequence of this procedure is that errors go on propogating themselves down the chain and are usually detected at a stage where rectifications/modifications become both costly and undesirable. The philosophy of continuous improvement implies changes at the initial stages with the aim of minimizing changes at later stages. To achieve this, it is imperative that there exists a strong communication between product developers and end-users.

It is worth mentioning here that "end-user" or "customer" does not imply only the "external customer" to whom the end-product is to be delivered, customer means both "internal" and "external customer". Every person involved in product synthesis is an element of the chain of "internal customers" which ultimately connects the vendor to "external customer". This link will help the designers to understand the customer requirements properly during the product-definition stage. The customer-vendor dialogue paves the way for defining priorities on requirements and evaluating trade-offs in various functional characteristics. Of course, initially the requirements are subjective in nature, and their objectiveness and level of transparency increases as the design process is further defined. Customer specifications must be used to establish more specific characteristics and features of the product. These must be then used for an explicit description of the related processes to allow CE to take place. The extent of "explicit description" will again be determined by the design stage.

Integrated, parallel product and process design is the key to concurrent design. CE approach as opposed to the sequential approach advocates a parallel design effort. The objective is to ensure that serious errors don't go undetected and that the design intent is fully captured. The above - mentioned integrated design process should have the following features:

- a) There must be a strong information sharing system, thus enabling the design teams to have access to all corporate facilities as well as work done by individual teams.
- b) Any design is necessarily an iterative process requiring successive redesigns and modifications. The CE process should ensure that the effects of a change incorporated by one team on other design aspects are automaticallty analyzed. Moreover, the affected functions should be notified of the changes.

- c) The CE process must facilitate an appropriate trade-off analysis leading to product-process design optimization. Conflicting requirements and constraint violations must be identified and concurrently resolved.
- d) All the relevant aspects of the design process must be recorded and documented for future reference.

The integration process discussed here is FUNCTIONAL integration in an organization. This binds the various functional areas viz. engineering, support etc., in the organization for greater efficiency of the whole venture. But, for this integration to be effective, it has to be accompanied by STRATEGIC and LOGISTIC integrations [7]. Strategic integration focuses on a company's business strategy. This strategy should tie decision making and other organizational policies together with the objective of realizing total quality management. Logistic integration is basically close coordination of the manufacturer with its customers and suppliers for cutting down logistic problems.

The realization of functional integration neccesitates a versatile and a flexible INFORMATION MANAGEMENT SYSTEM within the organization. The system must be domain independent and adaptable to both large and small industrial enterprises. Some essential and desirable characteristics of this system are listed below:

- a) The system should be tailorable to the needs of the specific organizations. It should be generic and at the same time modifiable to the requirements of the enterprise.
- b) It should have a diverse repository of organized knowledge which is easily accessible across the spectrum of product life-cycle disciplines.
- c) There must be an intelligent information distribution system which could provide information on a "need to know" and/or user-specified basis.
- d) It should have facility of interfacing with software tools and application databases existing in user's organization.
- e) The system should be capable of making the whole design team cognizant of the modifications done by a sub-group. In addition, it should have the ability to appraise the impacts of the modifications in a global manner i.e. on all the other design activities.
- f) Even though most of the activities should be automated, there must be provision for human intervention at every stage. Furthermore, a manual bypass alternative for autonomous activities should be provided.
- g) The system must support progressive refinement of product and process development from "design initiation" to the "design finalization" stage.
- h) There must be tools permitting rapid prototyping and testing, therefore paving the way for commercial production.

The information management system can therefore be visualized as composed of three basic blocks:
- i) data architecture,
- ii) information framework,
- iii) software tools and services.

These are discussed below.

3.1. Data Architecture

The primary purpose of data architecture is to organize the design and manufacturing knowledge of the enterprise to accomplish maximum exploitation of this data base. Due to the very nature of concurrent engineering, the diversity and quantity of data is immense. Therefore, only its structuring can result in rapid access, global transparency and fast information transfer [13].

Some main features of the architecture are enumerated below:
- a) Creation of data base packages that will enable an exhaustive storage of information on product-process descriptions and corporate resources.
- b) Providing sufficient flexibility to allow easy modifications of the data base. Moreover, there should be the capability of recording the "by whom" and "why" of the modifications done.
- c) The information must be organized into various sub-classes with explicit relations amongst them. These classes can be thought of as pertaining to: life cycle phase (design, manufacturing, support etc.), requirement attributes (performance, cost, time factor etc.), product type etc. Provision for extension of existing classes and addition of newer ones must be made.
- d) There must be a mechanism to ensure that the contents of the data base are quickly and conveniently accessible to the various product development teams.
- e) The data bases must be strongly protected against unauthorized use, and only identified users should have access to the system.
- f) There must be an extensive index to the contents of the data base, and record of applications software, available computing facilities and authorized users should also be there.
- g) There should be a facility for detailed documentation of an ongoing design effort. The idea should be not only to record the design data but the method of design as well. The method used for designing a present product can prove to be very valuable to future designers tackling a similar problem.

3.2. Information Framework

CE is dependent to a great extent on information network for managing the communications between various groups involved in product life cycle activity design. There is a need for an information framework capable of accomplishing the cooperation and coordination necessary for concurrent engineering.

Some of the key features of the framework are:
- a) The framework should be based on a set of standards and specifications so that it is general purpose. At the same time, however it should have attributes of adaptability to the special needs of an enterprise. Diligent efforts are required for developing a standard framework for CAD/CAM/CAE application tools. Moreover, the framework should be compatible with the existing hardware and software platforms.
- b) The framework should serve as a common environment for designers to perform a cooperative design effort. It should facilitate parallel execution of design activities.
- c) There should be mechanisms to highlight conflicts and notify designers of decisions and modifications. In addition, the framework has tools to analyze and evaluate global effects of local design modifications.
- d) The framework has intelligent information filtration and interpretation mechanisms so that experts of one domain can derive maximum benefits out of knowledge pertaining to other domains.
- e) The user has access to an assortment of tools and data bases across the network. He/she also has rapid data transfer facility at his/her disposal.
- f) The framework should have tools that can support multiple users as opposed to currently prevalent single user tools. This is very critical for group working which is inherent to CE approach.

3.3. Software Tools and Services

The CE information management system envisages to provide a whole range of software tools and services that will support an economical and an optimum product design. In addition to a multitude of CAD/CAE/CAM tools, there will be a host of other tools for project management, process planning etc. Services primarily include- friendly user interfaces, network monitoring and rapid prototyping.

Some important aspects of these tools and services are listed below:
- a) The application tools for CAD/CAE/CAM will be modeled to be domain specific and based on standard information framework. Domain specific tools will result

in a distinction of domain dependent and domain independent design characteristics of the product. This will help tremendously in enhancing of the design intent.
b) Tools will be developed which will enable project leaders to perform optimum task planning, monitoring and scheduling for timely project completion.
c) Friendly user interface modules will be implemented and this will enable the user to work in a fully interactive mode. Moreover, the users will have access to all tools from their individual work-stations.
d) A facility will be provided to keep account of the interactions amongst product developers on the network. This will monitor the information flow on the communication channels and diagnose faulty sub-systems.
e) A "first unit production" system will be established which will produce a prototype of the designed product with the desired quality requirements. Feature based design representations will be prepared which will be translated to tooling and manufacturing information. This information would be utilized by CNC (computerised numerically controlled) machines and robotic systems to perform actual manufacturing of the prototype.

4. Integration Levels In Concurrent Engineering

CE is a systematic approach to the integrated, concurrent design of products and their related processes, including manufacture and support. This approach is intended to cause the developers, from the outset, to consider all elements of the product life cycle from conception through disposal, including quality, cost, schedule, and user requirements. Thus, systematic integration of methodologies, processors, human beings, tools, and methods is one of the basic direction of concurrent engineering. This integration and developing tools that better accommodate, on the one hand, the iterative and evolutionary and, on the other hand, the concurrent nature of product design form the foundations of CE systems.

CE systems represent the evolution of a number of distinct paths of technological development including object-oriented programming, constraint programming, visual programming, knowledge-based systems, hypermedia, databases and information retrieval, and CAD/CAM/CAE. These technologies have reached a stage of maturity on their own so that it is now possible to define an overall unifying structure for product design. This structure caters to the needs of product developers for information that is relevant, easy to obtain, and helpful. Below, the architecture of CE systems is considered as four levels consisting of an object-oriented data base, an intelligent data base engine, a high-level interface and high-level tools. Within this architecture the integration of humans, their tools and methods, and processors to form a concurrent engineering environment for the development of products is treated in the context of cooperating knowledge bases. The presented approach is partially illustrated by selected solutions of DICE (DARPA Initiative in Concurrent Engineering).

The CE system architecture forms a new discipline which is responsible for providing the functionality of ready access to information throughout designing and manufacturing organizations. The CE data base four-level architecture is illustrated in Figure 4.1.

```
High Level Tools
    Data Presentation/Display
    Hypermedia Mangement
    Decsion Support
    Knowledge Acquisition
    CAD/CAM Applications

High Level Interface
    Multimedia
    Personal Database
    Note Taking
    Navigation Tools
    Object Editing
    Semantic Editing

Intelligent Database Engine
    Inference Engine
    Knowledge Manger
    Concurrency Manager
    Explanation Manager
    Object-Oriented Server

Object-Oriented Database
```

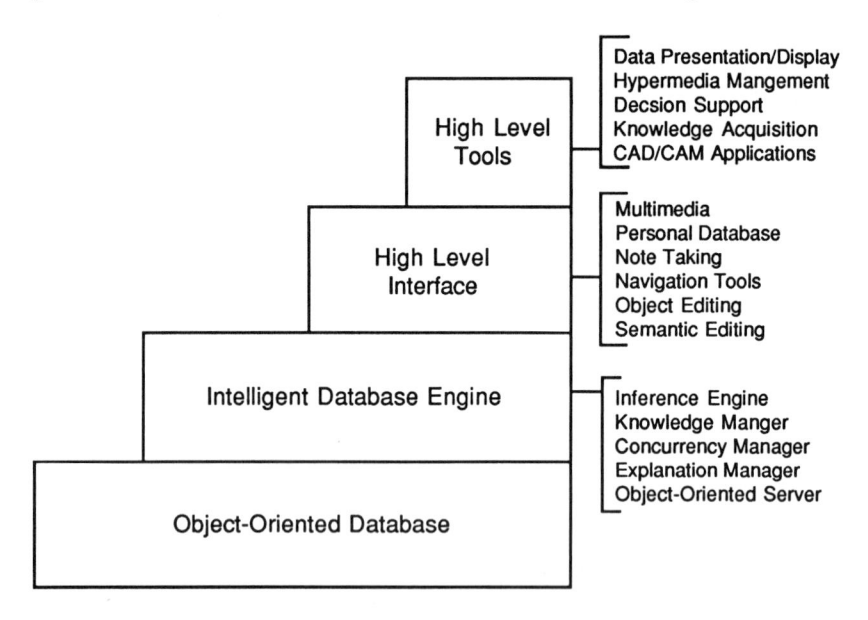

Figure 4.1. The CE System Architecture

The purpose of the object-oriented data base in CE, considered as the bottom level of the four level architecture, is to provide an engineering environment which supports real-time inspection and modification of enterprise data. For example, in DICE a very general data model, called the *Product Process Organization* (PPO) model [13], is introduced. The PPO model includes information about form and function of the product, process steps in manufacturing and design, all resources, all documentation, analysis results, major features, constraints, and so forth. The PPO data base combines techniques of relational data bases, knowledge-bases and object-oriented programming [17, 18]. The other three levels form a top-level subsystem, called the CE intelligent data base, that includes a data base engine, a high-level user interface, and high-level tools [14].

The second level, above the object-oriented data base level, is the *intelligent database engine*. The engine consists of an inference engine, knowledge manager, concurrency manager, explanation manager, and object-oriented server. For example, the DICE inference engine consists of the DICE Blackboard (DDB), Design Fusion, and domain-independent knowledge-based environments. The DBB incorporates a blackboard framework based on frame knowledge representation scheme [4,20] that allows for communication and cooperation between designers. Design Fusion is understood as a preliminary design system. By infusing knowledge of downstream activities into the design process in this system, designs that will satisfy the functional specification while taking into account the requirements of the total product life cycle will be created. This approach differs from post-design critiques where such knowledge is used to validate an existing design but is not part of the process of generating the design itself. Whereas Design Fusion provides a preliminary design, the DBB for Design Evolution provides a global workspace in which the evolving product development can be tracked. This global product development task is split, under the control of a Project Lead (PL), into local tasks being solved by relevant designers. Each such task can be solved under control of the DICEtalk knowledge-based system [15]. In the present architecture concept, DICEtalk splits, if necessary, the local task into subtasks and provides, if necessary, DICEtalk knowledge-based tools or standard CAD/CAE/CAM tools for the appropriate subtasks. DICEtalk can integrate a designer's single activity with DBB and appropriate tools in a homogeneous knowledge-based way.

The third level is the *high-level user interface*. This level creates the model of the task and data base environment that users interact with. Associated with this level is a set of representation tools that enhance the functionality of the engineering environment. The user interface is presented in two aspects. There is a core model that is presented to the user, and this model is based on hypermedia and visual programming techniques. In addition, there is a set of high-level tools which, although not an essential part of the core model, nevertheless enhances the functionality of the CE system for certain classes of user.

The fourth of these levels is the *high-level tools*. These tools provide the user with a number of facilities such as intelligent search, data presentation, data quality and integrity control, and automated discovery. They usually appear and work much as their stand-alone equivalents, such as spreadsheets and graphic representation tools, but they are modified so as to be compatible with the object-oriented data base and knowledge-base models. The CE system high-level tools contain CAD/CAE/CAM applications and tools of knowledge representation environment. On this level, we also consider intelligent system design tools that provide facilities for designing the object-oriented data base. These tools are somewhat peripheral to the central idea and functioning of the CE intelligent data base, but they are needed as part of the product development process.

4.1. Integration of Processors

Networked product developers may use different platforms appropriate for their tasks. In a general case, one developer can use more than one workstation, and there is a need to use engineering applications running under different operating systems. On the other hand, the coordination of complex tasks involving many humans and a long series of interactions requires a homogeneous operating system - a metaoperating system.

In the DICE project, such a metaoperating system is called the Concurrency Manager (CM). The CM consists of the Local Concurrency Managers (LCM) and the Coordinating Concurrency Manager (CCM). Each LCM is a local agent in each DICE workstation and the CCM is a master agent for local agents as illustrated in Figure 4.1.1. Roughly speaking, the CM, through its LCMs, provides connectivity, location transparency and network-wide access in the DICE heterogeneous environment [19].

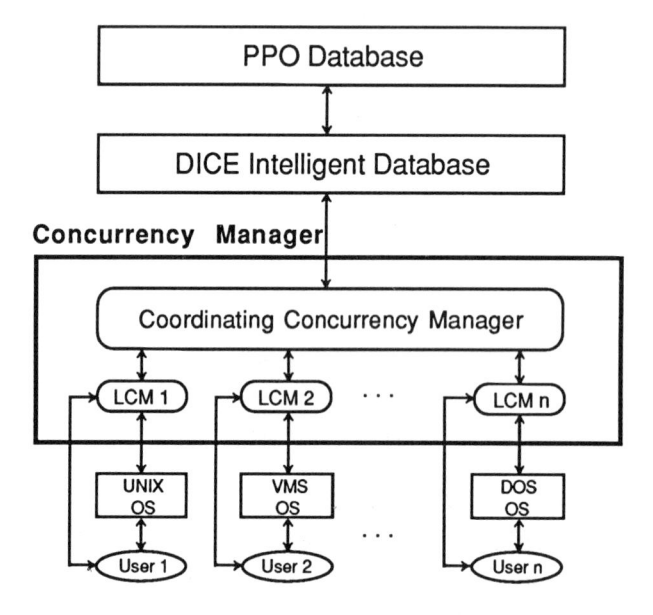

Figure. 4.1.1. The DICE Metaoperating System.

The LCM embedded in each processor takes away the responsibility for remote process communication from the designer application. Thereafter, when the application needs to send a message to a remote application or to a remote processor, it merely calls the LCM in its node. The LCM organizes a uniform interface and provides network-wide access to all the resources available to the product developer regardless of the actual location of the resources. On one hand, the LCM is an interface to the services, utilities, and persons associated with the entire DICE network, while on the other hand, the LCM is an interface to the local resources, services, tools, and clients. However, individual applications or product developers may choose to use an alternate utility to interact with other processes or users even while the LCMs are active; for example, the local operating system can also be used independently by LCM users.

4.2. Integration of Humans Beings, Tools and Methods

Many tools and methods used in product design and development are limited to algorithmic solutions. However, a number of problems encountered are not amenable to purely algorithmic solutions. Such problems are often ill-structured and an experienced engineer deals with them using judgement and experience. Artificial intelligence techniques, in particular the knowledge-based systems technology, offer a methodology to solve these ill-structured design problems. An ill-structured task is defined by a knowledge base that, when executed according to its inference method, needs user- specific knowledge relevant to the task being solved. All knowledge bases for subtasks related to the same task form a task knowledge-based environment. These knowledge bases can include algorithmic tools and methods which can be used as procedural knowledge of a knowledge-base system. The DICEtalk knowledge-base system [15] provides integration of algorithmic and knowledge-base tools and methods in a homogeneous knowledge-based way into a module called a *Knowledge Agent* (KA). The structure of KA is illustrated in Figure 4.2.1. Through the DICEtalk data structures, conventionally called the blackboard, all communication and coordination take place between knowledge-based, and algorithmic methods and tools. Thus, the DICEtalk KA can be viewed as the integration medium for all tools and methods used for solving a particular subtask.

A product developer can complete a task by invoking a collection of relevant subtasks, which can be solved by using appropriate KAs. A collection of DICEtalk KAs for the same task or product developer is integrated through the CM and is called the *Knowledge Module* (KM) as illustrated in Figure 4.2.1. Therefore, a KM is considered as the second level of DICEtalk integration of tools and methods for a single user. The blackboards of all KAs within the same KM form a unified object-oriented blackboard or, more precisely, a distributed dispatch-managing network [15,16].

In other words, the integration of a single user and his tools and methods can be viewed as the union of his judgements and experience with knowledge about his tools and methods. DICEtalk KMs provide a collection of standard high-level tools that complement the DICE intelligent database functionality. These tools may also be used by DICE knowledge-base system developers.

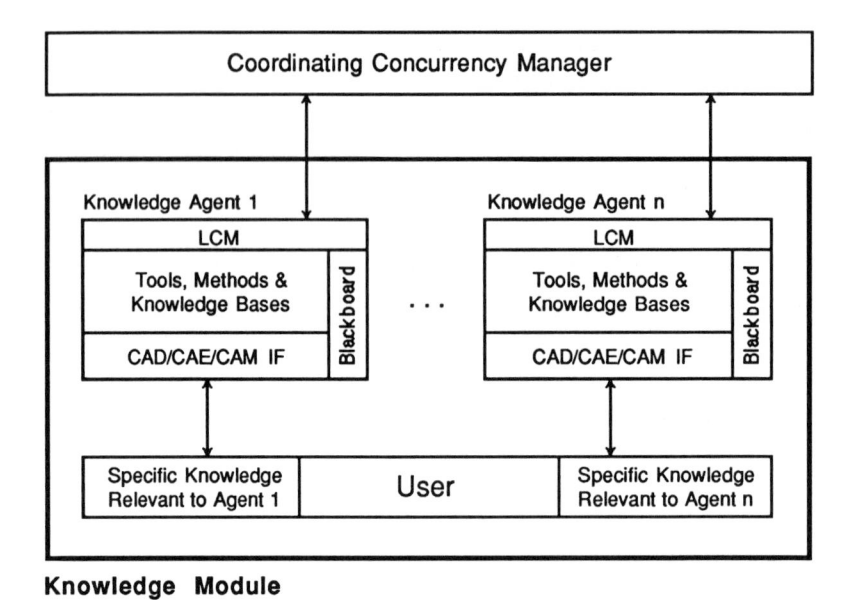

Figure 4.2.1. The Structure of a DICEtalk Knowledge Module.

4.3. Integration of Knowledge Modules

The DICE blackboard model [4,20] supports an object-oriented representation of PPO data that can be expressed and operated on in variety of ways. Product developers carry out their work on a part of the product by the current focus in the blackboard. They use their own experience and tools of their knowledge modules in their local workspaces to develop contributions to the evolving product development. In the DICE blackboard framework, two of the three blackboard panels, the design panel and the task panel, represent information needed by two closely coupled issues, that of global design and that of design tasks and fragments of the design being worked on. The third panel holds the assertions that are to be evaluated and possibly integrated into the design solution space. The blackboard enables designers to interact through the DBB interpreter and reach a consensus about tasks of the design before they are incorporated as part of the solution in the PPO data base as illustrated in Figure 4.3.1. Coordination of activities is supported by the Project Lead (PL) who disseminates local tasks to product developers. Solutions to local tasks are shared by product developers in the form of assertions. A final consistent solution configuration is approved by the PL before being checked into the PPO information model.

Thus, the DICE blackboard framework provides three distinct advantages:

a) it can be used to organize knowledge in a modular way (a collection of Knowledge Modules (KMs));
b) it can easily integrate different representation methods (frames, percepts [16], etc.) for different KMs;
c) it provides coordination in a distributed computing environment through the Concurrency Manager (CM).

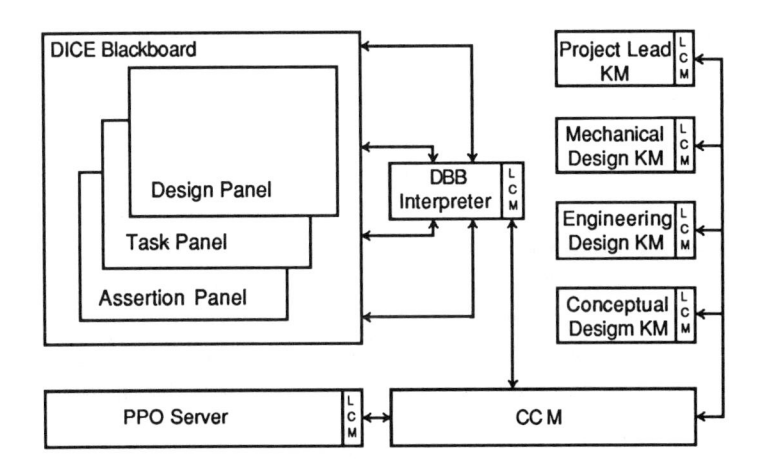

Figure 4.3.1. The DICE Blackboard System.

References

1. Suzaki, K.: The New Manufacturing Challenge - Techniques for Continuous Improvement . The Free Press, Div. of Macmillan Inc.
2. Reich, R.B.: The Quiet Path to Technological Preeminence. Scientific American, Oct. '89.
3. Evans, B.: Simultaneous Engineering. Mechanical Engineering, Feb. '88.
4. Munro, A.S.: Simultaneous Engineering- Questions with Answers. Proceedings of Conference on Simultaneous Engineering for Improved Product and Manufacturing Interface, Sponsored by SME, Oct. '88.
5. Wood, R.T: Overview of DARPA Initiative in Concurrent Engungeering. Symposium on Concurrent Engineering, May '89 at CERC, WVU.
6. Fabrycky, W.J.: Engineering and Systems Design: Opportunities for ISE Professionals. Proceedings of IIE Integrated Systems Conference, Nov. '89.

7. Crow, K.A.: Design for Manufacturability Its Role in World Class Manufacturing. Proceedings of Second International Conference on Design for Manufacturability. CAD/CIM Alert, Nov. '88.
8. Devol, F.E.: Design for Manufacturability: The Realities of Implementation. Proceedings of Second International Conference on DFM. CAD/CIM Alert, Nov. '88.
9. The Role of Concurrent Engineering in Weapon Systems Acquisition. IDA report R-338, Dec. '88.
10. Sarin, S.C.; Das, S.K.: Production Control Implications for CIMS. Proceedings of International Conference on CIM, May '88.
11. Five Ways to Meet Japanese Challenge. Newsweek, Oct. '89.
12. Bruck, R.E.: Building Supply Chains for Simultaneous Engineering. Conf. on Achieve Total Quality Through Simultaneous Engineering, sponsored by The Mfg. Institute - A Division of Institute for International Research. Oct. '89.
13. Cleetus, J.; Ueijio, W.: Red Book of Functional Specifications for the DICE Architecture, Concurrent Engineering Research Center WVU, Feb 28, 1989.
14. Parsaye, K.; Chignell, M.; Khoshafian, S.; Wong, H.: Intelligent Databases, Object-Oriented, Deductive Hypermedia Technologies, John Wiley , 1989.
15. Sobolewski, M.: DICEtalk: An Object-Oriented Knowledge-Based Engineering Environment. Proc. of the Fifth Int. Conference on CAD/CAM, Robotics and Factories of the Future. Dec. '90.
16. Sobolewski, M.: Percept Knowledge and Concurrency. Proc. The Second National Symposium on Concurrent Engineering, February 1990, Morgantown, West Virginia.
17. Hardwick, M. et. al., ROSE: A Database System for CE Applications, ibidem.
18. Sarachan, B.D.; Lewis, J.W., PPO Schema Management, ibidem.
19 Kannan, R., F.; Cleetus, K.J.; Reddy R., Distributed Computing with Local Concurreny Manager, ibidem.
20. Londono, F.; Cleetus, K.J.; Reddy R., A Blackboard Problem-Solving Model to Support Product Developement, ibidem.

Quality Design Engineering: The Missing Link in U.S. Competitiveness

H. BARRY BEBB

(Xerox Vice President, Retired)
Barry Bebb & Associates
Rochester, NY

Introduction

Past efforts to understand why Japanese Corporations deliver products with superior appeal to customers have often focused on the symptoms of what manufacturing delivers rather than the upstream Engineering Design fundamentals that determine what manufacturing can deliver. Detailed comparisons of Japanese and U.S. processes and practices utilizing Competitive Benchmarking concepts and techniques created by Xerox in the early 1980's [1,2] identify Engineering Design as a major competitive factor. In this article, these Competitive Benchmarking processes are used to demonstrate that Engineering Design process capability is the "missing link" in efforts to achieve world class competitiveness in mechanically intense systems found in most home, office, military, aerospace and factory products. Competitive benchmarking is also used to identify Quality and Engineering Design practices that need to be implemented to improve U.S. competitiveness.

Competitive Benchmarking of Product Development Processes

In the early-1980s, Xerox compared the contributions of engineering and manufacturing to the quality, cost and schedule outcomes of many commodities including plastic piece parts. American plastics suppliers proved to be a factor of two higher in cost than Japanese suppliers. About one-half of the short-fall was attributed to Xerox Engineering Design process capability and the other one-half to manufacturing process capability. The Japanese were simply out-designing U.S. industries: more functions per plastic part, fewer inserts, better manufacturability associated with mold designs, fewer finishing operations and so forth [3].

At about the same time that the Xerox study was in progress, Larry Sullivan of Ford Motor Company published the now well-known comparisons of Engineering Change Rates, Schedules and the role of <u>Quality Function Deployment</u> in comparable Japanese and American automobile corporations [4]. The results were disheartening. The best-of-best Japanese Competitive Benchmark Corporations developed new automobiles in one-half the time with one-half the people compared to Detroit corporations. This conclusion differs from the often cited comparisons of Japanese versus American automotive development schedules. Averages reference a collage of business needs and capability factors. Best-of-best benchmarking compares the fundamental capabilities of the world's best corporations. To evaluate basic process capabilities, best-of-best benchmarking processes must be employed.

Xerox 1981 profits plummeted to one-third of the 1979 levels [1]. New product offerings exhibited quality problems in the field which caused Customer Dissatisfaction. In addition, as David Kearns, the CEO of Xerox, often cites, the competition was introducing products at <u>prices</u> that were less than Xerox <u>manufacturing costs</u>. Continuing benchmarking studies compared Japanese and American manufacturing outcomes based on American designs and Japanese designs of the same product; these comparisons of American and Japanese Designs and American and Japanese manufacturing outcomes again showed that engineering design was a larger determinant than manufacturing of product cost and quality. <u>Synthesis of the myriad of published and Xerox internal benchmarking studies clearly establishes engineering design quality as a critical factor in achieving reduced time to market with high quality and low cost products [5]. These studies identify competitive differentiation in terms of</u>

<u>1) how engineers practice Engineering Design,</u>

<u>2) how managers supervise Design Engineers,</u>

<u>3) what contemporary design skills are required and trained and</u>

<u>4) what expectations are established by senior management.</u>

<u>These are not subjects that senior management of U.S. corporations often deal with.</u> <u>These are subjects that Japanese Corporations engage as a part of the normal course of business.</u> <u>As a result of these studies, I have concluded that Engineering Design is the Missing Link to U.S. competitiveness.</u>

A Concurrent Engineering Design Model

A model for organizing and communicating engineering design concepts is the "Bottle Model" in Figure 1. The center axis is the design set point. The "Bottle" curve that opens up from left to right over time depicts the product design iteratively progressing toward increasing latitudes of operation (robustness in Taguchi terms) [6]. Latitude defines the range of possible operating conditions in which the product can perform satisfactorily. The "Bottle" that closes down from left to right depicts the manufacturing process design dynamics of reducing variances (noises in Taguchi terms) [6].

<u>Engineering Design Quality</u> is defined as the difference between the latitude and the variance curves. The model differs from the more traditional optimization procedures in that it <u>treats reduction in variances and improvement in latitudes as co-equal design objectives</u>. While the concept of design latitudes is broadly understood, the concept of design variances is less familiar. Tolerancing provides a familiar example. "Loose" tolerances are easier to achieve in manufacturing than tight tolerances. Hence, "loose" tolerance designs reduce manufacturing variances.

Concurrent Engineering Design (or Simultaneous Engineering) is depicted in the "Bottle" model by the overlapping latitude and variance bottle shaped curves. Integration of engineering, manufacturing and other disciplines into teams is necessary to create the total set of skills needed to concurrently design for larger latitudes and reduced variances throughout the detailed design phases of Product Design and Manufacturing Process Design identified in the "Bottle" figure. The dotted lines delineating the four generic product development phases of Concept, Product Design, Manufacturing Process Design and Production are slanted to further emphasize the overlap and parallel execution of the various design tasks.

Good engineering design demands early attention to manufacturing requirements, and good manufacturing process design requires people skilled in manufacturing processes to participate in the very earliest Concept and Product Design phases of a product development program. Integration of engineering and manufacturing is essential to achievement of competitive product offerings. However, the notion of Concurrent Engineering Design must be extended to encompass Customers and Service representatives as full team members. The core set of functions that must be simultaneously engaged on product development programs from concept generation through launch include Customers, Product Planning, Engineering, Manufacturing (including Suppliers) Service, Human Factors, Industrial Design, Safety, Environmental Engineering, Distribution, Marketing and Sales.

The fundamental issue addressed by the notion of Concurrent Engineering Design is Design Quality prior to undertaking the build, debug, test and refine cycle for the first fully integrated Engineering Prototype. A benchmark single-cycle process for design, build, test and fix of a prototype is compared with a three-prototype-cycle process in Figure 2. At the start of detailed design, both examples show variances exceeding latitudes. In the benchmark single-prototype-cycle, functionally robust and manufacturable designs are produced by testing and refining models built to the first designs. American firms often initiate the first Engineering Prototype build prior to achievement of a Design Quality where latitudes exceed variances. In this circumstance, some tinkering and adjusting of each of the, say, 25 prototype models is required to make them operational. Thus, each model operates at a different set-point. Indeed, some of the units may never work satisfactorily and become relegated to carcasses from which parts can be scavanged. Clearly, a design that yields unit-to-unit variation is not sufficiently robust for quantity production; hence, a redesign and second build, test and refine cycle is required.

The curves depicting a three-prototype-design-build-test-refine cycle characterize chaos. When the first prototype build cycle commences, <u>the grander notions of improving design quality by increasing latitudes and reducing variances succumb to the tasks of assembling things, chasing parts, debugging things and so forth</u>. Experience and research [7] show that <u>bad first designs are irrecoverable</u>. No matter how many prototype iterations are undertaken, some portion of the first design decisions will persist through all iterations.

<u>Design Quality and the "Bottle" Model</u>

The collection of Taguchi methodologies provides a coherent description of the various elements of the "Bottle" model. The "Bottle" depiction of opening latitudes relates directly to Taguchi's orthogonal array characterization. Closing variances relates to Taguchi Manufacturing Process Paramenter Design and Taguchi Tolerance Design. The Design Quality can be evaluated continuously throughout the design process by relating the Taguchi Quality Loss Function to the difference between the latitude and the variance. While Taguchi provides a consistent characterization of the "Bottle" model, other powerful Engineering Design processes such as the myriad of Optimization methodologies described by Reklaitis, Ravindran, and Ragsdell [8] and Design for Manufacturability, Design for Serviceability and so forth [3] must continue to be employed to realize World Class engineering designs capable of yielding World Class manufacturing quality and customer satisfation.

Finally the design parameters characterized by the "Bottle" curves relate directly to Statistical Process Control (SPC) parameters utilized in manufacturing as illustrated in Figure 3. This linking of manufacturing SPC quality metrics and the upstream Product and Manufacturing Process Design "Bottle" parameters brings to mind the old adage that high quality and low cost cannot be sprinkled on in manufacturing and test. Quality, Cost and Time-to-Market are largely determined by the quality of the engineering design utilized for the first Engineering Prototype build, test and refine cycle.

Product Cycle Time and Continuous Improvement

Schedule is the most pressing problem confronting many U.S. corporations. Ralph Gomory and Roland Schmitt [9] distinguish between Idea Dominated Development and Incremental Development. Idea Dominated Development references reduction to practice and commercialization of new technologies. Incremental Development references the development cycle of the next automobile model, the next generation of RAM memory, the next generation of computers and so forth. The quality of Incremental Development depends on the quality of detailed design processes involving mature technologies. The quality of Idea Dominated Development depends on the quality of the technology reduction to practice.

Gomory and Schmitt argue that design improvements are only captured during a full product development cycle. For example, while manufacturing is producing 256K RAMs, engineers are developing the designs and processes for 1M RAMs. If the engineering teams are a year slower than the competition, their company will be non-competitive. A corporation might survive a one-year shortfall; however, if the next product design cycle is also a year slower, then the two year deficit would surely be fatal.

This thesis that incremental improvements are only captured during a full product development cycle <u>transforms schedule from a tactical "first-to-market" action into a strategic process for accelerating continuous improvements</u> as illustrated in the upper portion of Figure 4. This cycle time perspective of engineering improvement rates in engineering parallels the manufacturing learning curve perspective that relates the volumes achieved per unit time to the rate of improvement realized.

The lower portion of Figure 4 reflects the impacts of cycle time differences on the mature portions of the "S" curve paradigm [10]. Traditional treatments of the "S" curve paradigm conclude that investment in a technology is more rewarding during the rapid exploitation period, and as that technology matures, the rate of improvement slows, thereby reducing the impact of further investment [10]. The Japanese changed the rules [5]. <u>They invested in downstream development processes to achieve strategic level improvements in the quality, cost and schedule in very mature mechanical technologies</u>. Downstream product development process improvement investments are highly leveraged and typically modest in magnitude. As many Japanese corporations have demonstrated at the expense of their American competitors, the payback from such investments can be enormous. They taught the painful lesson that <u>competitive privileges derived from downstream Incremental Development improvements are often larger than those derived from upstream innovations</u>.

Change Agents for Continuous Improvement

Processes such as Design-for-Assembly [11], Taguchi methodologies [6] and Quality Function Deployment [12,13] have proven effective when they are utilized. However, benchmarking visits to a variety of large and small U.S. discrete manufacturing corporations revealed that while the benefits of utilizing these and other contemporary Engineering Design processes were known, they were not consistently implemented as part of the main stream product development process. In most of the large corporations, advocates have utilized the newer processes on a project or two and published the remarkable improvements achieved in technical journals; eventually, these advocates become frustrated because their efforts are not broadly propagated through their enterprises. There appear to be two <u>barriers</u> to the adoption of new Engineering Design practices. First is the normal enterprise <u>resistance to change</u>. The second is <u>lack of change agents</u>.

Competitive benchmarking clearly suggests that successful Japanese and U.S. enterprises invest in change agents in order to hurdle the barriers of resistance to continuous improvement. Various mixes of change agents such as quality specialist networks, mandatory and voluntary training programs, job rotational programs, cross-organizational teams and 'centers of excellence' are being utilized. Concerning the specific topic of continuous improvement in Engineering Design, universities, government agencies, and several corporations (including Xerox) have established 'Design Institutes' chartered to find, internalize, develop training materials, train the trainers and then train their technical communities in the world's best Engineering Design practices.

<u>Training is a proven change agent</u> for propagating new knowledge and expertise through an enterprise. Without training capability, there exists little prospect that improved processes developed either inside or

outside an enterprise can be pervasively propagated throughout an enterprise. While it is common for a researcher to learn from studying the works of previous researchers, it is uncommon for the many teams that make up large enterprises to uniformly adopt the best practices of other, often geographically remote, teams or from external developments. Overt actions must be taken to penetrate enterprises to the level of improving and evergreening the technical skills of the practitioners and the management skills of their supervisors. Centers of Excellence, such as 'Design Institutes', can synthesize best practices, often published in foreign languages, and provide uniform leadership and training throughout an enterprise.

Utilizing some portion of the above recommendations, Xerox has improved integrated engineering and manufacturing processes to achieve competitive parity in terms of the major quality and cost metrics. While Xerox has improved schedule by a factor of two, substantial additional improvements are required to achieve competitive benchmark levels. Xerox's dedication toward continuous quality improvement in the past decade contributed to Xerox's winning the Malcolm Baldrige National Quality Award for 1989.

Acknowledgements

The original concepts that led to the "Bottle Model" construct are due to Don P. Clausing, the Bernard M. Gordon Adjunct Professor of Innovation and Practice at the Massachusetts Institute of Technology, and Maurice Holmes, Vice President and Chief Engineer at Xerox. Ken Ragsdell, Associate Vice Chancellor of the University of Missouri-Rolla, has contributed toward improving Xerox engineering design capabilities throughout the decade of the 1980's. Professor Jack Dixon, Director of the Mechanical Design Automation Lab, and Mike Duffey, Associate Director of the Mechanical Design Automation Lab of the University of Massachusetts, contributed significant ideas and editorial perspectives to this article. Without the continued support and direct help of the above referenced people, this article would not exist.

REFERENCES

1. Jacobson, G. and Hillkirk, J., *Xerox American Samurai*, MacMillan Publishing Company, 1986.

2. Camp, R. C., "Benchmarking: The Search for the Best Practices that Lead to Superior Performance", a series in *Quality Progress*, 1989.

3. Bebb, H. B., National Academy of Engineering article based on my presentation within the forum "Responding to the Competitive Challenge, the Technological Dimension", published in The Bridge, Vol. 18, No. 1, Spring 1988, pp. 4-6.

4. Sullivan, L. P., "Quality Function Deployment", *Quality Progess*, June 1986, p. 39.

5. Abegglin, J. C. and Stalk, G., *Kaisha, The Japanese Corporation*, Basic Book Inc., 1985.

6. Taguchi, G., *Introduction to Quality Engineering, Designing Quality into Products and Processes*, Asian Productivity Organization, 1986 and *System Experimental Design, Engineering Methods to Optimize Quality and Minimize Cost*, UNIPUB/Krauss International Publications, 1987.

7. Jansson, D. G., and Smith, S. M., "Design Fixation", *Preprints, 1989 NSF Engineering Design Research Conference*, College of Engineering, University of Massachusetts, Amherst, June 11-14, 1989, pp. 53-79.

8. Reklaitis, G. V., Ravindran, A., and Ragsdell, K. M., *Engineering Optimization Methods and Application*, John Wiley and Sons, 1983.

9. Gomory, R. E., and Schmitt, R. W., "Science and Product", *Science*, 240, May, 1988, pp. 1131-32, 1203-04.

10. Dieter, G. E., *Engineering Design, A Materials and Processing Approach*, McGraw Hill, 1983.

11. Boothroyd, G., Poli, L. and Murch, L. E., *Automatic Assembly*, Marcel Decker, New York, 1982.

12. Hauser, J. R. and Clausing, D. P., "The House of Quality", *Harvard Business Review*, Vol. 66, No. 33, May-June 1988, pp. 63-73.

13. Cohen, L., "Quality Function Deployment: An Application Perspective from Digital Equipment Corporation", *National Productivity Review*, Summer 1988.

"BOTTLE" MODEL AND DESIGN QUALITY

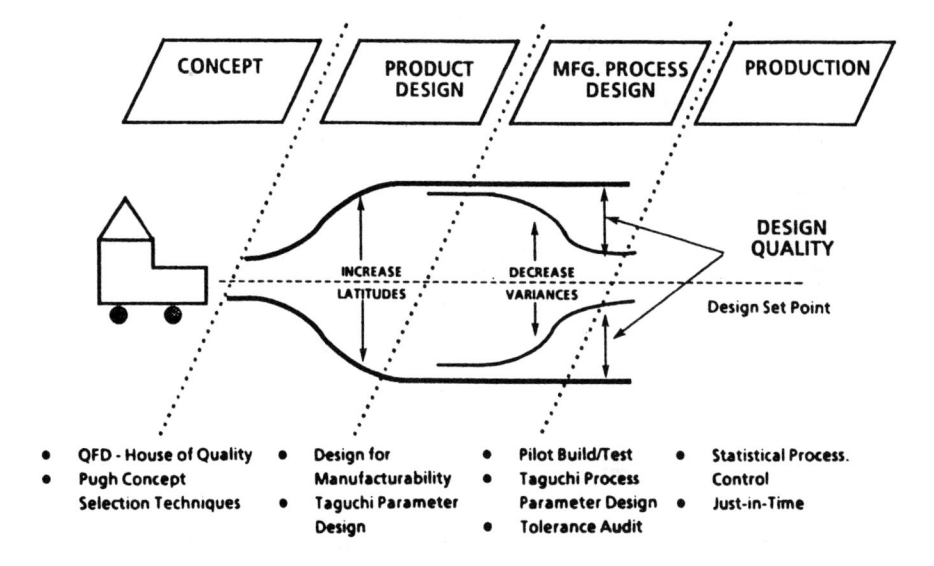

- QFD - House of Quality
- Pugh Concept Selection Techniques
- Design for Manufacturability
- Taguchi Parameter Design
- Pilot Build/Test
- Taguchi Process Parameter Design
- Tolerance Audit
- Statistical Process Control
- Just-in-Time

Figure 1

Product Design is the process of increasing engineering latitudes of operation. Manufacturing process design is the process of decreasing manufacturing variances. Latitudes represent "robustness against noises" in the Taguchi model and variances represent noises. The difference between latitudes and variances is a measure of design quality. Simultaneous engagement of Product Design and Manufacturing Process Design is frequently referenced as concurrent engineering.

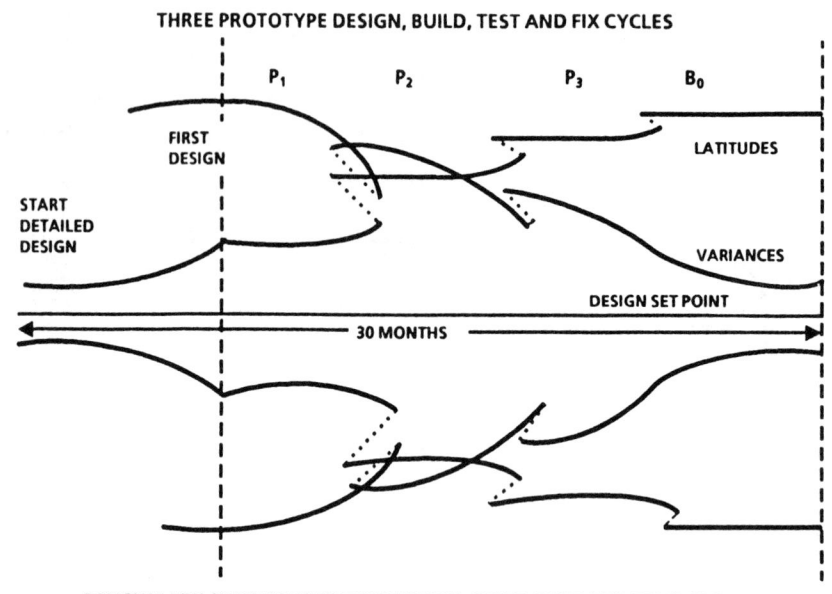

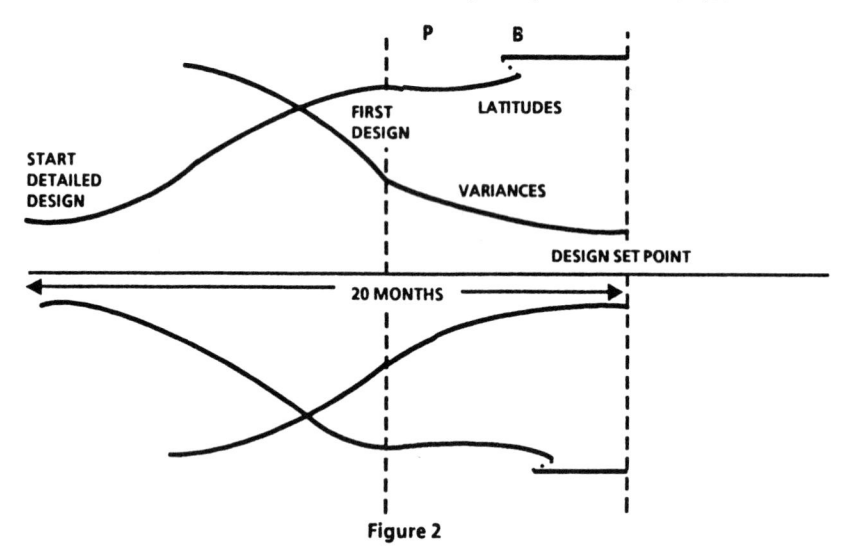

Figure 2

"Bottle" representations of three prototype cycle and single prototype development programs (See text).

RELATIONSHIP OF "BOTTLE" WITH COMMON QUALITY MEASURES

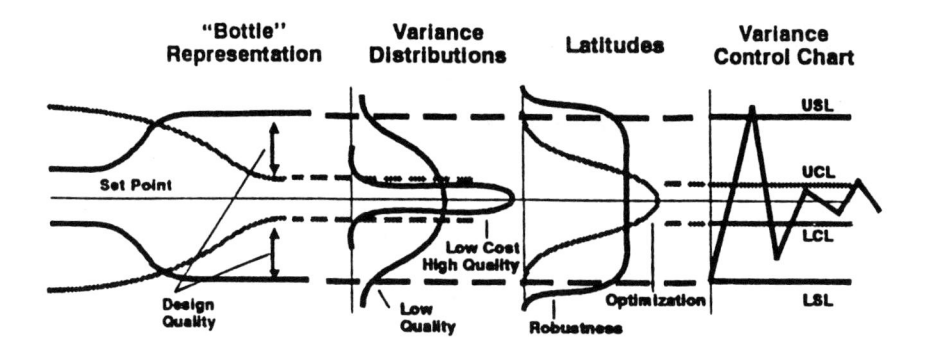

Figure 3

The relation of the "Bottle" representation to common quality measures such as Variance Control Charts and Variance Distributions is depicted. Performance latitudes are also shown in relation to the "Bottle" characterization. Performance values in <u>robust designs</u> do not change significantly for all operating parameters within the latitude window as illustrated by the flat latitude curve labeled Robustness. Performance optimization processes generally yield higher performance than robust design processes at the set-point; however, as the curve labeled <u>Optimization</u> indicates, the performance deteriorates more rapidly as parameters vary away from the set-point. Comparison of latitude and variance distributions yields an intuitive about the definintion of <u>Design Quality</u> as the difference between the latitude and variance "Bottle" representations. In addition, manufacturing Statistical Process Control factors (variances) are directly related to the up-stream Engineering Design outcomes.

THE RALPH GOMORY, ROLAND SCHMITT THESIS

1. IDEA DOMINATED DEVELOPMENT - U.S. IS GOOD AT
2. INCREMENTAL DEVELOPMENT - JAPAN IS GOOD AT!
   - CAPTURE IMPROVEMENTS DURING PRODUCT DEVELOPMENT CYCLE
   - MOST OF THE BATTLE IS INCREMENTAL DEVELOPMENT

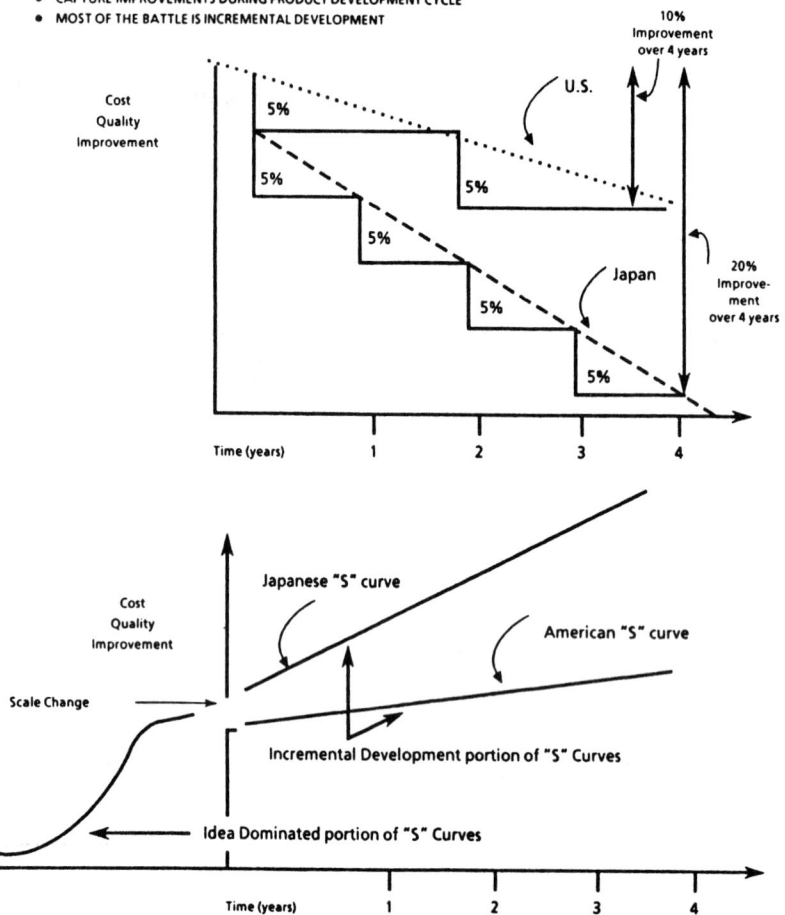

Figure 4
Incremental improvements are captured during Product Development cycles. Japan's success at reducing development cycle times increases rate of improvement. The lower portion of the graph depicts the impact on the traditional "S" curve paradigm. Japan has achieved enormous success by changing the rules of a mature technology such as mechanical rather than depending on breakthrough innovations.

New Activities in the Manufacturing Domain to Support the Concurrent Engineering Process

KEITH WALL and S.N. DWIVEDI

Concurrent Engineering Research Center
West Virginia University
Morgantown, WV

Abstract

Instilling productivity into the engineering process is addressed through the application of a concurrent engineering process. Parallel activities are designed into the engineering process providing high benefits in product value. Knowledge-based tools and methods along with new architectures for information handling improve the chance of successfully implementing a concurrent engineering process and further improving the productivity of engineering activities. This paper takes a look at creative activities that can be applied to the engineering process for the generic development of new products. These new activities are viewed from the manufacturing domain and seek to explore non-traditional roles in the early and late stages of product development.

A Concurrent Engineering Process

Concurrent Engineering (CE) is a process involving engineers of different disciplines who develop parallel project tracks and share relevant ideas and results. While traditional engineering processes employ a compartmentalized and serial approach toward developing new products, CE seeks to restructure traditional engineering activities in a way that progressively refines the product's definition while satisfying downstream life cycle concerns.

Process Engineering: A Migration Plan for CE

To understand the interworkings and relationships of the product development team, the process should be diagrammed in a format that highlights the individual activities and their communication needs. This establishes a blueprint of the process that can then be optimized for maximum efficiency. Serial and parallel relationships between activities will be obvious, and activities of low value to the process will be discovered. A first cut improvement exercise is to optimize the current process by weeding out the activities of low value. These activities are usually established to regulate the errors previously experienced. Before eliminating the low value activities, the root causes of the offending system activities must be identified, fixed, and made self-regulating. This is all part of a continuous improvement plan that is necessary prior to or in concert with the implementation of a restructured CE process. Process engineering is the next step to creating the process desired, and various standards and methods are available to the process engineer for performing this task. Nevertheless, a data flow diagram of the "as is" process is a good start for building an optimal process. Then the individual elements of the process can be visualized and communicated to the appropriate vendors for solutions in tools or architectures that the enterprise wishes to migrate toward.

Once diagrammed in sufficient detail, the task of improving and creating parallel process activities begins. Traditional downstream evaluations should be undertaken first to provide precursor examinations of conceptual designs. Incomplete data models can be evaluated in a statistical sense using designed statistical experiments and empirical design rules. Serial relationships can be considered customer-vendor models with each vendor satisfying the needs of the customer and each customer understanding the constraints of the vendor. This relationship aids the development of data flow requirements between activities. Parallel relationships are random in their communication needs and harder to model in the time domain. One activity usually has associative links with another especially when downstream integration activities exist. The random nature of communicating project results and the association of these results on parallel project activities make automation of the concurrent engineering process interesting.

This article makes a case for suggested new roles supported by the manufacturing function as a team participant in a CE development process. It is an attempt to define new contributions for a manufacturing professional in all phases of the product life cycle. In defining new roles, there must be some logic to adding activities even when project cycle time is not extended due to parallel processing. The new activities should adhere to prescribed conditions for acceptance and be of high value to the project. The following list describes many of the elements thought to be useful for an accurate model of a CE process. It can be used as a check sheet for screening and evaluating new and restructured process activities and their relationships. They are as listed here as goals of the process engineering task:

(1) Formulate better conceptual definitions that leverage competitive resources and focus on the customer's perception of value;

(2) Assess product risks early to minimize project time and/or cost while assuring product quality;

(3) Manage all life cycle constraints for design quality and product robustness;

(4) Promote interactive communications on an as-needed basis for efficient product development through information sharing;

(5) Capture design intent along with definition for better product representation late in the life cycle;

(6) Utilize old lessons learned and capture new ones so that the learning process is additive rather than recycled;

(7) Refine and improve design quality to reduce process noise and increase design tolerance;

(8) Make the process self-regulating to avoid low value activities (i.e. after-the-fact inspections and redundant approvals);

(9) Document the process model for future automation;

(10) Measure productivity improvements (project attainment vs. resources consumed);

(11) Automate stable processes for further enhancements in productivity;

(12) React dynamically to changes in the project requirements and resolve conflicts based on the customer's perception of value.

These can all be considered elements of improving white collar productivity with a focus toward automating product development.

Recommended Contribution Activities of the Manufacturing Engineer in the CE Process:

Before attempting to define process contributions, a generic description of the progressive phases of a product's life cycle will be discussed.

Requirements Definition:

The first phase is concerned with understanding the objectives of a new product. This usually encompasses an interpretation of the customer's needs put into terms of product performance and competitive attributes regarding project time, costs, and quality standards. A competitive definition of the objectives for a superior product is the end result.

Conceptual Design:

A phase where creative proposals for new products are offered up. Conceptual designs are high risk and vague in their descriptions because they offer new solutions that have yet to be tried by the enterprise. Some elements of the conceptual design may be well known to the enterprise and their risk can be classified. Typically, one of the aspects of the design or a unique combination of elements will be new to the experience of the enterprise and these will be assessed for further development risk. One or few conceptual designs will be started into the next design phase. The successful designs are chosen based on their promise of meeting or exceeding performance goals and are weighed against the risk that they will violate the constraints of the project at some future time.

Detailed Design:

Subsequent to the conceptual design phase is the Detailed Design phase, the phase in which validation and detailed definitions are developed for the product to be manufactured. Usually heavy in parameter testing and simulation, this phase attempts to minimize the risk of investing in a manufactured product and understanding the bounds of the products performance capabilities. A detailed product representation is the result of the detailed design phase.

Production:

This is full scale production where processes are optimized for efficient and effective operation. The production process must be competitive to produce high value products that have short inventory cycles, low cost, and high product quality assurance. Assumptions made earlier in the product development cycle concerning the producibility of the product are validated at this stage. Production resource levels are optimized and the manufacturing process is continually improved to incorporate alternate and emerging manufacturing processes. New producibility constraints are defined for use in the next product development cycle. The end result of this phase is a product that falls within the tolerances of its design and capabilities of it manufacturing process.

Support:

The support phase nurtures the product until retirement and is concerned with providing the customer with continued service of the product with reliable operation, and low cost maintenance. As the final phase, it continually tries to improve the product in these two aspects. Its overall objective is continuing customer satisfaction with the product and it accomplishes this by improving the reliability of the product and extending its life. The end result of this phase is the lessons learned about the overall customer satisfaction with the product. These feelings are archived and used to better develop similar products or product features.

A manufacturing engineer has a contributing role in all phases of the life cycle. A common misbelief is that the manufacturing engineer can only make critical interpretations of new designs. "Give me a design and I'll tell you why it won't work" and "it will cost too much to make it work" are the popular notions that design personnel have of the manufacturing input to the design problem. Many projects in concurrent engineering aim to develop design advisors that automate the ability to make critical judgement on new designs based on historical knowledge from a manufacturing data base. Others are satisfied to apply historical design rules about producibility when developing their product definitions in the attempt to Design for Manufacture (DFM). These design rules simply apply further constraints to the process of designing, but they make the problem narrower, less creative, and more complex. However, contributions can still be made to the up front development stage that are not critical in nature but rather add to the synergy of a creative definition process. The following is a proposal and defense of supporting activities that the Manufacturing Engineer can perform during all phases of the product life cycle. They are activities analogous to the tiger-team concept and other supported methodologies previously mentioned:

1) Competitive Manufacturing Resource Analysis (Requirements Definition)

Present competitive manufacturing techniques and resources that can provide an advantage in the value of the product. These techniques and resources could, all else being equal and in comparison to a competitor's product, provide a greater product value to the customer. The leveraging of this information could ease the requirements in a difficult feature of the design by allowing increased performance due to a particularly good manufacturing process or allowing project resources to be reallocated to more difficult challenges.

2) Resource Utilization for Conceptual Model Production (Conceptual Design)

Involving the manufacturing engineer in producing conceptual models aids the process in two ways. First, the manufacturing engineer performs a high level feasibility study and studies the design features represented by the model. Second, a conceptual design of the manufacturing process begins to take shape and the exercise can highlight sub-level problems brought out in the details of designing a representative manufacturing process. This can be especially true in assembly processes where assembly techniques can be determined from model construction.

3) Designing Statistical Experiments (Conceptual Design)

Knowledge-based reasoning can occur on the design features for which there exists historical attributes relating to the producibility of its product features. When creative design features emerge in a concept, there is still a need to quantify the manufacturing attributes of the feature even in a probability of occurrence sense. A well designed experiment for individual product features of unknown risk can benefit the Conceptual Design process by providing statistical probability data for risk assessment. The experiment allows the manufacturing engineer to further refine his manufacturing process design and gives better data for making cost and delivery estimates in the next phase of design. The ability of generating a more accurate cost and delivery estimate may be the most important contribution because so many decisions to proceed, modify, or kill are made based on attaining product cost goals. In real life situations, manufacturing engineers boost the cost estimate of products that employ features that don't have a sound basis for determining the historical effect on the manufacturing operation. Manufacturing cost estimates errors are always on the conservative side as are product performance estimates by designers.

As the process evolves into a more detailed design, the role of the manufacturing engineer gets more traditional. Cost estimates are made based on detailed definitions, and new processes are parameterized prior to production. If a good up front job of manufacturing process design is accomplished by the methods previously mentioned including some simulation and validation methods of the actual process, then the manufacture of a functional prototype can provide a vehicle for activities that involve characterizing process variance as defined below.

4) Process Characterization (Detailed Design)

This activity started in the earlier phase of the development with the employment of historical process data and designed experiments for new features and new processes. This step is a detailed process characterization by feature for the prescribed process. Its objective is to associate manufacturing process variance with feature tolerance. Design quality can be defined as the difference between the spread of the feature tolerance and the spread of the process control limits. Better design quality occurs when the feature tolerance range is greater than the control tolerance. Defining design tolerances is a detailed design task and it is up to the manufacturing engineer to get his process control capability data to the designer prior to establishing tolerances.

By participating in the up front design phases of the product's development, the manufacturing engineer now has insight into the needs of the designer. The production phase is a traditional role that focuses on the performance of the manufacturing organization for project performance attainment. In addition to the usual duties assigned, the manufacturing engineer now validates the assumptions about producibility and documents the lessons learned on new designs.

5) Building Attributes Descriptions of Product Features (Production)

In support of the next product development phase, the manufacturing engineer can now associate process performance with design features. Being involved in the process, he now knows the justification for choosing the product feature for the design and can suggest producible features from a product requirement need. This is an additional contribution that the manufacturing engineer can now contribute to the design process. Sometimes this knowledge can be valuable to the design group since the manufacturing engineer is exposed to a wider variety of designs from many different design groups. In this context, the manufacturing engineer acts as an integrating force for the different working design groups lending producibility knowledge and design alternatives derived from different projects.

When the products move into service, the manufacturing engineer must then react to the customer's.feedback. This is a traditional role. However, by working with support personnel, a plan can be laid out at the beginning of the production phase that will capture the processing data most needed by the support group.

6) Capturing Strategic Support Data (Support)

Manufacturing process data is often needed by the support organization to perform its duties efficiently. Such data can be used to determine fit-up dimensions to customer interfaces, to initiate and optimize product repairs and is helpful in analyzing weak areas in the product's reliability. An up front assessment of the data need for the support organization allows the manufacturing engineer to build a data collection scheme efficiently into the process.

In conclusion, this paper provides an outline for added value process activities that can be performed by the manufacturing participant in a concurrent engineering process. Most of the details of implementing these activities have been omitted, thus leaving this task for future work. A similar exercise performed for the Design, Support, and Project Management roles to determine value added activities in non-traditional phases of the product life cycle lays the framework for the CE process. The tools that enable a CE process do not make an enterprise more productive. Knowing how to be more productive in the engineering process and defining activities that contribute to the CE process allow product developers to leverage new technology tools in a productive way.

References

(1) DARPA Initiative in Concurrent Engineering (DICE), GE Aircraft Engines, Brochure, April 1990

(2) Red Book of Functional Specifications for the DICE Architecture, Concurrent Engineering Research Center WVU, Feb 28, 1989

(3) Winner, R. I., et. al., "The Role of Concurrent Engineering in Weapon System Acquisition" Institute for Defense Analysis, Report R-338, December, 1988.

(4) Taguchi, G., Elsayed, E.A., and Hsiang, T.C., Quality Engineering In Production Systems, McGraw-Hill Book Company, New York, 1989.

(5) Boothroyd, J. "Estimate Costs at an Early Stage," American Machinist, " Aug. 1988, pp. 54-57

(6) Introduction to Ptech, Associative Design Technology, brochure, Aug 15, 1989

(7) EIS Engineering Information Systems, "Vol I: Organizations and Concepts", Honeywell Systems and Research Center, October 1989.

(8) EIS Engineering Information Systems, "Vol III: Engineering Information Model Administrative Domain by TRW and ECAD Domain Model by Honeywell, Sensor and System Development Center," Honeywell Systems and Research Center, October 1989.

(9) Dhillon, Balbir S., Quality Control, Reliability, and Engineering Design, Marcel Dekker, Inc. New York, 1985

(10) Dehnad, Khosrow, AT&T Bell Laboratories, Quality Control, Robust Design, and the Tagucci Method, Wadsworth & Brooks/Cole Advanced Books & Software, 1989.

Chapter II
Intelligent Information Network

Introduction

Application of artificial intelligence tools at various design stages is described in the second paper in this section. For the synthesis phase, a rule-based component synthesis is presented, whereas for the evaluation phase, two inferential methodologies are presented. A broad perspective of conceptual design in a concurrent engineering environment is provided. Design of a Network Interface Unit (NIU) to integrate already existing equipment in a scenario of plant automation with LAN technology is presented in the next paper. The capability of a node on this network with an open-ended architecture is that shop floor equipment, process control devices, and equipment from a wide range of vendors having high incompatibilities through proprietary hardware and software can be integrated.

Data management has been a crucial issue in engineering design and analysis. The next paper describes a checker in an environment for application of software integration and execution (EASIE), where a number of independently developed programs exist to check data inconsistencies that are introduced to the database resulting from changes in data. In addition to presenting two efficient algorithms, the problem of flagging all the affected variables in the data set after having run the analysis programs on the modified data is also discussed in this paper. In the following paper, IGES translation for a mixed platform of CAD, CAM and CIM is discussed, and the future possibilities for solving the problem of data transfer are described. Methods involving fuzzy, set-based, single linkage cluster analysis and rank order clustering to develop a group technology cell formulation for machine components are the focus of the next paper in this section. The fuzzy rank order cluster alogorithm is capable of ranking machines with priority, and this enables one to deal with problems of machine capacity, exceptional elements, and bottle neck machines. The next paper describes the implementation and use of a design assessment tool in the DICE. The internal structure of the tool and the user interface are described in detail along with a brief discussion of work planned for the future. The last paper discusses integration strategies of local area networks connecting PCs with the DICE Network. Implementation of these strategies provides a relatively inexpensive method for constructing concurrent engineering systems in manufacturing industries.

Artificial Intelligence in Concurrent Engineering

A. KUSIAK and E. SZCZERBICKI

Intelligent Systems Laboratory
Department of Industrial Engineering
University of Iowa
Iowa City, IA

Summary

Design of a product progresses from description of needs and functions to detailed design of its components. To date a number of design methodologies have been developed. In this paper, features of various design methodologies are discussed. The impact of a concurrent engineering environment on the design methodology is presented. Concurrent engineering should benefit from artificial intelligence tools. Application of these tools at various stages of design is discussed.

Introduction

The complexity and the creativity associated with engineering design requires diverse problem solving techniques and expertise. Much of that can be accomplished with an artificial intelligence (AI) tools. Recent research efforts have resulted in a number of implementations of AI approaches in engineering fields related to design:

(a) design of printed circuit boards (Simoudis 1989, Steinberg and Mitchell 1984),

(b) design of computer software (Ellsworth 1989),

(c) architectural design (Flemming et al. 1988, Monoghan and Doheny 1986, Gero and Coyne 1987, Maher 1988),

(d) structural design (Grierson and Cameron 1988, Sriram 1985),

(e) mechanical engineering (Arora and Baenziger 1986, Serrano and Gossard 1988),

(f) chemical engineering (Banares-Alcantara et al. 1988),

(g) control engineering (Nolan 1986),

(h) engine design (Shen, Chew, and Issa 1989).

Careful study of the above and other applications shows that AI techniques should not replace a human specialist, but effectively perform specific tasks, that might form a critical or timely aspect of a larger problem. In other words, intelligent design systems should work with specialists rather then attempt to replace them. The latter has been considered while developing various design support systems reviewed in this paper. All the techniques presented were developed, or are under development in the Intelligent Systems Laboratory, Department of Industrial Engineering of the University of Iowa.

In the last two decades, attempts have been made to formalize the design process. Various methodologies have been developed. For example, a decision-making approach was proposed by Archer (1969) and recently expanded by Mistree and Muster (1989). System analysis and hierarchical decomposition was applied to design by Mesarovic, Macko, and Takahara (1970). An optimization approach was developed by Radford and Gero (1988) for architectural design, and by Arora (1989) for mechanical design. Design process was presented as an information processing activity by Koomen (1985) and Conant (1976). A knowledge-based approach was presented in Coyne et al. (1990). Pahl and Beitz (1988) and Hubka (1982) proposed a systematic approach for engineering design.

Design research has resulted in development of various models. A model that is widely agreed upon was proposed by Asimow (1962). The model includes three phases: analysis, synthesis, and evaluation. Since the model has been developed, the interpretation of its three phases has undergone changes. Nowadays, it is realized that the design phases should not be addressed in pure hierarchy or in ideal sequential order. Rather, they very often intermix and are carried out recursively. This is especially true when the evaluation phase is seen as the one that incorporates concurrency to the design process. In the paper, the three phases are briefly presented for the stage of conceptual design.

Analysis

The analysis phase, called also the formulation phase (Maher 1989) or specification phase (Pahl and Beitz 1988), is embedded in the space of needs and is to specify requirements and functions. The search of the specification space in conceptual design is qualitative and quantitative in nature. Thus it is impossible to use the existing search methods for the search because all of them use quantitative evaluation functions. A generalized search method to deal with qualitative and quantitative aspects of the search is required. A possible implementation of the generalized search is to use a set of production rules to guide the search. The generalized search algorithm is presented next (Kusiak, Park, and Szczerbicki 1990).

Step 1. Form one-element queue OPEN consisting of the root node.
Step 2. If the first element of the queue OPEN is a goal node, stop;
Otherwise, expand the first element of the queue OPEN generating its child nodes and remove the first element from the queue OPEN.
Step 3. If only one child node is generated, then add the child node at the beginning of the queue OPEN and go to Step 2; Otherwise go to Step 4.
Step 4. If qualitative knowledge is associated with child nodes, apply production rules to select nodes for further consideration. The selected nodes are added to the queue OPEN;
Otherwise, compute the cost of child nodes and add them to the queue OPEN.
Step 5. Sort the queue OPEN by estimated cost and go to Step 2.

In the algorithm a set of general production rules is used to decompose requirements into subrequirements. For example:

Rule 1
    IF   a requirement corresponds a function
    THEN   do not decompose it

Rule 2
    IF   a requirement does not correspond a function
    THEN   decompose it

Domain specific production rules for decomposition of requirements are also used. Examples of the domain specific production rules for design of a shaft coupling are presented next:

Rule 3
    IF   a requirement is "Design a shaft coupling"
    THEN   decompose it into "The nature of coupling is rigid"
    AND   "Coupling is able to transmit torque"
    OR   "The nature of coupling is flexible"
    AND   "Coupling is able to transmit torque".

Rule 4
    IF   a requirement is "Coupling is able to transmit torque"
    THEN   do not decompose it

Rule 5
    IF   a requirement is "The nature of coupling is flexible"
    THEN   decompose it into "Coupling allows for radial offset of the shaft"
    AND   "Coupling allows for axial offset of the shaft".

Examples of the domain specific production rules for selecting alternatives in conceptual design of are illustrated below:

Rule 14
    IF   a requirement is "Design a shaft coupling"
    AND   the nature of coupling is rigid
    THEN   select "The nature of coupling is rigid"
    AND   "Coupling is able to transmit a torque".

Rule 15
    IF   a requirement is "Design a shaft coupling"
    AND   the nature of coupling is flexible
    THEN   select "The nature of coupling is flexible"
    AND   "Coupling is able to transmit a torque".

The proposed generalized search has the following characteristics:
- It deals with both quantitative and qualitative aspects of the problem.
- It uses a set of production rules to guide the search.
- The computational effort to find the best solution is modest.
- New design knowledge can be incorporated into the existing knowledge base for conceptual design.

Synthesis

Design synthesis has been of interest to numerous authors. Maher (1989) applied synthesis to the domain of structural and foundation design. Chang and Tsai (1989) applied topological synthesis for design of gear mechanisms. A general synthesis strategy for mechanical design was proposed by Hoover and Rinderle (1989). Tsai and Freundenstain (1989) described synthesis of robot configuration. The synthesis of schematic descriptions was presented by Ulrich and Seering (1989). Hardware components were synthesized using design heuristics in Chin (1989). Synthesis is regarded as a creative process (see, for example, Hoeltzel and Chieng 1990) that is based on the general systems theory (Rapaport 1986, Mesarovic and Takahara 1989) and uses hierarchical structures introduced in Mesarovic et al. (1970) and more recently discussed in Oren (1984) and Zeigler (1984, 1987).

Kusiak and Szczerbicki (1990a) present a modeling approach to synthesis at the conceptual design level. Synthesis of components and models into an overall model of the designed object begins at the lowest level (level 1) of abstraction represented by the initial state of a model base. It is performed according to the algorithm presented next.

Step 1. Open a set MODEL_BASE consisting of all components generated at the specification and representation stage.
Set level = 1.

Step 2. Apply production rules to generate connections between elements in MODEL_BASE.

Step 3. If no connections are generated, stop;
Otherwise, match elements in MODEL_BASE into pairs using the existing connections.

Step 4. Define input and output variables for models generated by the matching process.

Step 5. Remove from MODEL_BASE all elements that have taken part in the matching process.

Step 6. Add to MODEL_BASE all models generated by the matching process.

Step 7. Set level = level + 1 and go to Step 2.

A set of production rules is used in the algorithm to generate connections between the elements stored in a model base. Production rules are structured to ensure that:

(i) an overall model of the designed object is generated in such a way that all components are included (the set of components at the initial state of a model base represents all functions specified at the specification stage),

(ii) all boundary inputs and outputs of the designed object are included in the synthesized model and they represent the only way that the designed object communicates with its external environment,

(iii) only physically and logically feasible matchings of components can be explored (i.e., only feasible variants of the design).

For example, production rules 5 and 6 make sure that components that have been defined as boundary elements at the representation stage create also the boundary of the overall model.

Rule 5

    IF   an element is an input boundary element
    THEN it can not accept an input from any other element

Rule 6

    IF   an element is an output boundary element

THEN   it can not provide an input to any other element

An object-oriented implementation of the above algorithm is presented in Kusiak, Szczerbicki, and Vujosevic (1990). The object-oriented programming paradigm is chosen since it simulates the designer's way of thinking during the conceptual design process. The system is implemented in Smalltalk-80 programming language and environment.

Evaluation

Evaluation can be seen as a process that involves interpreting a partially or completely specified design description for conformance with the expected performance (Maher 1989). Evaluation, during the early stages of design, is usually based on a subjective assessment of relevant criteria with incomplete information. In the absence of complete information it is not feasible to use an evaluation technique that requires a rigorous mathematical formulation of the problem. Hence, the methods that consider subjective multiple criteria dealing with uncertainty, imprecision, and incomplete information are recently used as design evaluation techniques. The relevance of imprecision and uncertainty to engineering design has been discussed in Wood et al. (1990). The concept of Pareto optimality was used in identifying the appropriate or best design solutions in Maher (1989). A model for design evaluation in simultaneous engineering using design compability analysis was developed in Ishii et al. (1989).

Concurrency evaluation refers to the integration of various activities within the broad scope of the product life cycle (Nevis and Whitney 1989). It is generally recognized as a practice of incorporating various life cycle values (concurrency attributes) into the early stages of design. Conceptual design is probably the most important phase in concurrent engineering and at the same time relatively unexplored. The concurrency attributes may include machinability, reliability, appearance, portability, safety, and so on. Different attributes are relevant to different design steps and thus the evaluation technique should be able to accept concurrency information and perform updating in a way that parallels availability of the information. Bearing this in mind, a methodology for concurrency analysis that uses the formalism based on Bayes' theory was proposed in Kusiak and Szczerbicki (1990). Bayes' theory was succesfully incorporated into inference engines (see, for example, Duda et al. 1981 and Forsyth 1984) and proved to be usefull in dealing with ill-defined problems. The Bayes approach helps to combine sample concurrency information with the prior information in evaluation of concurrency satisfaction during the conceptual design process. The probability associated with this prior information is called a subjective probability, in that it measures a designer's degree of belief in a given proposition. The designer uses his own experience and knowledge as the basis for arriving at a subjective probability (Kusiak and Szczerbicki 1990). Given a priori probability about the initial satisfaction of a certain concurrency attribute, the designer can modify this probability to produce a posteriori probability while moving through the design steps and adding some new components. The process of updating is repeated and each time the probability of the concurrency attribute being satisfied is shifted up or down using a Bayesian rule with a different prior probability derived from the last posterior probability. In the end, having collected all the available information, the designer can come to a final conclusion using the idea of upper and lower thresholds defining areas of acceptance and rejection as illustrated in Figure 1 (Kusiak and Szczerbicki 1990).

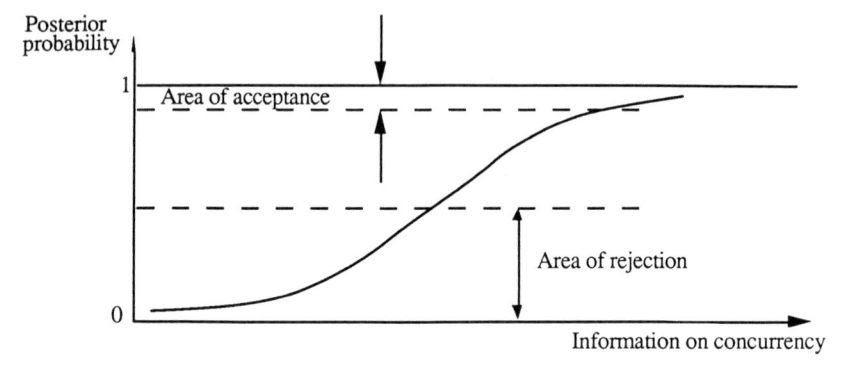

Figure 1. Upper and lower thresholds of the posterior probability for the attribute "reliability"

The main advantages of the Bayesian technique of handling uncertainty and incomplete information concerning the satisfaction of various concurrency attributes at the conceptual design level are as follows:
- it is based on probability theory which is easy to understand and accept, and which seems to mirror the natural way of thinking,
- it represents straightforward computational procedure that can be readily implemented,
- final posterior probabilities are not affected by the order in which the components are connected.

Above all, it represents a flexible technique which keeps the designer involved in the concurrency problem and thus, enables him to learn and benefit from the design process itself.

Bayesian approach is a "one-valued" approach as it represents the probability with one value only. This drawback can be overcome by the approach based on the Dempster-Shafer theory that was developed to create a "two-valued" approach (Garvey et al. 1981). Using this theory, the concurrency satisfaction can be represented by two values, a minimum and maximum likelihood. Uncertainty of the concurrency attribute is the difference between this two values. The likelihood of a hypothesis H about concurrency satisfaction is represented as an interval H[s(H),p(H)], where s(H) is the support for H (the minimum likelihood of H being true), and p(H) is the degree of plausibility of H (the maximum likelihood of H). The degree of uncertainty about the actual probability value of the hypothesis corresponds to the width of the interval u(H)=p(H)-s(H). Hence, if u(H) is zero the system becomes Bayesian. To clarify the above, the following examples are considered:

H[0,1]   no information at all about the concurrency attribute H,
H[0,0]   H is definitely false,
H[1,1]   H is definitely true,
H[0.25,1]   partial support is provided for H,

H[0,0.85]   partial support is provided for negation of H,
H[0.25,0.85] probability of H is between 0.25 and 0.85, that is, support is provided simultaneously for H and its negation.

The basic idea behind the application of Dempster-Shafer theory to concurrency evaluation is to have several knowledge bases that guide and evaluate the design from different perspectives, while the designer himself is responsible for making final decisions based on the model ratings and suggestions. This knowledge bases (knowledge sources) correspond to various life-cycle concerns that the designer should take into account. It is assumed that a knowledge source KS1 distributes a unit of belief across a set of concurrency attributes for which it has a direct evidence for support or negation. Thus a set of attributes, say $H1_1$, $H1_2$, $H1_3$, and so on has a total sum of belief equal to 1, i.e., $\Sigma m_1(H1_i) = 1$, where $m_1(H1_i)$ represents the portion of belief that KS1 has commited to the hypothesis of attribute H1, called the basic probability mass. If a second knowledge source KS2 is applied (or the opinion of another expert is taken into consideration) information is combined by computing the orthogonal sum as illustrated with the rectangle in Figure 2.

$$m_{1,2}(H1_i, H2_j) = m_1(H1_i) m_2(H2_j)$$

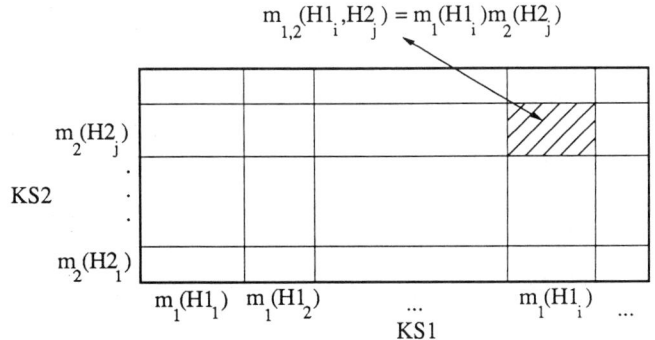

Figure 2. Composition of mass from knowledge sources KS1 and KS2

Figure 2 shows a vertical strip of measure $m_1(H1_i)$ commited to $H1_i$ by KS1, and a horizontal strip of size $m_2(H2_j)$ commited to $H2_j$ by KS2. The intersection of these strips commits $m_1(H1_i)m_2(H2_j)$ to the combination of $H1_i$ AND $H2_j$.

To allow for propagation of evidential information of various concurrency attributes, a number of rules is developed. As an example, the following rule is given (Szczerbicki 1990):

Rule 1
    IF $H1_1[s_1(H1_1), p_1(H1_1)]$
    AND $H1_2[s_2(H1_2), p_2(H1_2)]$
    THEN $H1[s(H1)p(H1)]$

where $s(H1)=\max[s_1(H1_1), s_2(H1_1)]$ and $p(H1)=\min[p_1(H1_1), p_2(H1_2)]$.

The Dempster-Shaffer methodology is an extention of the Bayes theory and as such has all the characteristics listed previously. Besides, the following may be considered as new advantages of this methodology:

(i) The precision of a hypothesis about concurrency satisfaction is clearly indicated by the difference p(H1)-s(H1). If the difference is small, then the knowledge about H1 is relatively precise, if large then correspondingly little is known. If p(H1)=s(H1), then knowledge of H1 is exact and reverts to point probability approach.
(ii) The amount of support for and against concurrency satisfaction is easily understood. The amount of support for H1 is at least s(H1) and the amount of support for the negation of H1 is 1-p(H1).
(iii) The methodology is developed on the basis of probability theory from which s(H1) and p(H1) are interpreted.

Conclusion

In the paper, three phases of conceptual design are discussed. Recommendations are made for implementations of various AI techniques suitable in design analysis, synthesis, and evaluation. The analysis phase is supported by a generalized search technique in design space represented by AND/OR trees. For the synthesis phase, a rule-based component synthesis is proposed. The evaluation phase is seen as one that incorporates concurrency in the design process. For this phase, two inferential methodologies are presented. The intent of this paper is to provide the reader with a broad perspective of conceptual design in a concurrent engineering environment.

References

Archer, B.L. (1969), The structure of the design process, in Broadbent, G. and Ward, A. (Eds), *Design Methods in Architecture,* Lund Humphries, London, pp. 76-102.

Arora, J.S. (1989), *Introduction to Optimum Design,* McGraw-Hill, New York.

Arora, J.S. and Baenziger, G. (1986), Uses of artificial intelligence in design optimization, *Computer Methods in Applied Mechanics and Engineering,* Vol. 54, pp. 303-323.

Asimow, W. (1962), *Introduction to Design,* Prentice-Hall, Englewood Cliffs, NJ.

Banares-Alcantara, R., Westerberg, A., Ko, E.I., and Rychener, M.D. (1988), The DECADE catalyst selection system, *Expert Systems for Engineering Design,* Rychener, M.D. (Ed.), Academic Press, Boston, pp. 53-92.

Chang, S.L. and Tsai, L.W. (1990), Topological synthesis of articulated gear mechanisms, *IEEE Transactions on Robotics and Automation,* Vol. 6, No. 1, pp. 97-103.

Chin, S-K. (1989), Synthesizing correct logic designs using formally verified design heuristics, *CAD Systems Using AI Techniques,* Odawara, G. (Ed.), North-Holland, New York.

Conant, R.C. (1976), Laws of information which govern systems, *IEEE Transactions on Systems, Man, and Cybernetics,* Vol. SMC-6, No. 4, pp. 240-255.

Coyne, R.D., Rosenman, M.A., Radford, A.D., Balachandran, M., and Gero J.S. (1990), *Knowledge-based Design Systems,* Addison-Wesley, Reading, MA.

Duda, R.O., Hart, P.E., and Nilsson, N.J. (1981), Subjective Bayesian methods for rule-based inference system, *Readings in Artificial Intelligence,* Tioga, California, pp. 112-129.

Ellsworth, R., Parkinson, A., and Cain, F. (1989), The complementary roles of knowledge-based systems and numerical optimization in engineering design software, *ASME Journal of Mechanisms, Transmissions, and Automation in Design,* Vol. 111, pp. 100-103.

Flemming, U., Coyne, R., Glavin, T., and Rychener, M. (1988), A generative expert system for the design of building layouts - version 2, *Artificial Intelligence in Engineering: Design,* Gero, J.S. (Ed.), Elsevier, New York, pp. 445-464.

Forsyth, R. (1984), *Expert Systems - Principles and Case Studies,* Chapman and Hall, London, UK.

Garvey, J.A., Lowrance, J.D., and Fischler, M.A. (1981), An inference technique for integrating knowledge from disperate sources, Proceedings of the 7th International Joint Conference on AI, Vancouver, BC, pp. 51-62.

Gero, J.S. and Coyne, R.D. (1987), Knowledge-based planning as a design paradigm, *Design Theory for CAD,* Yoshikawa, H. and Warman, E.A. (Eds), Elsevier, New York, pp. 339-374.

Grierson, D.E. and Cameron, G.E. (1988), An expert system for structural steel design, *Artificial Intelligence in Engineering: Design,* Gero, J.S. (Ed.), Elsevier, New York, pp. 279-294.

Hoeltzel, D.A. and Chieng, W-H (1990), Knowledge-base approaches for the creative synthesis of mechanisms, *CAD,* Vol. 22, No. 1, pp. 57-67.

Hoover, S.P. and Rinderle, J.R. (1989), A synthesis strategy for mechanical devices, *Research in Engineering Design,* Vol. 1, No. 1, pp. 87 - 103.

Hubka, V. (1982), *Principles of Engineering Design,* Butterworth Scientific, London, UK.

Ishii, K., Goel, A., and Adler, R.E. (1989), A model of simultaneous engineering design, *Artificial Intelligence in Design,* Gero, J.S. (Ed.), Springer-Verlag, New York, pp. 483-501..

Koomen, C.J. (1985), The entropy of design: a study on the meaning of creativity, *IEEE Transactions on Systems, Man, and Cybernetics,* Vol. SMC-15, No. 2, pp. 16-30.

Kusiak, A. and Szczerbicki, E. (1990), A Methodology for Conceptual Design of Mechanical Systems, Working Paper No. 90-05, Department of Industrial Engineering, The University of Iowa, Iowa City, Iowa.

Kusiak, A. and Szczerbicki, E. (1990a), Rule-based synthesis in conceptual design, Proceedings of the Third International Symposium on Robotics and Manufacturing, Vancouver, B.C.

Kusiak, A., Park, K., and Szczerbicki, E. (1990), A Generalized Search Approach to Design Specifications, Working Paper No. 90-05, Department of Industrial Engineering, The University of Iowa, Iowa City, Iowa.

Kusiak, A., Szczerbicki, E., and Vujosevic, R. (1990), Intelligent Design Synthesis: an Object-oriented Approach, Working Paper No. 90-15, Department of Industrial Engineering, The University of Iowa, Iowa City, Iowa.

Maher, M.L. (1988), HI-RISE: an expert system for preliminary structural design, *Expert Systems for Engineering Design,* Rychener, M.D. (Ed.), Academic Press, Boston, pp. 37-52.

Maher, M.L. (1989), Synthesis and evaluation of preliminary designs, *Artificial Intelligence in Design,* Gero, J.S. (Ed.), Springer-Verlag, New York, pp. 3-14.

Mesarovic, M.D., Macko, D., and Takahara, Y. (1970), *The Theory of Hierarchical Multilevel Systems,* Academic Press, New York.

Mesarovic, M.S. and Takahara, Y. (1989), *Abstract Systems Theory,* Springer-Verlag, New York.

Mistree, F. and Muster, D. (1989), Conceptual models for decision-based concurrent engineering design for the life cycle, Technical Report, Department of Mechanical Engineering, University of Houston, Houston, Texas.

Nevins, J.L. and Whitney, D.E. (1989), *Concurrent Design of Product and Processes,* McGraw-Hill Publishing Company, New York.

Nolan, P.J., (1986), An intelligent assistant for control system design, *Applied AI in Engineering Problems 1st International Conference,* Springer, Sothamptom, UK, pp. 473-481.

Oren, T.I. (1984), Gest - a modelling and simulation language based on system theoretic concepts, *Simulation and Model-Based Methodologies: An Integrative View,* Oren, T.I., Zeigler, B.P., and Elzas, M.S. (Eds), North-Holland, Amsterdam, pp. 3-40.

Pahl, G. and Beitz, W. (1988), *Engineering Design,* Springer-Verlag, New York.

Radford, A.D. and Gero, J.S. (1988), *Design by Optimization in Architecture, Building, and Construction,* Van Nostrand Reinhold, New York.

Rapaport, A. (1986), *General Systems Theory,* Abacus Press, Turnbridge Wells, Kent, UK.

Serrano, D. and Gossard, D. (1988), Constraint management in MCAE, *Artificial Intelligence in Engineering: Design,* Gero, J.S. (Ed.), Elsevier, New York, pp. 217-240.

Shen, S.N.T., Chew, M-S., and Issa, G.F. (1989), Expert system approach for generating and evaluating engine design alternatives, *Applications of Artificial Intelligence,* Trivedi, M.M. (Ed.), The International Society of Optical Engineering, Bellingham, Washington, pp. 532-543.

Simoudis, E. (1989), A knowledge-based system for the evaluation and redesign of digital circuit networks, *IEEE Transactions on Systems, Man, and Cybernetics,* Vol. 8, No. 3, pp. 302-315.

Steinberg, L.I. and Mitchell, T.M. (1984), A knowledge based approach to VLSI CAD - the REDESIGN system, Proceedings ACM IEEE 21st Design Automation Conference, Albuquerque, NM, pp. 412-418.

Szczerbicki, E. (1990), Informational uncertainty and decisionmaking models in cybernetic design of organization, *Systems Analysis, Modelling, Simulation,* Vol. 7, No. 7, pp. 252-265.

Tsai, L.W. and Freudenstein, F. (1989), On the conceptual design of a novel class of robot configuration, *ASME Journal of Mechanisms, Transmissions, and Automation in Design,* Vol. 111, No. 1, pp. 47-53.

Ulrich, K.T. and Seering, W.P. (1988), Synthesis of schematic descriptions in mechanical design, *Research in Engineering Design,* Vol. 1, No. 1, pp. 3 - 18.

Wood, K.L., Otto, K.N., and Antonsson, E.K. (1990), A formal method for representing uncertainties in engineering design, Proceedings of the First International Workshop on Formal Methods in Engineering Design, Manufacturing, and Assembly, Colorado Springs, Colorado, pp. 202-246.

Zeigler, B.P. (1987), Hierarchical, modular discret-event modelling in an object-oriented environment, *Simulation,* Vol. 49, No. 5, pp. 219-230.

Zeigler, B.P. (1984), *Multifacetted Modelling and Discret Event Simulation,* Academic Press, New York.

An Open Ended Network Architecture for Integrated Control and Manufacturing

K.C.S. MURTY and RAMESH BABU

Central Electronics Engineering Research Institute
Pilani, India

SUMMARY

Integrated plant wide automation has become popular with advances in LAN technology. Inter communication among plant machinery is being implemented with standard protocols and manufacturing message formats upto application level embedded in the machinery and equipment. However problem lies in integration of already existing equipment which are heterogenious with variety of non standard interfaces. This paper presents design aspects of Network Interface Unit (NIU) with specialised architecture by which such systems can be brought into an integrated network. The NIU has interface to standard network on one side and to non standard systems with nonstandard protocols on the other side. The NIU with open architecture is expandable and interfacable to a wide variety of plant machinery and process instrumentation. The node also acts as a cluster interfacing heterogenious types of devices and also takes up message translation activity to interface standard message protocols to proprietary ones. A LAN designed and developed with above approach was tested in a process industry. The paper presents approches and methodologies used and discusses implementation experiences of the Network.

INTRODUCTION

Manufacturing plants and process industries are getting highly automated with modern electronics. Distributed Digital controls and supervisory controls are being implemented rapidly into modern industrial processes. Turnkey solutions are getting available from leading manufacturers for plant wide automation. However, existing plants find problems in integrating heterogenious types of instrumentation as the LAN technology underlying in such systems, remain proprietary with the manufacturers. A total solution is not becoming practical due to incompatible systems already existing in plant with non standard communication mechanisms. Hence a need arises for the development of a Local Area Network optimised for filling the above gaps. In such designs some of the inherent features of the plant operations and processes can effectively be exploited to bring out optimised designs like slow changes in process parameters, limited and fixed number of nodes, limited distances etc. The above exercise was taken up for a typical sugar plant and the same can be applied to manufacturing environment also. The above design aspects are dealt in detail below:

2. SYSTEM REQUIREMENTS

An industrial LAN has the basic task of providing communication across autonomous plant devices. The foremost requirements are:
a)   Each node on the network should be autonomous. The entire network should not effected if one of the nodes is up or down. True distribution is possible through standard media access protocols. Token passing protocol is selected as it provides deterministic access to each node.
b)   Physical media of communication can be on twisted pairs with RS 422 protocol. Internode distances and noice environment in the selected plant would allow such media. However transmission through coaxial cable after FSK modem should be an option for harsh environment.
c)   Data Rates in the selected plant which are estimated to have a peak value of 32 Kb/s should be supported.
d)   Node tasks: The LAN node has to interface the communication channel on one side and the process or plant machinery on the other side. It is hence an interface unit (Network Interface Unit). The NIU should be able to be interfaced to any plant device. It should be able to cluster heterogenious types of such devices and communicate with them through separate drivers depending on the device characteristics. It should observe each such physical device as a logical device on the network. The NIU should coordinate and prepare messages to be routed to other devices on other nodes through front-end communications processor. The above requirements can provide easy interface of the network to any generic device. The NIU should also be able to directly interface with the process through sensors and actuators and control the process or operations directly. The control or data acquistion task should be able to be directly run on NIU. A combination of such clustering and direct control of the Process should also be possible.

3. HARDWARE ARCHITECTURE

The essential hardware (Fig.1) required to meet above requirements are a) Physical media b) Front end modem c) Communication processor d) Host Processor. Twisted pair cable is found optimum physical media with RS 422 interface. [1] For longer distances an FSK modem at 5MHz carrier is developed as an additional adapter connected over 75 ohms coaxial cable. [3]

The Network Interface Unit (NIU) has been designed to carry out the application and communication tasks. Keeping the system expandability requirements, the system has been designed as bus based. The node consists essentially two cards, the host processor and the communication processor. The Host processor is 8086 based standard card with necessry memory, interrupt controller and timer.

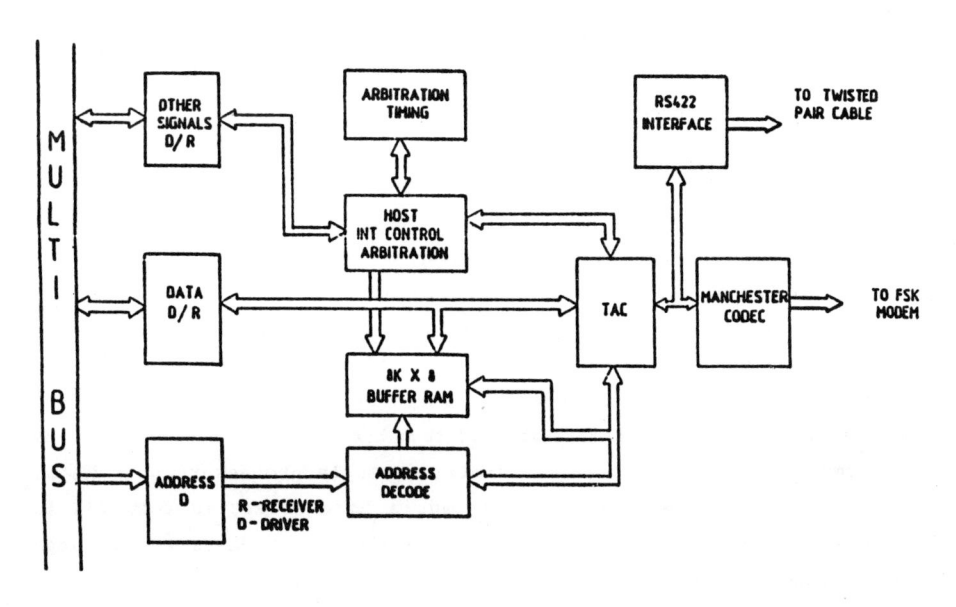

FIG. 1   LAN CONTROLLER ARCHITECTURE

The communication processor is a VLSI Token Access Controller(TAC). [2] The device accepts messages to be sent through a linked buffer chain and transmits to destination whenever token is passed to this node. Similarly after reception of data frame and CRC check, it places them in receive buffer chain. The device controls the Media Access (Token Passing Bus) mechanism, data and token frame generation. Buffer chains are accessed through in_built DMA. The Buffer memory is resident on Communication processor. This memory can be accessed by TAC and Host. Dual ports for the memory are designed and necessary arbitration for host and TAC access with priority to TAC are generated. The TAC interrupts the host through multibus interrupt. As the TAC has 8 bit data path, host transactions are also limited to byte wide. An additional data path from Multibus to TAC is also provided for controlling and observing the TAC by host. All operations on TAC are synchronised with a 16MHz clock. TAC outputs data and token control frames which are manchester encoded before transmission. Interface to control devices can be done through multiplexers or through available standard ports on host. Direct digital control can be incorportated by additional analog I/o cards and interface to process devices.

4. NETWORK SERVICES

The network services have been developed such that they can be directly utilised by the application without much effort. The software provides data link layer services with reliable data transfers. Keeping the system requirements and the quantum of data generation, no necessity for higher layers was found. The application software can directly utilise datalink services and incorporate itself any additional reliability requirements. As the communication overheads are small, a single processor can take up application and communication tasks and a cost effective solution is achieved. Faster responses to messages can also be achieved due to less communication overheads.

There are five major data structures in the communication software. The REG structure maps into registers of token access controllers and provides a means to acess them. The next structure is CB (Control Block) which maps into 16 locations in memory and controlled by TAC. CB consists of pointers to receive and transmit buffers. It also has counters updated by TAC to register different events. The third structure CTR is to keep an account of exceptions and other events of interest. The next two structures are definition of transmit and receive linked buffers. The main modules with parameters are listed below.

CHIP-DIAG	Carries out TAC diagnostics
ENTER_NET (Master)	Sets the station into network
WRITE_REG (Num, Value)	Writes into any TAC register a given value
READ_REG (Num)	Reads from a TAC register
DL_INIT (Myaddress, Priority, Xlimit)	Initialises the TAC and buffers in RAM
NODE_STAT	Provides access to node and network status parameters.
DL_RECEIVE	Transfers data of a pending received frame into a user buffer using pointer.
DL_TXMT (Packet_Ptr, Dest_addr, Source_addr, Packet_Len)	Transmits a given quantum of data.
FRAME_ASSEMBLY (Str_Ptr,Dest_Addr,Source_Addr, Packet_Len)	Assembles a frame for transmission from given information.
ADD_STATIONS	Allows addition of stations by first disrupting the logical ring and then regnerating it.
STAT_PROCESS (Ctr) :	Increments appropriate status counter
INTRPT	Interrupt handler for TAC interrupt on Receive.

5. IMPLEMENTATION

Three NIUs with above described hardware architecture were designed and Network services implemented in PROMS (Fig 2). One NIU is interfaced through a parallel port to process to acquire process parameters. The second NIU was used as a cluster to capture 24 parameters while a third NIU is interfaced to an IBM PC for supervisory functions like monitoring, data acquisition, storage and display. Third NIU is designed with application software to collect data frames from other NIUs, assemble and submit them to PC. The details of supervisory functions are not described here.

6. CONCLUSION

Network architectures can be tailored to the application needs and cost effective solutions can be derived. A node on the network with open ended architecture can integrate shop floor equipment, process control devices and equipment from wide range of vendors having high incompatibilities through proprietary hardware and software and take up direct control of plant devices and also cluster several such devices into a cell. Specifically, such solutions provide easy integration of existing instrumentation and derive the benefits of plant wide automation at marginal costs.

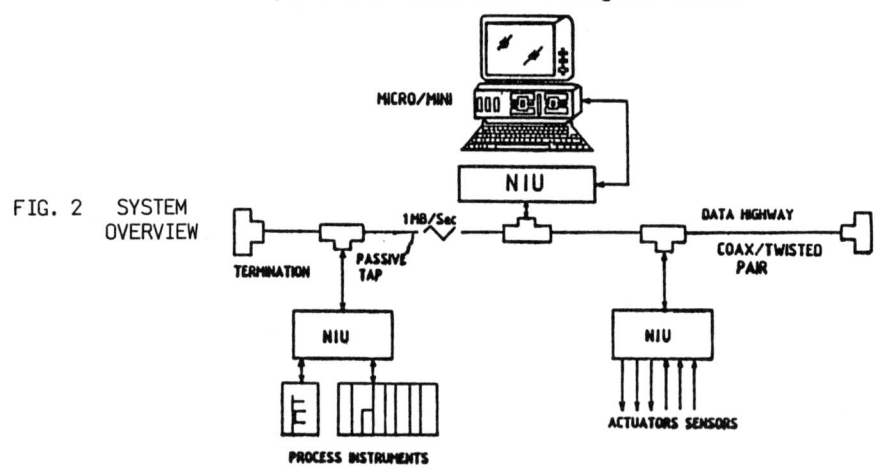

FIG. 2 SYSTEM OVERVIEW

7. ACKNOWLEDGEMENTS

The authors wish to thank Dr. W.S. Khokhle, Director, CEERI, for allowing to present the above work. The authors also acknowledge Dept. of Electronics, India for providing financial assistance for the above project.

8. REFERENCES

[1] IEEE 802.4/87/12. "Token Passing Bus Access Method and Physical Layer specification", Draft.G, July 1987.
[2] Wester Digital Corpn. "WD2511/WD2840 Technical Package", April 1984.
[3] R. Blauschild and R. Fabbri, "Two chip MODEM for FSK data Transfer", Elec. Comp. & App., Vol 6, No. 2, 1982.

A Data Base Inconsistency Checker for EASIE

K.H. JONES, S. OLARIU, L.F. ROWELL, J.L. SCHWING, and A. WILHITE

NASA Langley and Old Dominion University
Norfolk, VA

Abstract

One of the challenges facing the designers of an integrated computer-aided engineering system is to blend in a robust and efficient way a wide variety of independently developed design and analysis programs, each with its specific requirements for input and output. The Environment for Application Software Integration and Execution (EASIE) provides a methodology and set of utility routines to support building, maintaining, and applying computer-aided design systems consisting of large numbers of diverse, stand-alone analysis codes. Conducting the engineering activity usually calls for manually changing the contents of a given set of variables in the data base and for the subsequent execution of a set of analysis programs. As a result of the computations, other data base values will be altered. To support a high-productivity environment, a tool is needed to make the engineer aware of potential inconsistencies introduced in the data base as a result of any data changes. A technique is described for identifying, beforehand, the set of all the variables in the data base susceptible to inconsistency. We address this problem in general, and discuss how it can be implemented in EASIE.

1. Introduction

In the context of the increased complexity of space-age technology, computer-aided engineering systems are crucial for supporting sophisticated design methodologies and analysis techniques. In order to increase productivity, the many iterations inherent in the design process have to be supported by a user-friendly, quick-turnaround, computer-based design system. The Environment for Application Software Integration and Execution (EASIE) provides a methodology and a set of utility routines for a design team to build, maintain, and apply computer-aided design systems consisting of a large number of stand-alone analysis tools. Wilhite et al [5] describe the development history of several integrated design systems and give an overview of needed future enhancements. EASIE implements these needed capabilities by configuring the entire system around a central data base containing all the input and output variables for the integrated analysis programs. Utilities exist for constructing the data base schema, generating the routines to read from or write to the data base, incorporating the analysis programs into the data base, interactively reviewing and modifying values in the data base, incorporating the analysis programs into an interactive executive for easy selection and execution, and building menus and procedures to assist the process.

The relationships between analysis programs and their input and output variables are described by a *template* (Figure 1). An input template is, simply, a list of the input variables that are required for a particular program. Similarly, an output template is a list of the output variables which are to be inserted into the data base by a particular program. These templates,

---

1 Analysis and Computation Division, NASA Langley
2 Department of Computer Science, Old Dominion University
3 This work was supported by NASA under grant NCC1-99
4 Space Systems Division, NASA Langley

once built, are associated by name with the particular analysis program to which they relate. This concept of program input and output templates enables the development of a single generic interactive editor which can display the input and output variables for any selected program by requesting the appropriate template [1-4].

In typical design and engineering studies, use will be made of both commercially-available and in-house developed analysis tools. Among the techniques currently being used to combine independent analysis tools into design systems, *loose-coupling* of analysis programs is particularly attractive. The term loose-coupling describes a collection of programs, each of which has the capability to communicate with a central data base and which access common data but which may be executed in any sequence.

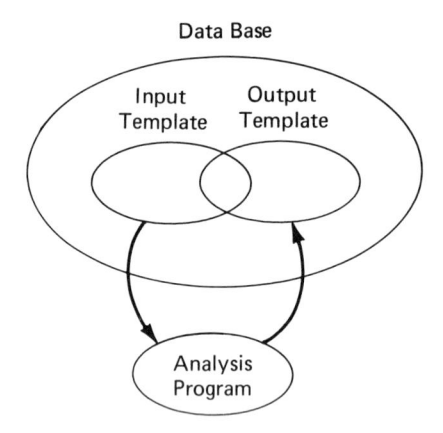

2. The Consistency Problem

The loose coupling of **EASIE** allows the designer to iterate through system components as desired. During this process, the designer will use engineering insight to define and possibly modify those input parameters which may also be outputs defined by other analysis modules. This technique helps decrease development time but may introduce inconsistencies into the data base. The purpose of this work is to define techniques to identify those inconsistencies. Parameters will be flagged in the data base so they can be queried at any time to determine the consistency status.

The remainder of the paper will formalize the essence of the design process described above and will consider the solution of the data base integrity problem in two cases. In the first case, the order in which the analysis programs will be executed is not known *a priori* but is, rather, determined dynamically during the design process, perhaps depending on some intermediate value. In the second case, the designer has a predetermined ordering of analysis programs that will be involved in the current analysis.

For the purpose of this work, a central data base is assumed, containing a set V of variables. In addition, a set A of loose-coupled analysis programs is also assumed. Every such program P is specified by an input template I_P containing the set of all the variables that are input to P, along with an output template O_P describing the set of all the variables that are output by P.

For definiteness, a variable is referred to as *in-out* if it is both input to some analysis program and output of some (possibly the same) program. A typical scenario inherent to the design activity, and arising in the process of verifying certain design hypotheses, calls for manually changing the contents of a set X of in-out variables in the data base and for the subsequent execution of a set B of analysis programs. Needless to say, this course of action will result in a number of variables in the data base whose values have been modified during the execution of the experiment. To help maintain the integrity of data in the data base, a tools is needed to identify *beforehand* all the variables whose values are potentially affected by this experiment. This capability would be useful in the very process of testing the design hypotheses since the engineer must become aware of the scope and of the potential impact of the changes performed. Of course, after the experiment is completed only those variables whose contents have actually been modified need be flagged in the data base for later reference.

A variable w in the data base is termed *potentially inconsistent* if there exists an order of execution of the programs in B which results in a modification of the contents of w. Similarly, a variable w in the data base is termed *inconsistent* if its value has actually changed during the experiment described above. It is useful to note that variables that are input variables but not in-out variables are never considered to be inconsistent. This is in accord with basic engineering principles that require a set of initial conditions to be input to the system under investigation.

In the remainder of this work it will be assumed that the set X contains in-out variables only. The procedure Prune_X described below will be executed as a preprocessing step for the purpose of removing from X all variables which are not in-out.

Procedure Prune_X;
{Input: a set X of variables in the data base;
Output: the same set consisting of in-out variables only}
0. **begin**
1. $I \leftarrow O \leftarrow \emptyset$;
2.   **for** all programs P in A **do begin**
3. $I \leftarrow I \cup I_P$;
4. $O \leftarrow O \cup O_P$
5.   **end**;
6.   return($X \cap I \cap O$)
7. **end**; {Prune_X}

It is easy to confirm that after executing Prune_X, the set X contains in-out variables only. Furthermore, with the assumption that the data base contains n variables and that there are, altogether, p application programs, procedure Prune_X runs in $O(np)$ time.

In general, the order in which the analysis programs in B are to be executed is dictated by the intermediate results of the experiment being conducted. Under these conditions is seems to be very hard to specify the set of inconsistent variables *without* actually running the analysis programs in B. However, the problem of identifying the set Y of all the potentially inconsistent variables in the data base without actually running the programs in B is solvable efficiently by using the input and output templates. In outline, the proposed algorithm does the following. To begin, the set Y is initialized to X. Next, for all the programs P in B with $I_p \cap Y \neq \emptyset$, the following actions are taken: the output template for P is added to Y and P is removed from B. (To justify this, it is useful to note that all the potentially inconsistent variables contained in O_P already belong to Y, making it unnecessary to add them again, at a later point.) This process is continued until either B is exhausted or else no remaining program in B uses a variable in Y as input. Note that the latter case can be easily detected using a simple boolean variable. The details of this simple algorithm are spelled out as follows.

Procedure Find_Potentially_Inconsistent(X,B);
{Input: a set X of variables and a set B of analysis programs;
Output: the set Y of all the potentially inconsistent variables in the data base}
0.   **begin**
1.   Y ← X;
2.   done ← false;
3.   **while** (B≠∅) **and** (**not** done) **do begin**
4.   done ← true;
5.   **for** all programs P in B **do**
6.   **if** $I_P \cap Y \neq \emptyset$ **then begin**
7.   Y ← Y∪O_P;
8.   remove P from B;
9.   done ← false
10.   **end** {if}
11.   **end**; {while}
12.   return(Y)
13. **end**; {Find_Potentially_Inconsistent}

To obtain a bound on the running time it is convenient to assume that the number of the variables in the data base is n (n≥0) and that B contains k (k≥0) programs to be executed. (Surprisingly, the running time of this procedure is independent of the number of variables in the set X.) The following statement establishes the correctness of the procedure and argues about its running time.

Theorem 1. *Procedure Find_Potentially_Inconsistent correctly determines the set of all the potentially inconsistent variables without actually executing the set B of analysis programs. Furthermore, the running time of the procedure is $O(k^2 n)$.*

Proof. First, let y be an arbitrary variable in Y. The proof of the fact that y is potentially inconsistent proceeds by induction on the number t (t≥0) of iterations of the while loop needed to add y to Y for the first time. If t=0, then y∈X and there is nothing to prove. Assume the statement true for all the variables added to Y in fewer than t iterations of the while loop. Since y belongs to Y, it must have been added to Y in line 7 of the procedure while the for loop in lines 5-10 was processing some program P in B. However, $I_P \cap Y \neq \emptyset$, implying that some variable z in I_P must have been added to Y in fewer than t iterations of the while loop. By the induction hypothesis z is potentially inconsistent, implying that y is also potentially inconsistent.

Conversely, let y be an arbitrary potentially inconsistent variable in the data base. To show that y must belong to Y, the proof proceeds by induction on the length t (t≥0) of the shortest sequence of programs in B which, when executed, result in a modification of the value of y. If t=0, then y∈X and line 1 in the procedure guarantees that y∈Y. Next, assume that every variable whose contents is modified by executing fewer than t programs in B is in Y. Let P be the last program in the sequence of t programs in B whose execution modifies the contents of y. This happened because some variable y' in I_P was already potentially inconsistent. By the induction hypothesis, y' belongs to Y. But now, when P is processed in line 5, y will be added to Y, as claimed.

To address the complexity, note that the while loop in lines 3-11 is executed at most k times. In each iteration of the while loop, the for loop in lines 5-10 takes $O(\sum_{P \in B}(|I_P|+|O_P|)) \subseteq O(kn)$ time, if the appropriate data structure is used to maintain sets. Consequently, the entire procedure runs in $O(k^2n)$ time, and the proof of Theorem 1 is complete. □

To be able to identify and flag the set of all the inconsistent variables in the data base when the analysis probrams in the set B have been executed, it is assumed that the integrated system has a logging capability. Specifically, it is assumed feasible to determine the exact order of execution of the programs during the experiment. Now identifying the set of all the inconsistent variables can be done efficiently by the following procedure. It is assumed, without loss of generality, that the analysis programs in B have been executed in the sequence P_1, P_2, ..., P_m, with all repetitions removed as a preprocessing step. Clearly, this can be accomplished by scanning the sequence once.

Procedure Find_Inconsistent(X,B);
{Input: a set X of variables and a sequence P_1, P_2, ..., P_m of programs in B;
Output: the set Z of all the resulting inconsistent variables}
0. **begin**
1.   Z ← X;
2.   **for** j ← 1 **to** m **do**
3.   **if** $I_j \cap Z \neq \emptyset$ **then**
4.   Z ← Z∪O_j;
5.   return(Z)
6. **end**; {Find_Inconsistent}

The following result argues about the correctness and running time of the procedure Find_Inconsistent.

Theorem 2. *Procedure Find_Inconsistent correctly determines the set of all the inconsistent variables after having executed the set B of analysis programs. Furthermore, the running time of the procedure is O(mn).*

Proof. To settle the correctness, it suffices to prove that a variable z is inconsistent after having executed the programs in B in the sequence P_1, P_2, ..., P_m if, and only if, z belongs to Z. First, assume that z belongs to Z. To show that z must be inconsistent it suffices to use induction on the number t (t≥0) of iterations of the for loop needed to add z to Z for the first time. If t=0, then z∈X and there is nothing to prove. Assume the statement true for all the variables added to Z in fewer than t iterations of the while loop. Since z belongs to Z, it must have been added to Z in line 4 of the procedure while P_t was processed. However, $I_t \cap Y \neq \emptyset$, implying that some variable z' in I_t must have been added to Z in fewer than t iterations. By the induction hypothesis z' is inconsistent, implying that so is z.

Conversely, let z be an inconsistent variable in the data base resulting from the execution of the sequence $P_1, P_2, ..., P_m$. To show that z must belong to Z, let j be the the smallest subscript in the range $1 \leq j \leq r$ such that after executing the sequence $P_1, P_2, ..., P_j$ the value of z is modified. The previous choice of j guarantees that $z \in O_j$ and so z will belong to Z after line 4 has been executed in the j-th iteration of the for loop.

To address the complexity, note that the for loop in lines 2-4 is executed m times and that each iteration takes at most O(n) time, for a total complexity of O(mn), as claimed. □

3. Conclusion

To support a high-productivity design and engineering environment, a tool is needed to identify potential inconsistencies introduced in the data base as a result of an experiment that involves
- manually changing the contents of some variables, and/or
- the subsequent execution of a number of analysis programs on the modified data.

We addressed the problem of identifying beforehand the set of variables which may result in inconsistencies in the data base as well as the problem of flagging the affected variables in the data base after having run the analysis programs on the modified data. Two efficient algorithms have been proposed. The associated tools are currently being implemented in EASIE. These tools will give the designer a handle on the scope and impact of the changes involved. In addition, the problem of flagging all the affected variables in the data base after having run the analysis programs on the modifed data is also discussed.

References

1. L. F. Rowell and J. Davis, The Environment for Application Software Integration and Execution (**EASIE**) Version 1.0 - Volume I - Executive Overview, NASA TM-100573, August 1988.
2. K. H. Jones, D. P. Randall, S. S. Stallcup, and L. F. Rowell, The Environment for Application Software Integration and Execution (**EASIE**) Version 1.0 - Volume II - Program Integration Guide, NASA TM-100574, July 1988.
3. J. L. Schwing, L. F. Rowell, and R. E. Criste, The Environment for Application Software Integration and Execution (**EASIE**) Version 1.0 - Volume III - Program Execution Guide, NASA TM-100575, April 1988.
4. L. F. Rowell, J. L. Schwing, and K. H. Jones, Software Tools for the Intergration and Execution of Multidisciplinary Analysis Programs, Old Dominion University, Tech. Report TR-89-04.
5. A. Wilhite, S. C. Johnson, and V. Crisp, Integrating Computer programs for Engineering Analysis and Design, AIAA Pare No. 83-0597, 21st Aerospace Sciences meeting, January 1983.

Database Exchange in the CAD/CAM/CIM Arena

BRUCE A. HARDING

MET/CIMT Department
Purdue University
West Lafayette, IN

Summary
Successful enterprise-level CIM environments share electronic information among all aspect of the organization. The process is relatively straightforward and well defined in procurement and other business related segments. However technical data sharing for CAD/CAM/CIM often require movement of complex coordinate geometry among more convoluted paths. The most common method for geometry movement, the initial graphics exchange specification (IGES), is fraught with potential problems. This paper discusses the IGES translation in mixed platform/mixed system CAD/CAM/CIM environments and suggests alternatives.

Problem Definition
With the proliferation of CAD systems today, especially stand alone microcomputer-based CAD, virtually every technical user may soon have a personal CAD system on their desk. This has also generated a movement away from single-source CAD systems to a multi-platform CAD environment featuring a number of CAD software and hardware systems. Each system often is chosen for specialty fiscal, power, niche, or ease of use rationales. This scenario, while enriching for end users, may cause severe difficulties organizationally when exchanging CAD data among multiple systems. Worse, exchange among users external to the organization may represent an even more diverse CAD user community.

Pre-dating microcomputers, CAD was primarily host-based with interchange among CAD systems rarely a concern because CAD was centralized and largely single sourced. However, in today's global marketplace a multi-platform CAD environment is becoming increasingly more common (see Figure 1). Typically CAD systems, both host-based and micro, store drawing databases containing graphical and non-graphical information in binary format to minimize memory utilization and maximize processing speed. Each CAD system utilizes unique algorithms to construct their proprietary database. Thus in their native database format, each CAD system is totally incompatible with any other CAD vendor, even if the two applications reside on the same physical platform. Additionally, few CAD programs generate consistent databases across hardware platforms running the same CAD application.

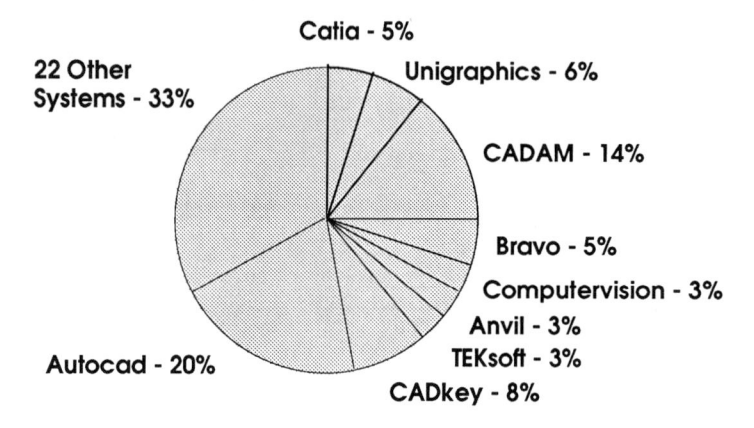

Figure 1. Typical CAD diversity in industry today. A sample of over 100 vendors to a major multinational corporation indicates that at least 36* different CAD systems are used by the vendors. Although all need to, only a few of the vendors success-fully exchange geometry data with the corporation. *31 shown. Some systems bearing the same name, but operating on different platforms, were combined for clarity.

Direct Translation

To facilitate database transfer among different CAD systems, two solutions have emerged. The first is the use of specific translators for direct conversion of one specific CAD database to another. Although optimal for specific systems, separate translators are needed for each combination of different CAD systems, and direction of conversion (A-to-B, B-to-A, etc.) for each pair of systems (see Figure 2). Additional combinations may be required if multiple versions of any one program are being used. Generally translators are platform specific because even like-logo CAD system tends to have slightly different database organizations on each platform or version supported.

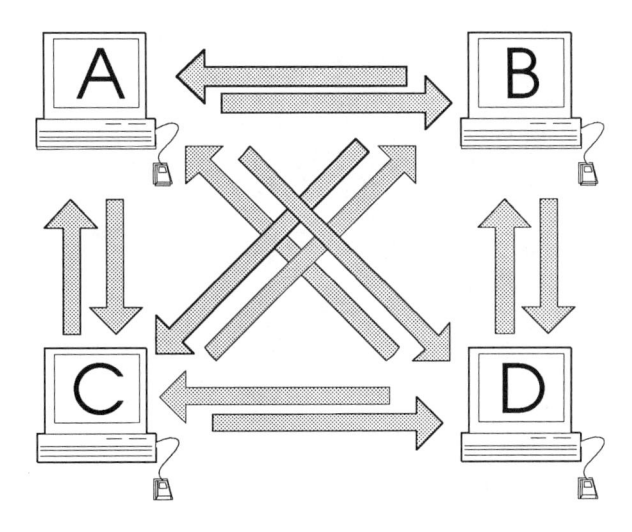

Figure 2. *Possible combinations of direct CAD database translation among 4 different CAD systems. 2 separate directional routines are required for each pair, resulting in 12 separate conversion routines for the four systems. Each routine must be updated separately as each CAD version is upgraded.*

Neutral File Translation

The second method of database conversion is via a "neutral" file format. Employing a neutral intermediary, each of 4 different CAD systems would require use and maintenance of only 2 translators — to and from the intermediate file (see Figure 3). Because each CAD vendor assumes responsibility for the translation to and from the neutral file, the process is platform and CAD version independent. This method of conversion is gaining in popularity because direct translators, while often more complete, are expensive and highly version and platform specific.

A number of neutral file formats have emerged. The most widely recognized and supported across both host-based CAD and micro CAD is the Initial Graphics Exchange Specification (IGES).

Each CAD system that supports IGES support entails bi-directional translators between the native CAD database and IGES format. Internally, the IGES database is ASCII text-based and readable by any other IGES translator.

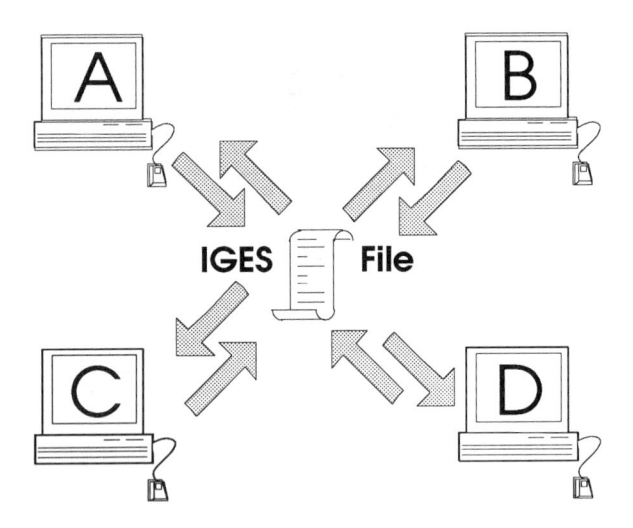

Figure 3. CAD database translation among 4 CAD systems using a neutral file format. Only 2 translation routines (bi-directional) are needed for each system.

IGES Problems

Although widely used, IGES is not a perfect method of translation. Depending on the source/designation CAD system, translational errors may arise. These errors may generally be traced to two phenomena. First, by nature, an intermediate file to be most broadly useful, must operate at the lowest common dominator of most CAD systems. Feature-rich CAD systems are not serviced as well as less capable systems. Secondly, IGES is a specification, a semi-structured suggestion, not a standard. As such, CAD vendors may augment the translation algorithms to enhance specific features; alterations which may benefit their features and totally ignore those of another CAD vendor. Thus translational problems may be manifest as the following:

- No equivalent element.
- Mutation of elements.
- Missing inter-element associations.
- Differing accuracy bases.
- Missing non-graphic or attribute information.

No Equivalent Element

Rarely do CAD systems use the same mathematical definition for graphic primitives. More advanced primitives such as ellipses, b-splines, ruled surfaces, symbols and surfaces of revolution may be missing entirely from some CAD systems. Elements lacking support at the destination end may be skipped entirely, or worse, mutated.

Mutation of Elements

A less capable destination translator may valiantly attempt to construct a reasonable facsimile of unrecognizable elements from the source. The facsimile, a best guess copy, could now be composed of different elements entirely, but which may be visually very difficult to detect. For instance, an ellipse failing to translate, may be mutated, but visually retain the appearance of an ellipse. However, on querying the geometry database directly, the ellipse has become a single polygon or even a set of 40 individual vector elements strung together to approximate the original element. In a CIM environment with integrated NC, such corrupted geometry might prove totally unusable

Non-geometrical information may be adversely affected also. For instance, text fonts (typefaces) in the source file but not resident or recognized at the designation end, may become unrelated vectors rather than the true ASCII characters they represent. Arrowheads, treated as a line attribute on some systems may become a set of three or more separate line elements on other systems. Associative dimensioning may cause unpredictable results in translation.

Missing Associations

A power feature touted by many CAD systems, is the ability to group elements and subsequently nest groups, forming complex geometric or boolean relationships. Some systems also use grouping functions to denote library symbols. Often during translation, group associations are lost; with them, a portion of the database's knowledge. Other object attributes, non-graphical information attached to the primitives, may be lost. These may include layer specifications, symbols, colors, naming conventions, line styles, line widths, tolerancing and other featured-based manufacturing data.

Differing Accuracy Bases

The base grid systems among CAD systems differ. In normal drawing operation, many CAD users 'snap to,' or attach to line intersections, tangents, equations of lines, surfaces, nodes, etc. During IGES translation database knowledge of the exact snap point as an element attachment point is lost. Rather, the intersection point is rounded to the nearest grid element. In high accuracy applications, this may generate a fatal error and is exacerbated in/from systems that operate in single precision rather than double precision.

Current Solutions

Although IGES has significant difficulties, a great number of users are successfully accomplishing translations among CAD system. To do so, they typically develop "work-arounds" based on trial-and-error experience. For example, when it is known that specific elements will be deleted or changed, users either avoid suspect elements altogether or build geometry out of alternative (and known) translatable elements. Groups are often maintained, or re-constructed, by placing element in different layers or in different colors (if maintained) and then re-grouped on the destination system. Loss of accuracy is minimized by avoiding snaps (lock-on) based on derived intersections.

The Future of Neutral Files
Ultimately, the elimination of proprietary databases, hence more data compatibility, would be the ideal solution. For the foreseeable future, current solutions will remain the only viable alternative. Direct translators will always be applicable where CAD multiplicity is limited, but neutral file systems will grow to dominate in the increasingly more common eclectic CAD?CAM/CIM environments. Under limited use now, PDES (Product Design Exchange Specification) will be the near term successor to IGES by embracing more extensive information than current versions of IGES. As the technology matures, artificial intelligence techniques coupled with translational systems will alleviate most limitations of current generation IGES/PDES translators. These techniques will check source database for occurrences of suspect elements and attributes, and interactively, or in batch processing, recommend or automatically modify CAD database information as necessary to insure translational integrity.

Formation of Machine Cells: A Fuzzy Set Approach

CHUN ZHANG and HSU-PIN WANG

Department of Industrial Engineering
State University of New York at Buffalo
Buffalo, NY

This research is directed toward developing a group technology (GT) cell formation methodology, with the inclusion of fuzziness commonly found in industry. Specifically, three studies are conducted: 1) The first study establishes a procedure, through which a non-binary machine-component matrix is constructed. This matrix forms the basis of the following two studies. 2) The second study applies the non-binary matrix concept to the single linkage clustering algorithm. 3) The third study focuses on a modified rank order clustering algorithm, which involves fuzzy set theory. The modified single linkage and modified rank order clustering algorithm are compared; the strengths and weaknesses are discussed.

Introduction

A large number of studies on forming machine cells have been reported, such as single linkage cluster analysis [4]. bond energy algorithm [5], rank order clustering algorithm [2], only to name a few. Although they are relatively different, these approaches share a common feature, that is, each utilizes a binary (two-value) machine-component matrix. In such a matrix, each matrix element takes a value of either "1" or "0". A positive matrix element means that a part will visit the corresponding machine.

The inherent problem with these two-value matrix approaches is that a part will visit only one machine for a particular shape. There is no way of specifying alternative machines for the feature of that part. However, in reality, a part feature can usually be formed on more than one machine. The inability of these approaches to include alternate machines has been the major driving force for us to pursue a different approach.

In this study, a non-binary matrix representation scheme for machines and components is employed. It is a generalization of the binary matrix concept. In a non-binary matrix, the value of each element denotes the "degree of appropriateness" that a part feature may be formed on a machine. A function is developed for each part feature to test the appropriateness of using a machine to make that feature. This function, in fact, can be viewed as a weighting coefficient, which reflects the "match" of a part feature and a machine. As it approaches unity, the degree of matching of a feature and a machine becomes higher. This non-binary matrix scheme offers a unique approach to the cell formation problem because it has more flexibility in grouping and clustering than the well known two-value (either 0 or 1) logic approach [3].

Non-binary matrices

Suppose that n parts and m machines are being considered for cellular manufacturing. A typical non-binary machine-component matrix is shown below:

$$N = \begin{array}{c|ccccc} & X_1 & X_2 & X_3 & \cdots & X_n \\ \hline Y_1 & u_{11} & u_{12} & u_{13} & \cdots & u_{1n} \\ Y_2 & u_{21} & u_{22} & u_{23} & \cdots & u_{2n} \\ Y_3 & u_{31} & u_{32} & u_{33} & \cdots & u_{3n} \\ \vdots & \vdots & \vdots & \vdots & \ddots & \vdots \\ Y_m & u_{n1} & u_{n2} & u_{n3} & \cdots & u_{mn} \end{array}$$

where

X_j is a part, $j = 1, 2, ..., n$;

Y_i is a machine, $i = 1, 2, ..., m$;

u_{ij} denotes the match between part j and machine i.

Match function between machines and components

The matrix elements are calculated from "match" functions between machines and components. The following steps are necessary to gain those values.

1. Define a match function for each machine-feature pair, using Zadeh's fuzzy sets theory.

2. Calculate the value of degree of match for each machine-feature pair.

3. Calculate the combined index of match for each machine-component pair, because a component usually has more than one feature. These combined indices go to the non-binary matrix for cell formation.

Zadeh's fuzzy sets theory is a generalization of a two-value mathematical concept of sets [9]. In a universe of discourse U, a fuzzy subset A of U is defined by a membership function $\mu_A(x)$. $\mu_A(x)$, the grade of membership of x in A, denotes the degree to which an event x_i may be a member of A, or belong to A. This characteristic function in fact, can be viewed as a weighting coefficient, which reflects the ambiguity in a set, and as it approaches unity, the grade of membership of an event in A becomes higher. For example, $\mu_A(x_i) = 1$ indicates strictly the containment of the event x_i in A. If on the other hand x_i does not belong to A, $\mu_A(x_i) = 0$. Any intermediate value would represent the degree to which x_i could be a member of A.

For example, if U is the set of required finished tolerances in some components, A is the fuzzy subset of tolerances that a machine can produce, then a possible match function $\mu_A(x)$ could be as follows:

$$\mu_A(x) = \begin{cases} 0 & x \leq a - a_2 \\ \frac{a_2 + x - a}{a_2 - a_1} & a - a_2 < x \leq a - a_1 \\ 1 & a - a_1 < x < a + a_1 \\ \frac{a_2 - x + a}{a_2 - a_1} & a + a_1 \leq x < a + a_2 \\ 0 & a + a_2 \leq x \end{cases}$$

where

$a=$ the average processing tolerance of the machine;

$a_1=$ the lower limit of processing tolerance of the machine;

$a_2=$ the upper limit of processing tolerance of the machine.

Determination of combined match indices

After all individual machine-feature match indices are calculated, we are ready to calculate the combined match index for each machine-component pair.

Let

$X_j = \{x_{j_1}, x_{j_2}, ..., x_{j_p}\}$ $(j = 1, 2, ..., n)$ be the fuzzy subset of features of component j,

$M = \{Y_1, Y_2, ..., Y_m\}$ be the fuzzy subset of machines,

w_k be the weight of feature k,

$\mu_{Y_i}(x_{j_k})$ be the degree of match between machine i and feature k of component j.

where

$n=$ the number of components,

$p=$ the number of features,

$m=$ the number of machines.

According to Zadeh's fuzzy sets theory, the combined match between machine i and component j is:

$$\mu_{Y_i}'(X_j) = \bigwedge_{1 \leq k \leq p} \{\frac{\mu_{Y_i}(x_{j_k})}{w_k}\}$$

By using the above equation, we are able to determine the combined match indices for all machine-component pairs, and to construct the machine-component matrix.

A Fuzzy Single Linkage Clustering Approach

The single linkage clustering approach is drawn directly from the field of numerical taxonomy, and is applied to the problem of group analysis by McAuley [4]. It involves a hierarchical process of machine grouping, in accordance with computed similarity coefficients.

To perform a single linkage analysis, several similarity coefficients have been developed, which are based on the binary machine-component matrix. However, they can not be applied to our problem directly, since they did not involve the fuzziness. We adopted this approach by defining a new similarity coefficient, combined with fuzziness.

The definition of similarity coefficient is based on the data in machine-component match matrix. It is defined as follows:

$$S_{ij} = 1 - \frac{\sum_{k=1}^{n} |u_{ik} - u_{jk}|}{m}$$

where
S_{ij} = the similarity coefficient for machines i and j.
u_{ik} = the degree of match between component k and machine i.
m = the number of components, whose entries for the two machines are not both zero.

It can be seen that the similarity coefficient reflects the degree of pairing relation between machines. After the calculation of similarity coefficients for each machine pair, machines can be grouped based on these coefficients, and a given threshold value. We can see, from the definition of the similarity coefficient, that this new algorithm is the extension of McAuley's similarity coefficient method [4]. The values of similarity coefficients will be the same as those obtained from McAuley's method, if all of the non-zero entries are 1's. By performing the above machine grouping and component assignment, the machine-component groups can be formed.

A Fuzzy Rank Order Clustering Approach

In the single linkage clustering method, the components need to be assigned to the formed machine groups. In some cases, it is very difficult to obtain a satisfactory result. In order to overcome this problem, we employed another approach - rank order clustering (ROC) algorithm, with inclusion of fuzziness. The ROC algorithm was originally developed by King [1,2]. This algorithm rearranges rows and columns in an iterative manner that will, ultimately, and in a finite number of steps, produce a matrix in which both columns and rows are arranged in order of decreasing value, when read as binary word [2]. The primary data of this approach is also the binary machine-component matrix. The basis of ROC algorithm is to cluster the positive entries (1's) into groups, and place them along the diagonal of the matrix.

With non-binary matrices, it can be seen that the diagonal blocks will most likely contain the majority of non-zero entries. Also in each block, the elements with larger weighted entries will be in higher levels.

The algorithm is as follows:

1. Assign weights to each column.
   The weight for column j: $w_j = 2^{n-j}$
   $j = 1, 2, ..., n$ $n =$ the number of columns.

2. Calculate the ranking value for each row.
   The ranking value for row i: $R_i = \sum_{j=1}^{n} w_j \cdot x_{ij}$
   $i = 1, 2, ..., m$ $m =$ the number of rows.

3. Rank the rows, in order of decreasing ranking value.

4. Rank the columns, using the same method as above.

5. Repeat the above 4 steps until no ranking change in rows or columns required.

Conclusions

In this paper, we discussed two methods - fuzzy sets-based single linkage cluster analysis and rank order clustering (ROC). Both are applicable to the machine-component grouping problem, with inclusion of fuzziness in production flow analysis. However, like the non-fuzzy case, the fuzzy ROC method has certain distinct advantages.

Unlike single linkage cluster analysis, which requires a secondary process of component allocation to the machine groups after machine grouping, the ROC performs both tasks simultaneously. In addition, the fuzzy ROC algorithm can rank machines with priority. This enables ROC algorithm to deal with the problems of machine capacity, exceptional elements, and bottleneck machines, which are normal in practical problems.

1. King, J.R., "Machine-component group formation in group technology: review and extension," *International Journal of Production Research*, Vol.20, No.2, 1982, pp.117-133.

2. King, J.R., "Machine-component grouping in production flow analysis: an approach using a rank order clustering algorithm," *International Journal of Production Research*, Vol.18, No.2, 1980, pp.213-232.

3. Li, J., Z. Ding, and W. Lei, "Fuzzy cluster analysis and fuzzy pattern recognition methods for formation of part families," *Proceedings of 14th NAMRC Conference*, 1986.

4. McAuley, J. "Machine grouping for efficient production," *Production Engineer*, Vol.51, 1972, pp.153.

5. McCormick, W. T., P.J. Schweitzer, and T.E. White, "Problem decomposition and data recognition by a clustering technique," *Operations Research*, Vol.20, 1972, pp.993-1009.

6. Wemmerlov U. and N.L. Hyer, "Research issues in cellular manufacturing," *International Journal of Production Research*, Vol.25, No.3, 1987, pp.413-431.

7. Wemmerlov, U. and N.L. Hyer, "Cellular manufacturing in the U.S. industry: a survey of users," *International Journal of Production Research*, Vol.27, No.9, 1989, pp.1511-1530.

8. Xu, H. and H.P. Wang, "Part family formation for GT applications based on fuzzy mathematics," *International Journal of Production Research*, Vol.27, No.9, 1989, pp.1637-1651.

9. Zadeh, L. A., Fuzzy Sets, Information Control, Vol.8, 1965, pp.338-353.

Design Assessment Tool

DANIEL M. NICHOLS and SUMITRA REDDY

Concurrent Engineering Research Center
West Virginia University
Morgantown, WV

Summary

This paper describes the implementation and use of a design assessment tool in the DICE (DARPA's Initiative in Concurrent Engineering) architecture. The internal structure of the tool and the user interface are described in detail along with a brief discussion of work planned for the future.

1 Introduction

Design assessment involves the ability to judge the quality of a design based on initial specifications such as customer requirements, safety regulations, and standards. Evaluating these criteria requires that they be known, and this information can come from the standards and guidelines adopted and constraints created by specific customer requirements. Assessment capabilities partly depend on the availability of domain specific tools to evaluate performance requirements.

Assessment, in the engineering domain, can concern only one part or a group of related parts that interests a designer. Assessment results come in the form of graphical and textual reports in response to assessments relating to that part or parts. This paper develops the idea of a design assessor into a tool that fits into the DICE architecture [4].

As future designs become increasingly complex, a project coordinator in a large design effort can lose substantial time sifting through a myriad of reports flowing in from a multitude of designers. A mechanism that can keep track of the information flowing throughout the design and can give reports about relevant information will prove to be a valuable tool to keep a designer in step with the progress of designs in the factory of the future.

This work has been sponsored by the Defense Advanced Research Projects Agency (DARPA), under contract No. MDA972-88-C-0047 for DARPA Initiative in Concurrent Engineering.

The design assessment tool (DAT) targets relevant information and monitors its progress throughout the design effort. It is a tool that keeps track of information flowing through the DICE Blackboard (DBB) which is used as a common place for designers to interact and reach consensus about the design and to provide visibility of the design[2]. The DAT provides graphical and textual reports about various aspects of the design. Additionally, aided by engineering applications registered in the DICE network, this tool can be used to provide external assessments of data.

2 Internal Structure

This section describes the design assessment tool from a programmatic standpoint. Assessment structure is discussed along with assessment instantiation. Collection and manipulation of data are presented next, followed by a discussion of the graphical and textual reports created from the data.

2.1 Assessment Structure

Blank assessments in the DAT, called *design assessment blank forms,* have a common structure. This structure implemented with LASER [1] and C is linked to the main DAT structure and has slots to hold the fields to monitor , the conditions to evaluate before storing any monitored data , a list of any variables that require values at the time of instantiation, a list of the tools and formulas to be applied to the raw data, and a list of the output parameters. Each of these fields has attached control information that organizes the underlying structure of the assessment.

2.2 Assessment Definition

A designer defines an assessment and makes it active by requesting an assessment type from a list of choices. Whenever this action takes place, a blank assessment *object* is instantiated and control information is created to complete the assessment structure. This control information concerns ownership of the assessment, the status of the assessment, the completion criteria specified by the designer, and information concerning the data monitoring routines.

Additionally, any variables present in the assessment type must be given values by the designer. These variables help to tailor the assessment to the specific user needs for the assessment type. Once these variables have been given proper values and the completion requirements have been specified, the user can either *save* or *send* the assessment. Saving the assessment allows the user to save the assessment for later use, and sending the assessment makes it active.

2.3 Assessment Activation

Once an assessment becomes active, real-time data collection begins. Data are collected by setting *demons* on the fields required for the assessment. Whenever objects are registered with the DAT, the design assessment tool compares the object's focus (what part(s) the object concerns) with a data dictionary that contains a list of foci and the assessments interested in them. If there is a match, then demons are set on the specific fields the assessment requires. The demon is then registered with the assessment to monitor the completeness of the data availability.

Whenever the information changes in a field where a demon is present, the demon becomes active, and it tests the condition associated with the field to determine whether or not to collect the data. If the condition evaluates true, the data are placed in a data object associated with the assessment. A time-stamp and the name of the object where the data came from is added as control information to the data field in the data object. The demon that *fired* remains active, but its status in the assessment changes from *unfired* to *fired*. This allows the assessment to know that at least one instance of this data has been gathered.

Active assessments periodically have their completion requirements tested by a completion monitoring routine. This routine evaluates the specific user completion requirements which center around time specifications such as daily or weekly requests for assessment results. If an assessment's completion requirements are satisfied, the data are sent to any external assessment tools identified in the assessment. The results from these tools are collected as data for the assessment report.

2.4 Assessment Results and Display

Assessment reports consist of both graphical and textual sections. Graphical sections can be in the form of x-y plots and histograms, and the textual section comes in the form of a formatted table (see section 3.2 for an example report). When defining the assessment, the user can specify the format of the report and what information should be displayed graphically. Users can also select multiple reports to appear as a group to compare several assessments at one time.

3 User Interface

The interface for activating an assessment and displaying a report is briefly described. The approach is to make it as Macintosh-like as possible using C and X-LIB [3], and allow the user to select the information required for an assessment from a list of choices.

The user invokes the design assessment tool by choosing the *assess* option under the main blackboard interface. A browser window showing the design assessments available and the reports generated is shown on the screen along with options for making an assessment, getting a report, and defining a new assessment (see figure 1). Only the options for making an assessment and getting a report are discussed.

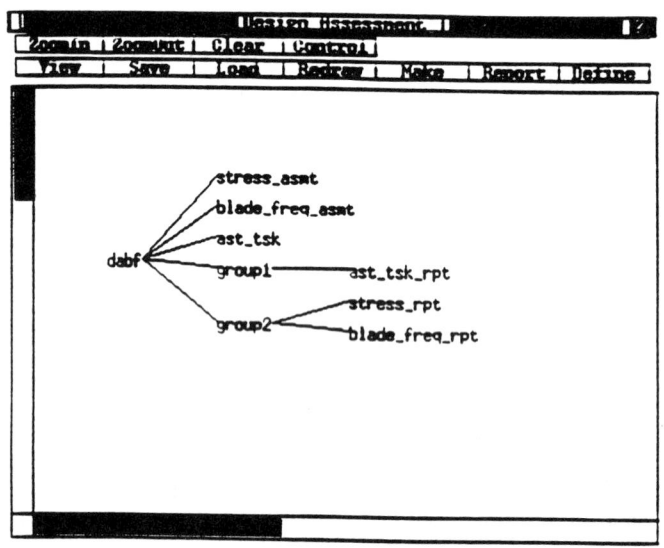

Fig 1. The main design assessment window showing the available assessments and reports and the design assessment functions.

3.1 Making an Assessment

To choose this option, the user first selects an assessment type and then clicks on the *make* button. This action causes a panel to open which displays information concerning the assessment's name, type, and description along with a list of all the external assessment tools used as part of the assessment. Additionally, there are panels requesting values for the assessment's parameters and completion requirements (see figure 2). Once the user successfully enters data for the required parameters and the completion criteria, there are several operations that can be performed.

The operations menu allows a user to *save, send, delete, cancel,* or *group* an assessment. Saving an assessment keeps the assessment in the DAT structure for later use ; sending an assessment makes the assessment active and data collection begins, deleting an assessment removes it from the DAT structure, while canceling an assessment deactivates it, but does not remove it from the DAT structure. Grouping an assessment identifies it as part of a group of assessments that are displayed together whenever the group is accessed.

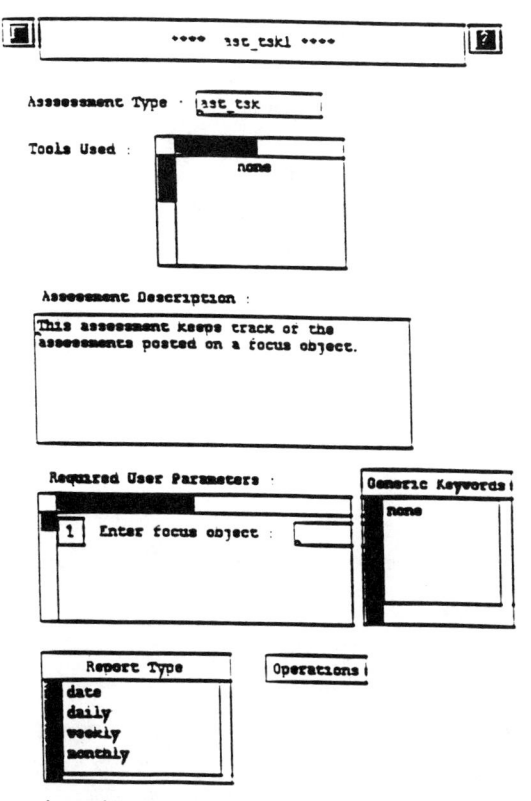

Fig. 2. An example make assessment menu. The user enters the focus object desired and the report type and then selects an operation.

3.2 Viewing a Report

One of the most useful features of the design assessment tool is the ability to convey a great amount of information in a condensed format. This information comes in the form of graphical and textual reports in a format the user specifies. To view a report, the user selects the report to be viewed and clicks on the *report* button (see figure 1). A report similar to that in figure 3 is displayed on the screen.

If more detailed information is desired, then the user can get it by clicking on one of the individual reports rather than on the group. The individual reports not only display the main section of the report, but also more detailed textual tables and/or graphs. These reports can be *saved*, *deleted*, or *printed* by selecting one of the operations in the report menu (not shown).

In the DICE architecture there is a module called the Electronic Development Notebook (EDN) whose role is to maintain the product development history and intent[5]. The DAT can function as a source of important information to be archived in the EDN, and we can provide an interface via EDN commands to extract the DAT assessment results and place them in the EDN with an appropriate index entry.

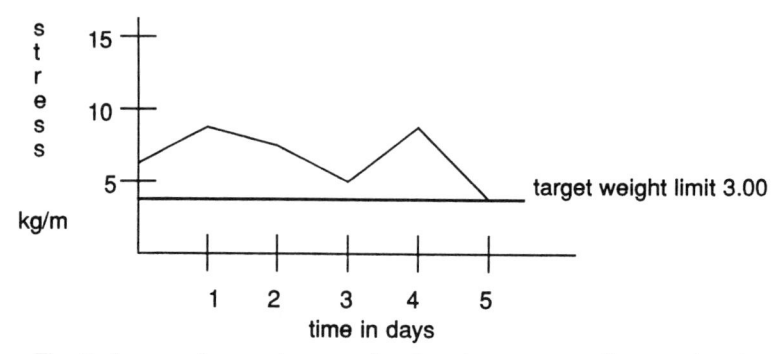

Fig. 3. An sample report comparing the stress across discrete time intervals to a target value.

4. Future Work

The initial prototype of the design assessment tool focused on information flowing through the DICE blackboard. While this information can provide assessments on the current state of the design, there is no facility to compare histories of previous designs or even a comparison of versions of the evolving design. The next stage of design assessment development will concentrate on interfacing with other knowledge sources in the DICE environment that can provide this information.

Additionally, the prototype can only simulate the execution of external assessment tools for more complex assessment capabilities. Research will continue on the integration of external assessment tools with the design assessor as these tools become available in the DICE environment.

5 Conclusions

This paper has explained, in outline, the initial prototype of the design assessment tool. With large amounts of data flowing in a design effort, this tool can help the project coordinator keep in step with the progress that is being made on the design. As has been shown, the DAT has the ability to monitor large amounts of data and condense the results into graphical and textual reports.

References

1. The LASER environment, LASER 2.0, Bell Atlantic Knowledge Systems : 1989.

2. Londono, Felix ; Cleetus, K. J. ; Reddy, Y. V. : A blackboard problem_solving model to support product development, Proceedings of the Second National Symposium on Concurrent Engineering. Feb. 7-9, 1990, Morgantown, WV, 91-110.

3. Nye, Adrian (ed.), Xlib reference manual for version 11 of the X window system. Newton, MA : O'Reilly & Associates 1988.

4. Red book of functional specifications for the DICE architecture. Concurrent Engineering Research Center WVU : 1989.

5. Uejio, Wayne H. : Electronic Design Notebook for the DARPA Initiative in Concurrent Engineering, Proceedings of the Second National Symposium on Concurrent Engineering. Feb. 7-9, 1990, Morgantown, WV, 349-361.

Support of PCs in Concurrent Engineering

NARESH C. MAHESHWARI and BRADLEY S. BENNETT

Concurrent Engineering Research Center
West Virginia University, Morgantown, WV 26506

Summary

The DICE (DARPA Initiative in Concurrent Engineering) Network is a collection of architectural services that support the usage of concurrent engineering (CE) practices in manufacturing organizations. Leading corporations depend heavily on supplier industries for up to 60 to 70 percent of their product components. These supplier industries connect PCs in their shops over local area networks (LANs). This paper discusses existing and future scenarios in PC-LANs and possible integration strategies and implementations using readily available commercial software packages. The PC user could exploit the advanced facilities provided by the DICE Network such as the Electronic Design Notebook, the DICE Blackboard, the sending of messages and pictures. At the same time users at Unix workstations could execute application on the PCs enabling access to thousands of applications available at a relatively cheaper price. The support of PCs in concurrent engineering provides a relatively inexpensive alternative in terms of hardware and software for manufacturing industries.

Introduction

DICE Network technology is being developed at the Concurrent Engineering Research Center to realize the immense benefits of concurrent engineering in U.S. industry. The DICE Network is a collection of architectural services designed to support CE practice in manufacturing organizations. The DICE architectural services include the Electronic Design Notebook[1], the DICE Blackboard[2], a Computerized Conference System[3], the sending of messages and pictures, and application integration using Wrappers[4]. The DICE architecture components run on Unix and VMS workstations connected through TCP/IP and supporting the X- Windows system. Leading corporations depend heavily on supplier industries for up to 60 to 70 percent of their product components. These supplier industries connect PCs in

*Acknowledgment:

This work has been sponsored by Defense Advanced Research Projects Agency(DARPA), under Contract No. MDA972-88-C-0047 for the DARPA Initiative in Concurrent Engineering(DICE).

their shops over local area networks (LANs). We plan to integrate these existing popular PC-LANs under the DICE Network with the least amount of disruption and cost. The leaders among PC-LAN operating system suppliers include: Netware from Novell; LAN Manager from Microsoft; Vines from Banyans Systems Inc.; LAN Server from IBM; and 3+ Open from 3COM Corporation. LAN Server from IBM and 3+ Open from 3COM Corporation are derivatives of LAN Manager from Microsoft. We shall discuss current and near-term integration strategies and implementations of PC-LANs in the DICE Network.

1.1 Novell's Netware

1.1.1 DOS Environment

As of today, DOS is the most common environment found on a Novell workstation. There are several ways to connect these workstations through TCP/IP over an Ethernet link to the Unix workstations.

1.1.1.1 Using WIN/TCP, WIN/ROUTE, and WIN/API from Wollongong Group Inc.

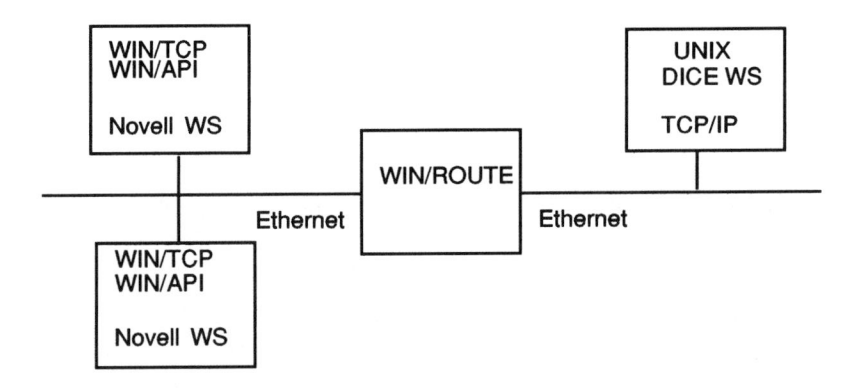

Figure 1: Connecting of Novell workstations using Wollongong Products

WIN/TCP for DOS from Wollongong provides TCP/IP protocol stack for PCs and supports telnet, ftp, and electronic mail applications. WIN/API for DOS provides a programmatic interface using Berkeley sockets. WIN/ROUTE for DOS provides Internet Protocols routing for TCP/IP based communication with PC-LANs. The connections are shown in Figure 1. Several Novell workstations installed with WIN/TCP can communicate using the single router.

In conjunction with X-Windows software such as "PC Xsight" from Locus Computing Corporation, the PC user can log on to any Unix DICE workstation and execute DICE software such as the Blackboard, Co-operate, and the Electronic Design Notebook. But more importantly, PC user can execute any DOS application on the Novell LAN and send the results to any DICE workstation user through the file transfer or electronic mail utility.

1.1.1.2 Using PC/TCP Plus from FTP Software Inc.

PC/TCP Plus from FTP Software Inc. has been the most widely used product to connect PCs to Unix workstations through TCP/IP. Refer to Figure 2.

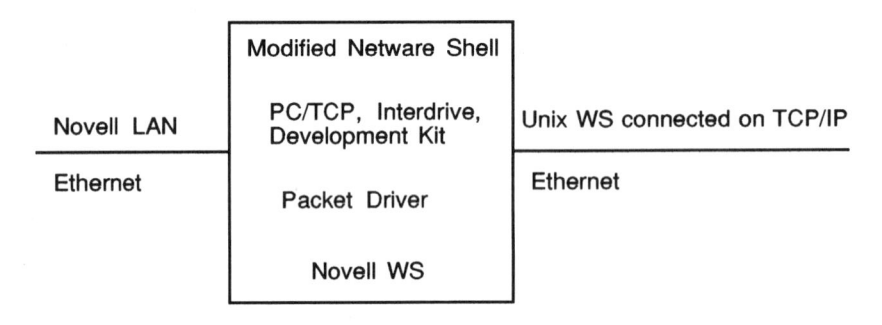

Novell LAN	Modified Netware Shell	Unix WS connected on TCP/IP
	PC/TCP, Interdrive, Development Kit	
Ethernet		Ethernet
	Packet Driver	
	Novell WS	

Figure 2: Connecting of Novell workstation using FTP Inc. Products

PC/TCP and Netware IPX share the same Packet Driver for the Ethernet card. These packet drivers for various Ethernet boards are available from Clarkson University. Modified IPX and Shell modules are available from Brigham Young University. PC/TCP Plus along with Interdrive provides TCP/IP protocol and NFS for PCs. It also provides telnet, ftp, rsh, rlogin, and Berkeley Socket library. A PC user can execute programs using the resources provided by a Novell Server and Unix DICE network resources. The results can be communicated to Unix workstations using NFS shared directories or through file transfer using ftp. In addition, the utilities can be developed to provide instant messaging between PCs and Unix workstations. Remote execution of programs on PCs from Unix workstations can also be provided. Again, the execution of DICE programs such as the Blackboard, Co-operate, and the Electronic Design Notebook can be provided by telnet and PC Xsight (X- Server) from Novell workstations.

1.1.1.3 Using PC/NFS from Sun Microsystems Inc.

PC/NFS is essentially a stand-alone PC product which allows PCs to be connected to Unix workstations LAN through TCP/IP over an Ethernet link. To support PC/NFS on a Novell workstation, the PC should have two Ethernet boards- one to communicate to Novell LAN and another to connect to DICE network through TCP/IP. We have developed utilities to provide instant messages from PCs to the users of Unix DICE workstations and remote execution of programs on the PC from Unix workstation. For details refer to the user's manual of the "DICE PC Server"[5]. These utilities have been developed using shared directories through NFS and rsh.

1.1.2 Windows Environment

Recently, Microsoft Corporation delivered Windows 3.0 for PCs. According to industry analysts, this is going to be the dominant environment for PCs over the next couple of years. It provides multi-tasking and larger memory for DOS applications. All the suppliers of TCP/IP products for PCs will be porting their respective products to Windows 3.x. It is also expected that suppliers of X-Windows on PCs such as Visionware Limited will be porting their products to Windows 3.x. Quarterdeck Office Systems has announced a X-Windows product on the top of Windows 3.x to be delivered in September, 1990. Integration of PCs into the DICE Network will become more attractive and effective because of multi-tasking facilities and more memory provided by the enhanced and protected modes of Windows 3.x.

1.1.3 Novell Netware386

Novell has announced intentions to support TCP/IP and NFS protocols in Netware386 by the beginning of 1991. This will permit the simultaneous access of the resources of Novell and Unix LANs by the Novell workstation.

1.2 LAN Manager from Microsoft Corporation and its Variants

The LAN Manager is closely identified with the OS/2 operating system. Hence, we will consider only the LAN Manager along with the OS/2 operating system. The following vendors are currently supplying TCP/IP products for the OS/2, LAN Manager, and Presentation Manager environments:

- i) TCP/2 from Essex Systems, Inc.
- ii) PC/TCP for OS/2 from FTP Software Inc.
- iii) IBM TCP/IP for OS/2 Extended Edition

The first two products use the Microsoft/3Com Network Driver Interface Specification (NDIS) and MAC-level drivers to support Ethernet adapters. These products provide the TCP/IP protocol and standard applications like telnet, ftp, etc. Support for NFS is expected in the near future. OS/2 is a multitasking operating system. Hence, the critical DICE modules can be ported under OS/2. One significant problem is getting X- Windows under OS/2 and Presentation Manager. One possible solution will be to use OSF/Motif instead of the X-Windows system for critical DICE modules. OSF/Motif and PM are supposed to be compatible. Another solution will be to port X11 under OS/2 with or without PM.

1.3 VINES from Banyan Systems

1.3.1 DOS Environment

The VINES workstation can be connected to a TCP/IP based Unix LAN by any of the following products to provide limited support of DICE modules:

- i) WIN/TCP and WIN/ROUTE from Wollongong Group Inc.
- ii) PC/TCP Plus from FTP Software Inc.
- iii) PC/NFS from Sun Microsystems

In addition, VINES Server based software can also be procured from Banyan Systems Inc. This software enables VINES server to act as Internet Protocols (IP) router and thus enabling IP traffic to be routed to the TCP/IP LAN of Unix workstations.

1.3.2 Windows Environment

VINES workstations under Windows will be capable of providing remote execution of applications from Unix DICE workstations, login to any Unix DICE workstation and execution of the DICE architecture services. Refer to section 1.1.2 for further discussion.

1.4 Conclusion

All major PC LANs can be connected to the DICE Network to enable the easy deployment of emerging DICE technology in supplier industries. The DOS environment provides minimal functionality for integration because of the lack of multitasking and program size limitations. The Windows environment extends the integration of PCs further under the DICE Network. OS/2 provides the functionality equivalent to a DICE Unix workstation.

References

[1] *W. H. Uejio*, "Electronic Design Notebook for DICE", Proceedings of the Second National Symp. on Concurrent Engineering, Feb 7-9, 1990, Pages 349-362.

[2] *F. Londono et al.*, "A Blackboard Problem-Solving Model to support Product Development", Proc. of the Second National Symp. on Concurrent Engineering, Feb 7-9, 1990, Pages 91-109.

[3] *Boris Pelakh et al.*, "Co-operate- Computerized Conferencing System" Concurrent Engineering Research Center, West Virginia University, March 1990.

[4] *P. J. Lohr et al.*, "An Application Integration Wrapper for DICE", Proc. of the Second National Symp. on Concurrent Engineering, Feb 7-9, 1990, Pages 241-256.

[5] *Bradley S. Bennett et al.* "DICE PC Server (DPS)- User Manual" Concurrent Engineering Research Center, West Virginia University, July 1990.

Note:

This article is based on the product literature supplied by the various vendors and trade journals.

Chapter III

Neural Network

Introduction

Neural nets are gaining prominence as a tool for solving a number of fuzzy type problems as an alternative method. The primary advantage of neural nets is their speed. In the manufacturing domain, the neural net concept has been applied to vision systems, robot path planning, pattern recognition, etc. The first paper deals with the employment of Artificial Neural Networks (ANN) to control a Molecular Beam Epitaxy process for growing thin film materials for semiconductors and to design a manufacturing system for group technology. ANN is applied as a pattern matching device for pattern directed scheduling for flexible manufacturing systems. The feasibility of an adaptive scheduler based on the integrated artificial intelligence framework is also presented. A neural network approach using the Adaptive Reasoning Theory (ART) paradigm to be used to classify the vectors obtained from the machine-part matrix is proposed in the next paper. In this paper, sensitivity analysis of the ART in recognizing the similarity between the vectors and preprocessing of these vectors is also analyzed. The last paper of this chapter discusses the application of ANN to nuclear reactor control. A two-layer feedforward neural network with back propagation as modeling paradigm for continuous process diagnostic is proposed. It is emphasized that a diagnostic neural network will be much faster than a diagnostic expert system.

Artificial Neural Networks In Manufacturing

KENNETH R. CURRIE

Department of Industrial Engineering
Tennessee Technological University
Cookeville, TN

Abstract

The topic of artificial neural networks (ANN) has received a great deal of attention among various groups of scientists and engineers as an alternative method of solving either intractable or "fuzzy" problems. In the area of manufacturing, ANN have been applied to vision systems, robot path planning, pattern recognition of data trends for use in analyzing production systems, and some recent research has been attempting to use ANN for feature recognition to facilitate CAD/CAM. This paper will highlight the scope of recent research accomplishments using ANN in manufacturing. Specific research in the control of a Molecular Beam Epitaxy process for growing thin film materials for semiconductors, and the design of a manufacturing system for Group Technology will be examined in closer detail. To conclude the examination of ANN in manufacturing, future trends and research areas will be discussed.

Introduction

The foundation of artificial neural networks can be traced to early research in neuroscience and neurobiology. Many of the advances in neural network development are directly linked to the biological constructs of learning, memory, and pattern recognition. The basic processing element of the neural network is the neuron which accepts electrical impulses or "firings" from other neurons through a network of fibers referred to as the synaptic conduit. Figure 1 gives a crude pictorial representation of a neural network as found in biological systems.

The biological network as shown in Figure 1 has been adapted in a very limited form to create an artificial neural network (ANN) that can be simulated on a digital computer. The simplest expression of a neural network is the parity network and is illustrated in Figure 2. The structure of the network is an input layer, a middle layer referred to as the hidden layer, and a final output layer. The parity network utilizes a supervised learning paradigm known as the generalized delta rule. The fact that the network requires supervised learning implies that a training set of possible input to output mappings are necessary to "teach" the network. Included in Figure 2 is a list of the input to output mappings by which the parity network has been

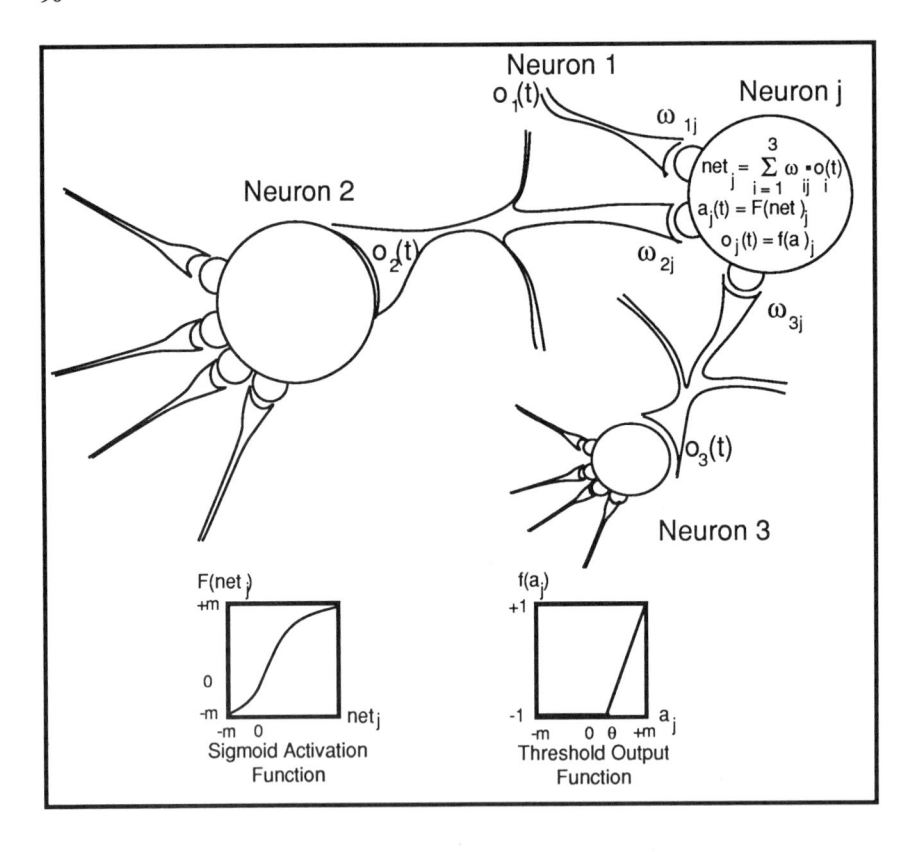

Fig. 1. Biological foundation of artificial neural networks
adapted from Rumelhart, McClelland, and et. al. [1]

trained. The neural network uses the training set one mapping at a time and adjusts the weights of the interconnections using the generalized delta rule to minimize the error between the output from the training set and the activation level of the output neuron. As the weights of the interconnections are changed the hidden layer seeks to find the appropriate dimension by which to map the input space to the output space. Supervised networks, also known as hetero-associative networks, are extremely useful in prediction of fuzzy, qualitative, and/or mixed variable patterns unable to be solved using classical methodologies. Another advantage of the network is the ability to be fault tolerant of input data that has either been improperly entered or is a new input mapping. The drawbacks of the hetero-associative networks are the necessity of a training set large enough to represent the mapping of the input space to the output space accurately, and the time necessary to train the network.

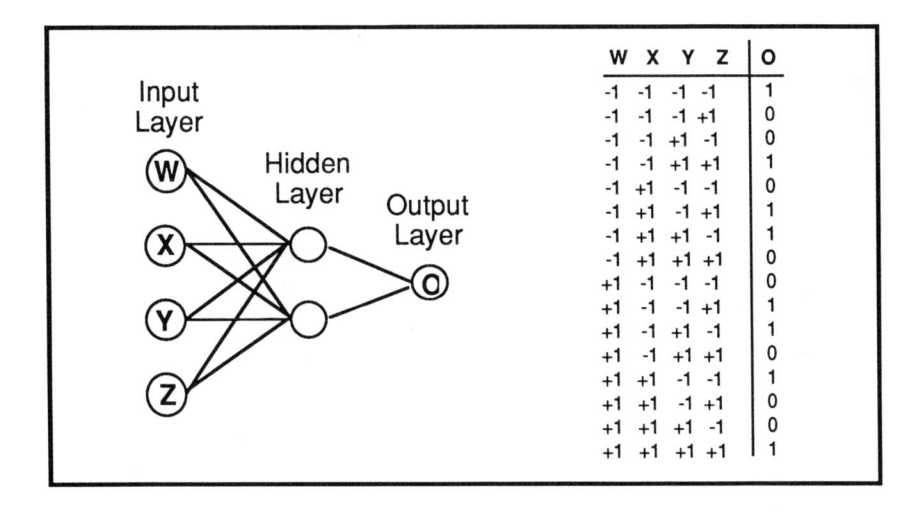

W	X	Y	Z	O
-1	-1	-1	-1	1
-1	-1	-1	+1	0
-1	-1	+1	-1	0
-1	-1	+1	+1	1
-1	+1	-1	-1	0
-1	+1	-1	+1	1
-1	+1	+1	-1	1
-1	+1	+1	+1	0
+1	-1	-1	-1	0
+1	-1	-1	+1	1
+1	-1	+1	-1	1
+1	-1	+1	+1	0
+1	+1	-1	-1	1
+1	+1	-1	+1	0
+1	+1	+1	-1	0
+1	+1	+1	+1	1

Fig. 2. Perceptron model of the Parity-4 problem

Another representation of an ANN is one which does not require a training set referred to as an auto-associative or self-organizing network. An example of an auto-associative network is the Hopfield Associative Memory. The Hopfield network works very much like human memory using sensory cues to associate similar objects or responses. As an example, a fragrance may immediately be associated in memory with a particular person and their attributes. The Hopfield Associative Memory uses cues to associate objects with similar attributes to the cue being used. Auto-associative networks are used in classifying pattern membership and in feature detection. There are limitations however on the capacity of associative memories, but adaptations to commonly used networks are being developed which enable larger capacities [2,3].

The use of simulated neural networks *is not* the end all solution to problems which are incapable of being solved using classical solution methodologies. It is however an alternative approach, another tool in the engineer's toolbox, that is extremely useful when adaptability and robustness are required.

Current Applications of ANNs in Manufacturing

The following review is by no means exhaustive, but is a sample of past, current, and future applications of ANNs in manufacturing. Applications of ANNs in manufacturing can be grouped into two broad categories: the first category is in machine control, and the second area is information processing of manufacturing data for decision making.

The most widely applied use of ANNs is in the control of robots. One of the first documented approaches to robotic control using neural networks was Albus [4] at the National Bureau of Standards (now the National Institute of Standards and Technology). In his four-part series, Albus details the biological and mechanical similarities of human movement, tracking, and grasping with that of the robot. Another article by Josin [5] develops a neural network that addresses the inverse kinematic problem when there are redundant degrees of freedom. Additional research into robotic control is being conducted at independent laboratories where information is proprietary. In a March 16, 1989 Wall Street Journal article, Julie A. Lopez reported on a robotic system under development by Martin Marietta Corp. for the Air Force that would use neural networks to fuse sensory inputs such as vision systems and sonar to control a robot. Neural networks are not restricted to control of mechanical equipment, but has also been used for adaptive process control. Articles by Widrow and Winter [6] and Passino, et.al. [7] represent a starting point for adaptive control using neural networks. Neural networks have demonstrated a reasonable measure of success in signal filtering and numeric to symbolic conversion as shown in these papers respectively.

The second application area of ANNs in manufacturing deals with processing of manufacturing information for decision making. Information processing is not limited solely to manufacturing. There are numerous examples where neural networks are used in financial applications as summarized by Humpert [8]. Two specific examples of information processing using ANNs in manufacturing will be discussed in detail to illustrate the possible breadth of applications neural networks will have in this area in the future.

Process Characterization Using an ANN

The author is currently using ANNs for rapidly characterizing a new technology for depositing thin film materials for semiconductors. Molecular Beam Epitaxy (MBE) is a state of the art technique for growing thin film epitaxial layers for semiconductor devices in the microwave/digital and opto-electronic areas. The MBE process is a very complex and unique process for growing epitaxial layers that conventional processes are unable to produce.However, there are stumbling blocks to the widespread use of MBE as a production tool such as contamination and lack of precise process control, resulting in low yields. Typically, the empirical characterization of the process takes several months to produce a single, new material type that achieves the desired optical, electrical, and molecular properties. The objective is the ability to self-improve process knowledge and thereby decrease the time necessary to find acceptable ranges for producing quality parts. The neural network will encode the knowledge of the MBE process in the network by adjusting the weights to account for the strength of the relationship between input and output variables. As new materials are characterized or as the process parameters shift, the network will be able to include the new processing information to create an improved knowledge base. The knowledge base will be dynamically changing as new materials are characterized, and as ex-situ material properties confirm process trends and patterns.

Manufacturing Cell Design Using an ANN

Group Technology exploits the similarities of part types to form part families and machine cells so that a particular family of parts is completely manufactured by a particular cell of machines. Previous research has focused on cell design as a mathematical exercise of partitioning a binary machine-component matrix to minimize the inter-cellular travel of parts between cells. In a paper by Currie and Creese [9] the topic of cell design was shown to involve interactions between both design characteristics and manufacturing characteristics. Current research is being conducted by the author to use both design and manufacturing characteristics in an auto-associative neural network to classify part families. The methodology uses an interactive-activation competition network as presented in the book by McClelland and Rumelhart [10]. An individual part is activated thus causing all of its attributes to be excited. The excited attributes will in turn excite those parts with similar attributes to the part initially excited. As an attribute receives an excitatory signal, all other dissimilar attributes receive inhibitory signals, thus the term competitive learning. The process of excitatory and inhibitory signals continues until an equilibrium state is reached with some parts having a high activation level and others with a suppressed activation level. This process is repeated for each part, and a part matrix can be assembled of the part activation levels relative to every other part. Then, using a matrix reorganization technique such as rank order clustering or a bond energy algorithm, the part activation matrix is partitioned into part families.

Conclusions

The applications of ANNs in manufacturing is quite diverse, but the potential for future application areas appears to be even broader. Consider the advances that ANNs have made in automatic target recognition with the ability to filter distorted images, detect features (regardless of orientation), construct objects from features, and then classify the objects. These same tasks are very similar to tasks required by vision systems, assembly operations, and the development of feature based design for CAD/CAM. There is also a great void in the research of optimization problems and scheduling and sequencing problems using ANNs. There are many specific applications of ANNs in manufacturing left to be explored, the determining factor is knowing when and how to apply them.

References

1. Rummelhart, D. E., Hinton, G. E., and McClelland, J. L. A general framework for parallel distributed processing. Parallel Distributed Processing, Volume 1: Foundations, eds. Rumelhart, McClelland, and the PDP Research Group. MIT Press (1986) 47.

2. McEliece, R. J., et. al. The capacity of the Hopfield associative memory. IEEE Trans. Infor. Theory. IT-33 (4) July (1987) 461-482.

3. Kosko, B. Bidirectional associative memories. IEEE Trans. Sys. Man Cyber. 18 (1) (1988) 49-60.

4. Albus, J. A model of the brain for robot control, part 2: A neurological model. Byte. July (1979) 54-95.

5. Josin, G. Integrating neural networks with robots. AI Expert. August (1988) 50-58.

6. Widrow, B. and Winter, R. Neural nets for adaptive filtering and adaptive pattern recognition. Comp. March (1988) 25-39.

7. Passino, K. M., et. al. Neural computing for numeric-to-symbolic conversion in control systems. IEE Cont. Sys. Mag. April (1989) 44-51.

8. Humpert, B. Financial applications using neural networks. Intel. Sys. Rev. 2 (1) (1989) 36-45.

9. Currie, K. R. and Creese, R. C. Justification of Cellular Manufacturing Using Multi-Attribute Part Family Loading - MAPFLO. Justification Methods for Computer Integrated Manufacturing Systems. Elsevier Press (1990) 203-219.

10. McClelland, J. L. and Rumelhart, D. E. Explorations in Parallel Distributed Processing. MIT Press. 1988 11-48.

Using Artificial Neural Networks For Flexible Manufacturing System Scheduling

LUIS CARLOS RABELO and SEMA ALPTEKIN

Computer Integrated Manufacturing Laboratory
Engineering Management Department
University of Missouri-Rolla
Rolla, MO

SUMMARY

The scheduling function plays an important role in flexible manufacturing systems. However, flexible manufacturing system scheduling is tremendously complex due to combinatorial explosion, technological constraints and goals to be achieved. Heuristics involving dispatching rules have been widely utilized to obtain good solutions. Nevertheless, it is difficult to identify any single dispatching rule as the best in all scheduling problems. Hence, a pattern-matching structure which can identify the "best" dispatching rule for a given problem will increase the probability of the generation of good schedules. In this paper, the utilization of artificial neural networks as pattern-matching devices for pattern-directed scheduling is reviewed. Several flexible manufacturing system scheduling problems and on-going research will be explained. Conclusions about the feasibility of an adaptive scheduler based on an integrated artificial intelligence framework and further research issues will be presented.

INTRODUCTION

Production scheduling may be defined as "the art of assigning resources to tasks in order to insure the termination of these tasks in a reasonable amount of time" [1]. According to French [2], the general problem is to find a sequence, in which the jobs pass between the machines, which is a feasible schedule, and optimal with respect to some performance criterion. Therefore, it is possible to define the FMS scheduling problem as an open deterministic dynamic job shop scheduling model. FMS scheduling problems belong to the class called "NP-hard". Several techniques such as mathematical programming and analytical models, dispatching rules and simulation, look ahead algorithms, and other unique approaches such as simulated annealing, and space filling curves have been applied to FMS scheduling problems [3]. Most of these existing FMS scheduling methodologies are not effective for real/time tasks. In addition, the knowledge of scheduling of FMS is mostly system-specific, not well developed and highly correlated with shop floor status, resulting in the absence of recognized sources of expertise [4]. This gives rise to problems on effective knowledge acquisition and knowledge representation schemes, and limits the development of generic knowledge-based expert systems for FMS scheduling. Consequently, an FMS scheduling system should have numerous problem-solving strategies. Each of these problem solving strategies should be selected according to the situation imposed by the scheduling environment. This calls for a pattern-matching structure to solve scheduling problems to select the problem solving strategy for a given scheduling situation.

Artificial neural networks (ANNs) are information processing systems motivated by the goal of reproducing the cognitive processes and organizational models of neurobiological systems. By virtue of their computational structure, ANNs feature attractive characteristics when applied as pattern-matching mechanisms for FMS scheduling. In the next sections, research assumptions, results of ANNs as pattern-matching mechanisms for FMS scheduling, and conclusions will be presented.

ANN FOR FMS PATTERN-MATCHING SCHEDULING

Assumptions. The assumptions used for this investigation are common to those utilized in simulation studies performed by other researchers [4,5]. These assumptions include the following:
- a) Processing times are deterministic
- b) Due dates are fixed
- c) There are no groups of similar machines
- d) The required resources to carry out the jobs have been identified
- e) The job routings are fixed and unique
- f) Pre-emption is not allowed
- h) Transportation times are negligible

<u>Performance Criterion Used.</u> The performance criterion which will be used is tardiness. Tardiness has been selected as the optimization criterion due to the importance of meeting due dates in industrial scheduling problems.

<u>FMS Dispatching Rules Selected.</u> The following rules were selected:

SPT (Shortest Processing Time)   CR (Critical Ratio)
LWR (Least Work Remaining)   EDD (Earliest Due Date)
SLACK (Minimum Slack Time)   S/OPN (Slack per Operation).

<u>Input Feature Space.</u> Appropriate ways need to be developed to represent the problem data to facilitate its understanding by the network. Without this key feature, the neural network will fail to learn the relationship with the desired efficiency and accuracy.

The problem of representing the number of parts and their routings was solved using a concept that is tied to FMS: Group Technology as the first dimension. The second dimension is the time remaining until the due-date. The third dimension is represented by the number of jobs that fall in the category defined by GT and the time remaining until due date. Each point is presented in the input vector (See Figure 1).

<u>Knowledge Acquisition Process.</u> Training examples should be generated in order to train a network to provide the correct characterization of the manufacturing environments suitable for various scheduling rules and the chosen performance criterion. In order to generate training examples, a performance simulation of SPT, EDD, LWR, S/OPN, SLACK, and CR for the same set of manufacturing situations was carried out (See Figure 1)-an example of a simulation problem is shown in Figure 2.

<u>Using Backpropagation.</u> The neural network paradigm utilized was backpropagation [6] and several scheduling problems were developed to test its respective performance. As an example, the case of a 10-job 4-machine scheduling problem will be presented. For this case an ANN using the paradigm specified above was developed.

This ANN developed, had a training set of 200 examples. The training session and its respective learning curve are shown in Figure 3. The training session took 35 hours on an IBM PS/2 Model 70. The ANN selected had 15 input units corresponding to the input feature vector size, 6 output units corresponding to each specific dispatching rule, and 68 hidden units in one hidden layer. The network was tested for 100 "new" examples and the performance of the network was found to be consistent (See Table I). These results were superior to those obtained using other dispatching rules.

Other Network Paradigms Investigated. After extensive research [7], backpropagation has been proven to achieve good performance. However, it has a number of limitations which include an inability to converge, long training times, and complex hidden units-hidden layers behavior. This encouraged the investigation of other network paradigms such as the Restrictive Coulomb Energy network [8]. In an RCE network, all examples of a pattern category define a set of points in the feature space that can be characterized as a region (or a set of regions) having some arbitrary shape. Consequently, an RCE network memorizes the pattern data. It has several advantages such as rapid system training, automatic node creation, and probability estimations to handle uncertainty.

A first attempt was made to utilize an RCE network using the NDS-500 from Nestor, Inc. to solve the 10-job 4-machine scheduling problem. The network was not able to learn and predict with an acceptable performance (less than 40% accuracy). These preliminary results indicate that the encoding of the input feature space used for backpropagation was not adequate for networks since RCE memorizes data patterns. A new encoding procedure was utilized, and the RCE network improved its performance (See Figure 4). This encouraged the development of new procedures which at this moment are an on-going research issue.

CONCLUSIONS

The major focus of this paper was the application of ANNs as pattern-matching mechanism for FMS scheduling. ANNs offer speed, learning capabilities, and convenient ways to represent manufacturing scheduling knowledge. On the other hand, several disadvantages of ANNs such as lack of explanations capabilities, and interpretation of declarative and procedural knowledge acquired, call for a hybrid approach to FMS scheduling. Therefore, procedural, declarative, and connectionist knowledge are all needed for a better knowledge representation of the FMS scheduling problem. This is only possible with the utilization of a hybrid inference process which would utilize inference-driven entities such as knowledge-based systems and connectionist-driven entities such as ANNs.

This Research was supported by a Grant-In-Aid of Research (SIGMA XI).

REFERENCES

1. Dempster, M.; Lenstra, J.; Kan, R.: Deterministic and Stochastic Scheduling: Introduction. Proceedings of the NATO Advanced Study and Research Institute on Theoretical Approaches to Scheduling Problems. D. Reidel Publishing Company, 1981, 3-14.
2. French, S.: Sequencing and Scheduling. Halsted Press: New York, 1982.
3. Kiran, A.; Alptekin, S.: A Tardiness Heuristic for Scheduling Flexible Manufacturing Systems. Proceedings of the 15th Conference on Production Research and Technology: Advances in Manufacturing Systems Integration and Processes. University of California at Berkeley, California, January 9-13, 1989, 559-564.
4. Wu, S.: An Expert System Approach for the Control and Scheduling of Flexible Manufacturing Cells, Ph. D. Dissertation, The Pennsylvania State University, 1987.
5. Gross, J.: Intelligent Feedback Control for Flexible Manufacturing Systems, Ph. D. Thesis, University of Illinois at Urbana-Champaign, 1987.
6. Rumelhart, D.; McClelland, J.; The PDP Research Group: Parallel Distributed Processing: Explorations in the Microstructure of Cognition. Vol. 1: Foundations. Cambridge: MIT Press, 1986.
7. Rabelo, L. C.: A Hybrid Artificial Neural Networks and Knowledge-Based Expert System Approach to Flexible Manufacturing System Scheduling. Ph. D. Thesis. University of Missouri-Rolla, 1990.
8. Reilly, D.; Cooper, L.; Elbaum, C.: A Neural Model Category Learning. Biological Cybernetics. Vol. 45, 1982, 35-41.

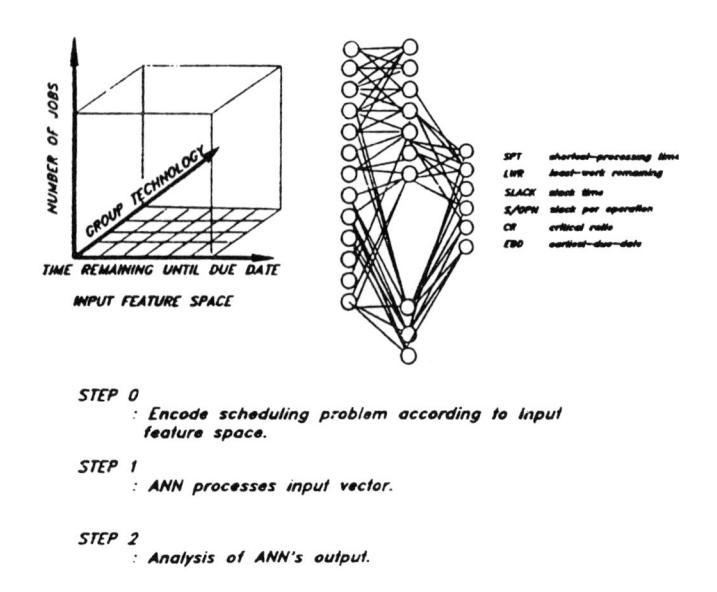

STEP 0
: Encode scheduling problem according to input feature space.

STEP 1
: ANN processes input vector.

STEP 2
: Analysis of ANN's output.

FIGURE 1. Using ANNs for FMS scheduling

SCHEDULING PROBLEM

JOB NUMBER	OPERATION NUMBER	READY TIME	DUE DATE	MACHINE NUMBER	UNIT TIME
1	1 (type 1) 2	0 0	13 interm. 13	1 3	3 4
2	1 (type 2)	0	12 interm.	3	5
3	1 (type 1) 2	0 0	6 very 6 near	1 3	1 8
4	1 2 (type 3) 3	0 0 0	30 very 30 distant 30	1 2 3	4 5 4
5	1 2 (type 4)	0 0	23 distant 23	2 1	8 5
6	1 (Type 1) 2	0 0	18 interm. 18	1 3	3 4
7	1 (Type 1) 2	0 0	12 interm. 12	1 3	1 8
8	1 2 (type 3) 3	0 0 0	28 very 28 distant 28	1 2 3	4 5 4

FIGURE 2. An FMS scheduling problem

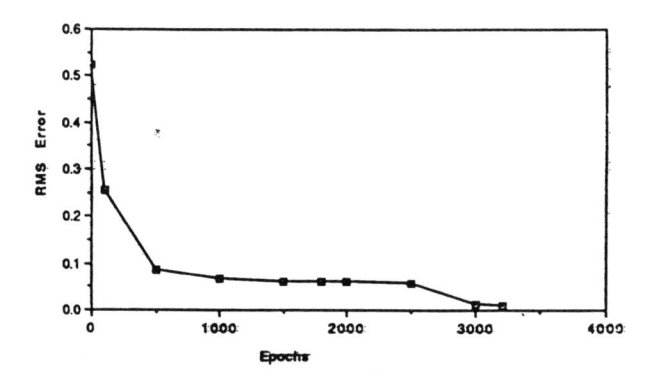

FIGURE 3. Learning curve

General Results Summary

Patterns	
Total	Processed
100	100

Identified		Uncertain		Unidentified
Correct	Incorrect	Correct	Incorrect	
74	26	0	0	0
74.0%	26.0%	0.0%	0.0%	0.0%

Size of NLS Memory	Total	Prototypes Exclusive	Overlapping
30324 bytes	886	879 99.2%	7 0.7%

F1=help   F10=detail   Esc=exit screen

FIGURE 4. RCE performance

Lowest tardiness frequency - 100 problems

SPT	LWR	SLACK	S/OPN	CR	EDD	ANN
12	11	60	62	53	50	83

Average total tardiness - 100 problems

SPT	LWR	SLACK	S/OPN	CR	EDD	DRC	ANN
82	83	72.9	72.6	73	76.3	69.6	71

DRC: Dispatching rules combined

TABLE I. Results for 10 job-4 machine problems

Machine-Part Family Formation Using Neural Networks

RAM HUGGAHALLI and CIHAN DAGLI

Engineering Management Department
University of Missouri-Rolla
Rolla, MO

ABSTRACT

Algorithms using analytic methods for the group technology problem of machine part family formation are relatively slow considering the typical large size of the machine-part matrix. In this study, a neural network approach that uses the Adaptive Resonance Theory (ART) paradigm to classify vectors obtained from the machine-part matrix, is proposed. The effect of varying the sensitivity of the ART in recognizing the similarity between the vectors and pre-processing of these vectors, on the formation of part groups is discussed. Initial test results obtained from the neural network application are compared with those of existing cell formation algorithms.

INTRODUCTION

The most important and exciting feature of neural networks, apart from their speed, is their ability to retain data, learn from past experience and adapt to new situations. These features of neural networks have already been applied to a number of real time situations with remarkable results.

Among the various applications of neural nets in manufacturing [1] is group technology, a concept that seeks to recognize and exploit similarities in physical and manufacturing attributes of the various elements involved in the manufacturing processes [2, 3]. A particular part's physical attributes can be recognized by a neural network, and on the basis of the attribute values, the network can group the part with a family of parts having similar attributes. This approach is a good alternative to existing classification and coding schemes which are tedious to apply.

Machine-part family formation [4], is a group technology problem for which neural networks can be used to obtain reasonable solutions. The objective here is to group similar machines and similar parts together to facilitate efficient planning and scheduling of the manufacturing processes. A matrix as in Table 1, representing the machine-part relations is called a machine-part matrix. Clustering of '1' elements in the matrix indicates the possible machine and part groups. Clustering algorithms developed for this purpose have been classified by Nakornchai and King as similarity coefficient, set theoretic, evaluative and analytic algorithms [4]. Also, a knowledge-based group technology system using a heuristic clustering algorithm was proposed by Kusiak [10]. All these methods, are sufficiently accurate and provide an excellent basis for comparison with the performance of the neural net approach.

The advantages of parallel distributed processing were, however not exploited by any of the methods mentioned above, due to the infancy of the technology. Hence, these algorithms can be painstakingly slow for machine-part matrices which are usually very large. Massive parallelization, as in neural nets, seems the only alternative to rapidly process such information and arrive at the result in a much quicker time.

GROUPING WITH THE ART PARADIGM

Among the different neural network architectures, the Adaptive Resonance Theory (ART) paradigm, is particularly attractive for this application due to its inherent classification scheme. It is now shown that an ART network [7, 8] can be applied efficiently to the cell formation problem.

An ART network accepts an input vector and classifies it into one of a number of categories depending upon the input vector's similarity with the stored patterns. A simplified version of the ART network is shown in Figure 1. Each neuron in the recognition layer is associated with a pattern that is stored as a set of weights to the inputs of the neuron. The input vector X is matched with all stored patterns, and the pattern that is closest to the input vector is adjusted to make it more similar to the input. The resultant vector C, obtained by a bit by bit AND of the input vector with the binary version of the stored pattern, must be similar to the input vector atleast to a 'vigilance' level, else the input vector is stored separately as an 'exemplar'.

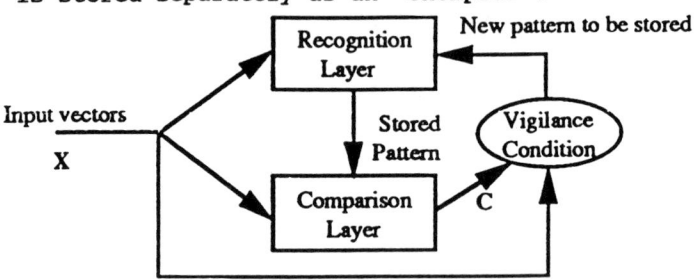

Figure 1. Simplified ART network.

Although Moon proposed an efficient algorithm [9] for the machine-part family formation problem, it is felt that the inherent binary vector classifying nature of the ART facilitates its direct application to this problem. In order to demonstrate how an ART can be used to form machine-part families, the architecture was modeled and simulated. Column vectors are first applied for sequential classification to form an intermediate matrix in which, similar column vectors are closer to each other relative to the original matrix. Then the row vectors in the intermediate matrix are likewise classified to obtain a final matrix. This process is depicted in Figure 2.

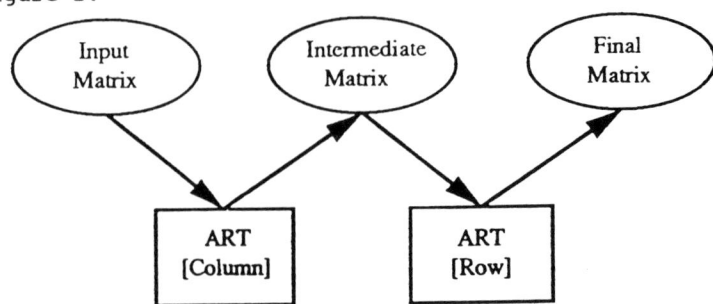

Figure 2. Column and row vector classification by the ART network.

Parts

	1	2	3	4	5	6			1	3	6	2	4	5
1	1						1	1					
2	1		1			1	2	1	1	1			
Machines 3			1	1			3		1			1	
4		1			1		5				1	1	
5		1		1			4				1		1
			(a)								(b)			

Table 1.(a) A 5 x 6 machine-part input matrix.
(b) Final matrix showing machine-part groups.

Consider the 6 part, 5 machine matrix in Table 1 (a) [6]. The column vectors (C1,..C6), that are first applied to the ART network are, 11000, 00011, 01100, 00101, 00010 and 01000. After, the classification of these vectors, the row vectors (R1,..R5), 100000, 101001, 001100, 010010 and 010100 are applied. The final matrix, obtained after classifying all vectors is shown in Table 1(b).

The ART network formed three part families (1, 3, 6), (2, 4) and (5). Although part # 5 should have been classified into the 2nd group, it was not, since the application of C4 before C5 lead to a resultant vector of 00001 being stored under group 2. This vector does not match C5 at all, hence a 3rd category had to be formed. A number of methods can be used to prevent such an event.

PERFORMANCE OF THE NEURAL NETWORK
The algorithm was studied with more realistic cases involving large machine-part matrices. The vigilance parameter plays a crucial role in classifying the input vectors properly. A very low vigilance value might lead to the classification of vectors that are barely alike, into one cluster, while a high vigilance value will form a larger number of clusters. Hence, the best solution from this method can be attained by starting with a low vigilance value and iterating the classification procedure for steadily increasing values of the vigilance level. It has been observed that using this procedure, the solutions of King's ROC2 algorithm for the

same cases can be closely matched. The performance of the neural network can be improved in a number of ways.

The main drawback of an ART for any application is that the number of 1's in any stored pattern under each category is non-increasing with the number of input vectors applied. This was demonstrated in the example presented earlier. Three approaches to counter this drawback are now cited.
1. If we pre-process the input matrix, so that vectors are applied to the ART in decreasing order of number of 1's, we can ensure, that the input vectors are always sparser in 1's than the stored patterns. This technique is particularly useful for large matrices that tend to have arrays of widely varying sparseness.
2. We can modify the ART algorithm so that the stored pattern under each category is adjusted only if the new input vector has more 1's. For example, if the stored pattern is 00011 or C2 and the input vector is 00110 or C3, we retain the stored pattern instead of adjusting it to 00100. This ensures that C5 is classified into group 2.
3. We can have another vigilance parameter to check the similarity of two vectors, with respect to 0's. Thus, if two vectors have to be classified into one group, they need to satisfy vigilances based on both 1's and 0's. The groups that are thus formed will contain parts that are more like one another.

With the help of these three techniques, the final matrix assumes an optimally clustered form. Machine cells and part families can be easily determined from the matrix. For the optimal clustering of machines or parts in case of ambiguity, algorithms developed by Wei and Gaither [6], or Seifoddini [5] can be used.

The size of the machine-part matrix has a direct bearing on the size of the neural net due to its parallel nature. Hence, the neural net, when realized on hardware, must be built for the maximum problem size expected to be solved. The parallelism involved, ensures that the time taken for any problem size will essentially be the same. Hence, a neural network classification system inclusive of additional but elementary processing techniques is still expected to be faster than any serial algorithm implementation.

CONCLUSIONS

Neural networks can be successfully applied to solve the machine-part family formation problem. The primary advantage of using neural nets is their speed. Neural net paradigms such as the ART can be applied in their basic form to arrive at reasonable results. The algorithms can be refined to achieve very accurate and optimal clustering. Research work to integrate the various techniques that optimize the performance of the basic ART is currently in progress. Yet another successful implementation of neural nets can hence be anticipated.

REFERENCES
1. Dagli, Cihan H.; Lammers, Scott; Vellanki, Mahesh : Intelligent Scheduling in Manufacturing through Neural Networks and Expert Systems, (to appear in Journal of Neural Network Computing 1991).
2. Hyer, Nancy L.; Wemmerlov, Urban : Group Technology in the US manufacturing industry: A survey of current practices. International Journal of Production Research. Vol. 27. 1989. 1287-1304.
3. Hyer, Nancy L. : Group Technology at Work. Society of Manufacturing Engineers. 1984. First Edition.
4. King, J.R.; Nakornchai, V. : Machine-component group formation in group technology: review and extension. International Journal of Production Research. 1982. Vol. 20. No. 2. 117-133.
5. Seifoddini, Hamid : A note on the similarity coefficient method and the problem of improper machine assignment in group technology application. International Journal of Production Research. 1989. Vol 27. No. 7. 1161-1165.
6. Wei, Jerry C.; Gaither, Norman : An Optimal Model for Cell Formation Decisions. Decision Sciences. 1990. Vol. 21. No. 2. 416-433.
7. Wasserman, Philip D. : Neural Computing - Theory and Practice. Van Nostrand Reinhold. 1989.
8. Carpenter, Gail A. : Neural Network Models for Pattern Recognition and Associative Memory. Neural Networks. 1989. Vol. 2. 243-257.
9. Moon, Young : An Interactive Activation and Competition Model for Machine-Part Family Formation in Group Technology. International Joint Conference on Neural Networks - Washington D.C. 1990. Vol II. 667-670.
10. Kusiak, Andrew : Knowledge-Based Group Technology. Expert Systems. Society of Manufacturing Engineering. 1988. 259-299.

Neural Networks in Process Diagnostics

SOUNDAR R.T. KUMARA and NAJWA S. MERCHAWI

Intelligent Design and Diagnostics Laboratory
Department of Industrial and Management Systems Engineering
Pennsylvania State University
University Park, PA

Abstract

Diagnostics and Control are critical to any continuous process set up. It is not only necessary that proper diagnostic conclusion be drawn but also with accuracy and speed. Artificial Neural Networks, which are biologically inspired are accurate, elegant and fault tolerant. In the contexts of known scenarios, they are faster in comparison with the case-based diagnostics systems. In this paper the authors propose a two layer, feed-forward Neural Network with back propagation as a modeling paradigm for continuous process diagnostics. This research is pursued in the domain of nuclear reactor control. Both the training and operational phases are described.

Introduction

The problem of diagnosis in industry is an important issue that has received much attention and created the new discipline of diagnostics. Diagnostics is concerned with developing procedures and mechanisms that detect failures in systems and suggest corrective action. As our systems get increasingly complicated, the diagnosis problem becomes more challenging. Artificial intelligence techniques have been used for cheaper and more automated diagnostics. In this paper, we look at the conventional diagnostic systems, identify problems with their implementation, and suggest that Neural Networks as a better candidate for diagnostic systems in terms of their accuracy and elegance.

Process Diagnostics: Conventional AI-based Approach

Diagnostic systems so far have been implemented as special cases of expert systems. This means that a knowledge base about the system is stored in a conventional computer. The knowledge base is a set of facts and rules about the behavior of the system. The inference mechanism (a running program) implements the actual diagnosis process. It inputs on-line data describing the current condition.

---

[1] Current address: Visiting Associate Professor, CSK chair, Research Center for Advanced Science and Technology, University of Tokyo, 4-6-1 Komaba, Meguro-Ku, 153 Tokyo Japan.

Then it searches through the rule space to identify the rules that apply to the current situation. It then asserts the conclusion of the rules which in turn can be premises of other rules. More searches are needed until no more inferences can be drawn. Search is a very time consuming process. Therefore expert systems are very slow. In process diagnostics, especially with continuous processes, speed is an important factor because some systems require a very high level of security and failures can be fatal.

The previous AI approaches to diagnostics have been either model-based or case-based. A typical approach described is based on Milne's theory of diagnostics [3]. First a knowledge base about the system is gathered. It consists of a set of parameters, their setpoint values, and a description of all possible failure scenarios in terms of relative parameter values with respect to their setpoint values.

The algorithm involves the following sequential phases:
1. Detect all out of bound values and the corresponding parameters (Contexts).
2. Identification of all the possible scenarios where these contexts can occur.
3. Identification of the priorities for these scenarios.
4. Select among the ones chosen in phase 3, the most likely one and
5. Suggest control actions.

We notice that this approach is time consuming. The time involved becomes intolerable with large problems where response time is critical. Moreover the system is incapable of making inferences in the event of partial or missing information.

Problem Description

We consider the Three Mile Island Nuclear reactor (TMI-II) example. It is not our intention to go through the details of operation here. Studies have revealed [Nann 88] that nine parameters play very important role in the nuclear reactor control. These nine parameters are described by $P_1, P_2, ...P_9$. From previous history we can identify different operational ranges for each of these parameters. Seven categories have been identified in this process. They are LLL, LL, L, N, H, HH, HHH. Where LLL and HHH are at the extremes of the spectrum denoting extremely low and extremely high values for the parameters (LLL: Extremely Low, LL: Very Low, L: Low, N: Normal, H: High, HH: Very High, HHH: Extremely High).

For each of the nine parameters definite values have been identified. We denote these as the setpoints. Table 1 gives the database of the set points. For example Parameter P_1 is normal if its value is 2150 (appropriate units). It is considered high if its value is 2250 etc.

There are twelve possible failure scenarios: $S_1, S_2, ..., S_{12}$. Through the practical experience or through quantitative derivations we will be able to set up a database describing the instance of each scenario. Table 2 describes the scenario database. The first line in this table is read: the lower the values of P1, P2, P4, P5, P7, and P8 higher the possibility of Scenario 1 to have occurred; the parameters P3, P6 and P9 can be interpreted as normal or irrelevant. The rest of the table is read in a similar manner.

Table 1. Setpoint Database from TMI-II Accident

Parameter	LLL	LL	L	N	H	HH	HHH
P1	1200	1900	2055	2150	2250	2355	2400
P2	45	150	200	222	240	260	280
P3	300	400	500	606	610	619	630
P4	800	850	900	940	1050	1070	1105
P5	10	30	45	160	170	180	190
P6	1	2	2.5	3	35	80	122
P7	800	850	900	940	1050	1070	1150
P8	10	30	45	160	170	180	190
P9	300	400	500	558	610	619	630

Given this data set the problem is to identify the parameters out of bounds and relate them to a particular scenario. Given this scenario, we can look into a corrections database and feedback the information. Nann et.al.[5] implemented this system through an SQL based relational retrieval scheme and intelligent feedback control. Some of the major problems related to such a scheme was the system did not perform very well in the event of missing or partial parameter values. In this paper we explore an alternative modeling paradigm for implementing this diagnostic system via Neural Network based architectures.

Neural Network Model for Diagnostics

Diagnosis is a matching process of input patterns with output patterns. The input patterns are the nine parameter values. The output patterns are the eleven failure scenarios and one normal operation scenario. Let us consider the TMI-II example described in the previous sections. The output patterns can be represented as binary strings. we need at least four bits to generate twelve different combinations to be assigned arbitrarily to the scenarios. One such assignment is: (S_1: 0001, S_2: 0010, S_3: 0010, S_4: 0100, S_5: 0101, S_6: 0110, S_7: 0111, S_8: 1000, S_9: 1001, S_{10}: 1010, S_{11}: 1100)

Table 2. Scenario Database

Scenario	P1	P2	P3	P4	P5	P6	P7	P8	P9
S1	Lower	Lower		Lower	Lower		Lower	Lower	
S2	Lower	Lower		Lower	Lower	Higher	Lower	Lower	
S3	Lower	Lower				Lower			
S4	Lower	Higher				Higher			
S5				Lower	Lower		Lower	Lower	Lower
S6			Higher	Higher	Lower		Higher	Higher	Lower
S7			Higher	Higher			Higher		Higher
S8			Higher	Lower	Lower		Lower	Lower	Higher
S9				Higher	Higher		Higher	Higher	Higher
S10				Higher	Lower		Higher	Lower	Higher
S11				Lower	Higher		Lower	Higher	Higher

To represent the input patterns, however, we need special attention. First, let us assume that external to the Neural Network we have hardware that inputs actual readings of parameter values and converts each to one of the seven states, LLL, LL, L, N, H, HH, HHH. Using comparator chips and necessary logic design techniques we can implement such a hardware at a reasonable cost. To represent seven states we need combinations of at least three bits. An example representation could be: (LLL: 110, LL: 101, L: 100, N: 000, H: 001, HH: 010, HHH: 011).

But such a representation is not appropriate for diagnosis on Neural Networks. This is because a Neural Network matches input patterns that look similar to the same output. In other words, the closer the patterns look, the more likely generate the same output and therefore mean the same thing. What is meant by patterns *to look close* is the following: A Pattern can be thought of as an image where a "1" in the pattern is a dark pixel and a "0" is a light pixel, arranged, say from left to right; patterns look close when their respective images look almost the same (differ only in a few pixels or bits). For the diagnosis process on Neural Networks, states of parameter values that have a close interpretation should look close to each other. For example, states HH and HHH have a closer interpretation and therefore their respective representations look closer. We have developed a representation scheme for the seven states based on the following criterion: States LLL and HHH have completely different interpretations and therefore their representation should look completely different. As the values of the parameters increase from LLL to HHH their corresponding patterns should start looking increasingly different from the representation of LLL and increasingly closer to that of HHH. The representation is:

(LLL: 000000, LL: 100000, L: 101000, N: 101010, H: 101011, HH: 101111, HHH: 111111).

To develop a training set for the Neural Network, an overall input pattern characteristic of each senario has to be derived. This pattern has to contain describtions of the states of all nine parameters, typical of the occurrance of the scenario. Therefore the overall input pattern for a scenario is the concatenation of the representations of the nine parameters, each parameter as described in Table2 using the following guidelines: When Table 2 reads "lower", the corresponding parameter is taken to be in state LL, when it reads "higher", the parameter is in state HH, and when the parameter is unaffected, it is taken to be in state Normal. States LL and HH are used to derive the patterns because it is assumed that they represent typical occurrances of the failure scenarios.

Diagnostic Neural Network

In this research we use a two layer, feed-forward perceptron with back-propagation.In this problem there are 9 parameters each one of them is represented by a string of 6 bits. Therefore the Neural Network will have a total of 54 input nodes. Since there are 12 output states (11 failures and one normal state), we can have 12 output nodes where each node represents a state, or we can have lesser number of nodes and have the states represented by a pattern of bits in the output nodes. In this preliminary implementation, we used 9 bits to represent the scenarios. The reason is that we tried to make the output patterns look as close as possible to the input patterns for faster convergence of the training algorithm. The output patterns are much shorter than the input patterns, so they cannot look that close. However, these input patterns are made of 9 groups of patterns, so we can compress these 9 groups of bits to just 9 bits, each group corresponding to one bit, as follows:

```
Group 000000      is compressed to   0
Group 111111      is compressed to   1
Group 101010      is compressed to   1 or 0, arbitrarily
```

The diagnosis process using neural networks consists of two stages: 1.Training and 2. operation.

Training Phase

We use a supervised training paradigm. A hidden layer of 27 nodes is used. Training is the process of determining the weight values associated with the connections in the network. These weight values are such that when the input patterns are presented to the network, the corresponding output pattern will be produced at the output thus identifying the scenario. The algorithm we used for training is the backpropagation algorithm. First, all weights are assigned a random number. Then each node in the hidden layer and then the output layer calculates its output and passes it through a threshold function. The threshold function we used is the sigmoid function which is of the form

$$F(x) = \frac{1}{1+e^{-x}} \tag{1}$$

Then the output of the network is compared to the target output and an error is estimated. This error term is the product of the derivative of the sigmoid function and the difference between calculated output and target output, for each node. In mathematical terms error at node i is given by

$$delta[i]=out[i]*(1-out[i])*(target[i]-out[i]). \tag{2}$$

The error term is propagated back from the output layer to the hidden layer and the weights are adjusted as to minimize the error. For further discussion of the backpropagation algorithm, see [6],[1]. Using this algorithm, it took only 27 iterations to train the network.

Operation phase

The objective of the neural network development is to use it for classifying the incoming values of the parameters into the appropriate scenarios. The system will be presented with readings of the nine parameters which now can be in any of the seven states. The output will be calculated using the set of weights derived during the training process. The resulting output pattern will be that of the fault scenario most likely to have occurred. At the current stage of development the system, given the 9 parameter values completely, effectively classifies them into one of the twelve scenarios. Due to the nature of the network architecture, even in the event of partial information about the nine parameters it is expected that the system will be able to identify the most appropriate scenario. However, under extreme circumstances, where the input patterns can not be classified into any of the output patterns the network will be able to generate new output patterns which directly lead to the identification of new scenarios.

Conclusions

The neural network architecture built in this research is quite encouraging. It is felt that diagnostic systems based upon neural network will perform better in terms of their identification of faulty situations, in terms of their speed and learning.

Results obtained with neural networks in other areas have been encouraging [1]. A diagnostic neural network will be much faster than a diagnostic expert system because once trained, its nodes process information in parallel and they are performing only simple calculations (additions and multiplications).

However, other important neural network issues will have to be studied more carefully, such as convergence during training, probability of error, and reliability.

References

1. Kamarthi, V. S.; Kumara, S.; Yu,F.; Ham,I.: Neural Network Based Architecture for Design Data Retrieval. Working paper Series, The Pennsylvania State University, Department of Industrial Engineering 1989.

2. Kumara, S.; Qu, L.; Lee, J.; Hicks, P.; Ham, I.: Sensor-based Real time Diagnostics for On-line Quality Control. To appear in the Journal of Wave and Material Interaction, 1989.

3. Milne, R.: Strategies for Diagnosis. IEEE Transactions on System, Man, and Cybernetics. SMC-17 (1987) 333-339.

4. Nann, S.: A Decision Support System for Control And Automation of Dynamical Processes. Unpublished M.S. Thesis, The Pennsylvania State University, Department of Mechanical Engineering 1988.

5. Nann, S.; Ray, A.; Kumara, S.: A Decision Support System for Control And Automation of Dynamical Processes. Journal of Intelligent Systems, to appear.

6. Rumelhart, D.E. ; McClelland, J.K.: Parallel Distributed Processing, Vols I&II Cambridge, Mass: MIT Press 1986.

Chapter IV

Knowledge-Based Engineering

Introduction

The application of Artificial Intelligence has spanned the engineering domain significantly within the last decade. Its importance in solving nonalgorithmic and fuzzy type problems as well as in providing technical aid have been realized. DICEtalk is such a knowledge-based environment that allows object-oriented programming to be built in engineering knowledge-based systems.

Data models for rigid body dynamic simulation and structural design sensitivity analysis are described in the second paper of this section. The data model is divided into a global data model and a local data model. The relation between the data models is also presented. In the next paper, the importance of manufacturing data in achieving the optimal production is presented, and an object-oriented data model is described. Knowledge-based evaluation for manufacturing is presented in the fourth paper.

Visual graphic knowledge representation for user interface design is the focus of the next paper. The design specification created by a designer together with the source code in the knowledge base are used by a source code generating engine to assemble the source code for the designed system. An intelligent path finder is described for a robot navigation in the next paper of this section. This system has the ability to augment its operational skill by means of interactive learning.

The next paper discusses a graphical user interface with an object-oriented knowledge-base approach, which requires an appropriate graphical user interface built on the modern principles of object and constraint-oriented programming, visual languages, and hypermedia.

In the next paper, a brief review and comparison of learning models of AI and automation is followed by the proposed structure of knowledge automation. The final papers of this chapter discuss the development of a knowledge based system for progressive die design, as well as for scheduling and selection of inspection level, respectively.

DICEtalk: An Object-Oriented Knowledge-Based Engineering Environment

M. SOBOLEWSKI

Concurrent Engineering Research Center
West Virginia University
Morgantown, WV

Summary

DICEtalk is a domain-independent knowledge based system implemented in the Smalltalk/V286 environment (Digitalk Inc.). It is being used in the DICE initiative to support both product and process evolution applications that are important ingredients to concurrent engineering. The DICEtalk high-level architecture includes a problem-solving engine, a high-level user interface, and high-level tools. The low-level architecture is described by more than fifty special different kinds of objects that are implemented in the system as Smalltalk classes. A problem-solving engine implements rule-based and knowledge source deduction with metaknowledge control, procedural attachment mechanism, and a broad interface to the Smalltalk environment. DICEtalk has more interactive windowing utilities and menus than one would normally expect to find in a knowledge-based system for an engineering environment. The emphasis in this paper is on the integration of declarative and procedural knowledge bases in the context of engineering tasks.

Introduction

Advances in computer hardware, software and engineering methodologies have led to an increased use of computers by engineers. In design, this use has been limited almost exclusively to algorithmic solutions such as finite element methods and circuit simulators. However, a number of problems encountered in design are not amenable to purely algorithmic solutions. These problems are often ill structured (the term *ill-structured problems* is used here to denote problems that do not have an explicit, clearly defined algorithmic solution), and an experienced engineer deals with them using judgement and experience. Knowledge-based programming technology offers a methodology to solve these ill-structured design problems.

Most knowledge-based systems have been developed for programmers. Their specialized knowledge description languages (frames, semantic networks) associated with programming languages (LISP, C) are often too difficult for use by engineers. The consequences of these relatively complex systems are long system development times with concomitant negative impacts on product cost, quality and supportability. A key idea behind the DICEtalk is the concept of object-orientation (high-level structured programming) providing procedural

---

*This work has been sponsored by Defense Advanced Research Projects Agency (DARPA), under contract No. MDA972-88-C-0047 for DARPA Initiative in Concurrent Engineering (DICE).

knowledge-based programming integrated with simplified English as declarative knowledge-based programming. The integration of both declarative and procedural knowledge bases is supported by an extended graphical user interface with broad interface to the Smalltalk environment, and through Smalltalk to other conventional languages, for example the C language, by user defined primitives. This integration has been successfully verified in the two knowledge-based applications: the Printed Wiring Board Manufacturability Advisor [2] and the Turbine Blade Fabrication Cost Advisor [CERC].

In DICEtalk a knowledge description scheme is based on a surface language, intermediate language, and deep language [4,5]. The last two languages are hidden from the user. The surface language sentences are considered as English-like sentences what allows engineers to create knowledge bases in natural way without learning specialized data description and manipulation languages. The expressions of intermediate language are logical formulas of the formalized percept language [3,4]. In this case language primitives of natural sentences and percept formulas are the same, i.e., a subject of a sentence and its complements, what allows to make conversion of natural sentences into percept formulas much easier and naturally. The deep language is Smalltalk-80 [1] for implementing structured objects that represent logical formulas at a computer level.

DICEtalk Architecture

The low-level architecture of the DICEtalk system is described by special kinds of objects that are implemented in the system as more than fifty Smalltalk classes. The DICEtalk high-level architecture includes a problem-solving engine, a high-level user interface, and high-level tools. The architecture is a result of integration of traditional approaches to databases with more recent fields such as object-oriented programming, visual programming, logic programming, and on-line information retrieval. The architecture forms the new discipline of engineering knowledge-based programming which is responsible for providing the functionality of ready access to information during design problem solving based on both declarative and procedural knowledge bases, and an extended interactive graphical user interface. The first level is the *problem-solving engine*. The problem-solving engine consists of an inference engine, knowledge base compiler, task dispatchers, task mangers, explanation manager, object-oriented knowledge server, and Smalltalk compiler. The inference engine implements rule-based and knowledge-source-based deduction with procedural attachment mechanism, metaknowledge control and a broad interface to Smalltalk. In DICEtalk the user interface is divided into two parts, each serving a different purpose. There is a core model that is presented to the user and in addition, there is a set of high level tools. Thus, the second level is the *high-level user interface*. This level creates the model of the knowlegbe-based problem-solving environment that users interact with. This model consists of the object-oriented representation of knowledge base along with the set of integrated tools for creating goals, facts, procedures, browsing, searching and problem solving. The third level are the *high-level tools* which enhance the functionality of the engineering environment for certain classes of users. These tools provide the user with a number of facilities such as inference control, graphical data presentation, low and high-level of inspecting and debugging.

Problem-solving engine

A problem-solving model is a scheme for organizing reasoning steps and domain knowledge to construct a solution to a problem. In other words, the central issue of problem-solving deals with the question: What pieces of knowledge should be applied, when, and how? A problem solving model provides a conceptual framework for organizing knowledge and strategy for applying that knowledge. The DICEtalk problem-solving model is object-oriented and based on a so called dispatch-managing problem-solving model. We can view a dispatch-managing model as a natural study of how group of individual solvers can combine to solve a goal (problem). The presented approach is to split the goal into simpler tasks and to solve each of these tasks by a so called *dispatch-managing module* (DM module). A dispatch-managing module consists of a task *dispatcher* and its *manager*. We suppose that tasks are not independent, i.e., they interrelate in some way.

The dispatch-managing model deals with problem-solving by separating a goal into a hierarchical structure of child tasks solved by DM modules. A dispatch-managing architecture of the problem-solving engine is made up of four basic components as shown in Figure 1:

1. The *knowledge base* (six panels)
   The main repository of goals, facts, procedures, and control advices.
2. *Supervisor* (master panel)
   The master DM module deals with solving the user defined goal. It creates the top DM module and controls a DM network activity.
3. The *DM network* (dispatcher and manager panels)
   The problem-solving tasks are organized into the hierarchical structure related to the current state of goal-solving. Each local DM module deals with local task-solving according to its local control strategy and a local knowledge-base taken as a subtask perspective of the knowledge base. Managers of local DM modules are responsible to its dispatchers for local strategy which decides what actions to take next for its dispatchers. Communication and interaction among parent DM modules and its child modules take place through its dispatchers.
4. The *working memory* (data panel)
   Subtasks are created by dispatchers and supplied for solving by its managers according to control advices. DM modules produce changes in the working memory in which lead incrementally to a global solution as a unification of all local

solutions. Parameters are user defined characteristics evaluated by dispatchers and then used as arguments of procedure calls.

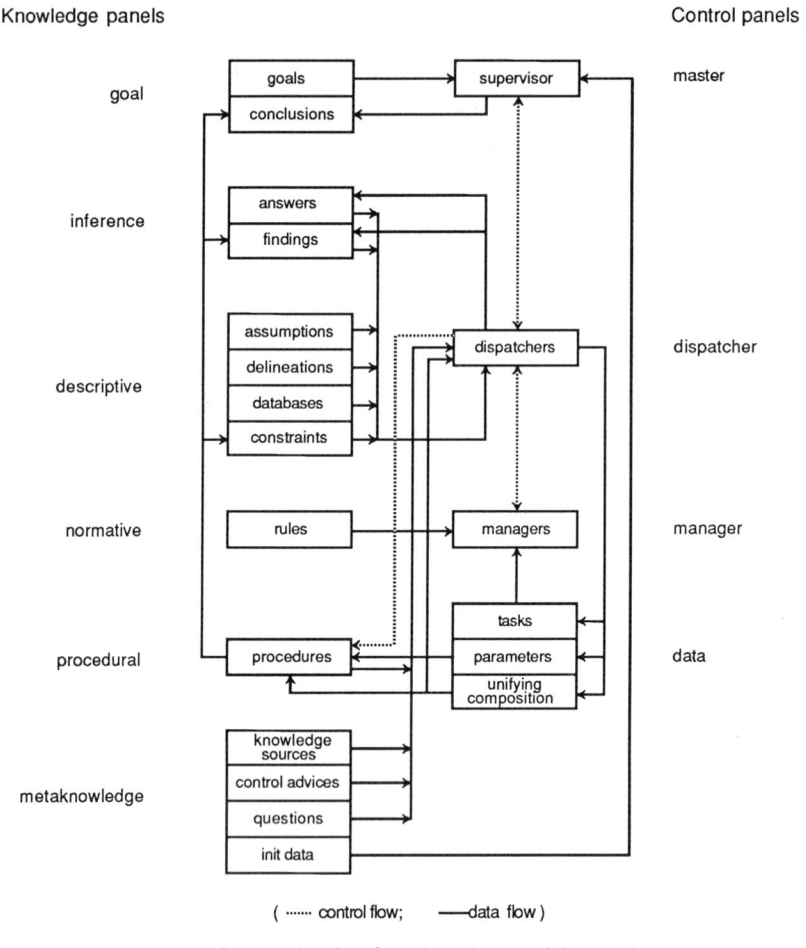

Figure 1. Schematic of DICEtalk problem-solving engine

The DICEtalk object-oriented framework of the presented model is implemented by three types of dispatchers: atomic-dispatcher, and-dispatcher, or-dispatcher, and three types of managers: task-manger, rule-forward-manager and rule-backward-manager. The problem-solving engine introduces an architecture that treats a declarative and a procedural knowledge base as one active object. Fundamental to this object-oriented integration is the notion of a DICEtalk knowledge base as an instance of a class created when new knowledge-based application is defined by a user. This new class, say **SubKnowledgeBase**, is a subclass of predefined **KnowledgeBase** class. Facts and goals of declarative knowledge base are stored in instance variables of **KnowledgeBase** class, whereas instance variables and user defined methods of **SubKnowledgeBase** class form a procedural knowledge base. Thus, this procedural knowledge base inherits declarative knowledge base from its superclass **KnowledgeBase**. Procedures defined as methods of procedural knowledge base can be called by inference engine when tasks to be solved contain messages send to *kb*, that refers to the current DICEtalk knowledge base. Let us consider the following rule (the Turbine Blade Fabrication Cost Advisor):

> *IF*
>> *cluster blade number is x1 and pattern assembly time is x2*
>> *and cluster dress time is x3 and cluster inspection time is x4 and [kb kpp for y]*
>
> *THEN*
>> *pattern wax cost is y*

ast1: parameter is Ncb, ast2: parameter is Tpa, ast3: parameter is Tcd, ast4: parameter is Tci

where *Ncb, Tpa, Tcd, Tci* are user defined parameters (for antecedent subtasks *ast*) for *cluster blade number, pattern assembly time, cluster dress time,* and *cluster inspection time* respectively. When this rule is used in backward direction and the task *[kb kpp for y]* is executed, the first thing that happens is that *kpp* is sent to the current knowledge base. The result of that message is used as the value substituted for the variable *y*, and then the inference engine returns the task *cost wax pattern preparation is y* appropriately instantiated as a finding. The message *kpp* can be implemented as the following Smalltalk method:

kpp
  | kppCost |
  kppCost := 1/Ncb value * (Rpl value + Tpa value + Tcd value + Tci value)
      * kc1Cost + kpCost.
  tbfCost := tbfCost + kppCost.
  ^kppCost asFloat

where *tbfCost, kc1Cost and kpCost* are instance variables (for other costs) in the procedural knowledge base, and *Ncb, Rpl, Tpa, Tcd, Tci* are parameters defined in the declarative knowledge base and treated as Smalltalk pool variables in the procedural knowledge base.

DICEtalk User Interface and High-Level Tools

The DICEtalk user-interface core model drives the functionality of the user interface with a DICEtalk data model. The model consists of a dictionary of declarative knowledge description language, declarative knowledge base, metadictionary of declarative metaknowledge description language, declarative metaknowledge base, and inference knowledge base connected by links into global knowledge base. The user interacts with these objects in a few fundamental ways, however the most basic concept within the core model of the user interface is that of movement and navigation. At each point in the interaction, the user is somewhere, that is at a particular object in the knowledge-based system. The user's understanding of this model is enhanced by the use of seven standard forms to represent objects during browse mode. Each form consists of fields that can represent text, list or graphic object. Since forms may contain more then one field, more objects then one object may be displayed on the same form. In the user-interface model, forms are dynamic, that is, the information is selected and pasted on to them at run time. With forms and fields are associated menus (icons, buttons) that navigate through the objects or execute functions and carry out interactive tasks. A form is the basic display unit. Most of the forms are so called browsers. A browser is a form consisting of at least two fields (forms), a list field and contents field. Selecting from the list field displays related information in the contents field. This contents may be modified, and eventually saved. Other forms are called prompters and are used to ask a user and then input to the system additional information. In the DICEtalk multi-windowing system each window represents a different form and forms may be overlapped and repositioned on the screen. The following browsers DICEtalk provides for editing and maintenance purposes: Dictionary Browser, Metadictionary Browser, Declarative Knowledge Base Browser, Procedural Knowledge Base Browser, and Declarative Metaknowledge Base Browser.

High-level tools supplement and complement the DICEtalk knowledge base functionality. These tools may be used by DICEtalk users as well as developers. DICEtalk tools provide a toolbox consisting of eight distinct tools, reflecting the facts that different applications have different needs and it is unnecessary to burden all applications and users with tools they do not need. DICEtalk provides the following browsers for presenting, maintenance and debugging purposes: the Attribute Accessibility Browser and the Attribute Tree Browser, both accessible from the Dictionary Browser; the Inference Browser, the Proof Tree Browser, the Subproof Tree Browser and the Task Consequent Browser, the last three accessible from the Inference Browser; two debugger windows: a Walkback window and a Debugger window.

References

1. Goldberg, A., and Robson, D. 1983. *Smalltalk-80: The Language and Implementation.* Addison-Wesley.
2. Padhy, S.K., and Dwivedi S.N. 1990. A Knowledge Based Approach for Manufacturability of Printed Wiring Boards. Proc. of the Fith Int. Conference on CAD/CAM, Robotics and Factories of the Future.
3. Sobolewski, M. 1987. Percept Knowledge-base Systems. In I. Plander (Ed.), *Artificial Intelligence and Information - Control Systems of Robots.* North-Holland.
4. Sobolewski, M. 1989. Percept Knowledge Description and Representation, *ICS PAS Reports* No. 663. Institute of Computer Science Polish Academy of Sciences.
5. Sobolewski, M. 1990. Percept Knowledge and Concurrency. *Proc. The Second National Symposium on Concurrent Engineering*, February 1990, Morgantown, West Virginia.

Data Models of Mechanical Systems for Concurrent Design

KIRK J. WU, FOOK CHOONG, and S. TWU

Center for Computer Aided Design and
Department of Mechanical Engineering
University of Iowa
Iowa City, IA

Mechanical systems data model for rigid-body dynamic simulation and structural design sensitivity analysis in a concurrent engineering environment are developed. Two levels of data model are proposed. The global data model serves as the communication basis for a variety of applications. The local data model extents the the global model to define a complete model for an application. Properties and contents of and relations between data models are discussed. The data models proposed define the semantics of models that can be used as the mechanical system schema for database applications.

Introduction

Much research has recently appeared on data model development, e.g. Refs. [1-4]. A data model is a formal device for describing a class of observed phenomena. Given an instance of the class of observed phenomena, a complete data model must be able to fully and unambiguously describe an instance of the observed phenomena, using its symbol structures and rules. The completeness property of a data model refers to the model's ability to provide necessary information for a particular purpose. Certainly a data model that captures only geometric information of a mechanical system is not adequate for dynamic analysis of mechanical systems. A data model must also be consistent, in that no extraneous information can be derived from the data model about the class of observed phenomena. For example, while an inconsistent data model may describe a real mechanical system, it may also allow description of an unrealizable mechanical system.

A data model, if it is defined to contain too many specific details or attributes, will be inefficient for use as a general model for applications that find these details of no importance. On the other hand, a data model model that is too general will force applications to extensively augment the model for their use. This will create duplicate information and hence the problem of keeping this duplicated information consistent. This suggests that in a concurrent engineering environment, where different applications coexist to analyze and evaluate the same product design and each application may view the same product differently, calls for a two-level data model, a global data model and a local data.

This paper focuses on data models for general mechanical systems that can be used in a simulation-based, concurrent design environment. However, attributes for defining technical aspects of simulation [5] itself and simulation results are not included in the paper. The purpose of this paper is to illustrate the use of a 2-level data model as the integration framework for concurrent design.

The major applications in the concurrent design environment that are considered in this study are rigid-body dynamic analysis (RBDA) [5] and structure design sensitivity analysis (DSA) [6]. Other major utility tools considered include a CAD system [7] and structure finite element analysis software (FEA) [8]. In a concurrent design environment, there is a Global Data Model (GDM) that defines a minimal set of physical, fundamental, and shared model parameters. The GDM will serve as an integrating agent of many, diverse applications, each of which may then have a very specific view of the GDM. The GDM is the common basis to derive and add application specific model information and to exchange and interpret shared model information. Local Data Models (LDM) are used to define application specific models, one for each application; i.e. DSA and RBDA. The data models that are presented in this paper are focused on the RBDA and DSA applications, which are rich enough to establish the feasibility of the proposed models to broad class of applications.

Design Scenario in Concurrent Engineering Environment

Figure 1 shows the flow and cycle of concurrent design that are consider in this data model development. Engineers who specialized in RBDA and DSA can extract fundamental and shared model information from the global model. They then add application specific model information to completely define models according to LDM and start analyses. After each analysis, results that need to be kept in global model for other applications will be stored. They may also post suggestions in a public domain for modifying models that may influence other designers' work, such as changing the location of a joint in a body of a mechanical system. Other designers may then need to carry out new analyses because of the modifications.

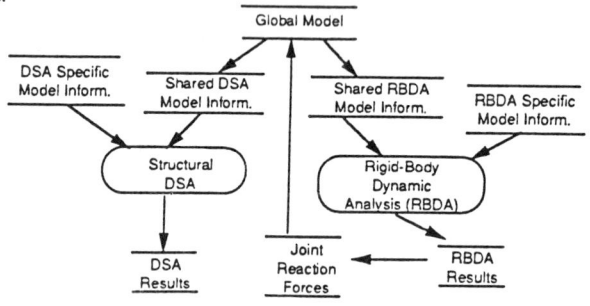

Figure 1 Design Flow in a Concurrent Engineering Environment

Global Data Model for Mechanical Systems

A GDM defines a minimal set of physical, fundamental, and shared characteristics of the model of mechanical systems for communication and consistency; The term "physical" is intended to mean to describe mechanical systems in physical terms. For this reason, the type of material will be kept in the GDM, instead of the density of the material. The term "fundamental" is intended to mean characteristics that cannot be extracted from other given data. Whether a parameter (characteristics, information, or attribute) is fundamental or not may depend on the modeling system. For instance, the mass of a part may be calculated from density and geometry, or it may be input by the user. In the former case, the mass is a derived parameter, while in the later case it is a fundamental parameter.

Shared data defines characteristics that are used by different applications directly or indirectly. For instance, the location of a hole in a part to insert the pin of a revolute joint is used in both RBDA and DSA. Through sharing model characteristics, consistent models across different application can be guaranteed. Characteristics that are needed only for a specific application will be stored in the LDM of the application. For instance, since Young's modulus of a material is only used by the DSA, but not RBDA, it should be stored in the local data model for DSA. However, the type of material of a part should be stored in GDM, since it is needed for both DSA and RBDA in defining the density of the part.

Figure 2 shows the contents and structure of the GDM for applications considered in this paper. Figure 2a shows that a mechanical system is composed of bodies, connectors, and subsystems. A subsystem is composed of bodies, connectors and its subsystems that are grouped for special functions. The superscript $^+$ represents one or more, and * represents zero or more. Two types of entity relationship (member-of and specialization) are used in the GDM, as shown in Fig. 2b. A kinematic or force connector

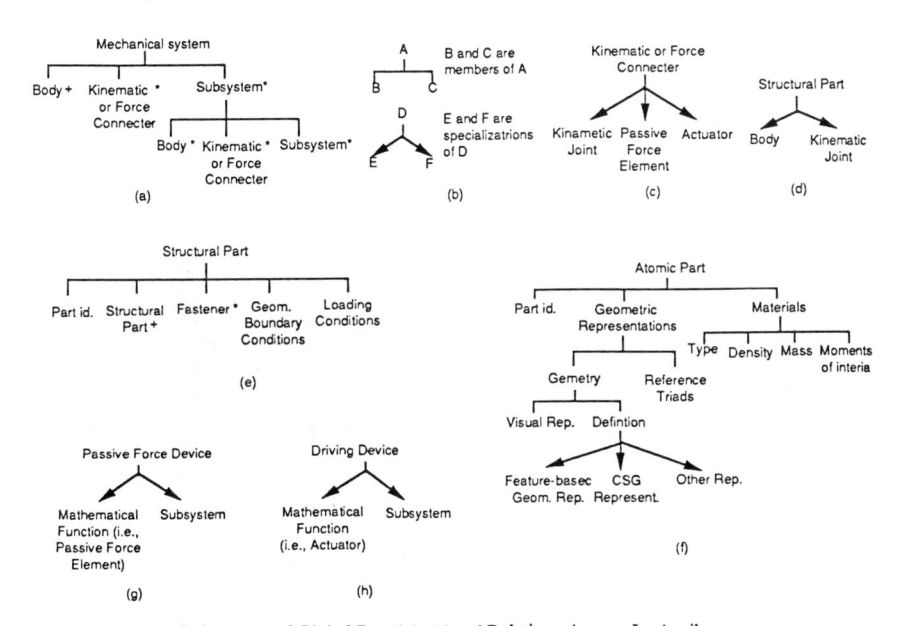

Figure 2 Structure of Global Data Model and Relations Among Its Attributes

defines the geometric constraints or force relationship between bodies, such as a joint, actuator, or a passive force element, as shown in Fig. 2c.

Figure 2d shows that both body and joint are viewed, by DSA, as structural parts. However, from a RBDA point's of view, they have different functions in a mechanical system. Between bodies there exist connectors only. A structural part is a solid entity that has geometry, material properties, loading conditions, and geometric boundary conditions. A structural part can be an assembly of several atomic parts that are fixed by fasteners, as indicated in Fig 2e. A fastener represents a general class of entities that fixes several structural parts together, such as a bolt and nut, rivet, weld, or adhesives [9]. An atomic part is a structural part that cannot be further divided from the modeling or analysis point of view. The attributes of an atomic part are shown in Fig. 2f. The geometry of an atomic part (created by CAD systems) is defined in terms the size and shape of the part, and reference triads that are used by dynamic analysis [5]. However, these two representations need to be consistent. The geometry of a structural part is the assembled geometry of its atomic parts.

A physical passive force device or an active driving device can be modeled as mathematical function or a subsystem, depending upon the interest, as shown in Figs. 2g and 2h. Some of the RBDA results must be sent back to the global model for DSA application; e.g. joint reaction forces, because these forces are the applied loading in structural DSA.

Local Data Model for Rigid-Body Dynamic Analysis (RBDA)

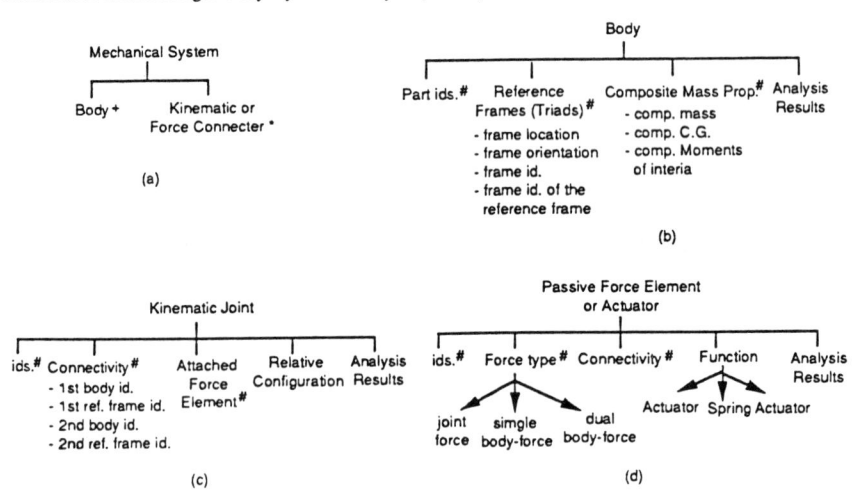

Figure 3 Structure of Local Data Model for Rigid Body Dynamics Analysis

In this section, only the parameters that are needed for describing mechanical systems for RBDA are discussed. A more complete data model for rigid-body dynamic simulation and animation has been provided in Ref. [10].

A LDM defines the semantics of the whole model used in an application. As shown in Fig.3, a local model contains model parameters that are obtained and/or interpreted form model characteristics defined in the global model (highlighted by a pound sign), such as locations of joint reference frames. A local model also contains additional parameters that are needed for a specific application, such as relative positions and orientations of a pair of joint reference frames.

Figure 3 shows the structure of the LDM for RBDA. In the LDM, a mechanical system is composed of bodies, kinematic connectors, and/or force connectors (see Fig. 3a). Bodies and connectors in subsystems are treated the same as other regular bodies and connectors in the local model. Most of the body information is obtained from the global model (see Fig. 3b). However, some model parameters of kinematic and force connectors are specified locally (see Fig. 3c and 3d). Most of the analysis results will stay in the local model. Only limited results will be sent to global model as discussed in the global data model.

Local Data Model for Structural Design Sensitivity Analysis (DSA)

Structural design sensitivity analysis investigates variations in system performance due to changes in the design parameters of structural parts. The overall procedure can be divided into four steps: geometry model generation, DSA design specification, structural finite element analysis, and design sensitivity computation. In order to carry out a structural DSA, data from a CAD geometry modeling system, structural finite element analysis, and design specifications are required.

The LDM for structural DSA, as shown in Fig. 4, consists of four kinds of representations that correspond to each step in the overall DSA computational procedure. All of attributes in Fig. 4 can be either obtained or derived from the GDM or defined as DSA specific data. The geometric representation for DSA is obtained form the GDM. The finite element representation, DSA design specification, and DSA computation are treated as DSA specific data.

In the finite element representation, material, boundary conditions, and loading attributes are either shared or derived data from the GDM. For instance, boundary conditions that are imposed at the finite element mesh points can be derived form geometric boundary conditions, as shown in Fig. 2.e. Other attributes in the finite element representation are finite element specific data. The finite element analysis results can be posted in a public place, for design evaluations. DSA design specifications include defining the design parameterizations and the performance measures. These are DSA application specific data. However, design parameterizations defined for the DSA application must be consistent with parameters defined in the GDM so that the parameterized design model can be used to communicate among all applications.

DSA computation is an integration process based on data associated with the geometric representation, finite element analysis, and DSA design specifications. The numerical integration scheme requires DSA specific data and can be specified inside the DSA computation code. Finally, DSA results that are needed for improved mechanical design are posted in a public area.

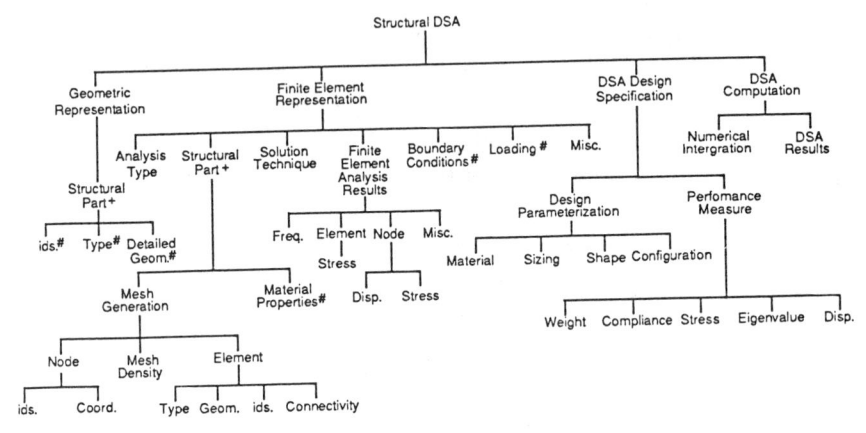

Figure 4 Structure of Local Data Model for Structural DSA

Further Research

A data model defines the static properties of the observed phenomena; i.e., a data model describes the necessary attributes and information and their hierarchical relationships. To complete the information framework for a mechanical system in concurrent design, the dynamic properties of the data must also be defined; i.e. a process model that describes how and when the attributes and information are generated is also needed. The connection between a data model and a process model is provided by viewing attributes that are defined in the data model as primitive or derived data. Primitive attributes are attributes whose values should be specified by designers in order to define an instance. Derived attributes are those attributes whose values are computed from primitive attributes and/or from other derived attributes. Computational requirements for calculating the value of a derived attribute define the process model. For example, joint force can be considered as a derived attribute. From the DSA point of view, joint force is just an attribute, but to the GDM, joint force is to be supplied by the RBDA.

A process model can, in addition to capturing the dynamic aspect of data, provide the communication framework for shared data between different applications. Much work is still to be done in formally linking the 2-level data model and a mechanical design process model, to complete the information framework for concurrent mechanical engineering design.

Conclusions

The concepts of global and local models are presented in this paper. The global model stores physical, fundamental, and shared information for many diverse applications. It serves as the integrating agent for different applications in a concurrent design environment. Criteria for determining whether a model parameter should be stored in the global or a local model have been proposed and illustrated by considering rigid-body dynamic analysis and structural design sensitivity analysis. The data model provides a basis for developing process models and for communication among applications in a concurrent design environment.

References

1. Kalay, Y. E. (ed.), MODELING OBJECTS AND ENVIRONMENT, John Wiley & Sons, New York, Chichester, Brisbane, Toronto, Singapore, 1989

2. Tomiyama, T., Kiriyama T., Takeda H., Xue D., and Yohikawa H., Metamodel: A key to Intelligent CAD System, Research in Engineering Design, Vol 1, No. 1, 1989

3. N. K. Shaw, M. Susan Bloor, A. de Penniton., Product Data Models, Research in Engineering Design, Vol 1, No. 1, 1989

4. Inmon, W. H., ADVANCED TOPICS IN INFORMATION ENGINEERING, QED Information Sciences, Inc., 1989

5. Haug, E. J., Computer Aided Kinematics and Dynamics of Mechanical Systems, Vol 1: Basic Methods, Allyn and Bacon, Newton, MA, 1988

6. Haug, E.J., Choi, K.K., and Komkov, V., Design Sensitivity Analysis of Structural System, Academic Press, New York, N.Y., 1986

7. Intergraph/Engineering Modeling System (I/EMS) Reference Manual, Intergraph Corporation, One Madison Industrial Park, Huntsville, Alabama, 35807-4201, 1988

8. DeSalvo, G.J. and Swanson, J.A., ANSYS Engineering Analysis System User's Manual, Vols I and II, Swanson Analysis System Inc., P.O. Box 65, Houston, PA, 1985

9. Jon R. Mancuso., COUPLINGS AND JOINTS DESIGN, SELECTION, AND APPLICATION, MARCEL DEKKER, INC. New York and Basel, 1986

10. J. K. Wu, Sung S. Kim, Sang. S. Kim, and K. J. Ciarelli., A Generic Mechanical System Data Model For Dynamic Simulation, Symposium of ASME Computers in Engineering Conference, Boston, MA, Aug 5-9, 1990, (in press)

Manufacturing Knowledge Representation Using an Object Oriented Data Model

RASHPAL S. AHLUWALIA and PING JI

Department of Industrial Engineering
West Virginia University
Morgantown, WV

Abstract

A common data base plays a key role in the development of a Computer Integrated Manufacturing System (CIMS) and a Manufacturing Data Base (MDB) is one of the necessary components of an integrated data base. A MDB needs to have information on part geometry features, operation processes, operating parameters, machine tools, cutting tools, jigs and fixtures. A MDB should also provide capabilities to model, store and manipulate manufacturing data in a manner suitable to the users. The commercial Data Base Management Systems (DBMS) are not widely used in manufacturing application primarily because the conceptual and internal modeling tools that they provide do not meet the requirements of a manufacturing user. This paper describes an object-oriented data model suitable for the manufacturing domain. In the object-oriented data model, it is easy to describe and maintain a manufacturing objects, such as parts, machine tools, cutting tools, fixtures, etc.

Introduction

The most important factor affecting the development and evolution of a computer integrated manufacturing system (CIMS) is a common data base [4] [5] which serves all functional areas in an enterprise. An effective CIMS data base is an invaluable tool that provides numerous design and manufacturing benefits. Experts feel that a CIMS Data Base Management System (DBMS) should have more capabilities than at present available from vendors to support large-scale scientific and business oriented data base [3] [4], or as Melkanoff indicated that no commercial system currently exists that can support all the requirements for a manufacturing database management system [8]. Current database management systems have been designed to deal only with alphanumeric data, but one of the major requirements of CIMS is the ability to store, transmit, and display graphical data integrated with text. These systems must be capable of supporting engineering analysis as well, which implies heavy numerical computations. Therefore, it is necessary to develop a specific DBMS suitable for a CIMS environment. However, there are many

---

Acknowledgements - This works has been sponsored by the Defence Advanced Research Project Agency (DARPA), under contract No. MDA972-88-C-0047 for DARPA Initiative in Concurrent Engineering (DICE).

difficulties in developing such a data base. The difficulties arise from a variety of requirements and constraints on the design and operation of the data base. Take the database size for example. Estimates are that a real CIMS database will be several orders of magnitude larger than the largest databases in use today [1]. In other words, such a database may require ten to a hundred gigabytes [8]. This implies that a CIMS database is unlikely to be centralized. On the contrary, the database is more likely to be distributed or consist of several sub-databases or local databases [2] [5] [8]. Because manufacturing is one of the main areas in a computer integrated manufacturing system and has a broad scope, from materials, part geometries, tolerances, to machine tools, cutting tools, operating parameters, it is critical to design a manufacturing data base. Such a data base can serve process planers, design for manufacturing engineers, facility administrators and other users.

Object-oriented Model

Traditionally, the data models used in data bases are relational, network, or hierarchical. In the formulation of relational data models, the mathematical theory of relation is extended to meet data base requirements. The properties of a relation is a mathematical set, and their representation is a table. The semantics of relationship types are represented by key, whereas many-to-many relationship types are represented by separate relationship relations. Network data models are based on tables and graphs. The functional restriction makes it impossible to represent directly many-to-many relationships types in a network data model. While the main restriction on the relationship types between tables is the functional link restriction in a network data model, a hierarchical data model imposes a further restriction on the relationship types. Such a restriction makes it difficult to represent many-to-many relationships types directly in a hierarchical data model.

In the manufacturing area, we can treat part, operation process, machine tool, cutting tool, fixture, and operating parameters as entities, so the relationships between each entity are mostly many-to-many and form a very complex graph. Figure 1 shows partial relationships. Furthermore, the manufacturing data is heterogeneous. In other words, the properties or attributes of same entities are neither consistent nor identical. For example, the operating parameters for drilling processes are feeds and speeds, given material type, hardness, material condition, tool material, hole diameter and length. However, the operating parameters for the turning process are feeds, speeds, and depth of cut, having given material type, hardness, material condition, tool material, and tool type. It is therefore difficult to represent such information by use of a relational, network, or hierarchical data

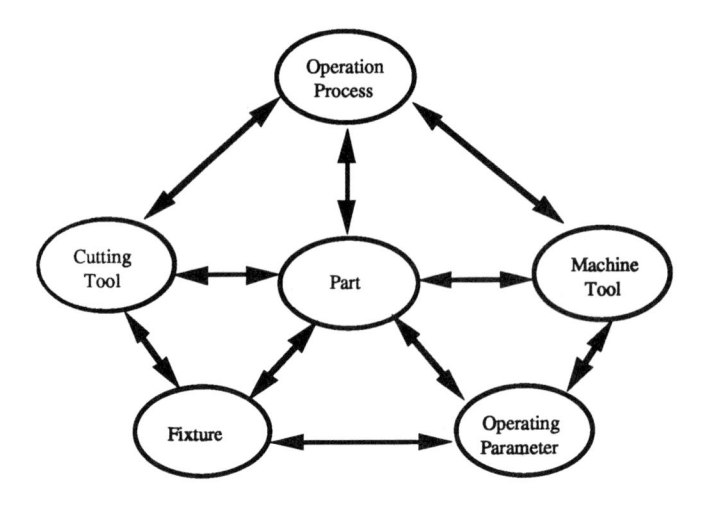

Figure 1. Partial Relationships
between Manufacturing Entities

model. In manufacturing, the data models needs to be object oriented rather than computer oriented.

In an object-oriented data model, the basic entity is that of an object. An object can loosely be defined as something of interest to the user. The attributes and relationships between entities in a relational data model are all properties of an object in an object-oriented data model. Objects can be compound, consisting of sets of objects. A compound object is called a class. The recursive definition of an object allows objects to be structured in many different ways. The resulting structure can be visualized as a complex graph that connects different objects, properties, and/or relationships at different times. The property of inheritance is a major advantage of object-oriented model over the three traditional models. An object can inherit common properties from its class, also the object can have its own specific properties, not shared by any other object or class.

Manufacturing Data Base (MDB)
The LASER system was used to develop a prototype object-oriented data base for manufacturing application. LASER, a system developed by the Bell Atlantic Knowledge Systems, Inc., provides a set of commands to create and manipulate objects and to specify relations between objects. It is a tool which helps to create models of concepts and situations that will constitute a database or a knowledge base [6].

The Manufacturing Data Base (MDB) has four modules: part, operation, facility and machinability. Part module deals with the basic information and manufacturing features of a product. The basic information of a part includes data such as part number or part identification, material, hardness, weight, etc. Manufacturing features are defined as round hole, nonround hole, slot, groove, gear, etc. A user inputs his or her part data through the part module. Operation module maps the manufacturing features into operation processes, i.e., it provides the capability of the manufacturing system. Facility module has data on available machine tools, cutting tools, fixtures (including jigs and auxiliary attachments). This module provides information on the availability of equipment in the manufacturing system. Machinability module provides the machining data such as operating parameters.

In MDB, part, operation process, machine tool, cutting tool, fixture, and operating parameter are classes. Each class can have many objects and each object can have many instances. For example, the cutting tool class has drill, reamer, etc. as objects. Each object can have many common properties. For example, the drill object has common properties such as: diameter, type, point angle, unit, etc. Say, we have two taper shank twist drills, one called drill87 with diameter 20 mm, morse taper number 2, flute length 156 mm and overall length 273 mm; another one being drill88 with diameter 43/64 inch, and morse taper number 2, and flute length 5 3/8 inches and overall length 9 1/4 inches. We can represent drill87 and drill88 as two instances of an object drill, as shown in Figure 2. In Figure 2, the relationship between a class and an object is represented by relation *isa* in LASER (*inverse_isa* is its inverse relation), and the relationship between an object and an instance is described as *instanceof* (its inverse relation is *instances*). The instance drill87 inherits all properties from its parent object drill, so the unit of drill87 is mm and point angle is 118 degree. However, drill88 inherits all other properties from object drill except the unit is inch rather than millimeter because unit is defined as inch in the instance drill88.

Other classes like part, machine, fixture, operating parameter can be described in a similar way. Such a representation is not only efficient, but also reduces the data redundancy. However, these do not represent the relation between instances of different classes. For example, we know that drill87 can be used in lathe2, while drill87 belongs to a class cutting tool and lathe2 is an instance of object lathe of class machine. Of course, drill87 does not have the properties of lathe2, and vice versa. In LASER, such relations can be represented by *partof* (its inverse relation is

```
{   cutting_tool
        inverse_isa #: drill reamer
}
{   drill
        instances #: drill87 drill88
        isa #: cutting_tool
        diameter:
        type:
        point_angle: 118
        operation: drilling
        unit: mm
}
{   drill87
        instanceof # : drill
        type: taper_shank_twist
        diameter: 20
        morse_taper_no: 2
        flute_length: 156
        overall_length: 273
}
{   drill87
        instanceof # : drill
        type: taper_shank_twist
        diameter: 43/64
        morse_taper_no: 2
        flute_length: 5 3/8
        overall_length: 9 1/4
        unit: inch
}
```

Figure 2   Drill Description in an Object-Oriented Model

parts). Relation *partof* can be used to relate objects without causing inheritance, so we can define that drill87 is a part of lathe3, as shown in Figure 3. Relation *partof* is not limited to relate different instances, it also can be used to describe the relationships between classes, objects and instances. By use of such a relationship, we can represent the complex knowledge data effectively.

Another feature of the MDB is demon application. For example, the operating

```
{   drill87
        instanceof # : drill
        partof * (parts): lathe2
        type: taper_shank_twist
        diameter: 20
        morse_taper_no: 2
        flute_length: 156
        overall_length: 273
}
```

Figure 3   The Relation between a Drill and a Lathe

parameters for drilling operation normally has the condition that the hole depth be less than twice the diameter [7]. Otherwise, different operating parameters should be used. For the twist drills example, if the hole depth is 3 times of the hole diameter, the speed and feed should be reduced by 10% of their normal values. In LASER, it is easy to realize this by use of demon, unlike a relational data model.

Conclusions

Several advancements have been made in the area of Computer Integrated Manufacturing. At the core of CIMS is a reliable database, it is believed that an effective CIMS data base does not exists so far. The existing data base models do not provide the capability to meet the requirements of a manufacturing data base. This paper proposes an object-oriented data model to develop a manufacturing data base. The object-oriented data model enables the representation of complex relationships required by the manufacturing system.

References

1. Appleton, David; "The State of CIM," Datamation, December 15, 1984, pp. 64-72.

2. Beeby, William; "The Future of Integration CAD/CAM Systems: the Boeing Perspective," IEEE Computer Graphics and Application, January 1982, pp. 51-56.

3. Beeby, William; "The Heart of Integration: a Sound Data Base," IEEE Spectrum, May 1983, pp. 44-48.

4. Groover, Mikell P. and Wiginton, John C.; "CIM and the Flexible Automated Factory of the Future," CIMS Series, Part 23, Industrial Engineering, Volume 18, Number 1, January 1986, pp. 75-85.

5. Krishnamurthy, Vishu; Su, Y. W.; Lam, Herman; Mitchell, Mary and Barkmeter, ED; "A Distributed Database Architecture for an Integrated Manufacturing Facility," Second Symp. Knowledge-based Integrated Info. Sys. Eng., May 1987. Integration of information systems: Bridging Heterogeneous Databases, edited by Amar Gupta, IEEE Press, New York, 1987.

6. "Laser manual," Bell Atlantic Knowledge Systems, Inc., 1988.

7. "Machining Data Handbook," Third Edition, Vol. 1 & 2, Machinability Data Center, Metcut Research Associates, Inc., 3980 Rosslyn Drive, Cincinnati, Ohio 45209, 1980.

8. Melkanoff, Michel A.; "The CIMS Database: Goals, Problems, Case Studies and Proposed Approaches Outlined," Industrial Engineering, Volume 16, Number 11, November 1984, pp. 78-93.

Knowledge-Based Evaluation of Manufacturability

SIPING LIU, VASILE R. MONTAN and RAVI S. RAMAN

Bell Atlantic Knowledge Systems
9 South High Street
Morgantown, WV

1.0 Introduction

Concurrent engineering enhances the cooperation among a group of product developers to reach the goal of a shortened design cycle for high quality, low cost products. The DICE (DARPA Initiative in Concurrent Engineering) project, in which our manufacturability evaluation model is a research task, aims to support a group of product developers with available computer and communications techniques.

It has been realized that design deficiencies from the point of view of manufacturing cause great time delays in the whole design procedure. Traditionally, a design may go between a designer and a manufacturing expert several times before it finally satisfies both the design requirements and the manufacturing constraints. Therefore, it is necessary to have a manufacturing expert (a human expert, a software package, or both) involved in this concurrent design team to answer questions related to manufacturing such as:

- Is this design feasible to be manufactured? If not, what are the main problems and how can they be corrected?
- How much will it cost to produce each artifact? Will the total cost exceed the budget limit?
- How much time will it take to manufacture the product? Will it be too long to meet the market requirement?
- Is the assumed factory suitable to manufacture this product? Or is it better to have it produced in a factory outside the enterprise?

The answer to these questions, especially those for cost and delivery time, has traditionally come from the manufacturing experts' estimation based on their past experiences and is often inaccurate, unreliable, and may be quite misleading in some situations. A more accurate analysis can contribute directly to lower cost and higher competitiveness in the market place.

Our solution is to use knowledge-based constraint evaluation tools to check complete or partial designs against the requirements and limits enforced by the manufacturing facilities. The result is a manufacturing procedure which may be used to generate a simulation model which reflects the machine tools layout in a certain factory. By running this model, both the designer and the project leader in this concurrent design team can get an estimation to the questions listed above at a much higher level of accuracy.

2.0 The Evaluation Facility

2.1 Data organization

The product design in the DICE environment is represented as features. This permits the system to work on a level higher than geometry data and is suitable for further knowledge-based analysis. Following this trend, we organize all manufacturing-related data in our system using LASER/KR™, a frame style knowledge representation language.

Design features are arranged in a feature-subfeature hierarchy, which is conceptually similar to a part-components organization. For example, a wheel can be considered as a feature with ring, spokes, and hub as its subfeatures. In our knowledge base, each feature is associated with the manufacturing processes which can be performed on the feature and each process is associated with all of the existing manufacturing equipment which can perform the processes. Each machine has certain operating conditions and a working range which must be checked before the machine can be selected.

2.2 Constraint evaluation

A constraint in our system is a logic expression which gives an easy way to represent the engineering knowledge needed to perform the evaluation. It can be attached to any object of interest. The activeness of each constraint is determined by the system status, as a human expert uses different criteria according to particular situations. For instance, cost will go up with an increase in machining accuracy, so the precision requirement has to stop at some point with a limited budget; if the product objective puts quality at a higher priority, cost constraints may be overlooked.

A sample constraint is "0.5 in. < hole diameter < 1.0 in." which may be attached to a drilling machine to specify its working range.

For each given design feature, the system selects all potentially applicable manufacturing processes by checking the constraints attached to all possible processes which are connected to the feature; then the system checks the constraints attached to all possible machine tools which can be used for the selected processes. As an example, suppose we have a hole as a feature on a working piece; from our knowledge base, we may find out that either a drilling process or a forging process can be used. However, a forging constraint may indicate that the process cannot be used for making holes deeper than a certain limit. Furthermore, various kinds of machines can be used for drilling. The system can single out the most suitable one according to the cost, speed, material, etc. This evaluation procedure goes through the feature-subfeature hierarchy and finally returns either an ordered manufacturing procedure for the product which satisfies all constraints or a report specifying which features can not be manufactured.

2.3 Knowledge-based simulation

Discrete event simulation has long been recognized as an effective method to obtain realistic results in a resource-sharing environment without actually having the circumstance. The manufacturing facilities (including machines and human workers) available for the production of the design can be described as such an environment. A manufacturing procedure suggested from the constraint evaluation naturally corresponds to a simulation model since it is a series of resource requiring and consuming activities. Combined with a description of one or more factory cell layouts, we can generate and run simulation models to answer most of the manufacturing-related questions we have raised before.

However, if we follow the traditional model building methods, as supplied by most existing simulation tools, we will end up requiring a simulation expert to generate models for each individual manufacturing procedure. This is obviously against the ides of offering an on-line evaluation system.

A knowledge-based system gives a natural and consistent data organization throughout different stages during the system execution. This makes it a lot easier for a knowledge-based simulation tool which is working in the same environment to adapt a process to a simulation model automatically. LASER/SIM™ is such a simulation tool in the LASER environment.

From the constraint evaluation process described in the preceding example, drilling may be suggested as the optimal method of making a hole in the current artifact; drilling will be recognized as an activity in a later generated simulation model simply because, as part of the pre-defined manufacturing knowledge, *drilling* has

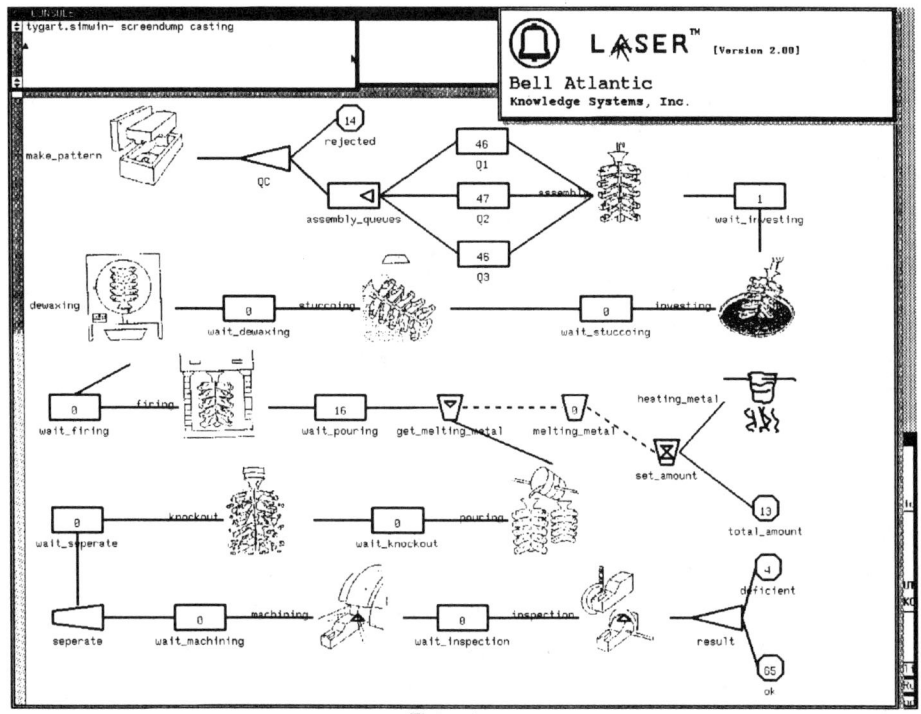

Figure 1

been related to the generic concept called *process* in our representation hierarchy and a process will be matched to a simulation server block. Some resource-related information, such as the number of a certain kind equipment, can also be used as domain knowledge as suggested in [1] to help the automatic generation of the simulation models.

Both the graphical display and the running result of a LASER/SIM simulation model generated for the casting process is illustrated in Figure 1. Some icons, such as *make_pattern*, represent the processes that a working piece will go through. Others may represent a quality inspection or an artifact queue, which are necessary in a simulation model to correspond to the real world. Similar models can also be generated for the machining procedures.

3.0 Conclusion

Similar efforts in design evaluation have been conducted in other organizations. Papers about them include [2] - [9]. Among them, Sanii[5]'s organization of manufacturing knowledge is quite like ours but its preprocessor can directly handle IG-

ES files; Marefat[2]'s work on shape understanding gives a possible way to find out constraints existing among individual features; Kusiak[4] did the detailed analyses at the machining pass level, which is not considered in our system since our output will not be used to drive a CAM system.

Even though there are a lot of differences, we would still like to compare our project with *Next-Cut* as introduced in [9]. The knowledge organizations in both systems are quite similar, but we have not put the lower level data for generating solid models in our knowledge-base. We and the Next-Cut designers have both realized the importance in evaluating the ramifications of a design. However, it seems that we have a difference on what kind of information is really needed by the design engineers. They considered that an accurate manufacturing process is too expensive and unnecessary to generate for the partial designs at each design step, therefore they concentrated on plan re-use for rapid re-planning based on the plan generated before; From our research, we concluded that even at the early design stage, it is still desirable to give the designers the choices of obtaining the manufacturing-related perspectives on the partial design at various details. This drove us to raise the idea of generating and running simulation models to offer more accurate information.

As to the knowledge-based simulation, Brazier [12] and Murray [13] have done some work on the automatic simulation model generation. But, applying this idea in a concurrent engineering environment requires matching similar concepts (such as process) among the data used for different knowledge-based software packages and is still a new field. Using a constraint evaluation mechanism to criticize a design from the manufacturing point of view can be effectively accomplished using a knowledge based system. It is interesting to note that the resulting manufacturing procedure can be naturally adapted to a simulation model in the system, provided that the system has already had enough knowledge about the factory layout. Running this model will yield some important design evaluation data at a higher accuracy.

Acknowledgments

We wish to thank Keith Wall of General Electric for his valuable suggestions and comments on our initial design.

This work has been sponsored by the Defence Advanced Research Projects Agency (DARPA), under Contract No. MDA972-88-C-0047 for the DARPA Initiative in Concurrent Engineering (DICE).

References

1. Davis, T. A.; Liu, S.; and Reddy, Y. V. *The Aquisition and Representation of Domain Knowledge.* Proceedings of the Summer Computer Simulation Conference, July, 1989: 585-589.

2. Marefat, M. and Kashyap, R. L. *Integrating design and automatic process planning by extracting machining features.* Technical Report TR-ERC 89-13. School of Engineering, Purdue University. November 1989.

3. Marefat, M.; Feghhi S. J.; and KashyapR. L. *IDP: Automating the CAD/CAM link by reasoning about shape.* Proceeding of The Sixth Conference on Artificial Intelligence Applications, March 1990: 138-145.

4. Kusiak A. *Knowledge-Based Process Planning.* Proceedings of the 1989 IIE Integrated Systems Conference and Society for Integrated Manufacturing Conference, 1989: 432-437.

5. Sanii E. T. and Davis R. E. *Computer Aided Process Planning Using Object-Oriented Programming.* Proceedings of the 1989 IIE Integrated Systems Conference and Society for Integrated Manufacturing Conference, 1989: 385-389.

6. Sycara K.; Roth S.; Sadeh N.; and Fox M. *An Investigation into Distributed Constraint-directed Factory Scheduling.* Proceeding of The Sixth Conference on Artificial Intelligence Applications, March 1990: 94-100.

7. Tong S. S. *Coupling symbolic manipulation and numerical simulation for complex engineering designs.* Conference on Expert Systems for Numerical Computing, Purdue University, December 1988.

8. Dixon J. R. *Artificial Intelligence and Design: A Mechanical Engineering View.* Proceedings of AAAI-86, 1986: 872-877.

9. Brown, D. R.; Cutkosky, M. R; and Tenenbaum, J. M. *Next-Cut: A Computational Framework for Concurrent Engineering.* Proceedings of the Second National Symposium on Concurrent Engineering. Februry, 1990: 179-196.

10. Dziedzic, R. T. and Raman R. S. *LASER^{TM}: Putting Expert Systems to Work.* AI Review of Products, Services, and Research. Vol. 2, August 1989: 63-66.

11. Reddy, Y. V.; Fox, M. S.; Husain, N.; and McRoberts,M. *The Knowledge-based Simulation System.* IEEE software, March 1986: 26-37.

12. Brazier, M. K. and Shannon R. E. *Automatic Programming of AGVS Simulation Models.* Proceedings of the Winter Simulation Conference. 1987: 703-707.

13. Murray, K. J. and Sheppard, S. V. *Knowledge-based Simulation Model Specification.* Simulation. Vol. 50, No. 3. March 1988: 113-119.

14. Singh, H.; Butcher, A.; and Reddy, Y. V. *Model Management in Knowledge-based Simulation.* Proceedings of Easten Simulation Conference. April, 1987: 11-14.

15. Liu, S.; Montan, V. R.; and Raman, R. *Soft Prototyping Facility in a Concurrent Engineering Environment.* Workshop on Concurrent Engineering Design, AAAI-90. July 1990.

Knowledge-Based Graphic User Interface Management Methodology

STEWART N.T. SHEN and JIH-SHIH HSU

Center for Artificial Intelligence and
Computer Science Department
Old Dominion University
Norfolk, VA

Summary

We have developed a methodology of using visual graphic knowledge base representation for user interface design and implementation. Expert knowledge of the underlying software system for user interface design and implementation is represented visually using workstations. User interface designers can easily understand and utilize the knowledge bases to design user interfaces. The design specification created by a designer together with the source codes in the knowledge base are used by a source code generating engine to assemble the source code for the designed system. The methodology is not limited to user interface design and in fact can be extended to the design and implementation of other types of software systems.

1. Introduction

Knowledge based systems have become very useful in preserving expert knowledge and in performing tasks that would otherwise require excessive human expert efforts. However, the application of knowledge based systems in creating and managing computer software systems is a relatively novel approach. Modern computerized systems require a very significant amount of effort in user interface design and implementation. From the survey results of numerous researchers, we can safely conclude that half or more codes of many computerized interactive systems just deal with user interface [1,2,3]. In a system that we have developed, the Graphic Knowledge Base System (GKBS) [4,5], we found that more than seventy five percent of the codes were devoted to user interface implementation.

Codes for user interface have some common aspects and can be treated in certain systematic ways so that much of the coding can be done in a very user-friendly manner, with the assistance of a computerized system, often called an user interface management system (UIMS). An UIMS can save a great deal of design and coding effort for the user interface in most interactive systems. UIMS's have become quite popular now just as database management systems, spread sheets, and expert system shells have. However, most UIMS's allow for the design and implementation of fixed-location based buttons and menus, as well as are not being knowledge based. In other words, they do not allow for the creation of context-sensitive menus and do not allow users to see any expert knowledge of the particular underlying software system. By a context-sensitive menu, we mean that a different menu is automatically associated with a different type of end-user created object or an empty space on the screen. In other words, in some application domain, an end-user can create arbitrary numbers of objects of certain types, and these objects are automatically associated with some specific menus. To our best knowledge, current UIMSs do not provide such capabilities.

Current UIMS's provide just pull-down menus to allow users to create user-defined menus or buttons. No knowledge of the underlying software system used to implement the user interface is explicitly represented and provided for users to see or to utilize. Thus the users can not learn from the UIMS about how the user interface is actually coded. A user can only do what a given UIMS supports, but can not become an expert on the underlying software system himself and thus can not implement anything that is not supported by the UIMS.

Our methodology uses graphic knowledge base representation for user interface design and implementation. A system has been designed for its implementation. The architecture and the operation of the system is illustrated in Figure 1. The knowledge engineer, who is also the domain expert of the underlying software system, utilizes the Graphic Knowledge Base Frame to build a graphic knowledge base. The user interface designer accesses the visual graphic knowledge base through the Graphic Knowledge Base Frame to create a design using the Graphic Scratch Frame. The Source Code Generating Engine accesses both the graphic knowledge base and the user-interface design specification to generate the codes for the user interface. An end user accesses the user interface to implement application problems.

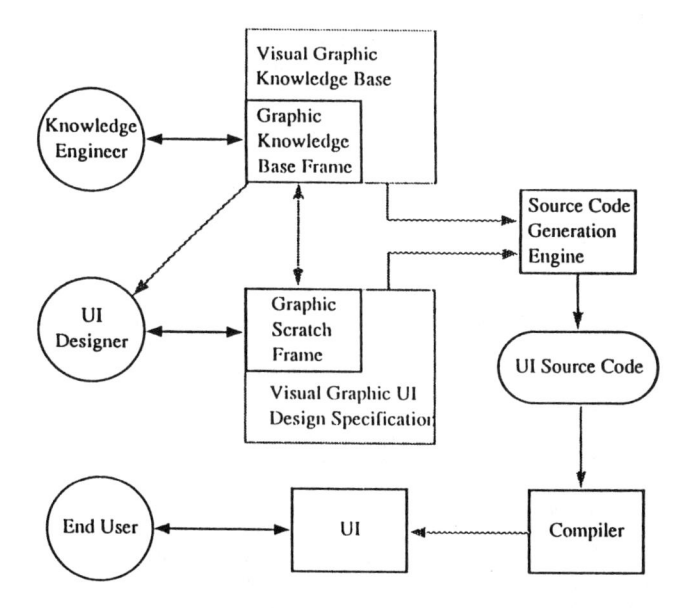

Figure 1. The architecture of the KBGUIM methodology

Our explicit representation of domain expert knowledge is different from other UIMS's. We allow deeper operations than creating just top level frames, buttons, and menus. We allow the automatic association of designer created context-sensitive menus to end-user created objects. In addition, the Knowledge Base Frame is not limited to representing only user interface domain knowledge. It also allows for the representation of other types of software system knowledge. Thus our methodology can also be applied to other types of software design and implementation.

2. Current User Interface Design Tools and Their Disadvantages

The current user interface design tools may be classified into four categories: user interface toolkit, language-based design tool, graphical specification design tool, and automatic creation design tool [6]. An user interface toolkit is a library of interaction techniques that can be called by an application program. This approach is less user-friendly and does not provide much user-dialogue control. A language-based design tool provides a special-purpose language for designers to specify the syntax of the desired user interface systems. There are many different forms of the languages that have been used for this purpose. In general, they are more difficult to learn. A graphical specification design tool provides a screen and allows designers to use a mouse and/or curser keys to move around the screen and to create menus at different locations on the screen. Such tools are mostly easy to learn but can support only a limited range of applications. An automatic creation design tool creates the interface from a given specification of the application's semantic procedures. The designer is allowed to modify the default parameter values to achieve different designs. While this last category of design tools is relatively new and quite promising, it does not provide a structural representation of the semantic procedures. As all the other current design tools, it also does not provide any insight into the underlying software system.

3. Basic Approaches in the KBGUIM Methodology

Our Knowledge Based Graphic User Interface Management methodology is based on several important basic approaches. We describe these basic approaches below.

3.1. Visualization

Cognitive scientists *Larkin* et. al. asserted that in many circumstances a diagram can be most easily understood and remembered and is worth ten thousand words [7]. *Topping* et. al. created a system called *Express* which may be categorized into the graphical specification design tool group [8]. They claimed that they built the system based on the concept that the display itself is the most natural form of specifying the desired image and is the most productive way to produce displays. *Ichikawa* and *Hirakawa* proposed the utilization of visual information in programming to overcome the complication and time-consuming problem in programming [9]. There are many other software systems that also adopt the visualization approach for its effectiveness.

Although quite effective in supporting software development, most software development systems using the visualization approach do not present any knowledge on the underlying software systems, which are typically some window systems. Thus the designers can become good users of the software development system as they are, but can not learn anything about the underlying software system. When they need to do anything that a particular software development system does not provide, they are simply stuck.

In our KBGUIM methodology, we present the relevant knowledge on the underlying software system to user interface developers in visualization. Thus a user interface designer can not only use our knowledge base to design and implement a user interface but also learn from it the workings of the underlying software system. Designers can become experts on the underlying software systems themselves. In cases when a knowledge base is insufficient for a particular design task, extensions may be easily made by the domain expert on the underlying software system, or may even be made by a user interface designer. We describe the visual graphic knowledge base representation in a separate section below.

3.2. Visual Graphic Representation

We adopt a form of visual graphic representation to represent the relevant knowledge on the underlying software system. In this representation, we use different visual graphic symbols to represent explicitly the fundamental knowledge objects needed in building user interfaces in a window environment, namely the SunView system of the Sun Microsystems. The objects are connected through different types of visual links to represent different types of relationships among the objects.

3.3. Visual Graphic Knowledge Base

The use of knowledge base technology in user interface design is quite novel. Rhyne suggested to incorporate developments in knowledge base system into the interface design process and into end user interfaces [10]. Folley used a frame-based expert system shell (ART schemata, or frames) in developing the User Interface Design Environment (UIDE) to assist user-interface design and implementation [11]. In Folley's UIDE, the user-interface design is still represented in a textual Interface Definition Language (IDL), which is not convenient for designers to visualize. The frame-based knowledge representation of UIDE is used only to denote the design of user interfaces. The representation of knowledge on the underlying software systems is ignored by current development environments.

Graphic representation use has gained popularity due to its ease of acceptability by humans. Some software requirements analysis methods, such as Data Structured Systems Development (DSSD) and Jackson System Development (JSD) apply graphic notations to represent information hierarchy in requirement analysis of software design [12]. These methods can be applied only manually, and thus are cumbersome and error prone for large systems. Later, automated analysis tools based on these methods became available. For example, SADT applies a graphic notation to support analysis processes [13]. The graphic notation used in SADT is composed of only one kind of box and arrow line. By supporting ten different object symbols and three different arc shapes, the KBGUIM methodology provides more descriptive power than SADT. Software through Pictures, a system developed by Wasserman, which also applies graphic representation and provides several graphic editors, does not support any means to represent the knowledge of the underlying software system [14].

development. We can consider that the KBGUIM methodology is more advanced since through the visualization of the design specification as well as the design process.

3.6. Automatic Source Code Generation

To facilitate the implementation of the designs completed by the designers, the KBGUIM system generates the source code for an user-interface design by composing the routines associated with the objects in the design specification into a complete program. These routines are retrieved from the visual graphic knowledge base by the source code generation engine and then put together in a specific manner according to the design specification. A knowledge engineer, who is an expert on the underlying software system and user-interface design is to provide the routines associated with the objects while building the visual graphic knowledge base.

There are some software development systems which are able to generate code skeletons. For example, the Structure Chart Editor of Software through Pictures, developed by Wasserman, aimed to generate a code skeleton in a special program design language [14]. However, even with the code skeleton generation process, developers still need to write nearly the entire code for each module. In order to further ease program development, the KBGUIM system directly generates most of the entire source code for a design.

4. KBGUIM System

The KBGUIM system is a prototype system to demonstrate the feasibility of our proposed methodology. KBGUIM system consists of three major components: Graphic Knowledge Base Frame (GKBF), Graphic Scratch Frame (GSF), and Source Code Generation Engine (SCGE).

GKBF applies visualization and graphic representation for knowledge engineers to create visual graphic knowledge bases. It is a visual graphic knowledge base environment. In this environment, a knowledge engineer can easily select desired objects and arcs, whose semantics have been defined earlier, to build the visual knowledge graph for a knowledge base. To complete a knowledge base, the knowledge engineer also needs to associate each object with the appropriate routine which will support the intended graphic operation. These routines are later retrieved by SCGE to compose the source code of the designed user interface.

GSF is an user-interface design environment which facilitates user-interface designers to specify desired user interfaces by applying flexible design and design in visualization. It allows the designer to direct the duplication, expansion, and/or elimination of objects as desired for specific designs. All these operations can be done conveniently by using the mouse and typing some names. While doing all these operations, the designer typically refers to the visual graphic knowledge base in the GKBF for guidance. In addition, GSF allows a designer to reuse a previous design specification by loading and modifying the old design.

SCGE is the source code generation engine. SCGE can recognize the user-interface design specification created in GSF. It first translates the specification into a two-dimensional graph, and then composes the source code by concatenating the routines retrieved from the visual graphic knowledge base created in GKBF.

Through GKBF, GSF, and SCGE, the KBGUIM system releases user-interface designers from working directly with the obscure underlying software system. It reduces dramatically not only the programming efforts, but also the depth of familiarity of the underlying software system otherwise required of the desinger. In the mean time, a new designer can quickly learn the workings of the underlying software system and become an expert in a short time.

5. Conclusion

It has been asserted that sometimes a picture is worth ten thousand words [7]. The KBGUIM methodology tries to take advantage of human's natural superior ability in accepting and understanding visual information. We propose the direct use of visual graphic representation of knowledge in the form of a visual graphic knowledge base. The modern workstation technology and window systems make our methodology feasible, effective, and pleasant to use.

The methodology allows the preservation of expert knowledge in specific domains, makes system design and implementation much more efficient, and assists designers to become experts in the underlying software systems

Besides being used for design specifications of user interfaces, the KBGUIM methodology is also used to represent the relevant knowledge of the underlying software system with visual multi-layer graphs. Figure 2 gives a graph representing the knowledge of how to declare a menu in SunView environment. The fact that the hexagon object and the ellipse object have the common object name, create_menu, indicates that create_menu can be specified recursively. The visual multi-layer graph allows knowledge engineers to clearly represent the knowledge on an underlying software system for a specific problem domain. As a result, a visual graphic knowledge base can be used to inform effectively the designers of the allowable components for user interface designs and the corresponding structures. This advising feature extends the capabilities of KBGUIM methodology to serve as an user-interface design expert system. The associated routines of all the objects in visual graphs are used to support automatic source code generation in the implementation of a design. Automatic source code generation will be discussed in detail in a later section.

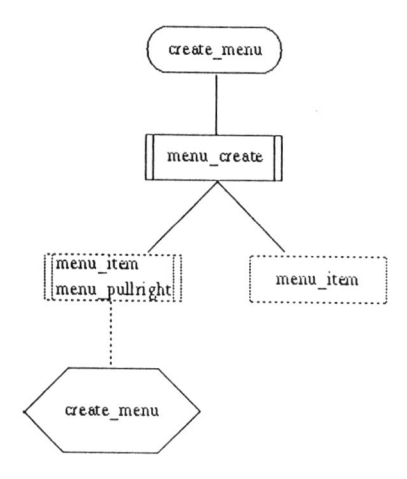

Figure 2. The knowledge graph of menu declaration in SunView

3.4. Flexible Design

During the user-interface design process, KBGUIM allows user-interface designers to directly retrieve appropriate knowledge from a visual graphic knowledge base. The knowledge retrieval is accomplished by the designer's freely expanding the expandable objects, such as those represented by ellipses, circles, double-rectangles, and hexagons of both mandatory objects, represented in solid symbols, and optional objects, represented in dotted objects. The flexible design supports a top-down approach in the user-interface design process. A user-interface designer starts the design process from declaring the outermost aspect of the desired user interface and then follows the knowledge graphs in the visual graphic knowledge base to further expand on the details. A previous user-interface design specification can be reused easily for other designs by making the necessary modifications.

3.5. Design in Visualization

The KBGUIM methodology allows a user-interface designer to design user interfaces by composing structures of objects which have associated routines that perform specific graphic interactions. This composition is achieved by automatic expansions at designer designated locations in the GSF. The designer uses the visual graphic knowledge base in the GKBF as a guide in carrying out his design process. The design specification in the GSF is expanded in front of the designer's own eyes as he directs the expansion.

Ichikawa and Hirakawa developed an iconic visual information processing language [15] called HI-VISUAL [9], which provided a navigation facility similar to the KBGUIM methodology's visual guidance. However, HI-VISUAL displays only a list of all candidate icons which can be used at each step of the program more speedily. In addition, the designers can interact with the knowledge engineers to prompt improvements in the knowledge base. After the designers become experts in the underlying software systems, they themselves can also readily modify the knowledge bases, and can thus soon go beyond the limitations of the originally supplied knowledge base. Current software design systems, including user interface management systems, do not provide separate knowledge bases and are thus quite inflexible.

The underlying software system needs not to be the SunView system and does not even have to be a window system. The methodology is not really limited to user interface management. Obviously, other types of system design and implementation can also apply this methodology.

References

1. Sutton, J. A.; Sprague, R. J., Jr.: "A Study of Display Generation and Management in Interactive Business Applications", IBM Research Report RJ2392(31804), Yorktown Heights, N. Y.

2. Smith, S. L.; Mosier, J. N.: "The User Interface to Computer-Based Information Systems: A Survey of Current Software Design Practice", Behaviour and Information Technology, Vol. 3, No. 3, 195-203.

3. Bobrow, D. G.; Mittal, S.; Stefik, M. J.: "Expert systems: Perils and Promise", Communications of the ACM, Vol. 19, no. 9, 880-894.

4. Shen, S. N. T. ; Liu, L.; Hsu, J.: "A Hyper-Graphics Methodology for Multidisciplinary Applications", 3rd Int. Conf. Expert Systems, Theory & Applications, Los Angeles, CA, Dec. 12-14, 1988, 111-114.

5. Shen, S. N. T.; Hsu, J.; "Knowledge Representation with Graphics, Texts, Procedures, and Property Lists", Proceedings of the Symposium and Workshop on Artificial Intelligence for Military Logistics, Williamsburg, VA, March, 1990, (to appear).

6. Myers, B. A.,: "User-Interface Tools: Introduction and Survey", IEEE Software, January, 1989, 15-23.

7. Larkin, J. H.; Simon, H.; "Why a Diagram is (Sometimes) Worth Ten Thousand Words", Cognitive Science, 11, 1987, 65-100.

8. Topping, P.; McInroy, J.; Lively, W.; Sheppard, S.: "Express-Rapid Prototyping and Product Development via Integrated, Knowledge-Based Executable Specifications", Proc., 1987 Fall Joint Computer Conf., Exploring Technology, Today and Tomorrow, 1987, 3-9.

9. Ichikawa, T.; Hirakawa, M.: "Visual Programming-Toward Realization of User-Friendly Programming Environments", Proc., 1987 Fall Joint Computer Conf., Exploring Technology, Today and Tomorrow, 1987, 129-137.

10. Rhyne, J.; Roger, E.; Bennett, J.; Hewett, T.; Sibert, J.; Bleser, T.: "Tools and Methodology for User Interface Development", Computer Graphics, Vol. 21, No 2, April 1987, 78-87.

11. Foley, J.; Gibbs, C.; Kim, W. C.; Kovacevic, S.: "A Knowledge-based User Interface Management System" CHI'88 Conference Proceedings, 1988, 67-72.

12. Pressman, R. S.: "Software Engineering A practitioner's Approach", McGraw-Hill Book company, 1987.

13. Ross, D.T.: "Applications and Extensions of SADT", IEEE Computer, Vol. 18, No. 4, April 1985, 25-35.

14. Wasserman A. I.; Pircher, P. A.: "A Graphical, Extensible Integrated Environment for Software Development", Proc. Second ACM SIGSoft/SIGPlan Software Engineering Symp. on Practical Software Development Environments, January 1987, 131-142.

15. Chang, S. K.: "Visual Languages: A Tutorial and Survey", IEEE Software, January 1987, 29-39.

Knowledge Augmentation Via Interactive Learning in a Path Finder

Q. ZHU
Department of Mathematics and Computer Science
University of Nebraska at Omaha
Omaha, NE

D. SHI and S. TANG
School of Engineering and Computer Science
Oakland University
Rochester, MI

ABSTRACT
We present an intelligent path finder which is capable of gaining and augmenting its operational skill in guiding a mobile robot navigating in unexplored environments. Rather than rendering the robot system to acquaintance with the specific environment, the robot is trained to acquire generic knowledge about path planning under various circumstances. The robot learns to determine a best direction of movement by means of interactive instruction in different environment situations . A pattern matching and state space transition scheme of learning is implemented.

1. INTRODUCTION
Path planning and collision avoidance in robot motion has been studied in two perspectives. One is to have the deploys and dimensions of obstacles in the environment known beforehand. Problem is to find a geometrically shortest path that enables the robot to accomplish its assigned task free from any risk of collision [1, 5]. This type of path planning is usually treated at a high level of motion control. It is not feasible for a robot navigating in an unexplored environment. The other is to have the robot equipped with active sensor devices. The robot is to response to the environment in real-time based on its identification of obstacles or barriers in the scene. A goal position of motion may be given, but the details of how to get to the destination is not specified before the motion. In this perspective, the problem of path planning is treated at a lower level of motion control [2, 3]. The robot determines its moving path towards the destination within the sight of view at every control cycle. The resulting path may not be globally optimal. However it allows the robot to proceed in the unexplored scenes.

When robot is operating via active sensing, it use automatically the information obtained and the knowledge about path planning to guide a search among alternatives for a good selection of a moving direction in the specific environment situation. Complexity and variations of the motion environment in real world preclude from programming a robot to possess a complete, error-free, and consistent knowledge base for path planning. To have the system adapt to the environment, one solution is to gain and improve the operation skill by learning in its running practice [7]. While totally automatic knowledge acquisition requires the use of complicated inference mechanisms, interactive learning provides a convenient way to compromise the sophistication and simplicity of such activity [8].

Learning systems that train the robot to acquire knowledge about the motion environment (obstacle and barrier settings) have been reported [4, 6]. After several times of travelling in the environment, the robot integrates the information gathered and tries to build an environment model in its internal representation so that a better travelling path can be generated thereafter. Drawback of this approach is that the learning results can only be applied to the same environment setting, and are not transferable to other unexplored situations. The learning scheme presented in this paper is different from that approach in the way that it is able to transfer the learned knowledge to any other similar or more complicated environment situations. The approach is to have the robot learn the path planning skills, rather than the specific environment setting. The robot does not simply repeat the motions in the same environment where learning takes place. It is supposed to applied the knowledge learned to anywhere else which bears fundamental similarity with the learning environment.

This paper is organized as follows: Section 2 describes the state space representation of the motion skill in path planning; Section 3 explains the interactive learning mechanism. Section 4 presents the learning system structure and computer simulation. Section 5 is conclusion.

2. STATE SPACE REPRESENTATION OF MOTION SKILL

We consider the path planning and motion control of robot as an asynchronous system. A "sampling - scene processing - path finding - maneuver" sequence is performed in every step of motion. A global goal is given in the robot motion task specification. Planning and adjustment of the path is done within the sight of view. The local path must be (1) consent with the task objective (reaching the destination); (2) collision free; (3) locally best (if not globally or sub-globally). Path planning using Configuration space (or Empty Space) and potential field [5] have been studied. Iyenger et al [4] have used spatial graph model and Voronoi diagram, where the entire motion environment is geometrically modeled. Rather than applying a geometric search in an environment map (which is time consuming), direction-driven approach of robot motion is employed in our research. The approach is simple and easy to be implemented for real time robot motion control.

While the robot is equipped with a sensor devise which has limited sight of view, it senses the local situation to form meta scenes. The environments of robot motion could be quite variant. We believe, however, that the meta scenes of the environments can be categorized into a finite number of patterns. Every environment is formed by a number of such meta scenes. The robot is trained to acquire knowledge about path planning under each of these meta settings with respect to different destination specifications. It then uses this knowledge to guide its motion in other similar or more complicated environment situations.

We use a state space representation to describe the pattern of meta scene, which is also the decision space of the robot path planning. Generally, the robot should make decision based on following attributes: (1) the position of global goal, that is, destination of the motion; (2) the positions of obstacles in scene; (3) relative directions and distances of the obstacles from the current position of robot; (4) possible directions to move and relative width of free space in each direction. The states are represented by a sequence of numbered digital pattern vectors, as

$$[d1, d2, d3, d4, t1, t2, t3, t4, ts1, ts2, ts3, ts4, ...].$$

The robot moves in a straight line with an upper speed limit at every control cycle. A state is changed by the motion action which is carried out by taking a step of movement in one of the geometric directions. The state transition sequence represents the motion of robot. Several factors determine the optimal choice of the operator in current situation: (1) Obstacles that cannot be passed through by the robot; (2) Distance from the current position of the robot to its destination of motion should be reduced; (3) Potential pitfalls should be avoided.

3. INTERACTIVE LEARNING

Learning can be defined as a process by which a system improves its performance by incorporating within its internal representation the information from the environment and use this information to make improvement of its knowledge base. The principle of learning can be explained by analog to human problem solving process. When a human problem solver encounters a problem, he first tries to classify his problem into subproblems for which he knows the solution. If he cannot, new method then must be constructed by applying some general problem solving technique. In constructing the new method, he must be carefully avoid certain pitfalls he has previously experienced. Method previously constructed may be used to solve subproblems. The new method is remembered so that it can be used to solve similar problems in the future. If any method, new or old, fails on a problem for which it is expected to work, the failure is examined and analyzed. As the result, the method may be modified to accommodate the new problem. Often the analysis of a failure can also be classified, and be remembered as a pitfall to avoid in the future. In any particular application, the environment, the knowledge base, and the performing task determine the nature of the particular learning problem and the particular functions that the learning elements must fulfill. This is the learning prototype adopted in our research.

We assume the robot has real-time capabilities for precise, fast, and highly interactive operations with a trainer when moving in a cluttered and evolving environment. The interactive capability allows the robot system to learn, augment, and improve its ability of finding (or deciding) a best path (a direction of motion) at every step for its motion towards the goal. The knowledge base in this system is a collection of move strategies represented in state space vectors. Interaction between the robot and trainer helps determining the set of rules that will bring out the best result for a particular group of cases. The algorithms involved allows easy implement and expansion of the knowledge base.

The motion planning knowledge can be expressed as predicates:

$$P: C_d \Rightarrow A_d,$$

where C_d is the condition (which is formed by the state space vector), A_d is the action of the predicate that specifies the motion path selection. Learning knowledge built in the system concentrates on the interactions with the robot and the manipulations of the motion skill knowledge (predicates). It makes inference and modification of the predicates. Problems of how to extract from previous experiences the strategy that is applicable to the accomplishment of the task or improvement of system performance in current situation are dealt with. Four operational schemes are applied [8]:

(1) *Learning Triggering Scheme*: An interactive interface is provided to trigger the learning process. The input to the system provides a learning instance that initiates a knowledge inference process.
(2) *Faulty Diagnosing Scheme*: It obtains feedback from a trainer and diagnosis the fault made in previous try to guide the modifications of existing decision rules accordingly. Supplementary information for learning is provided at this stage via interactive communication between the robot and trainer.
(3) *Knowledge Induction Scheme*: In knowledge induction, the causes of error pattern is identified by retracing the steps. The primary capabilities required are symbolic deduction

and heuristic search.

(4) *Knowledge Modification Scheme*: The modification of knowledge performs mainly a clause convolution operation. Various ways of symbolic manipulations are involved. The process augments the existing knowledge of path planning and constructs better ones that can be used to enhance the path decision skill in varying environment situations and tasks performed by the path finder.

We use S_d to denote the current measurement of the robot motion states (which is a description of local scene setting). S_d is attempted to match a C_d in the predicate set, where a decision is to instruct the robot in which direction it should proceed. Every component of state may have a "*donot care*" status, denoted as #, which can be matched by any value of the component. For deriving a learning result, an error pattern E_p is generated from the match between S_d and C_d, plus any supplemental information Q_s from the trainer. The Q_s can be a clause that negates the whole or part of C_d, or a description for the whole or part of the expected action A_d. An error pattern E_p captures the inconsistency of C_d with S_d, or P with Q_s. The modification of a predicate can be categorized into following four types:

(1) Rectifying the action of P:

$$(C_d => A_d) \cap E_p => (C_d => A'_d);$$

It sets up the actions of the predicates under certain conditions; The operation may simply add action to a null predicate, or form a complete condition-action pattern and add it to the rule set.

(2) Generalizing conditions of P:

$$(C'_d => A_d) \cap (C''_d => A_d) \cap E_p => (C_d => A_d);$$

This operation may merge rules and turn special values of conditions into "*donot care*" variables. It expands the scope of the applicability of certain predicates.

(3) Specializing conditions of P:

$$(C_d => A_d) \cap E_p => (C'_d => A_d);$$

This operation may drop condition components or add condition components of a predicate. It is to narrow the scope of the applicability of the predicates.

(4) Splitting P:

$$(C_d => A_d) \cap E_p => ((C'_d => A'_d) \cap (C''_d => A''_d));$$

This performs disjunction of conditions. It breaks down the condition part of a predicate p into two parts, and forms two new predicates that take each splitting part of the condition as their condition, and inherit, alter, or form new action parts.

4. SYSTEM ORGANIZATION AND SIMULATION

The robot system simulated is capable of detecting the presence of obstacles or barriers within its sight. When the robot proceeds, its view of the scene is continuously updated. For every new situation, the robot identifies its relative position with the goal, select a best path according to the local setting of obstacles and barriers, and proceed towards the destination directly or indirectly. It keeps a safe distance away from any obstacles or barriers.

The path finder applies an importance-dominated matching scheme. The rule set is organized in several groups in hierarchy. The first level only concerns the directional components. The second level deals with the relative positions of the robot with the goal. The third level deals with some more sepcific components, such as the width of path along the direction, the safe distance of robot path from the obstacles or along the barriers. When a robot motion situation is matched with a pattern at higher level, the components at lower level will not be tried. In this way the speed of robot motion control is fast. The similarity between two situations is measured by finding the best possible match according to what is more important as exhibited by the position of rules.

The control system is initialized with a null motion rule. For every step of motion, (1) the robot may consult trainer for the right way to move; (2) the trainer may interrupt the robot, instruct, or correct a move. The learning algorithm takes trainer's input, combining with the environment situation, robot motion state, and the current activating rule to make modifications of the rule set accordingly. An interactive communication interface of the simulation system is show in Fig. 1.

The skill of motion is accumulated in the learning process gradually. As the learning proceeds, the time of interruption for robot-trainer interaction decreases. when the time of interruption decrease, we say that the robot is trained into a deeper level. The capability of independent decision-making is accumulated and enhanced along with the used of different environment settings in the learning. With different levels of learning, the robot performs differently in each testing environment. When the robot is experienced with a sufficient amount of learning instances, it can operate in an environment all by itself without any further trainer intervention and consultation. The interactive learning system shows the trainer the formation of solution paths stepwisely. An example of graphical display of the robot motion environment is shown in Fig. 2, where the obstacles and barriers are shown as blocks. Goal position is marked by a black cross. The robot is shown circular with an arrow inside indicating its motion direction.

Main properties of our learning scheme is that the learning process is rather simple. Because of the state space representation of the knowledge, no complicated knowledge inference is involved. The interactive process allows simple manipulation of learning cases, easy to trace the rules and manage the modification of rules.

5. CONCLUSION

It has long been a goal of research to develop robot systems that could be taught rather than programmed. We present an intelligent path finder which has the ability to augment its operational skill by means of interactive learning. The system is used in planning a best path for robot motion in an unexplored environment while avoiding collisions with obstacles or barriers. The robot is not trained to repeat a path in the same environment, but to learning the skill of making decisions at different meta settings of the environment within its sight. Interactive learning provides a convenient way for the system to augment the capability of path planning and collision avoidance, and manage the creation, justification, and rectification of the motion control strategies. The ability of intervening user input in system's inner processes permits the implementation of knowledge inference efficiently. The method can be extended to the dynamic obstacle avoidance problems where obstacles in the environment are also moving rather than statically deployed.

REFERENCES

[1] Brooks, R. A., "Solving the Find-Path Problem by Good Representation of Free-Space", *IEEE Trans. Systems, Man, and Cybernetics*, Vol. SMC-13, N0. 3, pp. 190-197.
[2] Chattergy, R., "Some Heuristics for the Navigation of a Robot", *The international Journal of Robotics Research*, Vol.4, No.1, Spring 1985, pp. 59-66.
[3] Crowley, J. L., "Motion for an Intelligent Mobile Robot", *IEEE First Conference on Artificial Intelligence Application*, Danver, December 1984, pp. 51-56.
[4] Iyengar, S. S., et al, "Learning Navigation Paths for a Robot in Unexplored Terrain", *IEEE Second Conference on Artificial Intelligence Applications*, 1985, pp. 148-155.
[5] Khatib, O., "Real-time Obstacle Avoidance for Manipulators and Mobile Robots", *The international Journal of Robotics Research*, Vol.5, No.1, Spring 1986, pp. 90-98.
[6] Weisbin, C. R., et al., "Autonomous Mobile Robot Navigation and Learning", *IEEE COMPUTER*, June 1989, pp. 29-35.
[7] Zhu, Q., "Self-Learning Expert Systems", *Proceedings of the 1st Annual ESD/SMI Expert System Conference*, Dearborn, MI, June 1987, pp. 129-142.
[8] Zhu, Q., "An Interactive Refutation Learning Structure for Skill Acquisition in Knowledge-Based CAD Systems", *Proceedings of International Conference on CAD/CAM Robotics & Factories of the Future*, Southfield, MI, August 1988, Vol. 2, pp. 170-174.

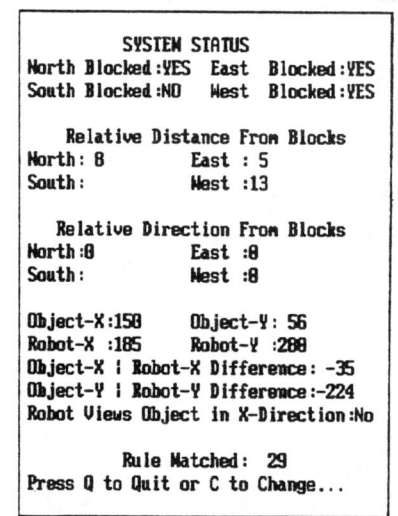

Fig. 1 Interactive communication interface of robot learning simulation.

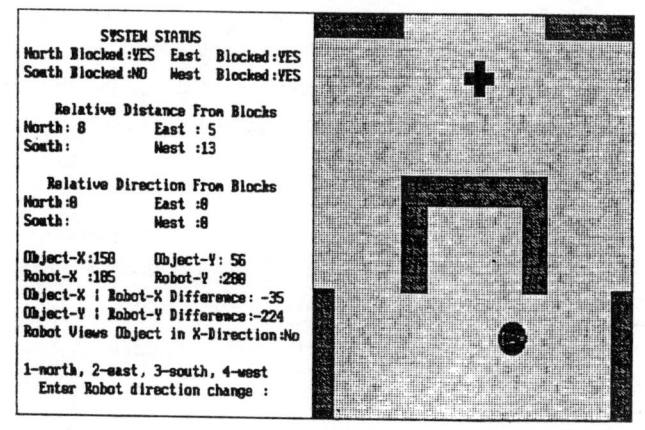

Fig. 2 Graphical display of robot learning simulation.

Graphical User Interface with Object-Oriented Knowledge-Based Engineering Environment

Z. KULPA, M. SOBOLEWSKI and S.N. DWIVEDI

Department of Mechanical and Aerospace Engineering and
Concurrent Engineering Research Center
West Virginia University, Morgantown, WV

Summary

One of the crucial problems in integrating machines and human experts in new-generation engineering design and manufacture systems is the effective man-machine communication. The object-oriented knowledge-based environment requires an appropriate graphical user interface built on the modern principles of object- and constraint-oriented programming, visual languages and hypermedia. The interface should provide semantic-rich graphical presentation of diverse domain-dependent data as well as a direct-manipulation interface easily adaptable to graphical symbology and conventions used in the particular application domain(s). To achieve these goals, the interface incorporates knowledge-base techniques both at the interface design stage and as a part of the interface manager. The interface will serve both the system designer (providing easy access to all the relevant mechanisms of the knowledge base and tools to create an appropriate domain-dependent end-user interface) and the end user (providing data acquisition from the user, system state presentation and control, and presentation of final results). The end-user interface design toolkit includes an extensive data and knowledge visualization subsystem, consisting of parametric, constraint-based image library, qualitative, quantitative (numeric) and structural data visualization modules, and interface-control specification module.

Introduction

User interfaces of expert systems are mainly based on a standard technique of so-called knowledge-base browsers of the kind proposed by [1, 7]. They are domain-independent tools aimed mostly at an expert system designer and allow for a textual (and limited graphical) access to the contents of the knowledge-base dictionary network, an actual state of the inference graph, etc. An end-user interface should, however, communicate information in a form compatible with graphical symbology and conventions used in the particular application domain. To easily create such an interface for every domain of a given expert-system shell application, the shell should include appropriate graphical interface design tools. Systems capable of graphical presentation of various data stored in knowledge bases were proposed quite long ago by [15, 9], though with a limited success, mostly due to the lack of powerful enough graphic synthesis paradigms and tools. These ideas and potential possibilities can be fully put into practice as a result of the development of novel techniques of object-oriented and constraint-

---

[*]Concurrent Engineering Research Center, West Virginia University, Morgantown, WV 26506

oriented programming [4, 11], visual languages and visual programming [6], improved user-interface design principles [5, 10, 14], and hypermedia approach to information presentation [13].The incorporation of all these developments in CAD systems will revolutionize this domain, making it possible to fully integrate machines and human experts in all the design and manufacturing cycle, from conceptual design to automatic manufacture and assembly. One of the crucial problems here is, inevitably, the effective man-machine communication.

In this paper, we present the general structure of a graphical interface based on these ideas and integrated with the object-oriented knowledge-based engineering environment built as an extension of EXPERTALK [1] and DICEtalk [2] systems.

Knowledge-based system architecture

Most knowledge-based systems have been developed for programmers. Their specialized knowledge description languages (frames, semantic networks) associated with programming languages (LISP, C) are often too difficult for use by engineers. The consequences of these relatively complex systems are long system development times with concomitant negative impacts on product cost, quality and supportability. A key idea behind our system (based on EXPERTALK [1] and DICEtalk [2] knowledge-based systems) is the concept of object-orientation (high-level structured programming) providing procedural knowledge-based programming integrated with simplified English as declarative knowledge-based programming. The integration of both declarative and procedural knowledge bases is supported by the graphical user interface with direct access to the Smalltalk environment, and through Smalltalk to other conventional languages, for example the C language, by user defined primitives. This integration (with the older version of the system, using conventional Smalltalk graphical interface) has been successfully verified in the two knowledge-based applications: the Printed Wiring Board Manufacturability Advisor [3] and the Turbine Blade Fabrication Cost Advisor [CERC, in preparation].

The low-level architecture of the system is described by special kinds of objects that are implemented in the system as more than fifty Smalltalk classes. The high-level architecture includes a *problem-solving engine*, a *graphical and application knowledge bases*, and *graphical user interface* (see Figure 1). The problem-solving engine consists of an inference engine, knowledge base compiler, task dispatchers, task mangers, explanation manager, object-oriented knowledge server, and Smalltalk compiler. The inference engine implements rule-based and knowledge-source-based deduction with procedural attachment mechanism, metaknowledge control and a broad interface to Smalltalk. The user interface is generally divided into two parts, each serving a different purpose. The first part creates the model of the knowledge-based problem-solving environment that users interact with. This model consists of the object-oriented representation of knowledge base along with the set of integrated tools for creating goals, facts, procedures, browsing, searching and problem solving. It is used mainly by the knowledge-base user solving his/her particular engineering problem (product developer). The second part provides the high-level tools which facilitate creation, modification and debugging of knowledge bases, both graphical and application knowledge base, for use by the application developer. These tools provide the user with a number of facilities such as inference control, graphical data presentation, and low and high-level of inspecting and debugging tools. Thus, the architecture is further complicated by the fact that the graphical knowledge base, used to control the graphical interface, can be simultaneously subject to modification with the high-level tools through the same graphical interface. To allow for this, appropriate safety and synchronization mechanisms are provided.

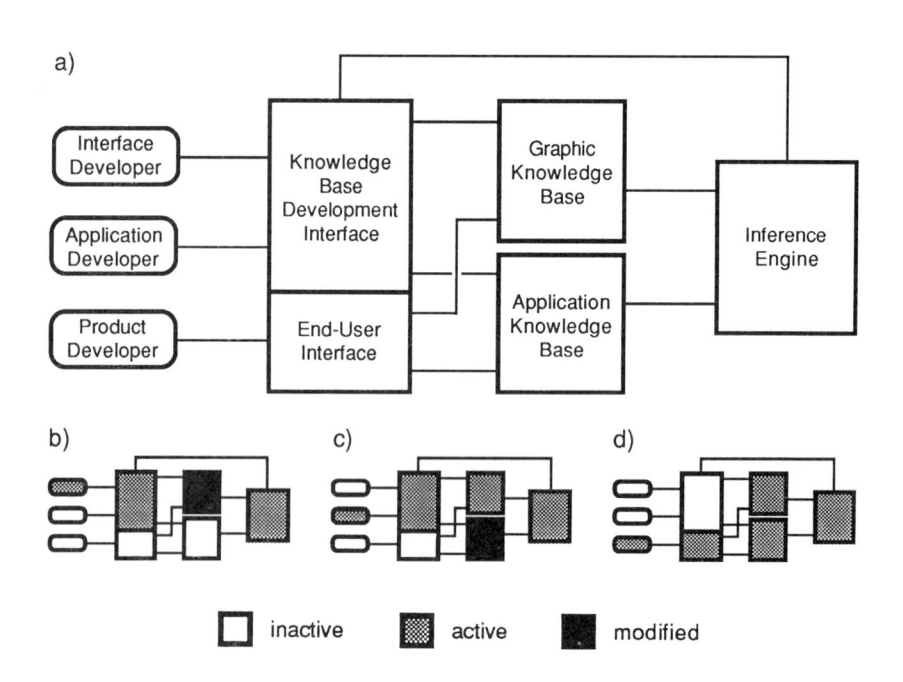

Fig. 1. Schematic of a system architecture
a) basic components; b) interface development mode;
c) application development mode; d) product development mode.

Graphical User Interface

The graphical user interface with a knowledge-based engineering environment should provide semantic-rich graphical presentation of diverse domain-dependent data, as well as a direct-manipulation interface incorporating user-domain terms, graphical symbology and action patterns. Moreover, to make true integration of the user and the machine possible, the user should be freed of the burden of over-specifying all the minute details of the input data again and again - the interface system should be able to supply by itself most of the "obvious" and repetitive data using context-dependent default values and intelligent guesses of user intent, based on the domain knowledge, the task at hand, and the user profile. To make this possible, the interface makes use of a graphical knowledge-base, both to help the application developer in construction of the application user interface, and to control the end-user interface manager during the problem-solving phase [8, 12].

The graphical interface of the application developer with the system should provide easy access to all the relevant mechanisms of the knowledge base: dictionaries, various sub-bases (declarative, meta, procedural and inference knowledge bases) as well as tools to create an appropriate domain-dependent end-user interface (with the help of dedicated graphical knowledge base). The end-user interface design toolkit contains an extensive data and knowledge visualization subsystem consisting of:
- parametric, constraint-based image library;
- qualitative data visualization module (handling of icons and other symbolic images);
- quantitative (numeric) data visualization module (creation of various types of graphs, charts, etc.);
- structural data visualization module (creation of networks, diagrams, maps, etc.);
- interface-control specification module.

The graphical interface of the end-user with the system provides mechanisms for:
- data acquisition from the user: stating graphically initial facts, answering system questions, etc.;
- system state presentation: intermediate results of the inference process and domain object simulations;
- system run control: stating graphically metaknowledge rules, choose search directions;
- final results presentation: stating conclusions, providing design advice, pointing out user design mistakes, plotting simulation results, and the like.

The graphical knowledge base is organized in a way similar to the application knowledge base - it is divided into declarative and procedural knowledge bases. Roughly speaking, the procedural graphical knowledge base contains methods for drawing and displaying graphical primitives (though sometimes as complex as graphs of functions or structural networks), whereas the declarative graphical knowledge base contains rules for selection of appropriate method of visualization of the given kind of data, e.g., type of graph, screen composition, and the like. The two bases are integrated such that attribute values used by the rules can be passed as parameters to procedures (in the procedural knowledge base), and the results of the procedures can be further used by the declarative knowledge base, e.g., as part of findings (inferences) originating from application of the rules. For example, a rule testing possibility to display some kind of graph on the screen might call a procedure to calculate display parameters of the graph, among others the area it would occupy on the screen, and then other rules might use this result to decide if this kind of graph fits properly into the current display, and what to do when it does not (e.g., suggest change of the graph type to use).

Conclusions

We presented the general structure of a graphical user interface with the object-oriented knowledge-based engineering environment. The interface is built around the modern principles of object- and constraint-oriented programming, visual languages and hypermedia. It provides semantic-rich graphical presentation of diverse domain-dependent data as well as a direct-manipulation interface easily adaptable to graphical symbology and conventions used in the particular application domain(s). To achieve these goals, the interface incorporates knowledge-base techniques both at the interface design stage and as a part of the interface manager. The interface will serve both the system designer (providing easy access to all the relevant mechanisms of the knowledge base and tools to create an appropriate domain-dependent end-user interface) and the end user (providing data acquisition from the user, system state presentation and control, and presentation of final results). The end-user interface design toolkit includes an extensive data and knowledge visualization subsystem, consisting of parametric, constraint-based image library, qualitative, quantitative (numeric) and structural data visualization modules, and interface-control specification module. With this kind of intelligent interface the true integration of the user and the machine becomes possible during all the design and manufacturing cycle, from conceptual design to automatic manufacture and assembly. This should be especially true within a concurrent engineering environment.

References

1. Sobolewski, M.: EXPERTALK: An object-oriented knowledge-based system. In: Plander, I. (ed.) *Artificial Intelligence and Information-Control Systems of Robots*, Amsterdam: North Holland 1989.
2. Sobolewski, M.: DICEtalk: An Object-Oriented Knowledge-Based Engineering Environment. *Proc. Fifth Int. Conference on CAD/CAM, Robotics and Factories of the Future*, 1990.
3. Padhy, S.K.; Dwivedi, S.N.: A Knowledge Based Approach for Manufacturability of Printed Wiring Boards. *Proc. Fifth Int. Conference on CAD/CAM, Robotics and Factories of the Future*, 1990.
4. Borning, A.; Duisberg, R.: Constraint-based tools for building user interfaces. *ACM Trans. on Graphics* 5 (1986).
5. Brown, J.R.; Cunningham, S.: *Programming the User Interface - Principles and Examples*. New York: J. Wiley 1989.
6. Chang, S.-K. (ed.) *Principles of Visual Programming Systems*. Englewood Cliffs, NJ: Prentice Hall 1990.
7. Clancey, W.J.: *Knowledge-Based Tutoring: The GUIDON Program*. Cambridge, MA: MIT Press 1987.
8. Faught, W.S.: Applications of AI in engineering. *IEEE Computer* 19 (1986) 17-27.
9. Friedel, M.: Automatic synthesis of graphical object descriptions. In: Christiansen, H. (ed.) *SIGGRAPH'84 Conf. Proc. (ACM Computer Graphics* 18, 3), New York: ACM Press 1984,.53-62.
10. Huang, K.T.: Visual interface design systems. In: Chang ,S.-K. (ed.) *Principles of Visual Programming Systems*, Englewood Cliffs, NJ.: Prentice Hall 1990.
11. Maulsby, D.L.; Kittlitz, K.A.; Witten, I.H.: Constraint- solving in interactive graphics - a user-friendly approach. In: Earnshaw, R.A.; Wyvill, B. (eds.) *New Advances in Computer Graphics (Proc. CG International'89)*, Berlin: Springer-Verlag 1989, 305-318.
12. Myers, B.A.: *Creating User Interfaces by Demonstration*. Boston: Academic Press 1988.
13. Parsaye, K.; Chignell, M.; Khoshafian, S.; Wong, H.: *Intelligent Databases: Object-Oriented, Deductive Hypermedia Technologies*. New York: J. Wiley 1989.
14. Thimbleby, H.: *User Interface Design*. Reading, MA: Addison-Wesley 1990.
15. Zdybel, F.; Greenfeld, N.R.; Yonke, M.D.; Gibbons, J.: An information presentation system. *Proc. 7th Intern. Joint Conf. on Artificial Intelligence,* 1981, 978-984.

Knowledge Automation: Unifying Learning Automation and Knowledge Base

A. CHANDRAMOULI and P.S. SATSANGI

Department of Electrical Engineering
Indian Institute of Technology
New Delhi, India

Summary
A brief review and comparison of learning models of AI and automaton is followed by the proposed structure of knowledge automaton. The behavior of knowledge automaton is summarized with simulation results for a simple case of single learning objective.

Introduction
The effort towards making intelligent computers/machines which had begun in 1940s [1] sets a pretty high goal of knowledge automation for the purpose of industrial/commercial use of AI. To explore the possibility of 'knowledge automation' we may frame the major research findings [2] into a premise "to exhibit intelligence and learning behavior some initial know_ledge is a must and similarly learning behavior is a must to update existing knowledge and/or to acquire new knowledge so that intelligent actions may be performed in new environments", which is a theoritical basis for the 'Knowledge Automaton' developed in this paper. As Learning is the key to change and to adapt to new environments, the learning models of AI and automaton are briefly reviewed and compared so as to integrate them into a model for knowledge automaton.

Models of Learning of AI & Automaton
Cohen [2] and Bolc [3] discussed various models of learning in AI field while Carbonell [4] has reviewed the main paradigms of machine learning currently being used in research. The primitive model of learning in AI research comprises of four components viz., environment, learning element, knowledge base, and performing element. In this Learning Model of AI [LMAI] the environment provides information to the learning element, which modifies the knowledge base so that the performing element can use the knowledge base to perform the required task. The per-

forming element feedsback the performance information to the
learning element, so that the experience gained may be used to
learn and be incorporated in the knowledge base of the system.

Bhakthavastalam [5] and Narendra [6] discussed the automaton
models of learning being used in learning automata theory. The
Automaton Model of Learning [AML] is an interactive feedback
system between the automaton and the environment. In this, the
automaton, a multi-action one, is the learning subject and the
environment would provide the object to be learned. Whenever
automaton selects an action, the environment would respond
either in a favorable manner or in an unfavorable manner. Based
on the response of environment and an updating scheme the
automaton may either repeat the same action or take another
action from among its action set. This process continues till
the automaton's behavior converges to or learns the
requirements of the environment.

Though the approaches of LMAI and AML are different, the historical relevence and the importance of finding alternate approaches to AI brings them together for comparing and exploring the possible integration of the two approaches. Environment
plays the key role in both LMAI and AML in the design and
classification of learning systems.

Learning element, performing element and knowledge base of LMAI
all put together are equivalent to the automaton of AML. Learning element and performing element, procedures, in LMAI have
the updating/learning scheme and selection of action by the
automaton as their counter parts in AML. The initial action
probability vector of the automaton in AML is similar to the
initial knowledge base in LMAI.

Learning takes place through feedback mechanism in LMAI as well
as AML except for the fact that their natures differ from
symbolic in LMAI to numeric in AML.

<u>Structure</u> <u>Of</u> <u>Knowledge</u> <u>Automaton</u>
Without going deep into neuro-physiological aspects, from the

layman's observations of reflexes of human being it may be observed that there are two levels of processing of information in the functioning of human being (human brain). One level of processing is locally at the functional unit and the other level of processing is globally at the seat of human knowledge, human brain, which takes into account all global information available and gives knowledge inputs to the functional unit.

From the basic premise and similarities of learning models mentioned in the preceding sections and the above observation of functioning of human being (human brain), the following structure, shown in figure.1, is proposed for the potential Knowledge Automaton which is evolved by unifying both the learning automaton and the knowledge base.

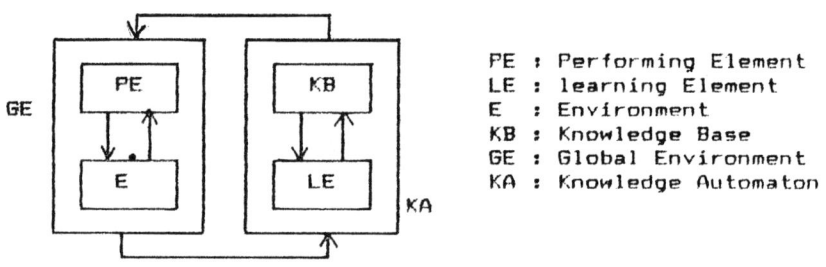

Fig.1 Structure of Knowledge Automaton

PE : Performing Element
LE : learning Element
E : Environment
KB : Knowledge Base
GE : Global Environment
KA : Knowledge Automaton

The components of the structure, viz, performing element (PE), environment (E), learning element (LE), and knowledge base (KB) can be considered as made up of the global environment (GE) consisting of PE & E and knowledge automaton (KA) consisting of LE & KB which are shown as two outer boxes in figure.1. Both global environment and knowledge automaton are structurally connected together in a feedback mechanism. Each component of the model is mathematically denoted as follows :
Performing Element (PE) : $\langle \alpha, P, T, U, K_p \rangle$
Learning Element (LE) : $\langle \beta, L, M, G_l, K \rangle$
Environment (E) : $\langle \alpha, F, U, G_l, f_l \rangle$
Knowledge Base (KB) : $\langle \beta, \emptyset, K, K_p, f_p \rangle$
where $\alpha : \{\alpha_1, \alpha_2, .., \alpha_r\}$ is the set of all performing actions of PE
$\beta : \{\beta_1, \beta_2, ..., \beta_s\}$ is the set of all learning actions of LE

PT : [p1, p2, ..., pr] is the performing probability vector of PE and pi(t) is the probability of performing action α_i being selected at the instant of time 't' and such that

$$\sum_{i=1}^{r} p_i = 1$$

L^T : [l1, l2, ..., ls] is learning probability vector of LE and li(t) is the probability of learning action β_i being selected at the instant of time 't' and such that

$$\sum_{i=1}^{s} l_i(t) = 1.$$

Gl : {gl1, gl2, .., glm} is the set of global learning objects.
U : {u1, u2, ...} is the set of responses elicited from the environment (E) by the performing element (PE).
K : {k1, k2, ...} is the set of knowledge responses to learning element (LE) from knowledge base (KB).
Kp : {kp1, kp2, ...} is the set of performing knowledge inputs to performing element (PE) from knowledge base (KB).
F : {F1, F2, ..., Fr} is the collection of random distribution functions which characterizes the environment (E).
Ø : {Ø1, Ø2, ..., Øs} is the collection of random distribution functions of sets of knowledge available which characterizes the knowledge base(KB).
T is the performance updating scheme in the sense it updates the performing action probability vector such that

p(t+1) = T(p(t), α(t), u(t), kp(t)).

M is the learning strategy modifier in the sense it modifies the learning action probability vector such that

l(t+1) = M(l(t), β(t), k(t), gl(t)).

fp is the global performing knowledge function.
fp : β X Kp ---> Kp i.e., kp(t+1) = fp(β(t), kp(t)).
fl is the global learning response function.
fl : U X Gl ---> Gl i.e., gl(t+1) = fl(u(t), gl(t)).

Due to space constraints a detailed discussion on these definitions is omitted here and the behavior of the knowledge automaton is summarized in the following section.

Behavior Of Knowledge Automaton

The behavior of the knowledge automaton can be explained in two ways : (1) If environment provides a global learning object, then the learning element (LE) elicits response from the knowledge base (KB) whether it is having the knowledge of the learning object or not; depending on this the knowledge base activates the performing element (PE) with performing knowledge inputs. The performing actions of PE elicit responses from the environment (E). (2) The knowledge base provides the performing knowledge inputs to PE which activates the action of PE that receives a response from the environment (E) which creates a new global learning object input to the learning element (LE) that feedsback to the knowledge base (KB). This is how it behaves in different situations according to the initiation either from the environment or from the knowledge base till it achieves the performance level to the satisfaction of the environment requirements.

Simulation Results

To illustrate the behavior of the knowledge automaton we have considered a simple case where the performing element (PE) and the learning element (LE) have two actions each. The knowledge base has two performing knowledge inputs and two knowledge responses corresponding to the PE and to the LE respectively. And the environment has two responses to the PE and a single global learning object to be learned. The environment provides the single global learning object as the input to the LE which initiates the functioning of the knowledge automaton.

Fig. 2.1 Knowledge Automaton

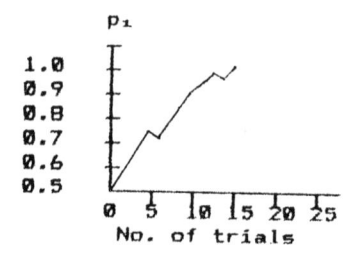

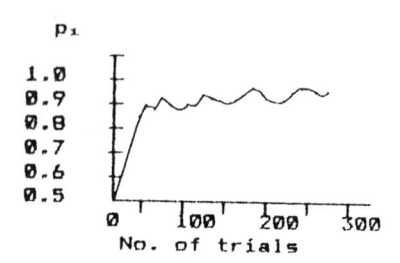

Fig. 2.2 Learning Automaton

Simulation results of this simple case of knowledge automaton and 2-action learning automaton [6] are plotted in figures.2.1 and 2.2. from which it can be observed that the performing action probability reaches unity with in a small number of trials. It explains that the provision of knowledge inputs improves the performing element actions to the satisfaction of the environment very fast.

Conclusions

The model presented in this paper is a starting point to develop theory and applications of knowledge automaton. Depending on Ø, the collection of sets of knowledge, one can classify different levels of knowledge and knowledge automata. The scheme 'M', learning strategy modifier, will be crucial to explain different ways of learning, from rote learning to learning by analogy. The possibility of integrating both the numeric and symbolic approaches may be explored. And the simulation results for the simple case illustrate that the performance level improves faster with knowledge inputs to the satisfaction of the environment.

References

1. McCulloch, W.S.; Pitts, W.H.; A Logical Calculus of the Ideas Immanent in Nervous Activity, Bull. Math. & Biophy. 5 (1943) 115-133.

2. Cohen, P.R.; Feigenbaum, E.A.; The Handbook of Artificial Intelligence Vol. III. London: Pitman Books 1981.

3. Bolc, L.; (ed.) Computational Models of Learning. Berlin: Springer_Verlag 1987.

4. Carbonell, J. G.; Introduction : Paradigms for Machine Learning. Artificial Intelligence 40 (1989) 1-9.

5. Bhakthavatsalam, R.; Purposeful Behaviour of Learning Automata in New Classes of Environments. Indian Institute of Science, Ph.D. Thesis. Bangalore. 1987.

6. Narendra, K.S.; Thathachar, M.A.L.; Learning Automata an Introduction. Englewood Cliffs, NJ: Prentice Hall 1989.

Developing a Knowledge Based System for Progressive Die Design

PRATYUSH KUMAR
Massachusetts Institute of Technology
Cambridge, MA

P.N. RAO and N.K. TEWARI
Mechanical Engineering Department
Indian Institute of Technology
New Delhi, India

Summary
The design of sheet metal dies for complex jobs is always a problem because of the principles involved. There is more to the practice rather than the hard engineering knowledge. An attempt has been made here to get the knowledge of the designers into the designing process as a set of rules, so that the system would be able to automatically make the necessary decisions and arrive at the optimum design. The system would also be able to give the requisite design details for successful production of the die.

1 INTRODUCTION

Designing a sheet metal die is always a challenging and time consuming task. A component, in practice, is essentially produced by combination of blanking, piercing and bending (forming) operations in a certain order. All these operations are interdependent and their nature and sequence has an influence on the total quality of the component produced. Any die that can perform all these operations simultaneously or sequentially has thus to be carefully and intelligently designed.

In this paper, an attempt has been made to concentrate on the piercing and blanking Progressive dies. It attempts at getting the optimum strip layout, the plan of strip development and a plan of die and punch blocks. This also selects the pilots dynamically, i.e., a different and the most suitable one for different stations.

2 PROGRESSIVE DIES

The progressive dies perform **two or more** operations simultaneously in a single stroke of a punch press, so that a **complete com**ponent is obtained for each stroke. The place where each of the operations are **carried out are** called stations. The stock strip moves from station to station undergoing the **particular operation**. When the strip finally leaves the last station, a finished component is ready.

At the start of the operat**lon, the** sheet is fed into the first station. After completing the operations at this station the **ram of the press** moves to the top and the stock is advanced from the first station move to **the second** station, while a fresh portion of the stock comes under the first station. The distance **moved by** the strip from station one to two so that it is properly registered under the stations is called advance distance. This advance distance should be the same between any two stations in sequence. Another variable called the feed distance is the amount of stock fed under the punch when ram comes for the next stroke. The feed distance may or may not be the same as the advance distance. The reason for the variance between the feed distance and advance distance, that is sometimes observed is because, often the sheet is overfed against a stop. The strip is, therefore positioned correctly under the punch by pulling it backwards with the use of pilots.

Progressive dies contain large number of stations. It is generally preferred to have piercing operation first in the sequence and a blanking or cutoff operation in the end to get the final component. Any of the pierced hole may be advantageously used as a pilot hole. If none of the pierced holes are satisfactory as pilot holes, then special pilot holes may have to be made in the scrap part of the stock.

2.1 PUNCH

The choice of the type of punch and its design depends on the shape and size of the pierced or blanked contour and the work material. For example, large cutting perimeters require large punches which are inherently rigid and can be mounted directly. However, smaller size holes require punches which may have to be supported during the operation, and therefore need to have other mechanisms to join it to the punch holder.

2.2 PILOTS

The pilots in progressive dies are used in order to bring the stock into the correct position for the succeeding blanking or piercing operations. The stock when fed manually is slightly over fed against a stop, provided by the pilot moving through the punched hole. The pilot when moves further down brings the sheet properly in position. The stock which is fed mechanically is slightly under fed, and the pilot pulls the stock for the correct registering. The main reason for this is that most mechanical feeders have provision for stock movement in only one direction.

The fit between the pilot size and the pierced hole determines the accuracy of the component produced. Too tight a fit between the pilot and pierced hole results in friction which would spoil the component as also, there would be excessive pilot wear. Depending on the type of work, the pilot size may deviate from the pierced hole size as follows:

0.050 to 0.100 mm	average work,
0.025 to 0.050 mm	close work, and
0.013 to 0.018 mm	high precision work.

The amount of over feed permissible is available in design books [10]. The length of the pilot should be such that it registers in minimum amount of time. Too short a pilot may not be able to register properly. Too long a pilot cause excessive friction which may be undesirable. The correct size of the pilot should enter the pierced hole fully, before any of the punches come into contact with the stock. The extra length of the pilot beyond the punch face may be approximately of the order of the sheet thickness or 1.5 mm, whichever is greater.

Pilots are in a way, perforator type punches. The die opening under the pilot should be larger than the pilot diameter by an amount double that of the clearance that would have been provided if it were a perforator. Sometimes, if the stock is misfed, then the pilot would act as a perforator and pierce the stock. If a large die opening is provided, then the material may be flanged rather than pierced. Hence, the die opening is restricted to the size as above.

2.3 STRIPPER

The stripper removes the stock from the punch after a piercing or blanking operation. The function of a stock guide is as the name implies, guide the stock through the various stations. The strippers are classified into two types: channel or box stripper and spring operated stripper or pressure pad.

2.3.1 Channel stripper

The channel stripper is simpler and easier to make. Fewer components are required for its construction and consequently, is economical. Also it is very rigid and therefore useful where large stripping forces are required. A channel stripper generally consists of a solid rectangular block of the same width as that of the die block for convenience. A slot or channel, slightly longer than the width and thickness of the stock is milled on one side of the block through which the stock is fed.

The height of the channel, H must be at least 1.5 times the stock thickness. For cases where the stock is to be lifted over a pin stop, the height may suitably be increased. The width, W of the channel should accommodate the stock width and any of its variation over the length of the die. Approximately 0.25 % of the die length would be able to take care of variations in most of the materials. If the stock is badly cambered, then the width may accordingly be altered. The clearance hole in stripper for the punch should be such that it is more than the size of the punch but not more than half the sheet thickness. More clearances should be provided for punching harder materials. The thickness of the channel stripper should be able to provide the necessary stripping force.

2.4 STOCK STRIP LAYOUT

Since, the components are to be ultimately blanked out of a stock strip, hence, precaution is to be taken while designing the dies for utilising as much of stock as possible. It is also necessary in progressive dies, to ensure continuous handling of the scrap on the die block, which means that the scrap strip should have sufficient strength.

The web of the stock to be left on the scrap strip, depends on the thickness and width of the strip, the contour of the blanked shape, and the type of blanking done. The parts with contours generally leave more material outside and therefore less scrap needs to be left on the stock strip. Those parts which have consecutive parallel blank edges, require slightly higher amount of scraps. The scrap values to be left on the stock are available from the design texts[5]. These values are for the progressive dies with single pass layouts. For single station dies, the scrap need not be as strong as for progressive dies, and hence these values can reduced by about 25 %.

In the case of two-pass layouts, the scrap to be left can be somewhat less than that of single pass layouts. Here also, for parts with curved edges, the scrap can be 1.25 T, where as for those with parallel edges, it is of the order of 1.50 T. If the part has partly curved and partly parallel edges, then 1.25 T would be enough for scrap strip.

2.5 STRIP DEVELOPMENT

Strip development refers to the choice of operations to be done in each of the stations in a progressive die. It is essential that the strip be properly developed since it will ultimately result in the best possible design of the die. Some of the principles arrived at based on the experience of the designers are as follows.

1. Always pierce the piloting holes in the first station. This helps in proper registering of strip for the subsequent stations.
2. If a number of the punched holes are very close, distribute them in more than one station so that the die block remains stronger. Similarly, holes nearer to the edge should be done in separate stations.
3. A complex contour should normally be split into a combination of simple shapes and punched out at a number of stations. This eliminates the expensive way of making a complex punch. It may also be possible that some of the simple shapes may readily be available commercially.
4. Use idle stations in strip development. Idle stations refer to the stations in a progressive die where no piercing or blanking is done. some of them may be used for simple piloting. A single idle station separate the corresponding punches by a distance equivalent to the advance distance, thus improving the strength of the die block, stripper and also the punch holder.
5. The hole nearest to the centre of the component will be the first punched hole.
6. The first punched hole should be away from the edges by at least 1.25 times the thickness of the sheet.
7. The diameter of the punched hole should be at least 2.5 times the thickness of the sheet.
8. If there are many such holes to be punched, select all those that lie in the same vertical line with the maximum X-coordinate value, i.e., those that are the right most holes in the piece-part.
9. If no such holes are available, then search for those holes which may not be farthest from the edges but still satisfy the above three criteria (6,7,8).
10. If no holes satisfying the above criteria, then punch hole(s) in the scrap area. The punched holes should be at least 1.25 times the stock thickness away from the piece-part edges.

    If the maximum dimension of the piece part to be produced is more than 250 mm, then two such holes are made in the scrap area. These holes are on the either side of the piece-part edges that are nearest to those stock strip ends that are parallel to the strip development direction.
11. If there are many holes that are to be punched, their sequence is determined by repeatedly applying the above stated conditions on the list of holes that are obtained by removing those which have already been punched.
12. The pilot hole would be the most recently punched hole for the next station.
13. The pilot for the last blanking operation is punched in the scrap area.
14. Select the internal contour to be blanked out first that has the smallest linear dimension. Others are sequentially placed depending upon their increasing linear dimensions.
15. The last operation is always the external contour blanking operation with external pilots in the scrap area.

3 USER MANUAL

Go into the AutoCAD (release 9.0 or higher) environment. Preferably on a virtual RAM disk, as otherwise the execution becomes painstakingly slow. The files required on the disk are all *Acad*.** files and *R*.SHX*, *M*.SHX*, *T*.SHX* files from the AutoCAD. Also copy *IN1.DWG*, *BIN.DWG* and *DIESYS.LSP* on the RAM disk.

Generate the piece part drawing with the following procedure:

1. Make first layer of the external contour as *OUTLINE*.
2. Draw outline contour as unfilleted polyline.
3. Draw this contour as a *Wide Run*.
4. Make all internal features in the layer *IN*.
5. Draw all polylines unfilleted.
6. If you want you can save this drawing.

An example is shown in Fig. 1.

Figure 1 Component shape

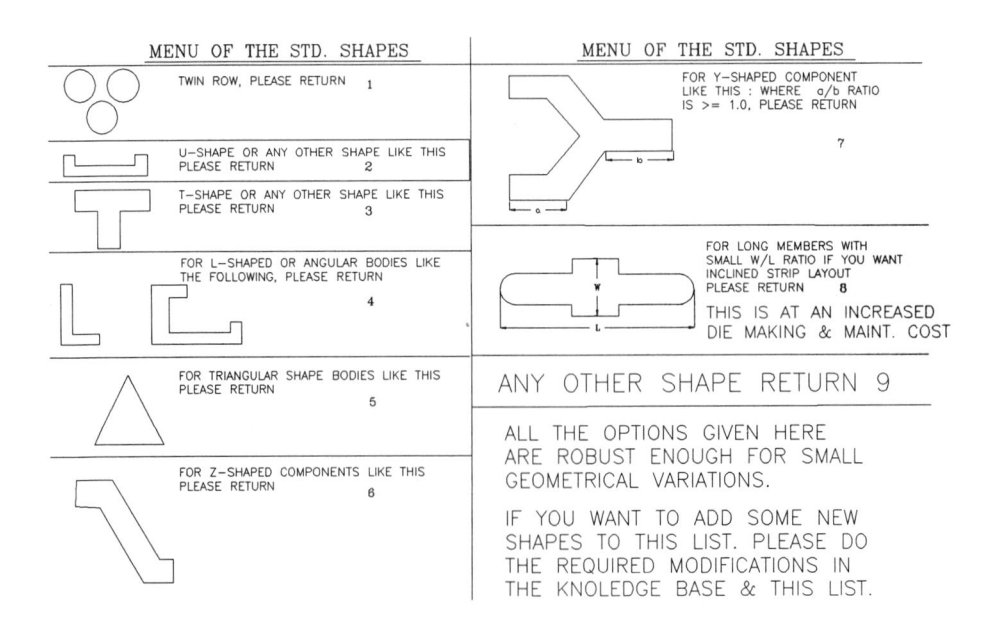

Figure 2 Options available for strip development

Interactive Run

After generating the piece part drawing load *DIESYS* in the following way:
Command: *(load "diesys")* return.

The menu options stored in the *IN1.DWG* file will be displayed to you (Fig. 2). Look at it carefully and find out whether any of the listed shapes match the given list of external contours. Give an appropriate option. Please enter the two extreme edges of the rectangle enveloping the piece-part drawing. This acts as an identifier for the piece-part. It will also ask you to give the other details as details such as stock thickness and the stock width.

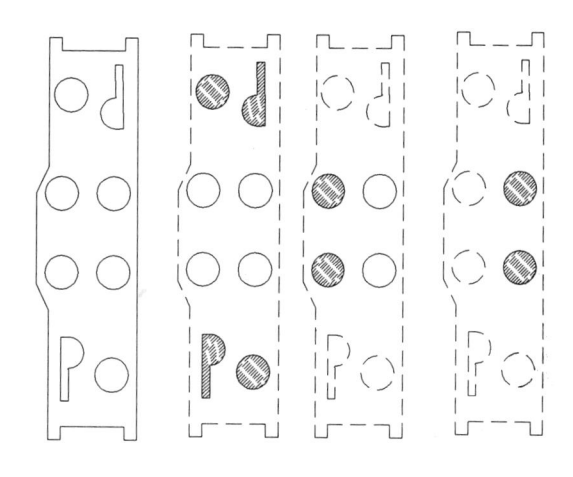

Figure 3 Stock-strip layout for the component

The next figure on the screen will give you the optimal stock-layout and will also give you an option to make any changes. Make any changes if you feel that the optimal layout given by the system is not appropriate. The next input required is the two extreme edges of the rectangle enveloping the oriented piece-part drawing. The default option is the previously inputed points. Based on these the system would generate the *Stock-Strip Development* (Fig. 3) and *Plan of the Die Block*. These two displays can be saved conveniently in appropriate files for taking Plotter Outputs.

REFERENCES

[1]   Hinman, C. W. : Press Working of Metals, *McGraw-Hill, New York, 1950*.

[2]   S M E : Die Design Handbook, *McGraw-Hill, New York, 1965*.

[3]   ASTME : Fundamentals of Tool Design, *Prentice-Hall, Englewood Cliffs, 1962*.

[4]   Ostergaard, D. F. : Basic Die Making, *McGraw-Hill, New York, 1967*.

[5]   Ostergaard, D. F. : Advanced Die Making, *McGraw-Hill, New York, 1967*.

[6]   Carlson, R. F. : Metal **Stamping** Design, *Prentice Hall, Englewood Cliffs, 1961*.

[7]   D.N.Ying : CPDDMS--An **Integrated** CAD/CAM System For Die Equipment Design and Manufacturing Application.

[8]   "AutoLisp"--Release 9.0 **Programmers'** Reference Manual, *1987*.

[9]   Jones, Franklin. D: Die design and die making practice, *Industrial Press, New York, 1957*.

[10]   P. N. Rao : Manufacturing Technology: Foundry, Forming and Welding, *Tata McGraw-Hill, New Delhi, 1987*.

An Expert System Model for the Use in Some Aspects of Manufacturing

R. B. MISHRA
Visiting Associate Professor

S. N. DWIVEDI
Professor
College of Engineering, West Virginia University, Morgantown, WV 26506

Abstract

Expert systems are widely used in the complex problems involving the symbolic representation of the problems rather than the computational simulation model. Manufacturing systems constitute many aspects of the processes such as scheduling, quality control, shop floor management, logistics, information management systems, etc., inviting the use of the expert system shell. This paper addresses the two problems, i.e., the scheduling in FMS and the inspection level in quality control to be represented in a framework (structure), i.e., EXPERT. The model is designed as a rule based system (i.e., IF and THEN rules) by incorporating different rules relating to the findings to the hypothesis. These findings are gathered through querries directed by control graphs. The model is implemented on a PC-XT using the Turbo-Prolog as the language tool. the confidence factor associates with the hypothesis reflects an attractive feature of the model and the ease of its implementation.

Introduction

Expert systems are computer programs, or a set of programs, using domain knowledge and reasoning techniques to solve problems normally requiring expertise caliber from human experts or an expert that uses rules for knowledge representation which is called a rule based system. The rule-based systems constitute the best currently available means for know how of human experts. Although no standard format exists for expert systems, they minimally comprise of knowledge base, infrence engine, and user interface. Knowledge can be stored in the knowledge base in any one of these several approaches: 1. Production rules,2. Semantic network, 3. Object attribute value, 4. Frames, 5. Logical expressions. Production rules take many forms. The Infrence Engine of a rule performs two functions: Inference and Control. Expert systmes are widely used now-a-days in the manufacturing process (1, 2, 3) and quality production (6, 7).

Due to the high grade of complexities involved in the operational level of FMS, the expert system models are being preferred over the simulation model. The FMS is basically a computer controlled production system capable of processing a variety of part types. The system consists of three main modules with numerically controlled manufacturing machines, an automated material handling system, and on-line computer control to manage the entire FMS. FMS is a very complex manufacturing system consisting of many interconnected components of hardware and software with many limited resourcs as pallets, fixtures, automated guided vehicles and tools. As the on-line scheduling problem of FMS is difficult to solve by mathematical models, an expert system model is developed involving the on-line scheduling problems in an FMS (8).

Light load, (ii) Machine state: Overload, Moderate load, Under load, (iii) Job Position: Overlate, Moderate late, and Normal late. The second level is to identify the criterial against which performance of the system are measured. The classified criteria are: Job computation times, Job due dates, Costs. The third level is the best way to satisfy the important critiera identified at the second level. This level has the decisions based on the first and second levels.

The FH rule, for example, is given below:

(Machine Utilization > .85, T)-----> (Machine Overload, .9)
(Machine Utilization > .85, F)-----> (Machine Overload, .65)

The above statements read that if the machine utilization > 85%, is true then the machine being overloaded has the confidence factor of 0.9, where as if the machine utilization >85% is false, the overload has the confidence factor of 0.65.

Similarly other rules are also formed regarding the throughput time for any job through system in part. For the time in hours, the slack value of job waiting ,the queue, Queue length-at the machine, blocking of the machine, etc. In each statement (hypothesis) is associated with a confidence factor ranging from -1 to +1.

The selection of inspection level is done based on the criticality of the component, application of the component, knowledge of manufacturing process, confidence level, number of components available for inspection, cost of inspection and cost of damage where a defective product is accepted. The four modules are as follows: (i) Criticality Module: The basic factors for the criticality requirements are reliability and connectivity of the component. The connectivity of the component is the number of sub-systems affected due to the failure of the component. The reliability assessment is subjective, no mathematical formulation is involved as though reliability is a vast and emerging field of study and research. Nevertheless, it is simply defined as the probability of not being failure of a component. Here only the qualitative (subjective) measure of the reliability of the component is taken into consideration.

The connectivity and reliability are qualitatively graded as Very, Very High (VVH), Very High (VH), High, Normal, Low, Very Low (VL), and Very Very Low (VVL). The examples of different rules based on these gradings are given below:

FF rules: (COMP-reliab., 0-0.8)------>(COMP-reliab., VVH)
(COMP-reliab., 0.8-0.6)---->(COMP-reliab.,VH)
(COMP-reliab., 0.2-0.4)---->(COMP-reliab., VL)
.
.
.

Similarly for the component connectivity are framed into FF rules as:

(COMP-CONNC., 1-1)------->(COMP-CONNC., VVL)
(COMP-CONNC., 16-20)---->(COMP-CONNC., High)

Application Part Module

Depending on the requirements and importance of the particular product, the different priorities are assigned to the product, keeping its appplication into consideration. Ten different users such as Missle, Space Vehicle, Aircraft, Ship, Office Equipment, Utility items, etc.

Another aspect of this paper deals with utilizing the expert system's framework (structure) for the selection of inspection level considering relevant factors. The Quality System in Manufacturing concerns with generating an effective and economic means for controlling quality and also to provide assurance about reliability, safety and performance requirements of manufactured products. In manufacturing, the inspection of raw materials, semifinished products, or finished products is an important part of the quality system. The selection of the inspection level is dependent on the relative amount of inspection based on criticality of the component and costs of inspection. The expert system framework thus developed is utilized in the selection of an inspection level for the given characteristics of the product.

This paper is divided into the following sections: Section 1 delas with the Expert System Framework. Section 2 deals with the formulation of the problem of scheduling in FMS and inspection level for the given product characteristics for Quality Control in Manufacturing. Section 3 covers the implementation of the structure for the problems and finally, Section 4 contains the discussion. The structure is implemented on PC-XT using Turbo-Prolog as the language tool.

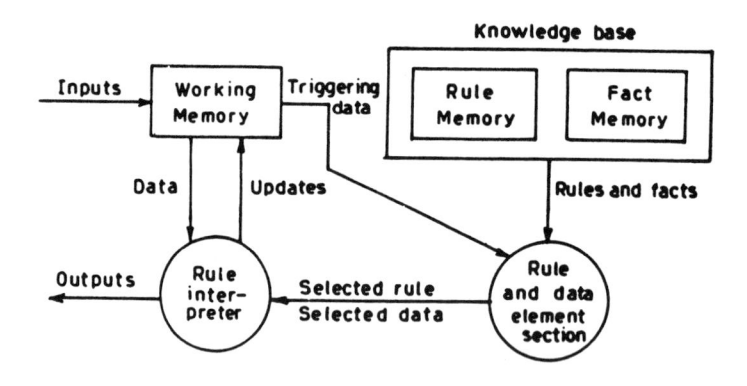

Figure 1. Rule-Based System

Framework of Rule-Based Systems (RBS)

The representation of a rule-based system is shown in Fig. 1. Consisting of various components such as (i) Knowledge Base; Dynamic and Static memory (ii) Inference Engine: rule interpretation, Query processor, etc. The design of an expert system consists of 10 steps as shown in Fig. 2, which are self-explanatory (5).

Several expert systems' (ES) development frameworks, based on early ES are now available and permit much quicker development of rule based systems. Examples of these commercially available packages are: (1) KAS which is derived from PROSPECTOR, (2) EMYCIN, MIKS 300, (3) EXPERT, derived from CASNET.

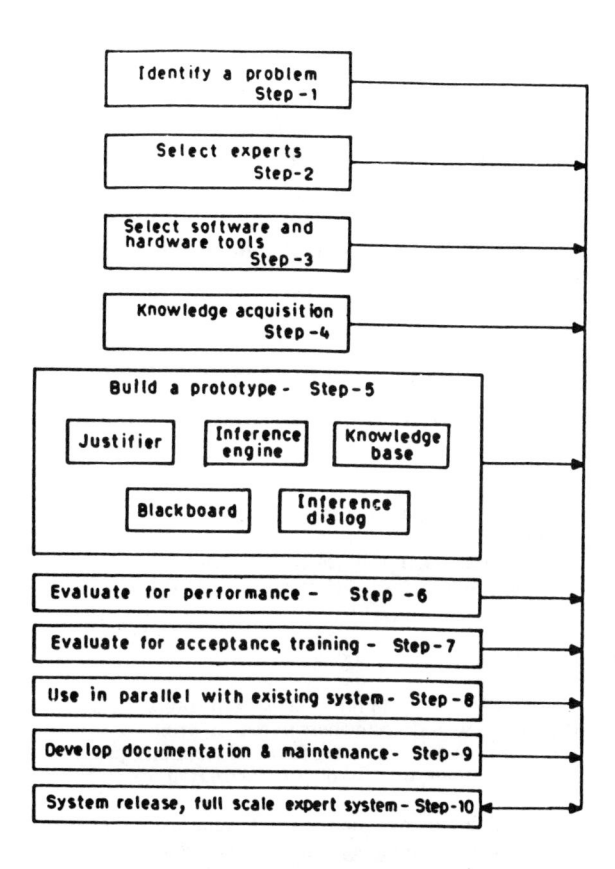

Figure 2. Design stages of an expert system.

The EXPERT framework is utilized here for designing the rule-based system for the above problems. The basic structure of an EXPERT model representation of knowledge consists of three sections: (i) Hypothesis, (ii) Findings, (iii) Decision rules. Hypothesis are the conclusions that may be inferred by the system. A measure of uncertainty is usually associated with a hypothesis. Findings are the specifications of the job, etc., and the Decision rules are three different types of rules such as: (i) FF rules: finding to finding rules, (ii) FH rules: Findings to hypothesis rules, (iii) HH rules: Hypothesis to hypothesis rules. These rules correlate one finding to another finding, finding to hypothesis and one hypothesis to another hypothesis. The findings are gathered by asking different types of questions such as: (i) Numerical, (ii) Multiple Choice, (iii) Yes/No (Binary) (4).

System Model

The FMS is based on a hierarchical approach consisting of three levels: (7). The first level is the classifications based on characteristics of the (i) System condition: heavy load, Moderate,

The following rules are framed in this module:

(Missle, T)------------>(Appl priority, 0.9)
(Space vehicle, T)---->(Appl priority, 0.8)
(Aircraft, T)----------->(Appl priority, 0.7)
.
.
(home applic, T)------>(Appl. priority, 0.2)
(utility, T)-------------->(Appl. priority, 0.1)

Cost Module

The Cost Module consists of different factors which contribut to the assessment of the cost of the product such as repair cost at inspection, cost of accepting a bad product, and cost of rejecting a good product. Apart from these criteria, the batch inspection level is also formulated based on the batch size and the cost ratio, which is defined as the ratio of inspection cost to damage cost.

(Good product, T)---->(Inspection cost, 0 8)
(Good product, N)---->(Inspection cost, 0.2)
(Good product, Y)---->(Inspection cost, 0.8)
(Good product, N)---->(Inspection cost, 0.1)

Incoming Quality Module

As some of the products do not fulfill the specifications required by the end user, leading to the increase in the percentage of defectiveness in the lot, produces significant effect on the level of inspection performed on the particular product. Hence, any product must satisfy the quality requirements, i.e., specifications, etc., for quality inspection.

The different modules are assigned four different levels, i.e., Special 1, Special 2, Special 3, Special 4, with the increasing weightage, i.e., confidence factor

(Special 1, T)---->(Inspection level, 0.8-.99)
(Special 2, T)---->(Inspection level, 0.6-0.8)
(Special 3, T)---->(Inspection level, 0.3-0.6)
(Special 4, T)---->(Inspection level, 0.1-0.3)

The FH rule for the inspection level of cost module are as given below:

(Cost Inspection Level, Z)---->(Good Quality, Z) & (Bad Quality, Z2)
& (Cost ratio, Z3) & (Batch Size, Z4)
& (Repair Cost, Z5)

Functional Findings:

$(Z = Z1 + Z2 + Z3 + Z4 + Z5)/5$

Similarly, the inspection level for other modulse are obtained together with the confidence factor calculated from functional findings in that module. The overall inspection level satisfying all the module requirements are given below:

(Component Inspection Level, Z)---->(Application Inspection, Z1)
& (Cost Inspection, Z2)
& (Quality Inspection, Z3)

Thus the component inspection level with confidence factor Z = (Z1 +Z2 +Z3)/3 is obtained.

Implementation

The model is implemented on a PC-XT due to the wide use of the PC-XT for implementation of many expert systems models now-a-days. The Turbo-Prolog is used as the programming language. The Prolog-version of the model represented in the previous section is shown for a particular inspection model in the form of control graphs (Fig. 3). Control graphs are directed graphs (having no cycles) in which all non-sink types nodes represent the queries asked. There are sink nodes which represents either success or failures. There is an edge connecting the two nodes if answer instantiation of one leads to asking of the other in atempt to establish some fact. The arcs are labelled with confidence factors. The confidence factor is summed over the entire paper and then divided by the number of edges to give the overall confidence factor for a particular success.

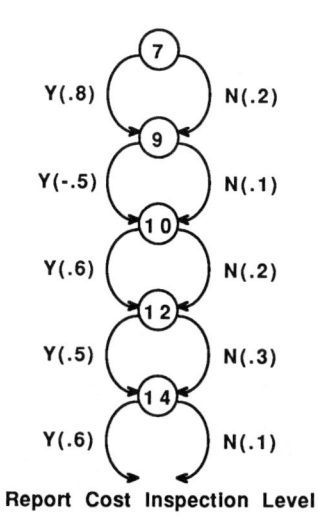

Figure 3: Cost Inspection Level Control Graph

The Prolog version of the Cost Inspection Level is shown below:

Cost Inspection Level :- Ask (7, Ans7), Good Quality (Ans7, CF1), Ask (9, Ans9), Bad Quality (Ans 9, CF2), Report Cost Inspection Level, Ask (10, Ans 10), High cost ratio (Ans 10, CF3) Cost Inspection Level Ask (12, Ans 12), Batch Size (Ans 12, CF4), Ask (14, Ans 14), repair cost (Ans 14, CF5).

Z = Z1 + Z2 + Z3 + Z4 + Z5/5

report ("cost of inspection level", Z)

Conclusions

The rule based system model has been developed based on the framework of EXPERT. The model starts its function by aksking a series of interelated questions as shown in the control graph (Fig. 3). Similar control graphs for each of the inspection modules and for the inspection levels are developed. The model covers quantitively the confidence in each of the hypothesis which it bears in carrying out the inspection modules. The inclusion of the qualitative assignment to the different clauses, i.e., statement such HIGH, VERY HIGH can be very much modelled using the Fuzzy Reasoning processes in this model. but the complexity of the implementation of the model would be increased. This model can be applied to the other manufacturing processes apart from the two processes: scheduling in Flexible Manufacturing Systems (FMS) and Inspection level in the Quality Control in Manufacturing systems are shown in this paper. The model is easy to implement on low cost PC-XT using 'C' as a programming language.

References

1. Stephen C-Y Lu, "Knowledge Processing for Engineering Automation", Proceedings of Advances in Manufacturing Systems Integration and Processes", David Adomfied (Ed), SME publishers, 1989, pp 455-468.

2. Kusia, A., "Designing of Expert Systems for Scheduling of Automated Manufacturing", I. E. July 87, pp. 42-46.

3. Bruno G., Elia A. and Laface P, "A Rule System to Scheduling Production", Computer Vol. 19, No. 7, 1986, pp. 32-40.

4. S. Weiss & C. A. Kulikowski", "A Practical Guide to designing Expert Systems", Rowman & Allanheld Publishers, N.J., 1984.

5. Turban E., "Review of Expert System Technology", IEEE Trans. Engineering Management, Vol. 55, 1988, pp. 71-81.

6. D. Chandra Reddy & S. S. N. Murthy, :Expert System Approach for Selection of Inspection Levels in acceptance", Proc., CAD, CAM, Robotics & Factories of the Future, Vol. III, TMH Publishers, 1989, pp. 519-527.

7. James r. Evans and William M. Lindsay, "A Framework for Expert System Development in Statistical Quality Control", Computers in Industrial Engineering, Vol. 14, No. 3, 1988, pp. 335-343.

8. Rambabu Kodali & S. S. N. Murthy, Knowledge Based System for Scheduling FMS", Proced., CAD, CAM, Robotics and Factories of the Future", TMH publishers, 1989, pp. 472-481.

Chapter V

Feature-Based Design and Manufacturing

Introduction

Feature-based design involves feature extraction or feature recognition as well as feature-based design. Feature-based approaches have been established as the link to CAD and CAM. Geometric interaction between features can cause a part to have several interpretations as a collection of features. To find the best process plan, the feature interpretation of the product needs to be alternated. The first paper of this chapter addresses this issue using feature algebra. The feature transformation system has been implemented and integrated with a solid modeler and a process planner. Feature extraction is the focus of the second paper. A new approach is described to recognize features using intermediate part geometry data and boolean operation, and a manifold dual model representation is used to define the part geometry. Protrusions and depressions are treated as lower features which are matched against templates to generate high level features for a given application; thus, recognition is done by matching instead of searching. The next paper describes a scheme to extract manufacturing features from the universal file of the I-DEAS software package. This technique is based on the same generic rules. Assembly feasibility is analyzed using features in the fourth paper. In addition, concave and convex properties of features for assembly are described and a new area optimization concept is introduced. The next paper deals with the feature-based evaluation of machinability and cost of product designs. Different modes of approach are analyzed for parts with complex geometry. The final paper discusses a feature based process planning system, which uses part features as medium of communication between geometry description and process planning functions. Components are defined by the user in terms of feature codes and their parameters. For process planning, the system supports the facilities for machine tool selection, setup planning, tool selection, and operation sequencing.

Using a Feature Algebra in Concurrent Engineering Design and Manufacturing

RAGHU KARINTHI and DANA NAU

University of Maryland
College Park, MD

Abstract

Concurrent engineering design requires the manufacturability aspects to be considered during the process of designing a part. One of the issues of interest in this context is that of geometric feature interactions. Due to geometric interactions among the various features of a machinable part, a machinable part could have several interpretations as a collection of features. In order to find the best process plan for a part; or in some cases to find a plan at all one needs to be able to generate alternate feature interpretations of a part, given only one of them. To address this problem, we have developed an algebra of features. By performing operations in the algebra one can obtain several feature interpretations of a machinable part given one feature interpretation. A feature transformation system based on the algebra has been implemented and integrated with our Protosolid[7] solid modeler and with our EFHA[6] process planning system.

1 Introduction

If the goals of concurrent engineering are to be achieved, CAD systems of the future must be able to communicate other modules such as those for process planning and fixturing. One of the primary problems in communicating between a solid modeler and a process planning system is the derivation of machinable features from the solid model. *Automatic feature extraction*, *design by features*, and *human-supervised feature extraction* are three of the commonly used approaches to solve this problem. Regardless of which approach is used for deriving the features, geometric interactions among the features can create situations where there are several possible feature representations for the same part. Since some of the feature interpretations may not be feasible and some others may be feasible but not optimal, it is important to have the capability to generate alternate feature interpretations in order to produce optimal process plans.

---

*This work was supported in part by an NSF Presidential Young Investigator award for Dr. Nau with matching funds from Texas Instruments and General Motors Research Laboratories, NSF Grant NSFD CDR-88003012 to the University of Maryland Systems Research Center, NSF Equipment grant CDA-8811952, and NSF grant IRI-8907890.

†Computer Science Department. Email: raghu@cs.umd.edu

‡Computer Science Department, Systems Research Center, and Institute for Advanced Computer Studies. Email: nau@cs.umd.edu.

For example, consider the part depicted in Figure 1. In this example, the part has been described as the part resulting from subtracting a rectangular pocket p_1, a rectangular pocket p_2 and a hole h_1, in that order, from a rectangular stock. Because of the interaction of the hole h_1 with the pocket p_2, the hole h_1 can be extended into the pocket p_2. Let h_2 be the extended hole. Let the extension of the hole h_2 into the pocket p_1 be h_3. Let the extension of the pocket p_2 into the pocket p_1 be p_3. The various possible feature interpretations of the part are $\{p_1, p_2, h_1\}$, $\{p_1, p_3, h_1\}$, $\{p_1, p_2, h_2\}$, $\{p_1, p_3, h_3\}$, $\{p_1, p_2, h_3\}$ and $\{p_1, p_3, h_2\}$. Of the above feature interpretations $\{p_1, p_3, h_2\}$ is not feasible from geometric considerations. From machining considerations, the larger pocket is made first, and hence the interpretations $\{p_1, p_3, h_1\}$ and $\{p_1, p_3, h_3\}$ are not good choices. Hence, the final set of features used for machining would be the pocket p_1, the pocket p_2 and one of the holes h_1, h_2 and h_3. The choice of h_1 or h_2 or h_3 cannot be be made without considering the parameters of the holes. The holes h_1, h_2 and h_3 have the same diameter but different depths. The holes h_1 and h_2 are located at the bottom of a pocket, and hence are more difficult to position than h_3 which is located on a face of the stock. However, they since their depth is less than that of h_3 they are less expensive to machine than h_3.

To address problems such as the one described above, we have developed an algebra of feature interactions. Given one valid interpretation of a machinable part as a collection of machinable features, all other valid interpretations of the part as other collections of machinable features can be derived through operations in the feature algebra. We have implemented a subset of this feature algebra dealing with rectangular solids, cylinders and counter sinks as the basis of a geometric reasoning system for use in communicating between a solid modeler and a process planning system.

In some previous work [1, 2], we developed a rather limited feature algebra for a small set of features, with the different kinds of features and their interactions all described as special cases. Since then, we have developed a unified mathematical way to describe features and feature interactions, so that our current feature algebra covers practically all features of interest to manufacturing. The current paper presents the highlights of this generalized feature algebra; the reader is referred to [4] for a more detailed presentation.

This paper is organized as follows: In Section 2, we describe the theory of the feature algebra. Section 3 describes a subset of the feature algebra that we have implemented and provides an illustrative example. Section 4 summarizes the work we have done on this topic.

2 The Algebra of Features

An algebraic structure [5] is a set, with one or more operations defined on the elements of that set. The feature algebra involves the set of all possible features (where *feature* is as defined below), and operations such as truncation and maximal extension (also defined below).

2.1 Domain

A *solid* is a compact, regular and semi-analytic subset of E^3. Let us see the scope and significance of the above definition. Regularity restricts a solid to be homogeneously three dimensional. Even parts with sheet metal components have a finite thickness, so this appears to be a very reasonable restriction. Since the solids that are considered are of finite dimensions, they are bounded and hence compact. The domain of semi-analytic sets covers practically all the shapes of interest to manufacturing. The reader may note that all planar polyhedra, cylinders, cones, spheres, tori and a variety of sculptured surfaces are encompassed by this set. It includes concave features such as T-slots, counter bores and counter sinks.

A *patch* is a regular, semi-analytic subset of the boundary of a solid. Figure 2 illustrates some examples of patches. Given a solid x and a patch p of x, the regularized complement of p is defined as $c^*(p,x) = b(\text{ps}(x)) -^* p$. From these definitions, we can prove the following lemma.

Proposition 1 *The set of all patches of a feature is closed under regularized union, intersection, complement and difference.*

A *feature* x is any pair $x = \langle \text{ps}(x), \text{patches}(x) \rangle$ such that $\text{ps}(x)$ is a solid and $\text{patches}(x)$ is a partition of the boundary of $\text{ps}(x)$ into one or more patches, each of which is labeled as BLOCKED or UNBLOCKED. For any patch p, we denote the label of the patch by label(p). The intended interpretation of a feature is as a solid $\text{ps}(x)$ that is subtracted from a larger solid, with the blocked and unblocked patches indicating boundaries of the patch that separate metal from air or air from air, respectively. For example, Figure 3 shows the BLOCKED and UNBLOCKED patches for the features h_1, h_2 and h_3 shown in Figure 1.

2.2 Operations on Features

This section describes the operations on features. In order to describe the propagation of patch labels we need to define additional concepts related to classification of a patch with respect to a solid. Due to lack of space, we describe only the first component: the set of all points of a feature.

2.2.1 Truncation

Given two features x and y, the operation truncation ($\mathcal{T}$) is defined as follows: $z = x\mathcal{T}y = \langle u,v \rangle$, where $u = \text{ps}(x) -^* \text{ps}(y)$. The second component is omitted from the discussion.

2.2.2 Maximal Extension

In order to define the maximal extension operation, we need to first define the infinite extension of a feature with respect to a patch. To keep the discussion simple, we present here the definition of infinite extension only for convex solids. For a more general definition, the reader is referred to [4]. Figure 4 illustrates some examples of infinite extension.

Let x be any feature and p_i be any point on x for which there exists a plane tangent to x at p_i. Then $H(p_i, x)$ is the closed half-space tangent to x at p_i that contains x.

The infinite extension of x with respect to the patch p is

$$\mathcal{I}_p(x) = \bigcap \{H(p_i, x) | p_i \in c^*(p, x).$$

Given two features x and y, and a patch p on x, the maximal extension of x in y with respect to a patch p (denoted by $x\mathcal{M}_p y$) is defined as follows:

$$x\mathcal{M}_p y = \begin{cases} \langle u, v \rangle & \text{if } \mathcal{I}_p(x) \neq \text{INVALID} \\ \text{INVALID} & \text{otherwise} \end{cases}$$

where $u = \mathcal{I}_p(x) \cap^* (\text{ps}(x) \cup^* \text{ps}(y))$.

2.3 Properties of the Algebra of Features

From the above definitions, a number of useful properties can be proved. Below are a few examples.

Proposition 2 *For any three features x, y and z, the following result holds:*

$$(\text{ps}(x) -^* \text{ps}(y)) -^* \text{ps}(z) = (\text{ps}(x) -^* \text{ps}(z)) -^* \text{ps}(y).$$

Proposition 3 *Given a feature x and a patch p of x, if $\mathcal{I}_p(x) = \text{ps}(x)$, then $\text{ps}(x\mathcal{M}_p y) = \text{ps}(x)$, for any feature y.*

3 Implementation

Based on the algebra of features, we have developed an algorithm that takes one feature interpretation of a part as input and outputs alternate feature interpretations. The operations, the properties and the algorithm have been implemented on a restricted domain consisting of rectangular solids, cylinders and counter sinks that have their planar faces parallel to the faces of the stock, such as holes, slots, shoulders, pockets, etc. Input to the algorithm is taken from a feature-based design system built on top of our Protosolid[7] solid modeler.

The implementation was done on a Texas Instruments Explorer II. The operations in the algebra are implemented as procedures in Lisp, and the algebraic properties (such as Propositions 2 and 3) as rules in Prolog. Whenever necessary, the features algorithm either asserts facts into the Prolog data base or queries it to determine the result of an operation on two features. If the query cannot be answered, then the procedures for the operations are invoked directly. Once the features algorithm has terminated, we have several possible feature interpretations of the part.

After these features have been produced, EFHA [6] (a successor to our SIPS process planning system[3]) can be invoked to make process plans for the features. These results give us an indication of which interpretation of the part is likely to be the easiest to manufacture.

3.1 Example

Let us illustrate the workings of the algorithm with the example discussed in Figure 1. In the beginning, we have exactly one feature interpretation of the part, viz. $\{p_1, p_2, h_1\}$. Due to the interaction between the features, in the first iteration we compute two additional features $h_2 = h_1 \mathcal{M}_a p_2$ and $p_3 = p_2 \mathcal{M}_b p_1$, where a and b are the top faces of h_1 and p_2 respectively. The additional feature interpretations added in this iteration are $\{p_1, p_2, h_2\}$ and $\{p_1, p_3, h_1\}$. During the second iteration we compute one more feature $h_3 = h_2 \mathcal{M}_c p_1$, where c is the top face of h_2. The additional feature interpretations added are $\{p_1, p_2, h_3\}$ and $\{p_1, p_3, h_3\}$. During the third iteration no new features are generated and hence the algorithm terminates after the third iteration. However, a new feature interpretation $\{p_1, p_3, h_2\}$ is added due to the interaction between the features h_1 and p_1. As discussed earlier, this combination is not feasible from geometric considerations.

For the example shown in Figure 1, the total processing time for computing the alternative feature interpretations was 6.35 seconds. In this example, the total number of interactions (the number of times an operator was applied) was 118, but only three of the interactions resulted in new features. For this example, 9 of these interactions were resolved via queries to the Prolog implementation of the algebraic properties, without invoking the procedures to compute the operations.

When the same example was run without access to the algebraic properties, the processing time decreased to 2.6 seconds. One reason for this result could be that the Lisp code is compiled and the Prolog runs interpretively. The other reason could be that since the features and interactions considered right now are rather simple, the procedures for the operations do not take a significant amount of time. With more complex features and interactions, access to the algebraic properties may turn out to be more useful.

The procedures for computing the operations take advantage of the nature of the features and interactions. Declaratively, the operations for this sub-algebra can be expressed in terms of set operations on solids. Therefore, another way to compute the operations would be to convert them into set operations on solids (computed by the solid modeler) and then test if the result of the operation is a new feature. Using this method, the time taken for computing alternate feature interpretations for the example in Figure 1 is 138 seconds. This shows that using the algebra is much more efficient than translating the operations into equivalent set operations. The reason is that if the result of an operation is not a valid feature there is no reason to compute the shape of the solid and then realize it is not a valid feature. Currently, we are working on a mathematical analysis of the computational time taken by both methods.

4 Summary and Conclusions

In this paper, we have summarized our work on an algebraic approach to handling feature interactions. We are currently working on extending the scope of the feature algebra to include additional operators to capture feature interactions. We believe this work will have utility not only for automated manufacturing, but also for other problems in geometric modeling and geometric reasoning.

References

[1] R. R. Karinthi and D. S. Nau. Geometric reasoning as a guide to process planning. In *ASME International Computers in Engineering Conference*, July 1989.

[2] R. R. Karinthi and D. S. Nau. Using a feature algebra for reasoning about geometric feature interactions. In *Eleventh International Joint Conference on Artificial Intelligence*, August 1989.

[3] D. S. Nau. Automated process planning using hierarchical abstraction. *Texas Instruments Technical Journal*, pages 39–46, Winter 1987. Award Winner, Texas Instruments 1987 Call for papers on Industrial Automation.

[4] D. S. Nau and R. R. Karinthi. An algebraic approach to feature interactions. Technical Report TR-89-101, Systems Research Center, University of Maryland, College Park, nov 1989. Submitted for journal publication.

[5] C. Pinter. *A Book of Abstract Algebra*. McGraw-Hill Book Company, 1982.

[6] Scott Thompson. Environment for hierarchical abstraction: A user guide, May 1989. Master's Scholarly paper.

[7] G. Vanecek Jr. *Set Operations on Volumes Using Decomposition Methods*. PhD thesis, University of Maryland, College Park, 1989.

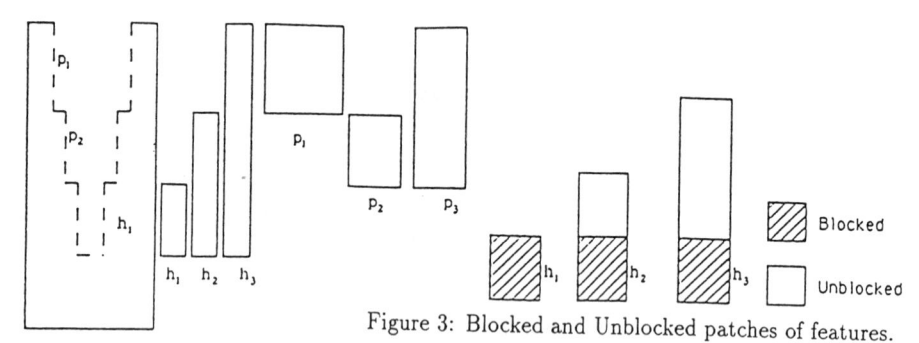

Figure 3: Blocked and Unblocked patches of features.

Figure 1: A part with two pockets p_1, p_2 and a hole h_1

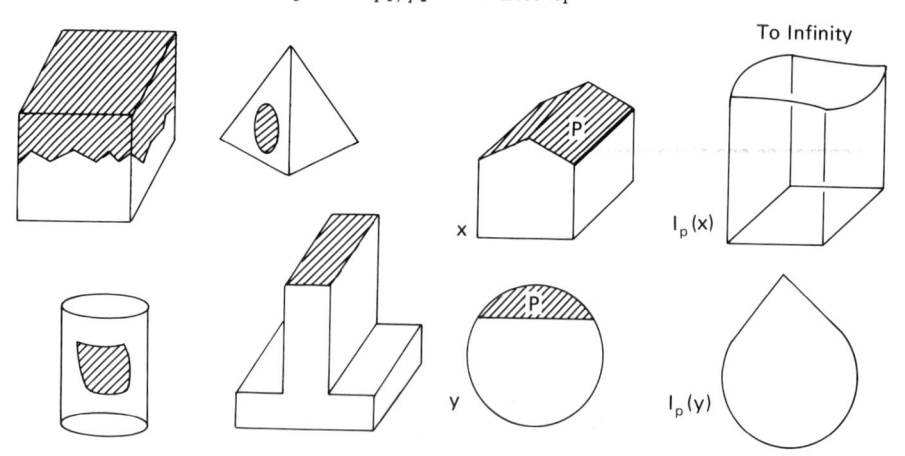

Figure 4: Examples of patches and infinite extension.

Figure 2: Some examples of patches (shaded).

Feature Recognition During Design Evolution

HYOWON SUH and RASHPAL S. AHLUWALIA

Department of Industrial Engineering
West Virginia University
Morgantown, WV

Abstract

This paper presents an approach to recognize features using intermediate part geometry data. Part geometry is defined by manifold dual model representation, namely, the boundary representation (Brep) and constructive solid geometry (CSG) tree. In the dual modeling system, the boundary of each primitive is split into two parts by a intersection edges. One of the split boundaries of a solid is joined with one of the two of the other solid to make a new solid. One of the two joined as the new solid is identified as feature by topological and geometrical properties. In this approach, features can be recognized by just matching a isolated set of faces (split boundary) with templates instead of searching. Moreover, the necessary volume (volume feature) can be generated based on the recognized feature (surface feature) using the split boundaries.

Introduction

Feature based design, automatic feature extraction and feature identification technique are generally utilized for the generation of shape features. In the feature based design approach, the features are predefined according to the given application and stored in a feature library. A part is then designed with features from the library. In the automatic feature extraction approach, the features are extracted from the completed part design. The feature extraction is performed by searching the same pattern with the templates specified over the part boundary data. In the feature identification approach the features are defined manually. A user identifies the features from the graphic display or blue print of a part. Most of shape feature related works can be categorized as one of these three approach.

The feature based design approach has some drawbacks: 1) the features of the part designed at the selected application mode can be meaningless for another application; 2) additional feature extraction or mapping process is necessary to generate the new features understandable by another application system; 3) In most cases this approach does not provide complete geometric information of part. One of the main drawbacks of feature extraction approach is the searching problem. The feature template in the feature knowledge base are tried out one by one to be matched with the same pattern of boundary data. Since both the feature and part boundary data are represented as graphs, the extraction procedure is expensive. In addition, since most of work

---

*Acknowledgements - This work has been sponsored by Defense Advanced Research Projects Agency (DARPA), under contract No. MDA972-88-C-0047 for DARPA Initiative in Concurrent Engineering (DICE)

in this approach only use boundary information by Brep or use unevaluated information by CSG tree, they need additional operation for the volume creation (volume feature) or evaluation procedure for the surface information. In the feature identification approach, the possible errors of feature generation are inevitable, and meaningless feature generation is possible because of human errors. The detail discussion is covered in [1].

Properties of shape features

A shape feature is defined as a set of contiguous faces or a volume based on the face set. A shape features are meaningful to an application. A set of faces has certain characteristic combination of topology and geometry so that the features for a feature library or a feature knowledge base are defined by the topology and geometry. The features in most mechanical parts are either protrusion or depression. Most of previous work on surface feature extraction can be classified as either "low level feature recognition" or "high level feature recognition". The former is to simply recognize the protrusion or depression, while the latter is to recognize the special pattern of the protrusion or depression by template matching. Both of the approaches use topology and geometry characteristic. Some of the previous work like [2] and [3] can be classified as low level feature generation, whereas [4] and [5] can be classified as high level feature generation.

This paper describes a methodology to recognize feature as the part is being modeled. The low level features, protrusions or depressions, are recognized first and then the recognized features are matched with the specific template to generate the high level features.

Low level feature recognition

Figure 1 shows boolean operation on two blocks. A block is subtracted from another block to make a "notch" feature. The details of modeling procedure of dual representation (Brep and CSG tree) modeling system are as follows: The intersection curves between boundaries of the two blocks are computed, making a edge loop as shown by heavily lines in Figure 2(a). The intersection curves split each boundary into two part and the resulting boundaries are classified as *AinB, AoutB, BoutA* and *BinA* according to the 4-way classification shown in Figure 2(b). An 8-way classification would include *AonB+, AonB-, BonA+* and *Bon-* [4]. From the components of the 4-way (8-way) classification, the result of boolean operation can be computed as follows:

$A \cup B = AoutB \otimes BoutA \otimes (AonB+)$ $\otimes$:gluing(joining)
$A \cap B = AinB \otimes BinA \otimes (AonB+)$
$A - B = AoutB \otimes (BinA)^{-1} \otimes (AonB-)$

The final model of the example is obtained by gluing the $(BinA)^{-1}$ on *AoutB* according to set operation (-) as shown in Figure 3(a).

In Figure 3(a), the boundary $(BinA)^{-I}$ glued on the final part seems the feature "notch" as extracted on most of previous works. If the split part $(BinA)^{-I}$ is recognized as protrusion or depression with respect to the neighbor, it can be a low level feature. The depression can be characterized by a set of contiguous faces which are adjacent to each other by concave edges and bordered by convex edge loop, while the protrusion is by set of faces adjacent to each other by convex edges and bordered by concave edges. The intersection edge loop as shown by the heavy line in Figure 2(a) is a convex border of the $(BinA)^{-I}$ in a new solid in figure 3(b) according to the following properties:

• Property1: *The intersection line of each boundary plane of two primitives solid is classified as either a convex for boolean operations intersection and difference, or concave for boolean operation union.*

• Property2: *If the two solids intersect each other in planes, and if one edge of the intersection loop is concave or convex, the rest of the edges of the loop have the corresponding convexity or concavity properties with the edge identified according to Property1.*

In the next step, the inside of the intersection loop, the 3 faces of $(BinA)^{-I}$ in the new solid are identified as concave by the following "Property3" because the face normal of the $(BinA)^{-I}$ is changed as it joins with AoutB.

Property3: *1) Any portion of the surface boundary of convex solid is convex or planar. 2) If the face normal of the portion is changed to the opposite direction, the convexity property is changed.*

Thus the part, $(BinA)^{-I}$, of the new solid is recognized as the "depression" low level feature. In general, one of the classified components in Figure 2(b) is a low level feature according to boolean operation. Here the proof of Property1 is introduced. For the proofs of the other properties refer to [7].

Proof of Property1: When the two solids intersect each other in planar surfaces, fa and fb in Figure 5(a), four divided parts represented by intersection of half-spaces, (A1, A2, B1, B2), are generated because the planar surface of one solid divides the other solid into two part as shown in Figure 5(b) and Figure 5(c). In such a case, each divided one is a convex portion of a solid because the planar surface (180°) is divided into two part less than 180°. The conclusion can be drawn as shown in Figure 5(d). The result of the intersection and difference operation is convex because the result part is only one part of divided one which is convex, while the result part of union is concave because the result part is joined part with a convex part, A1 or B1, and plane, A or B.

High level feature generation

In the previous high level feature generation research, the feature template in the feature knowledge base are tried out one by one to be matched with the same subgraph of the boundary data causing expensive searching problem like the NP-complete problem. In the proposed approach, the high level feature recognition is reduced to just matching procedure without searching, because an isolated set of faces shown in Figure 4(a) is generated during design evolution, which is recognized as a low level feature. The isolated and recognized low level feature is simply matched with the templates in the feature knowledge base to recognize the specific pattern to result in "high level feature". The Figure 4(b) and (c) show the template of feature "notch" and recognized high level feature "notch".

Conclusion

This paper presents a procedure, where the features are recognized as the part evolves. The procedure is based on using intermediate boundary data and boolean operations. The protrusions or the depressions of a part are recognized as low level features. These features are then matched with the template to generate the high level features for a given application. In this approach the features are recognized by just matching, not searching. The recognized features can be listed and the part can be represented by the integrated geometry and feature data structure. In addition, the necessary volume (volume feature) can be generated based on the recognized feature (surface feature) using the split boundaries. The non-uniquness of modeling like CSG tree representation is the critical issue in this approach. The scope of feature geometry is limited to the primitives and their boundary type defined by the modeling system.

Reference

1. Suh, Hyowon and Ahluwalia, R. S., "Features in CAD/CAM/CAE", Report, Industrial engineering, West Virginia University, 1990.

2. Floriani, L. De and Bruzzone, E., "Building a Feature-based Object Description from a Boundary Model", CAD, vol 21, no 10, Dec 1989, pp602-609

3. Nnaji, Bartholomew O. and Liu, Hsu-Chang, "Feature Reasoning for Automatic Robotic Assembly and Machining in Polyhedral Representation", int. J. Prod. res, vol 28, no 3, pp517-540

4. Joshi, S. and Chang, T. C., "Graph-based Heuristics for Recognition of Machined Features from a 3D Solid Model", CAD, vol 20, no 2, March 1988

5. Safier, Scott A. and Finger, Susan, "Parsing Features in Solid Geometric Models", European Conference on AI-90, 1990

6. Mantyla, M. "An Introduction to Solid Modeling", Computer science press, Inc., 1988

7. Suh, Hyowon, Miller, J.E. and Ahluwalia, R. S., "Concavity and Convexity Properties of Intersection of Solids", Under preparation, CERC, West Virginia University, 1990.

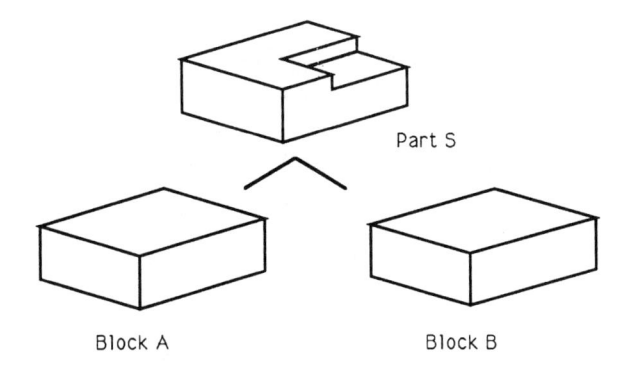

Figure 1. Boolean operation of two blocks

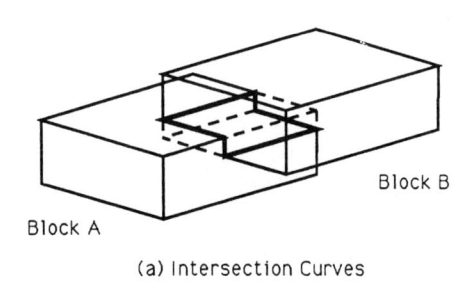

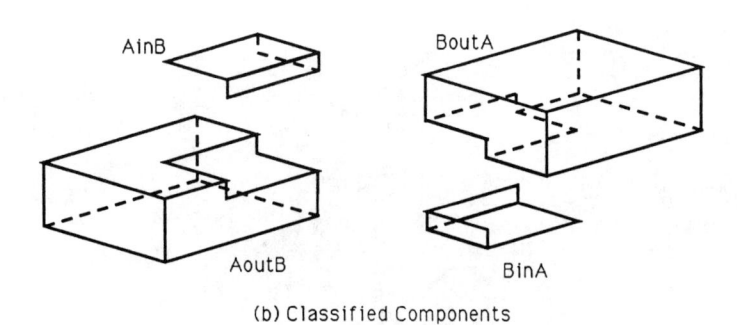

(a) Intersection Curves

(b) Classified Components

Figure 2. Boundary classification of two blocks

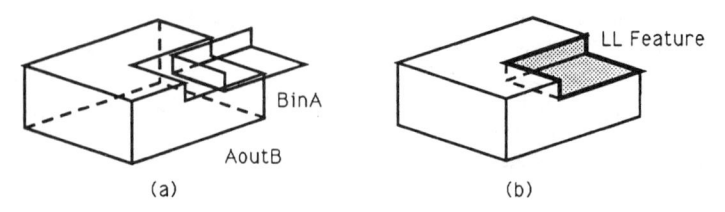

Figure 3. Low level feature generation

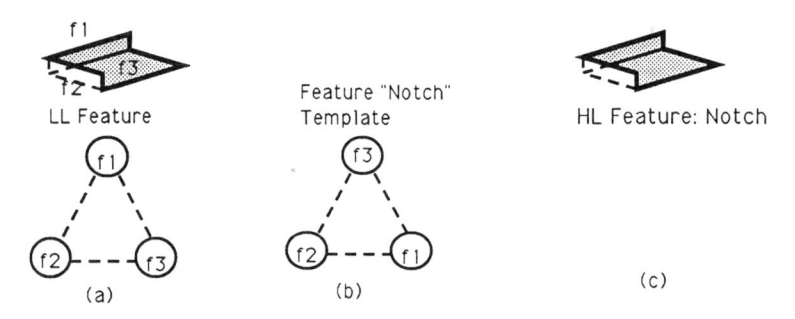

Figure 4. High level feature generation

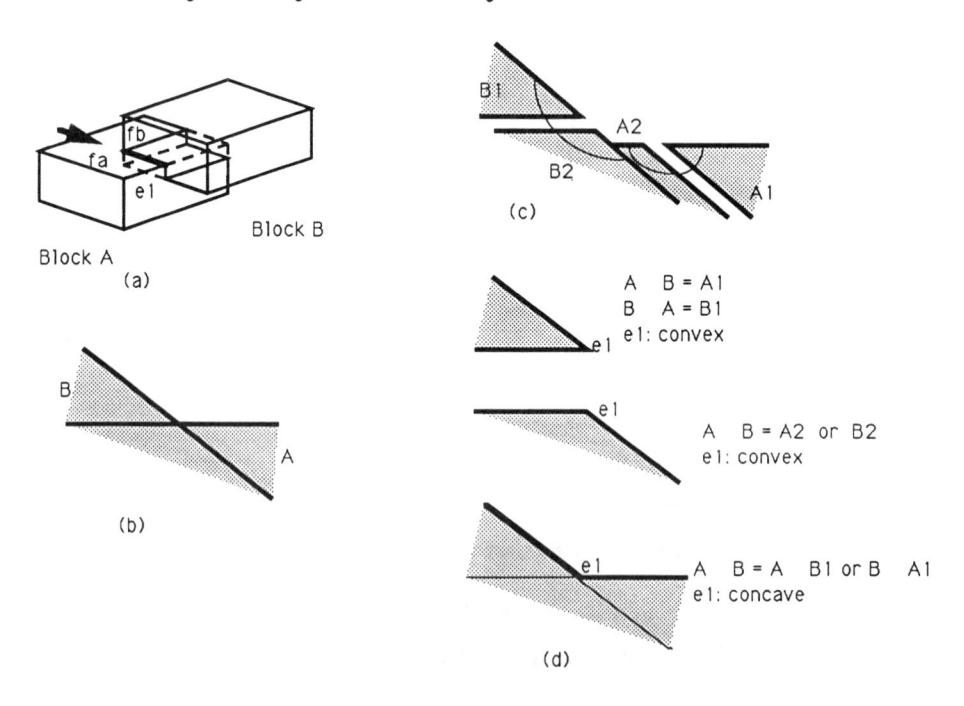

Figure 5. Convexity of Solid Intersection

Extraction of Manufacturing Features from an I-DEAS Universal File

JONG-YUN JUNG and RASHPAL S. AHLUWALIA

Department of Industrial Engineering
West Virginia University
Morgantown, WV

Abstract

I-DEAS, a Computer Aided Design (CAD) software package, generates a Universal file containing geometry and topology information of objects. The Universal file also has the history of boolean operations resulting in an object. This paper presents a scheme for extracting manufacturing features from the Universal file. The feature extraction technique is based on rules such as a boolean difference of a cylinder and a block, at the proper position, results in a hole. The paper presents an example of generating manufacturing features for machining applications.

Introduction

Integration of Computer Aided Design (CAD) and Computer Aided Manufacturing (CAM) is critical to achieving higher levels of productivity in small batch manufacturing. Small batch manufacturing accounts for 80% of all manufacturing in the U.S. Currently, there are islands of automation which are not appropriately linked. Typically, a CAD system provides part topology and geometry information, which is critical for generating manufacturing plans. Part geometry can be divided into geometric entities called features. Features have been defined in many ways, a broad definition being that it is a region of interest on the surface of a part [1]. A manufacturing feature would have meaning with respect to manufacturing operations. This paper describes manufacturing features extraction from a Universal file generated by the I-DEAS CAD software.

Many research results have been reported in the area of feature extraction. The FEATURES [2] system is based on logic programming to extract high level knowledge in the form of part feature definitions from stored part description. It uses ROMULUS, which is a boundary representation scheme. Exhaustive search is required to match features, computational time is therefore significant when many features exist. Attributed Agency Graph (AAG) [3] reduces computational time. It defines convexity or concavity for every adjacent two features, the system is limited to polyhedral features. Augmented Topology Graph (ATG) [4] describes shape features using a

---

[1] Acknowledgement- This work has been sponsored by the Defence Advanced Research Project Agency (DARPA), under the control No. MDA972-88-C-0047 for DARPA Initiative in Concurrent Engineering (DICE).

graph grammar, which relates the connection of concave and convex. ATG recognizes features by parsing against the graph of objects. Woo [5] uses volumetric decomposition scheme. Alternating Sum of Volume (ASV) uses the idea of volume equality between a cavity and sum of each feature primitive volume. ASV utilizes decompositions by using form features for structural analysis. This system recognizes cavity to transform volumetric designs of part into descriptions for NC machining. ASV has no semantic information with respect to volumes.

Geometric Modeling

There are several types of graphic representation schemes in geometric modeling. The most widely used are wire frame, boundary representation (B-rep), and constructive solid geometry (CSG). Wire frame is mostly used for two dimensional drafting. CSG tree is used for three dimensional solids. The CSG scheme uses regularized boolean operation of primitives, which has subtraction, union, and intersection operations. Figure 1 shows tree structure of CSG and its non-uniqueness. CSG can build the same model using different tree structure. B-rep scheme represents the oriented surface of objects in data structure which are composed of vertices, edges, and faces [6]. Figure 2 shows a non-manifold object, which is not considered as a valid object and does not satisfy Euler equation. The equation is used to check validness of objects in B-rep.

Universal File in I-DEAS CAD

The I-DEAS CAD package was developed by SDRC [7], it uses CSG representation scheme for the solids. I-DEAS generates a model file, a program file which contains user input command, and an Universal file. The Universal file contains all information on geometry and topology of objects. The file consists of several data sets and each of them having an identification number. The data sets are arranged by the identification number. Data set 407 in the universal file represents an object facet, which provides information on each facet and vertex point label [7]. Data set 408 is for object history tree control structure. It gives tree nodes and leaves information. In the object history data set, nodes are written as negative numbers and leaves are written as positive numbers. Data set 409 describes object history node which contains information on types of major and minor operation, type of transformation matrix, and label of the first and second child object. Data set 410 represents object history leaf. It has information on type of leaf, type of transformation matrix and its value. Data set 431 is for object faces. It gives label number and number of facets. Data set 433 gives minimum and maximum coordinates of an object. Data set 450 represents object header which contains surface area, volume, density, center of gravity, and principal axis of an object. Data set 520 gives coordinates of each points of an object. Figure 3 shows a section of Universal file with an object option for a cylinder. On the second line in the data set 450, it explains that the cylinder is one body composed of three surfaces, 32 points, and 18 facets. The next line represents the center point of the cylinder as the origin of coordinates. The other data is extra information for the cylinder. As shown in the Figure 3, all data sets are separated by '-1' in the fifth and sixth column.

Manufacturing Feature Extraction

This paper considers manufacturing feature extraction for milling operation. Raw material is assumed to be a block. It is assumed that only subtraction among regularized boolean operation is used to build an object. Union operation can be used, but we get extra lines on the object surfaces which are connected to each other after the operation. Unnecessary lines can cause confusion to the user. Manufacturing features such as holes, pockets, and steps are identified. These three features are most common for mechanical parts. Subtracting a cylinder from a main block at the proper location results in a hole. The dimensions of the cylinder and the main block come from data set 410. The data set also gives a matrix for transformation of the cylinder. If the height of a cylinder is larger than the height of a block and both sides of a cylinder are outside of a block, then the object is a through hole. If one side is in the block and other side is out side of the block, then the object is a blind hole. A blind hole and a through hole may look similar to the designer, however the manufacturing process is different. A through hole can be easily made by the drilling operation, but it is difficult to make a blind hole without additional manufacturing operations. A pocket may be a polygon, but it is assumed to be rectangular. The condition for pocket is that of subtracting a rectangular block from the initial block. The dimensions of the two blocks come from data set 410. The dimensional information after an operation comes from data set 409. The factor distinguishing a blind pocket from a through pocket is same as the holes. For the pockets, corner radius is a key factor in deciding manufacturing processes. Relatively large radius is easy to manufacture, but very small radii are much more difficult. A step is considered if a block whose width is larger than the initial block and both sides of the block are outside of the initial block. The subtracted block should be on one edge of the initial block. A step has freedom for manufacturing because there are many tool approach directions. If all conditions for each manufacturing feature are satisfied, we can extract them from the Universal file. Figure 4 shows each manufacturing features and location of substraction operation on the initial block.

Conclusion

The extraction of manufacturing features from geometry data is a difficult task. There are many feature extraction scheme, none of them have been successful for a general case. The manufacturing feature extraction scheme presented in this paper is not general either, however we suggest a scheme which uses topology and geometric information from a Universal file. We divide holes into blind holes and through holes. We differentiate pockets with relatively large radius for corners from pockets without corner radius. A designer can create a blind pocket without a corner radius, but the part would be extremely difficult to manufacture.

References

[1] Pratt, M.J. and Wilson, P.R., requirements for the Support of Form Features in A Solid Modeling System, Report No., R - 85 - ASPP - 01, CAM-I Inc., Arlington, TX, 1985

[2] Henderson, M.R., Extraction of Feature Information from Three Dimensional CAD Data, Ph.D. Thesis, Purdue University, West Lafayette, IN, 1984

[3] Joshi, S., CAD interface for Automated Process Planning, Ph.D. Thesis, West Lafayette, IN, 1987

[4] Safier, S.A. and Finger, S., "Parsing Features in Solid Geometric Models", European Conference on Artificial Intelligence-90, 1990

[5] Woo, T.C., Feature Extraction by Volume Decomposition, Proceedings of Conference on CAD/CAM in Mechanical Engineering, MIT, Cambridge, MA, 1982

[6] Hoffmann, C.M., Geometric and Solid Modeling, Morgan Kaufmann Publishers, Inc., CA, 1989

[7] I-DEAS Level 4 User Guide, SDRC, Milford, OH, 1988.

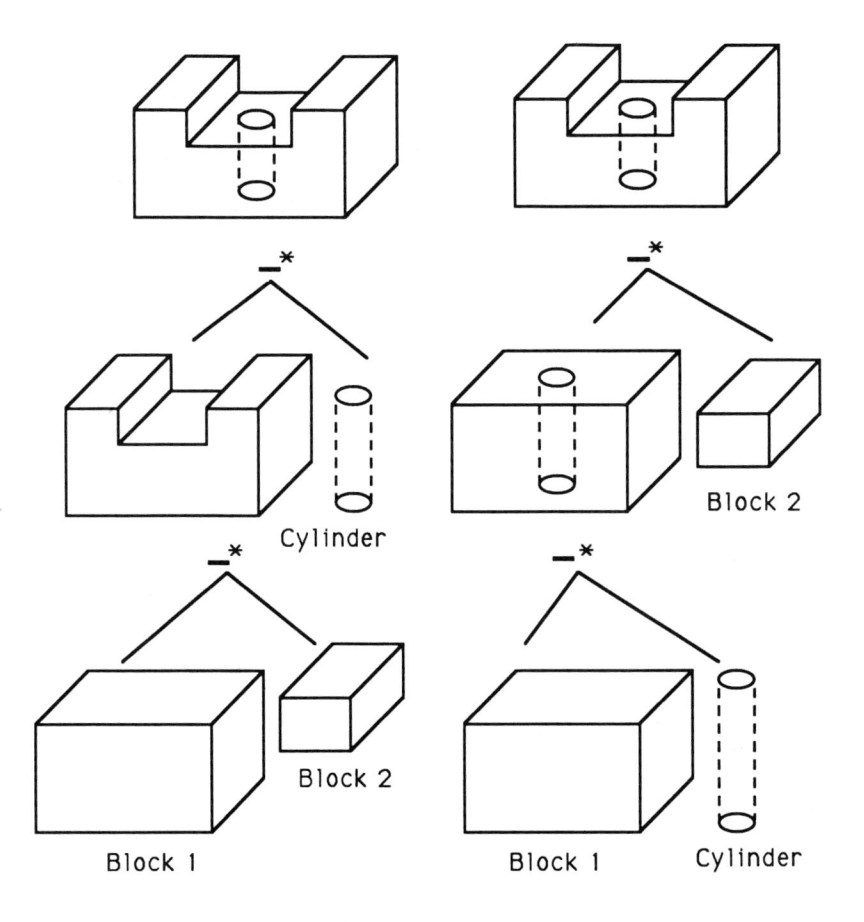

Figure 1 CSG tree structure and ununiqueness

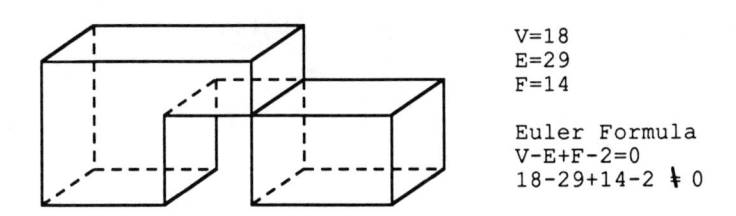

V=18
E=29
F=14

Euler Formula
V-E+F-2=0
18-29+14-2 ≠ 0

Figure 2 Non-manifold object

```
    -1
   450
        1CYLINDRR                                        1
        1           3           32          18
 0.00000E+00  0.00000E+00  0.00000E+00
-1.00000E+00 -5.00000E+00 -1.00000E+00  1.00000E+00  5.00000E+00
 0.00000E+00  0.00000E+00  1.00000E+00  0.00000E+00  0.00000E+00
 0.00000E+00  0.00000E+00  0.00000E+00  0.00000E+00  0.00000E+00
 0.00000E+00  0.00000E+00  0.00000E+00  0.00000E+00  0.00000E+00
 0.00000E+00
    -1
    -1
   520
        1    1.0000000E+00   5.0000000E+00    0.0000000E+00
        2    9.2387950E-01   5.0000000E+00   -3.8268346E-01
        3    7.0710677E-01   5.0000000E+00   -7.0710677E-01
        4    3.8268343E-01   5.0000000E+00   -9.2387950E-01
        5    0.0000000E+00   5.0000000E+00   -1.0000000E+00
        6   -3.8268352E-01   5.0000000E+00   -9.2387950E-01
```

Figure 3 A part of an Universal file for a cylinder

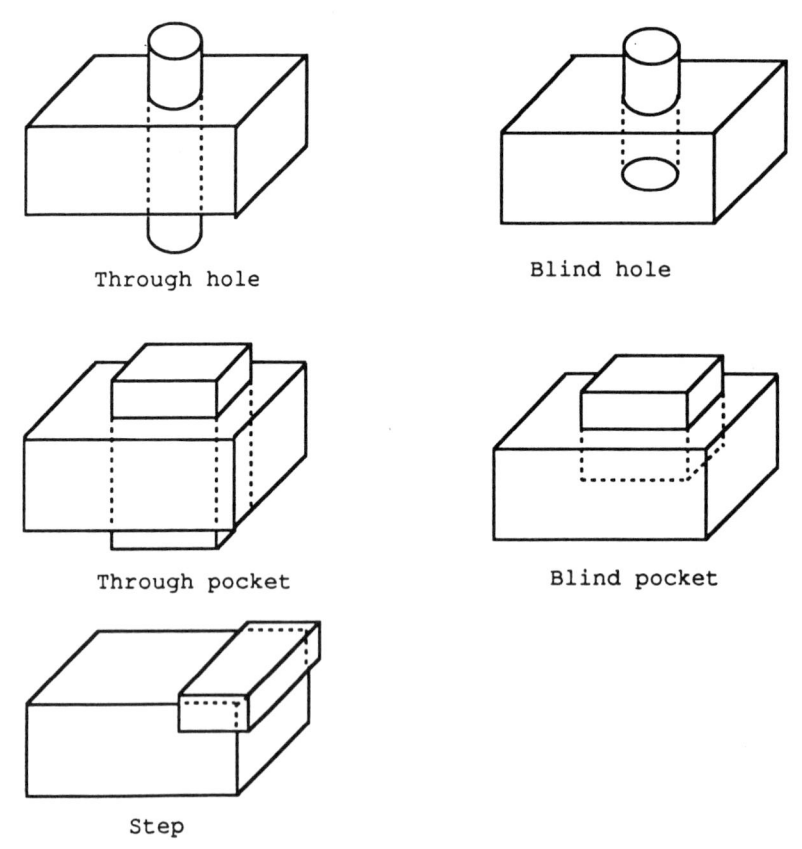

Through hole

Blind hole

Through pocket

Blind pocket

Step

Figure 4 Location of two primitives for substraction operation

Feature Based Design Assembly

SISIR K. PADHY and SUREN N. DWIVEDI

Department of Mechanical and Aerospace Engineering
West Virginia University
Morgantown, WV

Abstract

Assembly has been an important part in any production environment. Automation of assembly has been a major research issue in the last couple of years to achieve accurate assembly. The gross manipulation can be controlled but the problem of fine motion is very complex to handle because there is always uncertainty exists the problem becomes more complicated. The analysis of a assembly can be done by use of features effectively.

Features are gaining prominence in manufacturing processes as a potential aid to link CAD and CAM. An assembly can be modelled by features which can represent the design intent of the process. In this paper the contact and geometric effects of features in assembly are analysed. Geometrical representations schemes for the features are presented. Different contact strategies are considered and an area optimization concept is also presented to optimize assembly.

Introduction

Oflate there is considerable interest in the assembly automation and use of robots in the assembly tasks. As the need is growing for the complete automation of the factory of the future, fine manipulation of objects assembly, the contact problem and part mating theory becomes prominent in the assembly research domain. The fine motion planning the objective is to find a path that attains a position and orientation involving contact among components. Force control and compliance controlled assembly schemes for robots are proposed and found quite successful. But due to the contact uncertainties and errors due to control and sensing creeps in to the problem and make it more complicated. Contact formation can be viewed as different way in which a feature be in contact with another feature. The shape of the feature and contact information decides the assembly operation and the planning. This paper presents the feature-based approach for assembly.

Geometry Controlled Assembly

The contact analysis of features and their shape effect is the crucial step to understand the assembly (only the process of mating involving components, no adhesives, no welding and no fasteners) automation. A feature is defined as *a geometric entity that can be assembled*. Sub features are defined as a part of the the feature that makes contact with the another feature. The prediction of contact forces and motion of features with other surfaces is a very complex task when friction is accounted. The problem of determining contact forces are solved only for simple cases when friction is present. Moreover there is no simultaneous considerations of normal and tangential and torsional load although they can be modelled by hertz and coloumb law.

Feature Geometry

Feature is defined to be a volumetric shape that is assemblable in this analysis. Each feature occupies a space in the world and their spatial relation is the key issue in the assembly. Features can be classified in to certain categories. They can be treaded rigid, articulation and elastic. In this paper only rigid features are considered. All this features can be in contact formation and can be classified in to six types. They include:

a. Face ~ Face (FF)   b. Face ~ Edge (FE)

c. Face ~ Vertex (FV)   d. Edge ~ Edge (EE)

e. Edge ~ Vertex (EV)   f. Vertex ~ vertex (VV)

With these notations the configuration space (CS) for an assembly configuration for two features will be

$$CS = \begin{bmatrix} (F_{ij}, F_{mn}) & (F_{ij}, E_{mn}) & (F_{ij}, V_{mn}) \\ (E_{ij}, F_{mn}) & (E_{ij}, E_{mn}) & (E_{ij}, V_{mn}) \\ (V_{ij}, F_{mn}) & (V_{ij}, E_{mn}) & (V_{ij}, V_{mn}) \end{bmatrix}$$

With the contact formation whether assembly is possible between two features can be analysed using feature reasoning. This is described in subsequent paragraphs.

Feature Reasoning

Assembly can be analyzed by feature reasoning. The feature reasoning can determine simple features that are likely to be mating sites during assembly. It is likely that a feature with concave attribute will be assembly site for a feature with convex attribute. In otherwords a component with concave feature may be a possible mating surface for a component with convex feature. A feature is convex or not can be determined by Simon's algorithm. The criteria for convexity as as follows:

A set S in the space (N dimension) is convex if for any two points p_1 and p_2 such that $\{p_1, p_2\} \in S$, and any scalar value β such that $0 \leq \beta \leq 1$. Then if the point $p = \beta p_1 + (1-\beta)p_2$ and $p \in S$ then the set S is convex.

The geometrical interpretation of a convex set implies that a straight line drawn between any two points in the set lies entirely within the set. A hollow cylindrical feature is a concave set while the solid cylindrical feature is a convex set. It is also quite consistent that the solid cylinder may be inserted in the hollow cylinder for assembly. However this method does not constrain the assembly of two concave features. For instance according the definition a slot is a concave feature and a T-section is also a concave feature and the assembly between this two feature is possible.

Theorem 1.: *Assembly is possible between two concave features only if one concave feature contains a convex feature within it while the other concave feature has a concave sub-feature corresponding to the convex sub-feature.*

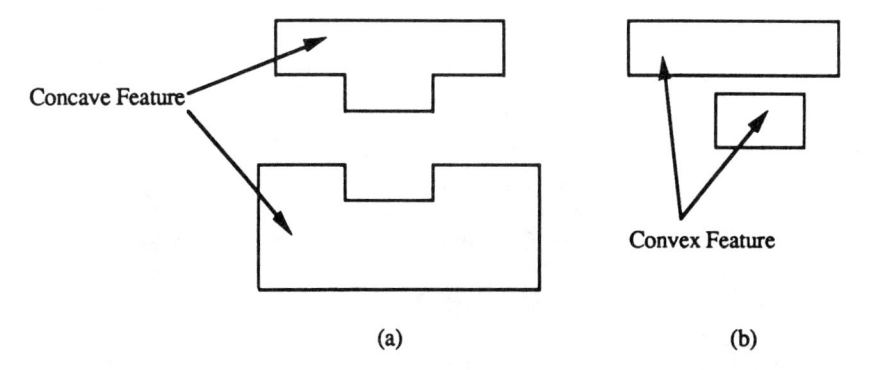

Figure 1. (a) Two Concave Feature; (b) T-feature consists of two Convex features

Proof: Consider a T-feature (feature 1)and a rectangular solid with a slot (feature-2) (Fig. 1). Now the feature 1 has two rectangular bars and these bars are convex. As the assembly is done one bar fits in the slot of the feature 2.

Corollary 1: A concave feature contains both concave and convex sub features.

Corollary 2.: A convex feature only and only contains convex sub-features.

Theorem 2.: Assembly in between two feature is possible if atleast one sub feature surface of one feature has one to one correspondence with the subfeature of the other feature with respect to a world frame.

Proof: Consider the ball and socket joint of a human body (Fig.2). The socket has the concave feature which is expressed as a concave function $\psi(x)$ and the ball has a convex feature which is expressed as a convex function $\Phi(x)$.

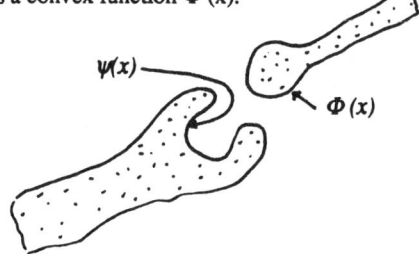

Figure 2. A Ball and Socket Joint.

Now for assembly to be possible the following condition has to be true:

$$\psi(x) = \lambda \Phi(x)$$

Now   if $\lambda < -1$; there is loose fit between the features.

if $\lambda = -1$; there is normal fit between the features.(Same Symmetry Group]

if $\lambda > -1$; there is interference between the features.

The minus sign is introduced to signify that the concave feature is opposite of that of the convex feature. With λ value of minus 1 the sub features have one to one correspondence while with value less than minus 1 the $\psi(x)$ is a function whose geometry is larger than that of $\Phi(x)$ and can encapsulate $\Phi(x)$. The -1 value correspondence to the bodies having same symmetry group.

The importance of this analysis can be appreciated if the robot has to assemble by intelligent reasoning using couple of parts. The best way is possibly to instruct or build an expertise with in the robot cell to assemble a triangle to a triangular slot or hole rather than to give all the combination. As indicated earlier it is the subfeatures met not the bodies or features. If there are two pair of features with m and n number of subfeatures with potential mating capabilities then the combination will be a matrix of m by n i.e. the total number of combinations will be m multiplied by n. So if the subfeatures are recognized it will be easier to see a relation between them for probable assembly.

Area Optimization Concept

The assembly consists of a number of component and the components form a spatial reflation with each other to form the assembly. The total area of the the components is the sum of all individual area before assembly.

$$A_c = \Sigma A_{oi}$$

However after assembling the components the area of the assembly is less that that of without assembly. For example considering a peg in the hole case the assembly is optimized when the peg has traversed the whole depth. In other words if the inner surface of the hole is excluded from area calculation in the assembly then the hole has a perfect assembly. However the final desired shape of the assembly is the first criteria and the area optimization should be done keeping that sense. This method can differentiate between a loosely fit assembly to that of the fit assembly. Another aspect is the minimization of number of component to optimize the area of the assembly.

Conclusion

The contact and geometric aspects of features in the assembly are discussed. The separation cones for the uncertainty in fine motion are discussed. It is observed that the concept is very useful to achieve fine motions to perform a precision assembly. The basic theorems for the features to qualify for assembly are proposed and proved. Symmetry group concept is presented that can eliminate the random search for an automated assembly by robot. An area optimization concept is suggested to determine the optimum assembly.

References

1. Popplestone, Robin J., Liu, Yanxi and Weiss, Rich.:"A Group Theoretic Approach to Assembly Planning," AI Magazine, Vol. 11, No. 1, pp 82-97,Spring 1990.

2. Vijaykumar, R.and Arbib, Michael A.:"Problem Decomposition for Assembly Planning", IEEE Trans, 1987.

3. Rocheleau, D.N. and Lee, K.:"System for Interactive Assembly Modeling", CAD,Vol 19, No. 2 ,pp 65-72, March 1987.

4. Padhy, Sisir K., Sharan, R. and Dwivedi, S.N.:" Design with Features at Conceptual Stage", Proc. Second National Symposium on Concurrent Engineeriing, Morgantown, WV, Feb. 7-9, 1990.

Feature Based Machining Analysis and Cost Estimation for the Manufacture of Complex Geometries in Concurrent Engineering

B. GOPALAKRISHNAN and V. PANDIARAJAN

Department of Industrial Engineering
West Virginia University
Morgantown, WV

Abstract

This paper describes the feature based evaluation of product designs based on machinability and the estimation of costs. The concepts related to the designation of features in products is described. The differing modes of approach are examined for simple parts and parts with complex geometries. The methodologies for performing machining analysis and cost estimation are described.

Introduction

Product designs are accomplished by describing a product in terms of coordinate points in a three axis framework, mostly in a computer aided environment. The geometry, defined in this manner is supplemented with other product attributes such as the material and quality characteristics, to fulfill the design requirements. The design then undergoes substantial testing and simulation to determine its strength, endurance, and other desirable characteristics as it functions in the designated environment.

Concurrent engineering can be termed as the systematic approach to the concurrent design of products and their related processes, so that considerable reductions in time and cost can be achieved at both levels [4]. In the domain of concurrent engineering, product design and process planning can no more be performed in a sequential fashion, as the influence of process planning activities is mandatory for the production of satisfactory product designs.

Acknowledgement
This work has been sponsored by the Defense Advanced Research Projects Agency (DARPA), under contract No. MDA972-88-C-0047 for DARPA Initiative in Concurrent Engineering (DICE).

Concurrent Engineering Approach

The President's Blue Ribbon Commission on Defense Management has discovered that even in weapon systems, the product development times are too high, and the functional aspects of the system are not up to expectations [3]. This indicates the need for concurrent engineering in many diversified industrial applications.

Concurrent engineering can also be defined as the merging of the efforts of product designers and manufacturing engineers to improve manufacturing processes and products with concurrence, constraints, coordination, and consensus as the primary ingredients [2]. Some aspects of computer aided concurrent engineering have been developed specifically in terms of an analysis of performance and cost [1], the consideration of reliability, maintainability, and supportability being the main focus of the research. These system level concepts would be successful only if the design is satisfactory in terms of manufacturing and assembly constraints.

Product Design and Manufacturing in Concurrent Engineering

The evaluation of product designs based on manufacturability, specifically machinability, is mostly with regard to their geometry and material characteristics. Complex geometries are difficult to machine, and machinability then depends upon the capabilities of the existing machine tools, and the skill of the operators. The quality of the finished product may not be satisfactory for the machining of complex geometries, irrespective of the sophistication of machine tools available.

The probable solution to this problem is to suggest alternate product designs which are more machinable. This means the alteration of the product geometry by the manufacturing engineer, in order that certain areas which are difficult to machine are rendered machinable.

The material characteristics affect machinability profusely, since hard materials produce high tool failure rates and lesser control over machining processes. These "difficult to machine" materials pose increased machining costs and higher production times. The evaluation process should be able to provide a cost estimate for machining which would reflect the extent to which material characteristics have played a role in product machinability.

The cost estimation of the design should be sensitive with respect to the level of the product design. Preliminary designs do not offer as much information about the product as detail designs, and cost estimation would obviously be less accurate for the former as compared to the later.

The evaluation of product designs, which is termed as the machining analysis, is a function of the product geometry, as described above. Product geometry, if grouped into product features, offer a new dimension in the way products are described. A product feature may be defined as a distinguishable entity in the product, which conveys a sense of form to it. For example, a cylinder of a certain diameter and a certain height may be described in terms of coordinate points on a CAD system, or could be alternatively described in terms of features such as center on the top face circle, center on bottom face circle, cylinder height etc. In each case, the features would have to be qualified in terms of their values, whether symbolic or numeric.

Products with Simple Features

A product with simple features, such as holes, pockets, grooves, and slots, can be subject to machining analysis by evaluating the product features for machinability and by estimating the costs of machining. Sharp corners, holes with large L/D ratios, deep slots, and other such features cannot be machined easily. In such cases, the modifications to the features may be suggested to the designer, such as increasing the radius of corners, decreasing the L/D ratios of holes, decreasing the depth of slots etc. These modifications to the design should however be acceptable to the designer, based on requirements apart from machinability.

Empirical formulae may be used to estimate costs for the machining of the individual features, using several processes and machines or using one NC (Numerical Control) machine and one setting. The length of the tool path plays an important role in the function for cost estimation in all cases.

Design of a Product with Complex Geometries

Products with complex features, such as the turbine blade, are designed in sections, using a B-spline curve to represent the profile cross section. The surfaces are all sculptured, and one cannot designate any commonly known feature types to the object. There are no grooves, pockets, holes, or other well defined features which can be attributed to airfoil shape of the turbine blade. In this case, machining analysis is difficult to perform, in regard to the suggestion of machinable alternatives and cost estimation. At the preliminary design stage, the form of the product being designed is mainly in terms of coordinate points. For instance, for the design of a generic airfoil, the suction profile would be in terms of points on the camber line and the two enclosing profile curves. Two such suction profiles, placed at a distance apart, forms the basis for the shape of the airfoil. At this stage, not much information is available on the final airfoil shape, material, dimensions, or the quality attributes.

The blade is designed in sections, each section being created in the CAD system as a closed B-spline curve, generated from a series of coordinate points. The created sectional profile is extruded along one axis to create the solid section of the turbine blade. Other sections with differing sectional curves and extrusion characteristics are combined using Boolean operations to create the final version of the blade.

For the purpose of analysis, we consider one particular section of the blade. This sectional curve has been extruded for a length of 2 units. The design was conducted using the TRUCE system (GE Proprietary software for solid modeling) on the VAX workstation.

Methodology for Providing Manufacturable Alternatives

The modification of the B-spline curve was considered for the generation of a possible alternate design which might have better manufacturability than the preliminary design. It was the intent to generate an alternate profile not far in shape from the original design, yet being more machinable, in terms of production cost.

Extreme convexities and concavities in the shape of the profile curve mean that as the tool travels along them, the feedrate has to be reduced for providing stability to the cutting tool. Slower feedrates mean higher machining times and higher machining costs. To reduce the extreme concavity or convexity between any three consecutive points on the profile curve, the angle between the surface normals for the first two points and for the second and third point was determined, and reduced by moving the middle point suitably.

The angle between surface normals was considered as a barometer for the measurement of concavity or convexity between any three points on the profile curve, and input by the user, as far as an acceptable maximum value is concerned. It is obvious that more the value for this angle prescribed by the user, less modifications will be done to the profile curve. This process is repeated for all the points in sets of three and several iterations are performed until steady state is reached regarding the stability of the coordinate points.

The angle between surface normals, the direction of the normals, the distance between surface points, and the number of surface points were found to be useful features in the domain of machining for complex geometries. A feature based machining analysis was effectively performed using these features. Thus, there is a world of difference between the designation of features for products with simple geometry as opposed to those with complex ones.

The machining time for the modified balde profile was much lesser as compared with the original design in most cases. This verification was accomplished using NCS (GE 5 axis NC Simulation Software). The designer at the preliminary or detail design stage should consider the modified profile as an alternative and perform tests to determine its feasibility and acceptability. This process when performed using the methodology described above, is likely to result is a robust, manufacturable design.

Methodology for Providing Cost Estimates

Cost estimation can be done in an approximate fashion at the preliminary design stage, and at a more accurate level at the detail design stage, as more information becomes available. The length of the tool path can be determined by the length of the profile curve, generated from the features, multiplied by the number of passes made to remove the material completely, and to generate the profile along the solid surface. The length of the tool path is an indicator of the cost of milling at the preliminary design stage, while the failure of the tool and its associated costs are an additional factor at the detail design stage, with the workpiece material being known. At the preliminary design stage, tool costs may be considered by identifying various workpiece materials and determining the costs for each one of them.

The system named PDEM has been developed using FORTRAN 77 on the VAX workstation. This system enables the modification of designs to present manufacturable alternatives, and estimates production costs. In testing the system, the results have been commendable, in that machining times have been substantially reduced with minimal alteration to the general form of the original design. Although the dimensions may have been changed, the general form of the design has been retained in all cases considered.

Conclusions

Concurrent engineering is an exciting concept that is certain to provide tremendous benefits when implemented. Manufacturability is one of the key issue that the designers need to be informed, and the aspects discussed in this paper pave the way for successful incorporation in this regard. The methodologies discussed for the evaluation and the generation of alternate designs, as well as cost estimation are bound to lead to robust designs of products with low developmental times and costs. The discussion regarding the designation of features for complex and simple geometries provides an important step in the direction towards feature based design and manufacturing.

References

1. Fabrycky, W. J., Design for the Life Cycle, Mechanical Engineering, Vol. 109, No.1, Jan. 1987.
2. Stauffer, R. N., Simultaneous Engineering : Beyond a Question of Mere Balance, Manufacturing Engineering, (1988), pp. 43-46.
3. Sullivan, L. P., Quality Function Deployment, Quality Progress, (1986), pp. 39-50.
4. Winner, R. I., Pennel, J. P., Bertrand, H. E., and Slusarczuk, M. M. G., The Role of Concurrent Engineering in Weapons Systems Acquisition, Report R-338, Institute for Defense Analysis, 1988.

Use of Part Features for Process Planning

S.K. GUPTA, P.N. RAO and N.K. TEWARI

Mechanical Engineering Department
Indian Institute of Technology
New Delhi, India

Summary

Proposed paper discusses a feature based process planning system. This system uses part features as medium of communication between geometry description and process planning functions. Component is defined by the user in terms of its geometry and features to be machined. Component database is expressed in terms of feature codes and their parameters. For process planning, the system supports the facilities for machine tool selection, setup planning, tool selection and operation sequencing. Tool selection module makes on-line dialogue with user and informs the status of tool search. For operation sequencing, besides functional requirements, two type of feature interactions are also examined. Examination of these interactions helps in taking the corrective actions during part program generation and hence generating optimum operation sequences. Design of the system is based on the integrated database approach and is of direct relevance to many of the integrated CAD\CAM activities.

1. Introduction

Of late the concept of using form features (shape elements) for integrating CAD\CAM is receiving attention of researchers. Increasing attempts of integration required a common communication medium between design and manufacturing activities and features were found to be suitable for both the applications. Features represent a collection of entities in an intelligent form that match the way engineers think and hence provide information at a higher conceptual level than the purely geometrical representation like lines, arcs, and text used by current CAD/CAM systems. Features allow both design and manufacturing engineers to perform their separate tasks using a language they have always shared, the language of features - "holes", "slots", "pockets". [1,2,3]

Currently there are two major schools of thought as to the kind of CAD data to use as the input. One approach is to take a general CAD model (in order to provide complete and unambiguous data, solid models are used) and develop an interface to recognize the manufacturing features from this model. The advantage here is that a solid modeler can be used for design. However the recognition is a non trivial problem, as it is extremely hard to recognize complex features.[4]

The other approach uses a specially designed CAD model incorporating shapes that are immediately recognizable by the manufacturing planner. These familiar shapes or features are designed around manufacturing operations. Such a feature based design environment limits the designer to the available manufacturing features, which are by definition feasible for manufacturing. By using a feature model, one can ensure that the part designed are manufacturable. In the latter approach where a component is modelled in a dedicated environment (also called as feature based design) offers certain distinct advantages over feature extraction type systems. Currently most of the feature recognition systems employ pattern matching for feature recognition, therefore the system is able to recognize only limited number of features (only features whose pattern is programmed can be identified), hence flexibility available at front end (design or modelling) is limited. Since most of the general purpose modelling systems overlook the tolerance information, this is to be separately appended to the database after feature extraction before these feature can be directly used for manufacturing applications. This causes severe bottleneck in the way of achieving integrated CAD/CAM systems.[2,3]

2. Overview of System

The CAPP system, presented here, is for generating process plans for the prismatic components. Some of the important characteristics of the system are the following :

1. Complete system is based on a set of manufacturing features, which can be tailored to specific requirements and all modules of the system are utilizing a central database which is created interactively by the user and is expressed in terms of manufacturing features.
2. Throughout its operation the system makes dialogue with user. Hence user gets the insight of the system and also the necessary assistance for trouble shooting.

This system consists of eight modules. First module is master module while other modules are used for part description and various process planning functions. Various system modules are described below:

2.1 Description Of Part Geometry

Part description information is presented in component database (central database used by all other modules for their operation) which includes part description as described in the following five sections:

(i) External Geometry : External geometry describes the basic shape of the part to which features to be machined are attached. System supports parallelopiped shapes and extruded solid shapes with polygon bases.

(ii) Internal Geometry : Internal geometry describes the internal shape element of the component. System supports blind cavities (not having opening on any of the surfaces, holes and other small accesses are considered exceptions) and open cavities (having major openings in one or more surfaces).

(iii) Work Surfaces : Surfaces, needing any kind of machining are called work surfaces. On these surfaces, locations of all manufacturing features are defined in local coordinate system.

(iv) Locating Elements : This is an optional information. If certain pre-machined surfaces or features are known, user should specify these surfaces as locating surfaces (also called datum surface).

(v) Features to be machined : The system supports a set of basic manufacturing features. At present, the list of basic features is prepared, based on the survey of common prismatic parts. Some of the basic features are shown in Fig. 1. These basic features may be combined to form complex features.

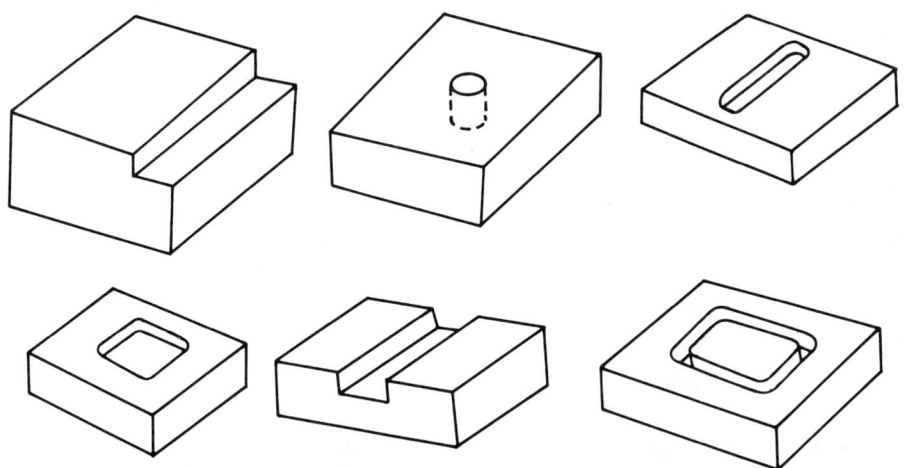

Fig. 1 Some of the available basic features

2.2 Related database

Besides the component database, the system uses two additional databases. First one is Machine Tool Database, consisting of two separate files. The first file has information pertaining to the machine tool capabilities or functional details while the second file is status file consisting of the current status of machines (information related to machine tool availability). Second database is tool database. System supports a set of tools to be commonly used on a machining center.

2.3 Machine Tool Selection

The selection procedure involves three stages.
STAGE-1: Selection of machine tools based on availability
STAGE-2: Selection of machine tools based on functional constraints
STAGE-3: Optimization

2.4 Setup Planning

Setup planning activity includes sequencing the order of work surfaces machining, identifying the locating elements for all the work surfaces and, if necessary, identifying the qualifying surfaces. This module uses a hybrid approach, making use of backward planning as well as forward planning. Program generates all possible sequences of machining the work surfaces. Using backward planning approach, each sequence is tested and best sequence is identified based on the minimum qualifying surfaces criteria. If all the surfaces are not satisfying the locating element constraints, then by forward planning approach program identifies the qualifying surfaces for the machining of job. This part considers the best sequence of machining the work surfaces and identifies the qualifying surfaces for all the work surfaces, which were not satisfying the locating element constraints.

2.5 Tool Selection

If a required tool is not found for a particular feature, then System informs the user about the tool it could not find and the name of the feature and its location. If no tool could be found for a particular feature, user has the option to either change the parameters of the feature or procure the necessary tool and add this information to the tool database. For all the features, tools are selected by backward planning. When a tool is searched for a feature, the program first searches tool for the final operation needed to produce that feature. The required pre-operation is returned by all successful search. Until a search returns no pre-operation, program continues searching tools for the same feature. When a successful search returns no pre-operation, the system proceeds for the next feature. If a search for a tool fails, user is informed and further search for that feature is aborted and the program continues tool search for the next feature.

Tool can be searched in database by three types of search calls:
1. This search returns the tool of specified key parameter (diameter in most of the cases). For example, this type of search call is used for the drills and other hole making tools.
2. This search returns the tool having greater or equal key parameter than specified value. For example, this type of search call is used for face mills, shoulder mills etc.
3. This search returns the tool having lesser or equal key parameter than specified value. For

example, this type of search call is used for groove making tools.

2.6 Operation Sequencing

Based on the geometric location and mutual interaction of the features, system generates the sequence in which to machine the features on a particular work surface. For sequencing the operations, three major considerations are used by this module :

(i) Feature type : Features like Plane face are to be machined first.

(ii) Geometric location of feature : Features are sorted into different layers depending upon their Z-coordinates of location. Features with same Z-coordinate (location) are considered in same layer. Various layers are arranged in increasing order of depth of layers (Z-Coordinates) with respect to work surface. Features lying in the layer with least depth are machined first.

(iii) Mutual interaction of features : Among the features lying on the same layer two type of interactions are considered :

(a) Features overlapping with each other : If two features are overlapping with each other (to form complex feature), then the following sequencing rules are followed :

RULE 1: Feature with larger depth is to be machined first.

RULE 2: If depth of both the features is same then feature with higher area is to be machined first.

This kind of interaction, explained in Fig. 2, provides information about the common area (in order to avoid repetition in machining) of the two features.

(b) Features lying too close to each other (features are not overlapping but wall thickness between the two is very less):

For this kind of interaction following rule is followed :

RULE : Feature with higher area is to be machined first.

This kind of interaction is explained in Fig. 3. Second type of interaction provides information to help in deciding the machining parameters of the second feature in order to avoid the deformation of thin wall between the two features during machining of second feature.

3. Programming Languages Used and Operating Environment

Part Description Module is developed inside AutoCAD (Release 9.0) using its internal programming language AutoLISP. Setup Planning Module is developed in TURBO PROLOG (Ver 2.0). Master Module and rest of the modules are developed in Quick Basic (Ver 4.0).

4. Discussion and Conclusions

Special consideration has been given to provide a user friendly interface to the system. At present, the system is making use of limited number of features to model the typical prismatic components to be machined on a vertical machining center. Work is under progress to make the available features compatible with CAM-I report.[5] Setup planning and feature interaction capabilities are also being enhanced to make the system more general purpose.

During the last decade, trend has shifted from interfaced automation to integrated automation. Feature based design systems help in bringing the manufacturing considerations into the design phase, and the coordination and integration between design and manufacturing is partially built-in, and thus the need for process planning function is partially eliminated. [6]

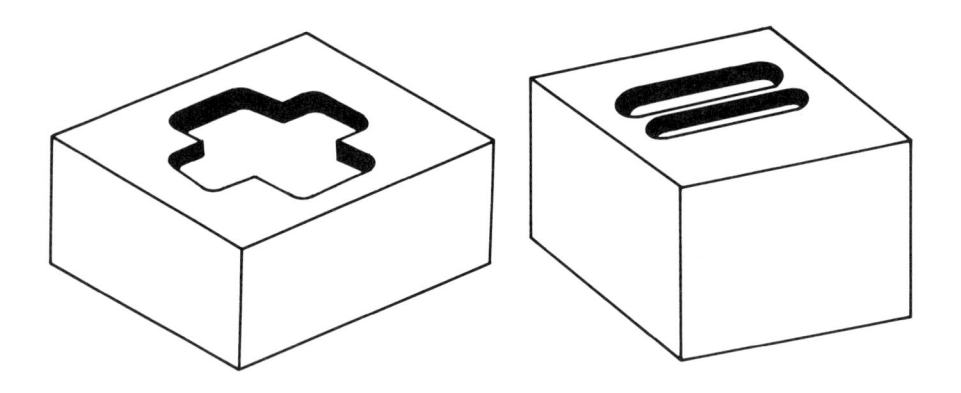

Fig. 2   Overlapping features

Fig. 3   Features lying too close

REFERENCES
1   Klein, A.: A Solid Groove Feature Based Programming of Parts. Mechanical Engineering, March 1988, p. 37.

2   Clark, A.L. and South, N.E.: Feature Based Design of Mechanical Parts. Proc. AUTOFACT, Nov. 1987, Detroit, p. 1-69.

3   Drake, S. and Sela, S.: A Foundation for Features. Mechanical Engineering, Jan. 1989, p. 67.

4   Gindy, N.N.Z.: A hierarchical Structure for form features. Int. J. Prod. Res., 1989, vol. 27, No. 12, pp. 2089-2103.

5   Butterfield, W.R., Green, M.K., Scatt, P.C., and Stoker, W.J.: Part Features For Process Planning. Report C85, 3,CAM-I Inc, 1987, Arlington, Texas.

6   Ham, I. and Lu, S.: Computer Aided Process Planning, The Present and the Future. Annals of CIRP, 1988, vol. 37/2, p. 591.

Chapter VI
CAD and FEM

Introduction

A new method is described for 3-D curved object-recognition of curved and planar surfaces by matching the scene and model surface primitives. The method employs conversion of orientation and position information into Euler parameter, eigenvalues, and translation parameters. The matching algorithm not only establishes the identity of the object but also provides the rotational and translation components which are used for robot manipulation. In the second paper of this chapter, an automatic algorithm for evaluation of model characteristics for unsymmetric turbine blades is developed with a few parameters outside the conventional procedures. In addition, the airfoil profile is generated using cubic splines. A computer-based life prediction method is discussed in the next paper to estimate the fatigue life in the initial design phase for material selection. Composite shell theory has been incorporated in the program to perform stability analysis for various loadings.

Model Based 3-D Curved Object Recognition Using Quadrics

M. HANMANDLU, C. RANGAIAH
Department of Electrical Engineering

K.K. BISWAS
Dept. of Computer Science and Engineering
Indian Institute of Technology
New Delhi, India

Summary

A new method is presented for 3-D curved object recognition by modeling the scenes and models as quadrics. The parameters of quadrics are converted into features which include Euler parameters accounting for the orientation of the surface. The matching process consists of finding a transformation of the model surfaces and devising a measure of consistency such that the error between the feature spaces of the model and the scene is a minimum.

INTRODUCTION

Model based methods are widely used in robot vision[1,2] as these lessen the requirement of multiple views for recognition by way of knowledge stored in models. Different methods in this category vary in the manner of representation of surfaces and of matching scheme chosen. Fisher[3] gives a detailed account of methods related to 3-D object recognition. Our treatment of the problem follows the guidelines in [4,5] but differ from it in the representation where we use quadrics as opposed to planes and in matching where we use Euler parameters instead of normals and depths.

REPRESENTATION AND MATCHING

In this method a curved surface is represented by the quadratic equation or a quadric[6]:

$$ax^2 + by^2 + cz^2 + dxy + exz + fyz + gx + hy + iz + j = 0 \qquad (1)$$

In matrix form, this becomes

$$Y^t D Y + E Y + j = 0 \qquad (2)$$

The unknown parameters can be obtained by least squares method. The location (ie., surface origin) is obtained from (2) after translating the variables in Y such that E=0. Accordingly, we have

$$Y'^t DY' + k = 0 \tag{3}$$

where k is a translation parameter which is related to parameters in (1). The definitions and computation of various vectors and matrices in equations (1) to (3) are refered to [6],[10].

The orientation is obtained from a matrix P which makes D diagonal in (2). The diagonal values of D are called eigenvalues. P has eigenvectors corresponding to eigenvalues of D as its columns.

The segmentation of the surfaces is done by finding equidepth contours (EDC) which yields step, semistep and roof edges[7]. The surfaces enclosed within the edges are modeled as quadrics. Planes are treated as a special case of quadrics. Similarly, all the model surfaces are treated as quadrics. The matching consists of finding a transformation which includes both rotation R and translation T. The rotation is made with respect to an axis of reference system and is to be followed by the translation. We need to rotate the orientation of the model to match with that of the scene. Similarly, we need to translate the model surface origin to match with that of the scene. In terms of these operations, the error measure is chosen as:

$$g(M) = \sum_i ||P_i - R P'_i R^t||^2 + w ||k_i - k'_i - T^t D'_i T||^2 \tag{4}$$

where w is a weighting factor. P_i, P'_i represent the orientations of scene and model surfaces respectively. These are treated as rotation matrices in the sequel. D'_i matrix belongs to the model surface.

For the calculation of R and T we have to minimize the sums

$$F_r = \sum_i ||P_i - R P'_i R^t||^2 \text{ and } F_t = \sum ||k_i - k'_i - T^t D'_i T||^2$$

respectively. The minimization of the first sum gives R. Since R bears a nonlinear relationship with rotation angle, we use quaternion approach to simplify the minimization. A quaternion is a 4-D vector and represents a finite rotation of an axis v in space by an angle ∅. It is defined as [8]

$$\alpha = a_0 + a_1 i + a_2 j + a_3 k = a_0 + a \tag{5}$$

Where ai (i=0,...3) are real parameters. A quaternion yields Euler parameters when $\|\alpha\| = 1$. Interms of these parameters R can be easily found in[8]. An important result is the multiplication of two quaternions :

$$\alpha \ (x) \ \beta = (a_0+a) \ (x) \ (b_0+b) = \alpha^+ \beta = \beta^- \alpha \tag{6}$$

For definitions see [8]. Using the above result, we can replace the rotation matrices Pi,P'i and R in the following by their equivalent Euler parameters L_i^+, L'_i^-, l_r respectively.

$$\|R\|^2 F_r = \sum_i \|P_i R - R P'_i\|^2 = \sum_i \|L_i^+ l_r - L'_i^- l_r\|^2$$

where we have multiplied Fr by $\|R\|^2 = 1$ for applying the result (6). The procedure to find Euler parameters corresponding to a rotation matrix is given in Paul [9]. The application of the result leads to [10]

$$F_r = l_r C l_r^t \tag{7}$$

where $C = \sum_i L_i L_i^t$ (a symmetric matrix)

with $L_i = L_i^+ - L'_i^-$

The minimum eigenvalue of C gives the error measure corresponding to rotation (e_r). The eigenvector corresponding to this minimum eigenvalue will yield Euler parameters from which we can obtain R.

The minimization of second sum is done by least squares method. In order to simplify the minimization here, we transform the equation such that D_i^2 becomes diagonal. In that case, the translation parameters are easily obtained. After the transformation with $PT = \hat{T}$, the second sum becomes:

$$\hat{F}_t = \sum_i \|k_i - k'_i - D'_i \hat{T}^2\|^2 \tag{8}$$

Taking
$$d_i = k_i - k'_i \tag{9}$$

$$\hat{W}_i = [\hat{D'}_1, \hat{D'}_2, ... \hat{D'}_i]^t$$

with $\hat{D}'_i = [D'_{i1}, D'_{i2}, D'_{i3}]$ = Diagonal elements of D'_i will lead to [10]

$$\hat{F}_t = ||d - \hat{W}\hat{T}^2||^2 \tag{10}$$

where $d = [d_1,..,d_N]^t$ is a N vector and $\hat{W}$ is a N x 3 matrix. Applying the least squares method to (10) will yield the solution

$$\hat{T}^2 = (\hat{W}^t\hat{W})^{-1}\hat{W}^t d \tag{11}$$

The resulting error is given by

$$\hat{e}_t = d^t(d - \hat{W}\hat{T}^2) \tag{12}$$

Actual translation matrix T can not be obtained from(11). We have to use another method given in[10] for that computation. But for recognition this poses no problem as we can still use the measure $\hat{e}_t$. The errors e_r and $\hat{e}_t$ are involved in the tree search procedure. The tree search[11] consists of numbering all the model and visible scene surfaces. Each number has a feature associated with it. For the purpose of matching, we take a scene surface and compare with all the model surfaces after transforming them. The model which yields minimum error is taken as the match for the scene. This procedure is continued for all the surfaces. We choose a specified error for backtracking. The result of transformation gives a rotation matrix, from which we can find the orientation of the object and a translation matrix, from which we can find the location of the object with respect to global coordinate system origin.

RESULTS OF IMPLEMENTATION

The proposed method is implemented on the range map of the object whose segmented map and edge map are shown in Fig.1a,1b respectively. Model is shown in Fig.1c with numbers indicating surfaces. The feature data consisting of Euler parameters, Eigenvalues and translation parameters is presented in Table 1 for a typical model surface and a scene surface.

We have made an exhaustive search for the first three levels and identified a few partial solutions with g(M) as the measure which consists of e_r and $\hat{e}_t$. In the initial phase, we have used a weighting factor of 2. The results for 3 levels are presented in Table2.

Table 1: Feature data of the model and the scene

Euler parameters	Eigenvalues	Translation parameters
0.4811 -0.7634 0.0963	0.2567 -0.3490 11.1810 -0.3490	-0.104993
0.6926 0.9364 0.3687	-0.3461 -2.5821 20.3535 0.0287	0.000048

Table 2: Partial solutions

Solution	Partial matching	Error measure
1	(1,2,4)	1.35 *
2	(1,3,7)	5.63
3	(1,6,7)	9.70
4	(2,1,3)	1.70 *
5	(2,5,3)	3.15
6	(2,6,5)	7.29
7	(3,1,6)	1.73 *
8	(3,2,7)	2.98
9	(3,6,7)	8.15

Table 3: Complete solutions

Solution	Matching	Error Measure
1	(1,2,4,6)	3.14
2	(1,3,7,6)	2.19 *
3	(2,1,3,6)	12.14
4	(2,5,3,6)	20.62
5	(3,1,6,2)	18.29
6	(3,2,7,1)	20.16

*selected solution

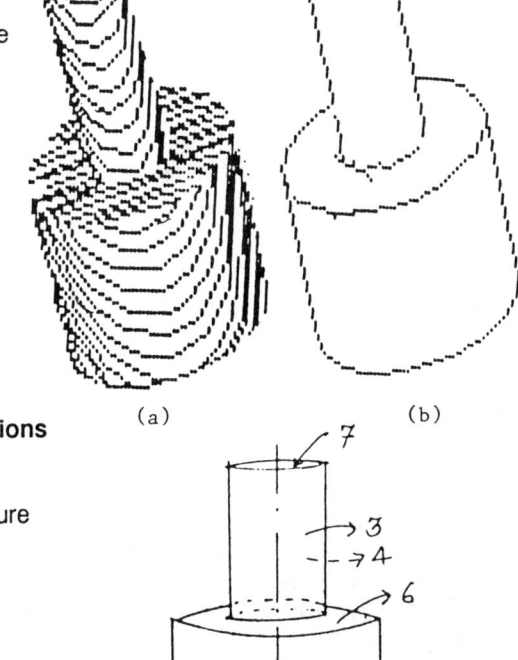

(a)   (b)

(c)

Fig.1 Details of scene and model

Based on these initial selected solutions, the search is extented to the 4th level,where we have used w of 6 to give more weightage for the translation error. Since we have four visible surfaces, our search concludes at the 4th level. At the end of 4th level,the matching results appear as in Table 3.

Table 3 shows that the final solution selected eventually consists of visible surfaces 1,3,7,6, viz.,bottom front cylinder surface,top front cylinder surface,top plane and middle plane respectively as identified in the model surfaces.

CONCLUSIONS

A new method has been presented for 3-D object recognition of curved and planar surfaces by matching the scene and model surface primitives. Knowledge of the object model in the form of quadrics has been assumed. It has been shown that by converting orientation and position information into Euler parameters,eigenvalues and translation parameters, the matching algorithm not only establishes the identity of the object, but will also provide as an important by-product, the rotation and translation components of the object so often required during robot manipulation.

REFERENCES

1. Chin,R.T.; Dyer,C.R.: Model-based recognition in robot vision, in ACM Comput.Surveys 18,1(1986) 67-108.
2. Besl,P.J.; Jain, R.C.: Three Dimensional object recognition, Comput. Surveys 17,1(1985)75-145.
3. Fisher,R.B.: From Surfaces to Objects:Computer Vision and Three Dimensional Analysis,John Wiley & Sons Ltd.,1989.
4. Faugeras,O.D.;Hebert,M.: A 3-D recognition and positioning algorithm using geometrical matching between primitive surfaces, in Proceedings,8th IJCAI,(1983)996 - 1002.
5. Faugeras,O.D.;Hebert,M.: The representation, recognition and locating of 3-D objects, The Int. J. Rob. Res. 3(1986)27-52.
6. Hall,E.L.;Tio,J.B.K.;McPherson,C.A.;Draper,C.S.;Sadjadi,F.A.: Measuring Curved surfaces for robot vision, IEEE Trans. Computer, (1989)42-54.
7. Wani, M.A.; Biswas,K.K.: A parallel algorithm for range image segmentation, communicated to IEEE Trans. Pattern. Anal.Mach. Intell. PAMI,1989.
8. Chou,J.C.K.;Kamel,M.; Quaternions approach to solve the kinematic equation of rotation, Aa Ax = Ax Ab of a robotic manipulator, in Proceedings, IEEE Int. Conf. Rob. Automation ,(1988) 656 - 662.
9. Paul,R.P.;Robot Manipulators : Mathematics, Programming and Control, MIT Press 1981.
10. Hanmandlu,M.;Rangaiah,C.;Biswas,K.K.:A new matching technique for 3-D object recognition communicated to Int.J.Comp.Vision,Graphics&Image Processing(1990).
11. Shapiro,L.;Haralick,R.: Structural descriptions and inexact matching,IEEE Trans. Pattern. Anal. Mach. Intell.,PAMI,3(1981)504 - 519.

Finite-Element Model for Modal Analysis of Pretwisted Unsymmetric Blades

N. T. SIVANERI and Y.P. XIE

Department of Mechanical and Aerospace Engineering
West Virginia University
Morgantown WV

Abstract

A quick and automatic algorithm for evaluation of modal characteristics for unsymmetric turbine blades is developed. Unlike conventional procedures, only a few geometric and material properties are needed in the present approach as input for a finite-element program by which the modal analysis is carried out. A cubic spline interpolation method is generated to prescribe the profile of airfoils so that related section properties are computed by means of numerical integrations. The finite element analysis, in this study, employs a special beam element with fifteen degrees of freedom, and uses a large-deflection small-strain Euler-Bernoulli beam formulation. Coriolis acceleration, centrifugal forces, warping effects, and rotary inertia are considered in the dynamic analysis. The effect on natural frequencies and modal shapes as changes to blade geometric and stiffness characteristics are analyzed. The method developed in this study can be utilized in the computer-aided design (*CAD*) for such blades.

Introduction

A designer of turbine blades needs knowledge of natural frequencies and mode shapes in order to avoid resonance and flutter. This knowledge becomes particularly important in the analysis of the dynamic behavior of thin and relatively long rotating blades. A relatively simple beam element is often adopted for the finite element model. The conventional methods of vibration analysis treating the blade as a uniform, straight, and cantilever beam in rotating or non-rotating state were presented by several authors [1-4]. The dynamic characteristic parameters, such as rorating frequencies and mode shapes, were obtained in the undeformed equilibrim system. Several kinds of beam elements with shear deformations and warping effect, which are extremly useful in the finite element analysis of turbine blades, have been studied [5-8]. In the existing algorithms, section properties such as area and mass moments of inertia are required as input for a computer program to evaluate the modal characteristics of a turbine blade.

The computer program developed in this study contains a library of airfoils that requires only the airfoil number as input in the case of standard airfoils. If the airfoil is not a standard one, the user needs to give several points in terms of x-y coordinates in the profile of the airfoil. The so-called piecewise cubic spline interpolation method [9] is developed to automatically generate the shape of the section. The area and moment of inertias are then calculated by means of numerical integrations. The analysis is carried out in dimensionless quantities so that the deflections and natural frequencies are normalized respectively by the length of the blade and the rotational speed of the turbine. The Campbell diagram is applicable to the geometrically similar blades since the natural frequencies are made nondimensional by the rotating speed. Also the centrifugal stiffening, Coriolis acceleration, warping, and rotary inertia are considered in the analysis.

Formulation

The turbine blade treated in this study is rotating at a constant angular velocity Ω. The Cartesian coordinate system x, y, z is set up to the undeformed blade at a precone angle β_{pc}.

The equations of motion are obtained from Hamilton's principle:

$$\Omega = \int_{t_1}^{t_2} (\delta U - \delta T - \delta W) dt = 0 \qquad (1)$$

where δU and δT are respectively the first variation of the strain energy and the kinetic energy, and δW is the virtual work contributions. At the constant rotating speed which is the case of this study, the δU, δT, and δW are independent of the time derivations. Hence, Eq. (1) can be expressed as:

$$\Omega = \delta U - \delta T - \delta W = 0 \qquad (2)$$

The energy expressions δU, δT, and δW are nondimensionalized by using $m_0 \Omega^2 R^3$ where m_0 is a reference mass per unit length. Terms that would be zero for the symmetric section are included in the first variation of strain energy, so that the dynamic analysis of nonsymmetric airfoils can be carried out.

Automatic generation of section properties

As a procedure of automatic generation, the program developed in this study requires the user to give several breakpoints of the x coordinates and their corresponding y coordinates so that the profile of airfoils can be found by means of a piecewise cubic spline interpolation method, and corresponding section properties are calculated by using traditional numerical integrations.

In cubic spline interpolation procedure, one determines $y_i(x)$ to interpolate the function between each two breakpoints x_i and x_{i+1}, so that at connected points of the polynomials, the function values, first derivatives, and second derivatives are continuous. Each cubic piece $y_i(x)$ is defined as:

$$y_i(x) = c_{1,i} + c_{2,i}(x - x_i) + c_{3,i}(x - x_i)^2 + c_{4,i}(x - x_i)^3 \qquad (3)$$

where the coefficients $c_{j,i}, j = 1, \ldots, 4, i = 1, \ldots, N$.

Generally, well spaced breakpoints are necessary in order to interpolate the airfoil profile accurately. Once the cubic piece $y_i(x)$ is known, the airfoil geometric properties can be evaluated. It has been found that the cubic spline interpolation is a highly effective procedure in modeling the airfoil profile. Further details on the definition of the profile center-line can be seen in reference [10].

The finite element discretization

The element forces can be obtained by using Hamilton's principle that is discretized as:

$$\Omega = \Sigma \, \Omega_i = 0, \qquad i = 1,\ldots, n. \tag{4}$$

where n is the element number. The assembly of element matrices results in global equations of motion in terms of the nodal degrees of freedom. The equations of motion are transformed to the modal space giving the equation as:

$$[M]\{\ddot{q}\} + [C]\{\dot{q}\} + [K]\{q\} = \{Q\} \tag{5}$$

where [M], [C], and [K] are the mass, damping and stiffness matrices, respectively. q is the modal displacement vector, and Q is the force vector. The displacements u, v, v', w, w' and ϕ are zero at the root. The details of the spatial finite element discretization are given in reference [11].

Results and discussion

(1) Standard symmetric airfoils

Two standard turbine blades (section 1 and 2) considered have length of 30", chord of 3", cross section of NACA 0009 and 0021. They are made of aluminum, and rotating at the speed of 10,000 rpm. It is evident that the change of thickness has very little effect on the rotating natural frequencies of the blade. Similar conclusions can be drawn for the mode shapes.

The tendancy of the natural frequencies with the increase of pretwist has been studied. The tip twist angle varies from 0° to 20°. Here again the effect of higher twist are to reduce the natural frequencies.

(2) Unsymmetric airfoils

Fig. 1 gives three typically unsymmetric airfoils, which are Sections 3, 4, and 5. The coupling of flap *(W)* and edgewise *(V)* displacements occurs for unsymmetric sections. For the first flap mode the change of twist angles at the blade tip does not significantly affect the natural frequencies. However, this effect may not be neglected for the second flap mode, as shown in Fig. 2.

For the first torsion mode, the increase of taper ratio and twist angle causes decreasing of the natural frequencies.

A designer of turbine blades can either change the tape ratio or the tip twist angle to alter the natural frequencies of the rotating blade with symmetric or unsymmetric sections. It is felt that increasing the tape ratio while keeping the section area constant could constantly decrease the value of natural frequencies. However in the case of changing pretwist of the blade, after reaching a certain value of the tip twist angle, the natural frequencies of the first flap mode almost remain constant.

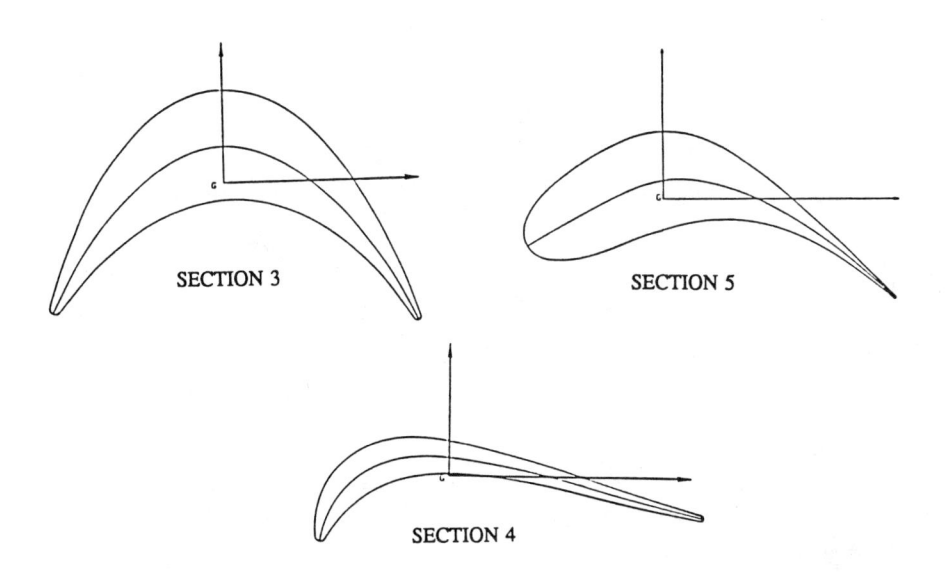

Fig. 1 Unsymmetric airfoils

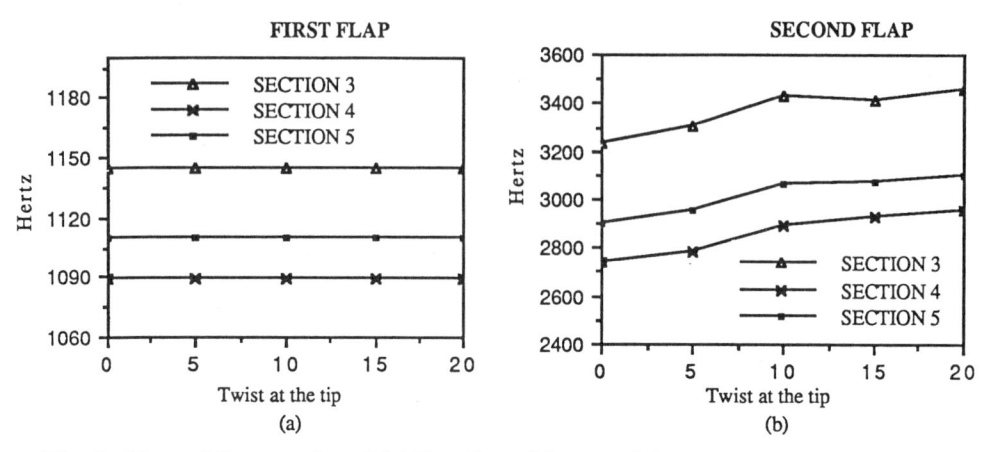

Fig. 2 Natural frequencies of (a) first flap; (b) second flap

References

1. Putter, S. and Manor, H. *Journal of Sound and Vibration*, Vol. 56, No. 2, (1978), pp. 175-185.
2. Straub, F.K. and Friedmann, P.P. *Vertica*, Vol. 5, (1981), pp. 75-98.
3. Rao, J.S. *Shock and Vibration Digest*, Vol. 12, No. 2, (1980), pp. 19-26.
4. Hodges, D.H.; Ormiston, R.A.; Peters, D.A. *NASA TP 1566*, (1980).
5. Krenk, S. and Gunneskov, O. *Journal of Applied Mechanics*, Vol. 52, No. 2, (1985), pp. 409-415.
6. Abbas, B.A.H. and Irretier, H. *Journal of Sound and Vibration*, Vol. 130, No. 3, (1989), pp. 353-362.
7. Yuan, F.G. and Miller, R.E. *AIAA Journal*, Vol. 26, No. 11, (1988), pp. 1415-1417.
8. Noor, A.K.; Peters, J.M.; Min, B.J. *Finite Elements in Analysis and Design*, Vol. 5, No. 4, (1989), pp. 291-305.
9. Conte, S.D. and Carl de Boor, "Elementary Numerical Analysis", McGraw-Hill Company, (1980).
10. Carnevale, E. and Zunino, E. *Engineering Software II*, R.A. Adley, Editor, CMC Publications, (1981), pp. 747-757.
11. Sivaneri, N.T. and Chopra, I. *AIAA Journal*, Vol. 20, (1982), pp. 716-723.

Computer Based Life Prediction Methodology for Structural Design

T. L. NORMAN, T. S. CIVELEK and J. PRUCZ

Department of Mechanical and Aerospace Engineering
West Virginia University
Morgantown, WV

Abstract :

Advanced concepts in structural design must include information on geometric and economic constraints, as well as static strength and fatigue life of the structure under long term service conditions. In this research, computer based life prediction methodology on the structural component level has been developed to estimate the fatigue life for material screening in initial phases of design. Estimates are based on available stress-life data of coupon test from advanced composites, such as epoxy resin reinforced with unidirectional carbon fibers, and metals. A cylindrical shell is chosen as the test component consistent with other preliminary Concurrent Engineering (CE) efforts. In addition to fatigue life estimates, composite shell theory has been incorporated to perform stability analysis for any combination of axial, torsional, flexural and internal pressure loading. To meet the needs of user friendly interactive design, the software is so structured that trade-off studies can be performed based on fatigue life, relatively compared specific strength, stiffness and weight.

Introduction :

In structural design, engineers are encouraged to optimization structural components for maximum material savings and weight reductions. Reduced weight is a factor of great importance especially in all forms of transport where reductions in weight result in greater efficiency, energy savings and larger payload. During the design phase, it is essential to answer fundamental questions regarding the structural strength, stiffness, and fatigue life. Fatigue analysis however is usually considered in the late stages of the design process. As a result, modifications to the original design are required which can be costly. Since costs of investigating new applications of advanced materials are expensive, complete design analysis and fatigue life prediction are necessary to reduce time and labor consumption, cost and increase the efficiency. Thus, there is a great need for a computer based fatigue-life prediction methodology on the structural component level. The software developed in this research makes use of available S-N data to estimate the fatigue life of approximately defect free thin walled cylinders, subjected to various loading conditions. Both metals and composites are considered. This code is designed to permit the user to perform trade-off studies by determining the effects of varying certain parameters such as geometry, material and composite material fiber orientation .

System Architecture :

The computer program, written in Turbo-Basic consists of the following sections (Fig. 1.): A *Material Selection Database* containing information about mechanical properties of certain metals and advanced composites ; *Structural & Geometrical Parameters* such as diameter, length, thickness, number of plies, ply orientations and ply thicknesses to be specified ; *Loading Condition* including any combination of axial, torsional, flexural and internal pressure loading to be specified ; *Static Analysis* for strength and stability analysis of composites and metals ; *Fatigue Analysis* to determine the fatigue life based on S-N data ; *Trade-off Studies* to perform quick comparisons of fatigue life, component weight and relative specific strength and specific stiffness .

Method of Analysis :

The software is designed to perform both static and fatigue analysis of the selected materials combined loading conditions.

Static : Analysis for strength and buckling of isotropic and anisotropic cylindrical shells is performed using classical thin shell theory of small deflection. For laminated composite shells, first ply failure is chosen as an indication of failure. Maximum stress and strain criterion are used in the analysis. Flugge's differential equations of equilibrium [1] are used to derive the characteristic equation for buckling under combined loads . End constraints are neglected and the cylinder is assumed to be long. For metals, failure is determined by Von Mises failure criteria. Buckling strength of the isotropic thin walled cylinders subjected to combined loads is determined according to the design curves under various loadings [2]. For each individual case of loading, buckling charts are simulated and interaction equations are used to predict failure.

Fatigue : For practical applications, it is convenient to have only one fatigue life which characterizes fatigue behavior of the material at the applied stress . Therefore empirical relations for predicting life, such as the stress-life (S-N) curves, are given as an estimation. Fatigue stress-life curves of Steel and Aluminum under completely reversed loading case are shown in [3] and [4], respectively. Since composite materials are inhomogeneous and frequently anisotropic, the fatigue processes are very complex. To completely characterize composites under fatigue loading, a very large database consisting of stress-life data is necessary since data must be generated for each different lay-up and material system. Experiments have been performed for a number of graphite/epoxy systems, loading conditions and lay-up geometries [5-7]. Stress-life curves for various As4/epoxy and T300/epoxy laminates are presented in [8], however, limited data is available for structural components such as composite cylinders. Thus predictions for structural components must either be estimated from available coupon test data, or determined experimentally. Our method in this preliminary study is based on estimation from coupon data as discussed below :

Assuming a general two dimensional state of stress, both mean and alternating components due to the fluctuating combined loading are computed in terms of the normal stresses (S_x) and (S_y) and shear stress (S_{xy}). Principal mean (S_m) and alternating (S_a) stresses are calculated using equations for biaxial state of stress [3], as follows :

$$(S_m)^2 = (S_{xm})^2 - (S_{xm}S_{ym}) - (S_{ym})^2 + (3S_{xym})^2 \qquad (1)$$
$$(S_a)^2 = (S_{xa})^2 - (S_{xa}S_{ya}) - (S_{ya})^2 + (3S_{xya})^2 \qquad (2)$$

The full reversed fatigue stress, S_f, is determined using the modified Goodman equation [4] as :

$$(S_a / S_f) + (S_m / S_u) = 1 \qquad (3)$$

where S_u is the ultimate tensile strength and S_m and S_a are determined from equations (1) and (2), respectively. The value for the fatigue stress is entered on the stress-life diagram of the material to determine the fatigue life of the component. It should be noted that this analysis assumes approximately defect free materials in the absence of significant stress concentrations. Under these conditions, it is assumed that the component behaves according to the results of the coupon test. This we feel provides a good approximation to fatigue life estimates in preliminary design.

Demonstration :

To demonstrate the function of the software the following example is shown. From the material selection database (Fig.2.), three materials are chosen : 1) T300/5208 with lay-up configuration of $[0/45/90/-45/90/45/0]_s$; 2) Aluminum; and 3) As4/3502 with lay-up configuration of $[0/45/-45/90]_{2s}$. Positive directions of the applied loads are chosen as illustrated in Fig. 3. Applied combined loading conditions are presented in Table 1. Two cases are considered because of thickness differences between the composite laminates. For Case 1 T300/5208 and Aluminum are chosen while Case 2 considers Aluminum and As4/3502. The dimensions of the cylinder are entered as having a diameter (D) of 2540 mm, a length (L) of 1905 mm. The thickness (H) for the first case is 1.778 mm consistent with 14 plies of T300/5208 and in the second case it is 2.032 mm consistent with 16 plies of As4/3502.

Results of the fatigue life prediction are shown in Table 2. Since the predicted fatigue life for the third material, As4/3502 is determined to be greater than the limits of the available stress-life data (10e7) given in[8], it is represented by ">10e7" as shown in Table 2. Although two different cases are considered, for comparison purposes loading conditions are so arranged that the predicted fatigue life of the second material, Aluminum, is same in each case. Specific strength and stiffness, weight (based on the volume of the chosen structural component), and fatigue life are compared relatively in Fig. 4.

Discussion :

A system has been developed to estimate fatigue life and perform static analysis of a structural component subjected to combined loading. A thin walled cylinder is chosen as the component and fatigue life estimates are based on the available stress-life data for certain conventional and composite materials. The primary assumption here is that the structural component is approximately defect free absence of significant stress concentrations, and that it behaves according to the results of a coupon test. This should provide a good preliminary estimate for material screening purposes. For most applications however the database will need to be improved to include the effect of complex geometries. To provide and improve the necessary database, an experimental program to characterize a number of baseline component types should be developed. This data could be used to improve the present algorithm so that fatigue life estimates could be made for more complex geometries.

Acknowledgement -- This work has been supported by DARPA under Contract No. MDA972-88-C-0047 .

References :
1. Cheng. S. ; Ho, B.P.C. : Stability of Heterogeneous Aeolotropic Cylindrical Shells under Combined Loading. AIAA Journal, Vol. 1, pg 892-898, 1962.

2. Bruhn, E.F. : Analysis and Design of Flight Vehicle Structures. Tri-State Offset Company1973.

3. Shigley, J.E. ; Mitchell, L.D. : Mechanical Engineering Design. McGraw-Hill, Inc. 1983.

4. Bannantine, J.A. ; Comer, J.J. ; Handrock , J.L . : Fundamentals of Metal Fatigue Analysis . New Jersey : Prentice Hall 1990.

5. Kremple, E, ; Niu., T.M. : Graphite/Epoxy $[45/-45]_S$ Tubes . Journal of Composite Materials , Vol. 16, pg 172-187, 1982.

6. Yang, J.N. : Fatigue and Residual Strength Degradation for Graphite/Epoxy Composites. Journal of Composite Materials, Vol. 12, pg 19-39, 1978 .

7. Foley, G.E. ; Roylance, M.E. ; Houghton, W.W. : Life Prediction of Glass / Epoxy Composites Under Fatigue Loading. Albany, Newyork : ASME International Conference In Life Prediction Methods 1983.

8. Tsai, S. W. : Composites Design ,4th Edition . Dayton,Ohio : Think Composites 1988.

Table 1 : Loading Conditions for Case 1 and Case 2

		Maximum Load		Minimum Load	
Load Type		case 1	case 2	case 1	case 2
Axial Load	kN	100	100	10	10
Pressure	kN/mm^2	.00075	.001	.0005	.0009
Moment	kN*mm	4000	4000	200	200
Torsion	kN*mm	5000	5000	500	500

Table 2 : Fatigue Life Prediction

		Fatigue Life, cycles	
	Material	case 1	case 2
1	T300/5208	9e5	-
2	Aluminum	2.0e4	2.0e4
3	As4/3502	-	>10e7

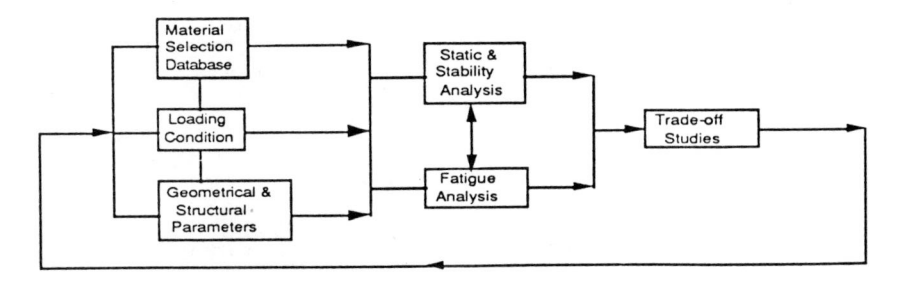

Fig. 1. Overview of the System

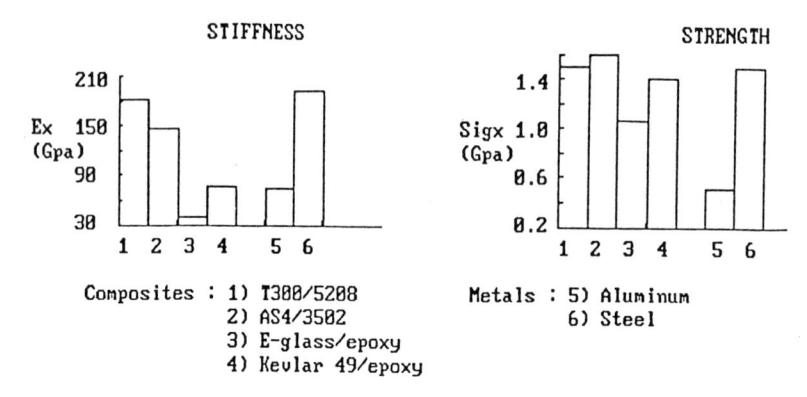

Fig. 2. Material Selection

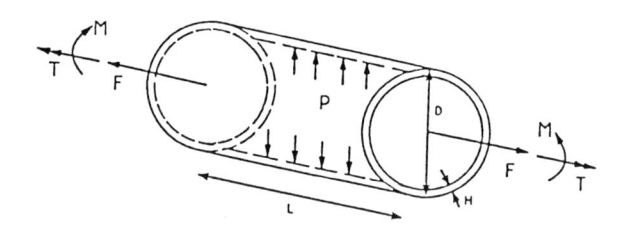

Fig. 3. Circular Cylinder with Applied Load Orientations

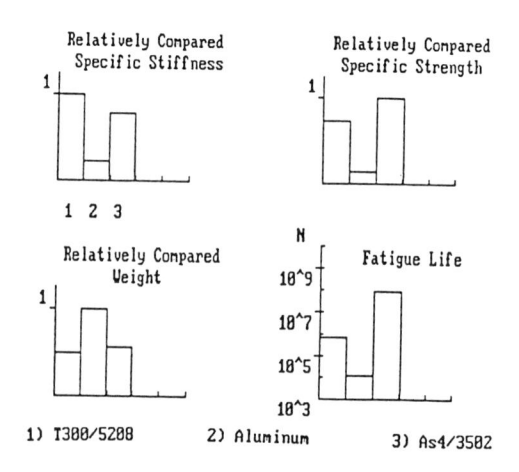

Fig. 4. Comparison of the results from design analysis

Chapter VII

Process Modeling and Control

Introduction

This chapter discusses aspects related to process modeling and control in the concurrent engineering environment. The first paper in this chapter provides an overview of the market, trends, and processing of superalloys and related materials used in the manufacturing of gas turbine engines in the 1990s and beyond. The second paper describes the application of the Finite Element Method (FEM) for the simulation of metal forming processes, and provides a rational methodology for designing and optimizing these processes. This paper reviews two general approaches - the flow formulation and the solid formulation used in describing the deformation mechanics of metal forming. The third paper discusses Concurrent Product and Process Development (CP/PD), which combines marketing, finance, design, engineering, manufacturing, purchasing and suppliers in the development process from concept initiation to customer delivery using computer data base technology. The fourth paper lays a foundation for the construction of computerized concurrent engineering systems by analyzing and outlining the design process of mechanical components in a way that readily lends itself to computer application. The fifth paper discusses modeling, analysis and performance evaluation of CIM systems using PETRI nets. The sixth paper describes the concepts of concurrent engineering and the modules of integrated production systems as well as implementation factors for proper communication and integration at both the design and production stages. The seventh paper illustrates the methodology for successfully adopting a rule-based expert system with process knowledge for control of the turning process. The eighth paper of this chapter outlines the design and development of an expert system which will enable the selection of milling processes for the manufacture of a product. It describes data and knowledge acquisition methods, examines inference procedures, and identifies effective expert system design methods. The last paper deals with forging a die design with artificial intelligence.

Processing of Superalloys in the 1990 s

F. ROBERT DAX

Director, Engineering/Technology
Cytemp - Powder Products,
Pittsburgh, PA

This paper will provide an overview of the market, trends, and processing of superalloys and related materials used in the manufacturing of gas turbine engines in the 1990's and beyond.

As we all know, the order backlog for new commercial aircraft and engines is the largest in history and is expected to be full through the 1990's. As of August, 1990 the backlog through 1998 for commercial aircraft is over 7200, and worth over $406 billion. This unprecedented backlog will keep all segments of the aerospace community busy and allow for the development of larger, more fuel efficient and more powerful engines. Figure 1 shows how the major components of an engine have changed since 1980 and will change by 1999. Figure 2 shows the change from 1960 through 2010 in the make up of an engine by material groups. It is from these two figures that the processing of superalloys will be directed in the 1990's.

TRENDS

There will be little change in the substitution of new materials or in processing forms in the manufacture of gas turbine engines in the 1990's as shown in figures 1 and 2. The dramatic switch from forged and ring rolled products to castings that occurred in the 1980's will end and new materials won't be substituted in quantity for nickel and titanium based alloys until at least 2000. The trends shown in figure 3 which allow engines to increase performance will be achieved by advancements in the current manufacturing processing for existing alloys and material systems. Even though new materials offer the high temperature properties required for the gas turbine engines of the future, the ability to process these and the present materials economically is important. Thus, to improve engine performance by substituting the new materials, improvements must be also be made in the processing of these materials which will be the "key" trend in the 1990's. In addition to the substitution of new materials, improve processing of existing materials can lead to better performance.

The following six process areas will be discussed:

1. Airfoil Casting
2. Powder Metallurgy
3. Rapid Solidification Rate Powder
4. Superalloy Melting
5. Forging
6. Intelligent Processing of Materials

AIRFOIL CASTING

In the 1980's castings grew from nineteen percent of the composition of an engine to 42 percent. This dynamic growth is mainly the result of the substitution of large structural castings for rolled rings and forgings. In the 1990's the primary change in the casting of airfoils will be in the continued growth of directionally solidified and single crystal cast airfoils replacing equiaxed airfoils. In high performance engines, such as those used by the military, improvements in strength can be gained by using directionally solidified and single crystal cast alloys. Today directionally solidified and single crystals airfoils make up about 2 percent of all airfoil castings. This will increase to at most ten percent by the year 2000. Yields are still poor and will be the primary area in which processing technology will focus.

These two processes rely on an increase in the thermal gradient and solidification rate in the processing of airfoils. Figure 4 shows this relationship. Airfoils processed by one of these methods have improved creep resistance, the main property related to the failure of superalloy cast airfoils, by the minimization or elimination of grain boundaries.

In addition to creep resistance, the fact that grain boundaries have been eliminated in single crystals allows for the use of alloys with improve strength by removing those alloying elements that are part of the composition that increase grain boundary strength. Thus cobalt, carbon, boron, zirconium, and hafnium can be removed from the composition. Though, these alloys strength the grain boundaries they lower the melting temperature. Thus, their removal allows the new alloy to be used at higher operating temperatures.

POWDER METALLURGY

During the 1990's the use of Powder Metallurgy (P/M) will continue to grow as more disks are converted from conventional, cast and wrought, superalloys to P/M. In addition, as Near Net Shape technology improves the cost disadvantage of P/M will decrease, allowing the use of P/M parts in even more applications. Figure 5 shows the three primary means of manufacturing P/M based superalloys. These are: 1) direct extrude, 2) Hot Isostatic Pressing (HIP), and 3) Consolidation at Atmospheric Pressure (CAP). The main advantages of P/M over cast and wrought are 1) better mechanical properties, 2) more consistent properties, 3) more homogenous microstructure, and 4) a potential cost savings in eliminating material and process steps, especially in near net shape manufacturing.

Figure 6 shows a comparison of properties for CAP plus rolled to cast and wrought for Alloy 718. This figure illustrates both the higher mean property values of P/M processing and the more consistent property data. It is this second attribute of P/M that is important to engine designers since they determine the life of an engine based on the lower control limit of the property value. Thus, even though the mean property values are only slightly higher, 2 to 3 percent, than cast and wrought, because the lower standard deviation of these values is significantly higher, 10 to 15 percent, the life cycle costs of an engine can be significantly reduced.

Figures 7 and 8 show how P/M can reduce the costs of producing parts and a relative ratio of potential cost savings comparing the conventional cast and wrought production practice with three P/M manufacturing methods. The main savings in the three methods, shown in figure 8, is realized through the use of near net shape molds and the elimination of the closed die forging steps, as shown in figure 7.

RAPID SOLIDIFICATION RATE POWDER

The three primary rapid solidification rate processing methods are shown in Figure 9. All three methods produce a very homogeneous microstructure similar to conventional P/M methods. Centrifugal atomization can be used to produce powder, as the Pratt & Whitney RSR method does or it can be used to build-up or deposit a layer of material as is done by the Osprey process.

Melt Spinning produces a thin strip of homogeneous material that can then be used as a coating, layered- both single and multilayered- to produce the desired properties. Its advantage over conventional powder is that the strip can be easily handled and is already fully dense.

The third Rapid Solidification Rate processing method is the use of a Laser to localize melt the surface of a material and then have that surface rapidly quenched to produced a homogenized surface, similar to plasma spray disposition of powder.

SUPERALLOY MELTING

Through the 1970's and 1980's the principal method used to remelt superalloys has been the Vacuum Arc Remelt (VAR) process. This processing method helps to produce a more uniform structure and to remove impurities from the initial Vacuum Induction Melted (VIM) ingot. However, in the 1990's the Electro Slag Remelt (ESR) process, and the similar processes of the Electron Beam/Plasma cold hearth refining processes will increase their importance in the remelting of superalloy ingots.

The ESR process has been used for nearly as long as the VAR process but its use has been limited because of problems associated with its control. Essentially, the ESR process is similar to the VAR process except that it uses a slag layer to cover the molten ESR ingot and remove impurities from the melted electrode. The slag skin, a layer of slag one-eighth to one-half inch thick formed during the remelt process on the surface of the remelted ingot, lowers the cooling rate of the ESR ingot, which increases the segregation within the ingot. The VAR process uses a vacuum to remove gases with impurities floating to the surface of the melt pool and solidifying on the outside surface.

Figure 10 shows a comparison of defect size among the VAR, ESR and ESR + VAR practices. The ESR + VAR practices is an attempt to use the best of both practices. However, during the VAR practices, the impurities that remain in the ingot after the ESR step are recombined to form larger impurities. Because the molten slag makes direct contact with the molten drops from the superalloy ingots, more impurities can be removed then by the VAR process. The slag layer not only produces cleaner material but also causes problems in controlling the process. In the ESR process the current passes through the slag layer which effects it. In addition, the composition and depth of the slag is constantly changing. Finally, the slag is forming a layer on the surface of the remelted ingot which effects the heat transfer and cooling of the ingot.

Two relatively new processes which are now being commercialized are the Electron Beam and Plasma cold hearth refining processes. Both processes are similar except for the type of torch /gun used for melting the superalloy. Figure 11 shows a schematic of the Electron Beam Process. The electron beam is used to melt the initial charge. The molten charge then flows into a water cooled copper hearth. No refractories are used, thus minimizing pickup of contaminating particles. Impurities present in the charge materials, float to the surface. A dam bridges the surface of the hearth which prevents the impurities from floating into the pouring spout. All this takes place in a vacuum chamber which also removes gases produced during the melting operation. In addition, the electron beam continues to control the temperature and melt rate of the molten metal, both behind and in front of the dam. The molten metal then flows from the spout into a water cooled copper crucible, similar to the VAR process.

FORGING

During the 1980's ring rolling and forging lost considerable market share to casting in the manufacturer of gas turbine engine components. This trend is expected to stop in the 1990's with the improvements in near net shape processing in conjunction with isothermal forging. Figure 12 shows the material savings possible in using Isothermal forging of near net shape parts. Frequently, the billet material used in isothermal forging is made from powder because of its superplastic properties. The high cost of powder, relative to cast and wrought material, emphasizes the savings generated in using isothermal forging.

INTELLIGENT PROCESSING OF MATERIALS

During the 1990's the primary advancement in superalloy technology will come not from new alloys, but rather from improvements in process technology. The areas that improvements will be made will include process development, process modeling, and process control. By combining these three areas and the use of computer technology, the 1990's will begin to see the development of the intelligent processing of materials (IPM).

Figure 13 shows a schematic a IPM controller. The essentially difference between this and regular process control is the inclusion of a predictive material process model. Figure 14 shows an example of such a model. These models frequently relate the metallurgical state of a material to the processing parameters. In figure 14 a process map relating the deformation of a titanium alloy is shown. Deformation in this case is related to strain rate or the speed at which the material is forged. Frequently, a three-dimensional plot is used which also relates strain. In all material forming operations, deformation or forming occurs at different rates and temperatures at different places in the material, the resultant structure will be non-uniform. Thus, by comparing the deformation in the material with a process map which relates structure and/or properties to the deformation or processing conditions, the process can be designed and controlled to produce the material with the desired structure and properties.

Figure 15 shows how forging process variables such as lubrication, die temperature, and ram velocity are related to the forging process model. This model is then used to predict the metallurgical and geometrical quality and whether the forging press is capable of forging the desired part. Presently, this analysis is conducted offline and an expert system is used to control the material forming system. However, as advances in computers continue in the 1990's, the Intelligent Processing of Materials will be used real time.

SUMMARY

The 1990's will see a continued increasing demand for superalloys as the commercial aerospace market expands and replaces old aircraft. In addition, larger commercial and more powerful military engines will be required. These economic factors will enable the superalloy industry to make improvements in the processing and the control of these processes. It is in these areas that superalloys will make the greatest advances in the 1990's.

249

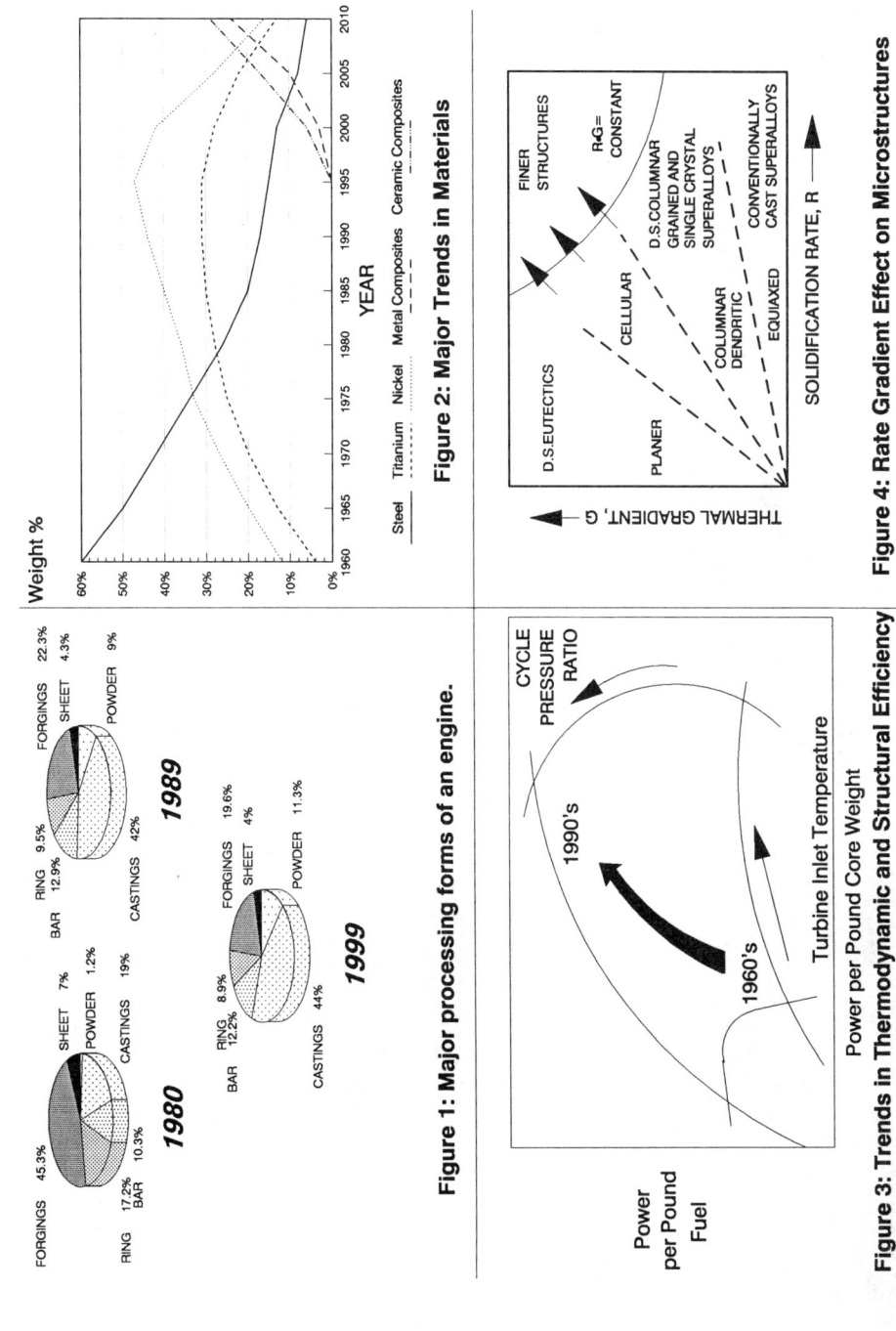

Figure 1: Major processing forms of an engine.

Figure 2: Major Trends in Materials

Figure 3: Trends in Thermodynamic and Structural Efficiency

Figure 4: Rate Gradient Effect on Microstructures

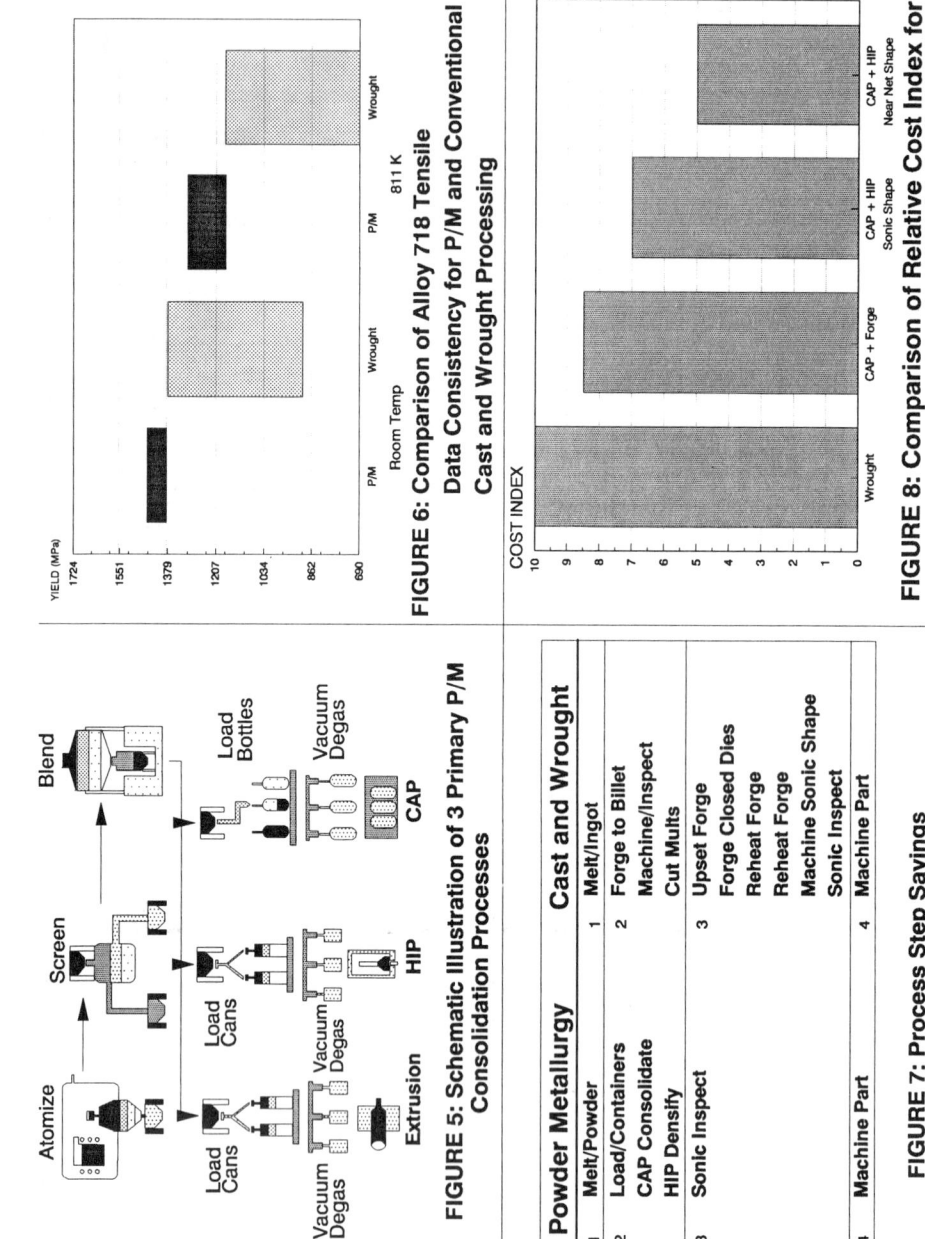

FIGURE 6: Comparison of Alloy 718 Tensile Data Consistency for P/M and Conventional Cast and Wrought Processing

FIGURE 5: Schematic Illustration of 3 Primary P/M Consolidation Processes

	Powder Metallurgy		Cast and Wrought
1	Melt/Powder	1	Melt/Ingot
2	Load/Containers CAP Consolidate HIP Densify	2	Forge to Billet Machine/Inspect Cut Mults
3	Sonic Inspect	3	Upset Forge Forge Closed Dies Reheat Forge Reheat Forge Machine Sonic Shape Sonic Inspect
4	Machine Part	4	Machine Part

FIGURE 7: Process Step Savings

FIGURE 8: Comparison of Relative Cost Index for Superalloy Produced via Conventional Wrought Process versus 3 P/M Processes

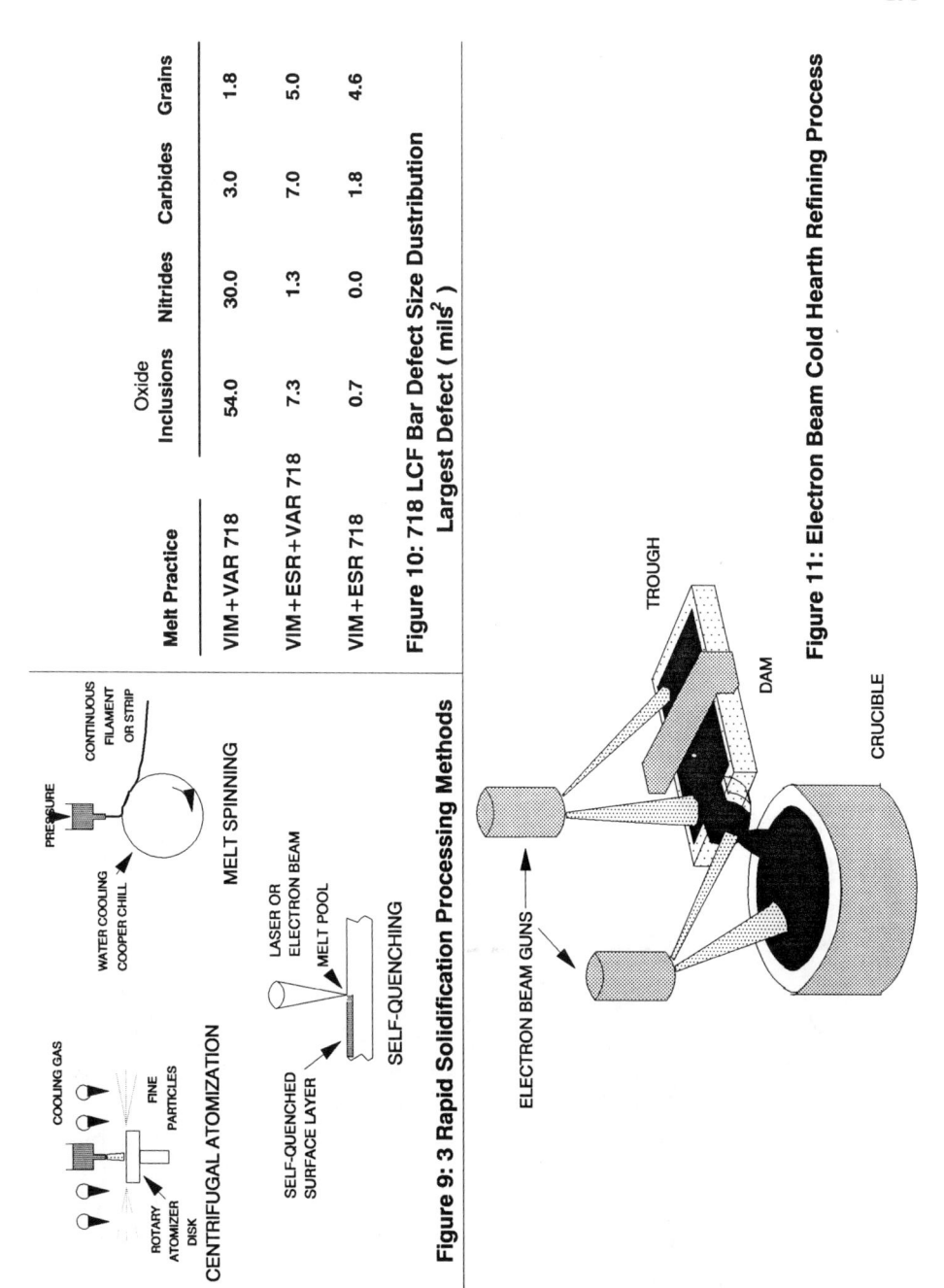

Melt Practice	Oxide Inclusions	Nitrides	Carbides	Grains
VIM+VAR 718	54.0	30.0	3.0	1.8
VIM+ESR+VAR 718	7.3	1.3	7.0	5.0
VIM+ESR 718	0.7	0.0	1.8	4.6

Figure 10: 718 LCF Bar Defect Size Dustribution Largest Defect (mils^2)

Figure 9: 3 Rapid Solidification Processing Methods

Figure 11: Electron Beam Cold Hearth Refining Process

252

Figure 12: Isothermal Forging Process Cost Reduction

Figure 13: Intelligent Processing of Materials - Controller

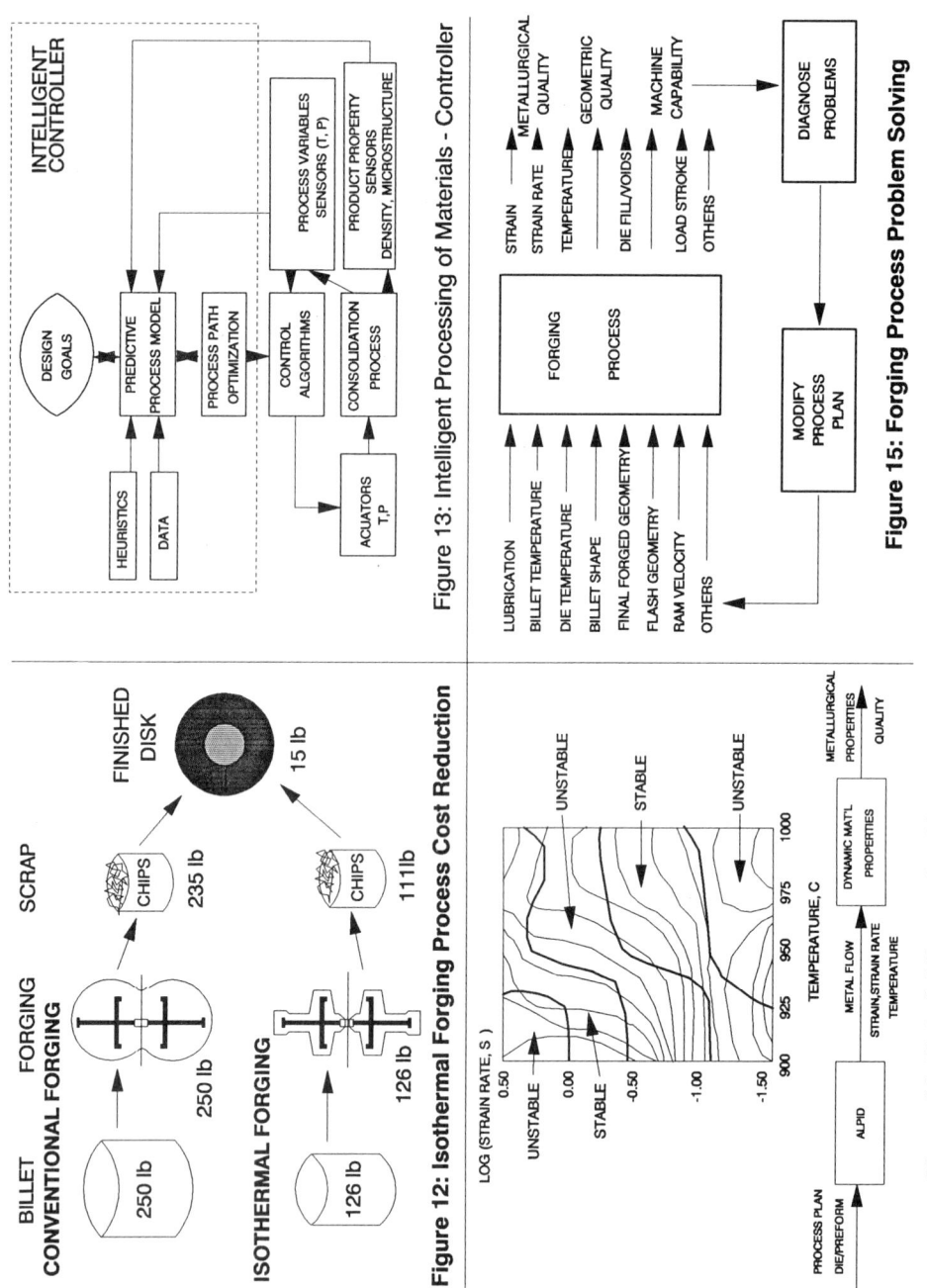

Figure 14: Material Processing Map

Figure 15: Forging Process Problem Solving

Application of the Finite Element Method in Metal Forming Process Design

SHANKAR RACHAKONDA and SUREN N. DWIVEDI

Concurrent Engineering Research Center
West Virginia University
Morgantown WV

ABSTRACT

The application of the Finite Element Method (FEM) for the simulation of metal forming processes has provided a rational methodology for designing and optimizing these processes. This paper reviews two general approaches - the flow formulation and the solid formulation - used in describing the deformation mechanics of metal forming. Some considerations related to the constitutive behavior of the material, the contact between the die and the workpiece are described. These considerations must be taken into account in an appropriate fashion while conducting the finite element simulation. The problem of severe grid distortions that is encountered during the simulations is discussed in detail with particular reference to remeshing. Results obtained using a finite element package based on the flow formulation are presented. Some of the future directions in this fast growing area of research are also discussed.

INTRODUCTION

The advent of FEM in the 1960's has provided a means to obtain detailed analytical solutions to highly complicated design problems in diverse fields such as solid mechanics, fluid mechanics and heat transfer etc. This method can be applied to metal forming problems in order to provide a rational methodology for process design activities which include preform design, die design and forming load prediction. The traditional approach to process design in the forming industry is based on empirical rules and guidelines and actual production trials. However, by the application of FEM, various metal forming processes can be simulated and designed accurately by obtaining a detailed description of the metal flow, load requirements and the variation of important process variables such as accumulated strain, strain rate and temperature. Thus the trial-and-error approach which is common in the industry can be avoided.

One of the first applications of FEM to metal forming problems was an analysis of plane-strain and axisymmetric flat punch indentation using a small strain elastoplastic formulation [1]. Considering the large plastic deformations and strains which occur in metal forming and questions regarding the solution accuracy, a formulation based on rigid-plastic characterization of the material behavior was developed [2]. In this formulation the elastic part of the deformation was neglected and the incompressibility constraint required by the above assumption was imposed by a Lagrange multiplier technique. It was further generalized to rate sensitive materials using a rigid-viscoplastic characterization and a Penalty Function Method was used to impose the incompressible constraint [3]. Since this approach is analogous to non-newtonian fluid flow analysis it was termed the "Flow Formulation".

It was further enhanced to include the temperature effects which are typically found in hot forming operations [4], and was specialized for sheet forming applications [5]. Quite recently, the flow formulation was extended to three dimensional analyses of some simple forging operations [9]. Parallel to these developments, several advances were made in implementing large deformation, large strain capabilities in the elastoplastic formulation, so that greater increments in deformation are permissible [7]. Since this formulation is based on the traditional elastoplastic or elastic-viscoplastic characterization of solids it has come to be known as the "Solid Formulation". Several variations of this formulation have been proposed and it is still an actively researched area.

Because of its simplicity and capabilities to take into account larger incremental deformations easily, the flow formulation has gained a wider acceptence in the metal forming community. A general purpose package for two dimensional metal forming simulation called ALPID was developed based on the above formulation which is capable of simulating bulk forming processes with arbitrarily shaped dies [5]. However, it is not capable of predicting the residual stresses because the elastic strains are neglected. Several examples of the application of FEM in metal forming process design are available in open literature including a rather novel application for preform design in forging and shell nosing [8]. In this paper, some important aspects of finite element simulation which require careful consideration are briefly discussed and a simulation example is presented.

CONSIDERATIONS RELATED TO MATERIAL BEHAVIOR AND FRICTION

The behavior of the material under deformation process conditions is an important input to the finite element packages used for metal forming process simulation. An accurate material model is needed for an accurate prediction of press loads and distribution of process variables. Material models typically consist of constitutive equations which relate the flow stress of the material to the accumulated strain, strain rate and temperature. These constitutive equations are developed by conducting compression tests on the given workpiece material. In addition, one of the important objectives in metal forming process design is achieving the specified product integrity which is largely dependent on the workability of the material. If the material model is also capable of analytically determining the workability of the material, such considerations can be incorporated in the finite element simulation. In this context, a novel methodology called "Dynamic Materials Modeling" [12], has received much attention. With this methodology it is possible to develop processing maps which indicate safe regions for deformation.

The flow of the material in the die cavity is greatly influenced by the friction at the tool-workpiece interface. In order to reflect this very important feature in the finite element simulations, a fundamental understanding of friction and lubrication phenomena is needed. This aspect is currently receiving much attention and efforts are being aimed at developing detailed frictional models. Currently, simplified frictional boundary conditions are being incorporated in the simulations by considering a constant sliding (or Coloumb) friction for cold forming applications and sticking (or shearing) friction for hot forming applications. A constant value of the friction factor which is determined from the friction calibration curves of the specified lubricant is input to the finite element packages. Friction calibration curves for different lubricants are generated by using experimental methods such the ring test or the spike test.

Accurate workpiece material models and friction models are two of the most important ingredients needed for a succesful application of FEM in metal forming process design. They emphasize the importance of experimentally generated accurate data that must be used in computer-based simulations. The development of such databases for various workpiece materials and lubricants is critical for the advancement of FEM-based design methodology in metal forming.

REMESHING

In order to simulate the metal flow that occurs during forming operations by FEM it is necessary to update the mesh continuously as the deformation progresses. On account of this the finite element mesh becomes gradually distorted effecting the predictive capability of the elements. It is then necessary to remesh the arbitrarily shaped intermediate configuration of the workpiece and transfer the field variables on to the new mesh using accurate interpolation algorithms. An inaccurate interpolation can introduce very large errors in the solution, and hence this aspect requires serious attention. During the simulation of certain complicated forming processes a number of remeshings can be necessary. Typically, remeshing is a time consuming process and makes the simulation exercise an arduous and time consuming task, especially in three dimensional cases. Several approaches are currently being pursued in the research community, in order to alleviate the problems associated with remeshing. One of the approaches involves providing automatic mesh generation capabilities in the finite element packages so that the time involved in generating the new mesh is drastically reduced. However, automatic mesh generation schemes for arbitrary shaped objects in three dimensions are still under development. An alternative approach is the use of an Arbitrary Lagrangian-Eulerian (ALE) finite element formulation which offers the possibility of improving the geometry of distorted elements and automatic mesh adaptation [10]. Another adaptive simulation scheme was also reported which involves continuous refinement and optimization of the finite element mesh as the deformation progresses [11]. Error estimates and indicators are used for measuring the error in the given mesh and regions where mesh refinement is needed are identified.

SIMULATION OF AN L-SECTION FORGING WITH ALPID:

All the above mentioned material and frictional considerations were properly taken into account in conducting an isothermal closed-die plane strain forging simulation of an L-shaped section, figure 2(a), using ALPID. The objective of the simulation was to predict the deformation requirements for the process in order to obtain the final cross section starting from a rectangular cross-section using a finisher punch. Some of the aspects that were closely watched include die filling, metal flow and load requirements. As a first step a proper model of the billet material, Al2024, was created by developing processing maps and optimum processing conditions were chosen. One such map is shown in figure 1. Friction calibration curves for the lubricant used, MoS_2, were generated and an interface sticking friction factor of 0.35 was chosen. The bottom die was held stationary and the top die was moved down with a velocity of 0.035 in/sec. The billet was 1.0" long and 0.5234" wide.

Figure 2(b) shows the initial orientation of the workpiece and dies. Figure 3(a) shows an intermediate stage during the simulation when the workpiece was remeshed and figure 3(b) shows the fully filled die cavity. In this particular case, a maximum load of 56 kips for forging the L section was predicted. Since the die filling was achieved with reasonable load requirements, it may be concluded that preform design is not necessary for forging the given L section. However, such a recommendation can be acceptable provided suitable forging presses are available and and several other factors such as cost, time and effort involved in preform design are taken into account.

CONCLUSIONS AND FUTURE WORK

The various aspects of metal forming simulation by FEM discussed in this paper clearly demonstrate that it is a viable process design technique but requires attention to many details. Some of the issues which need to be addressed by researchers in the future include enhancing three dimensional simulation capabilities, developing automatic meshing capabilities and increasing the speed of computation. Finally, the metal forming industry must be educated in the benefits of using these simulation techniques for solving their process related problems.

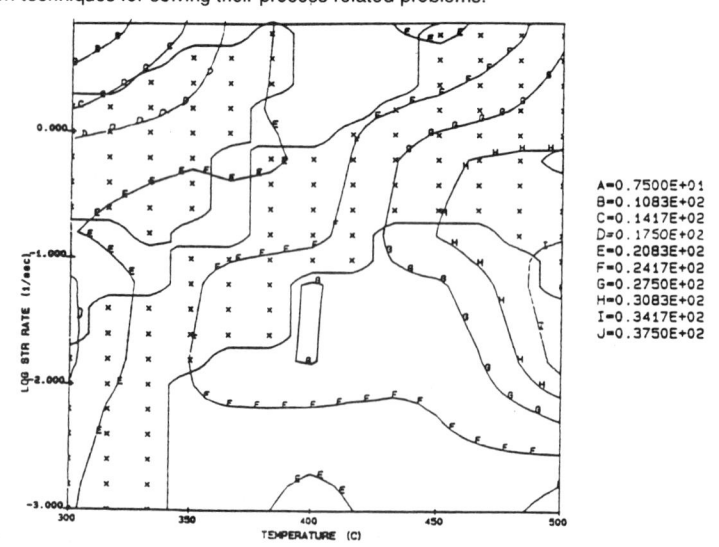

Figure 1. Stability map of Al2024 at a strain of 0.1

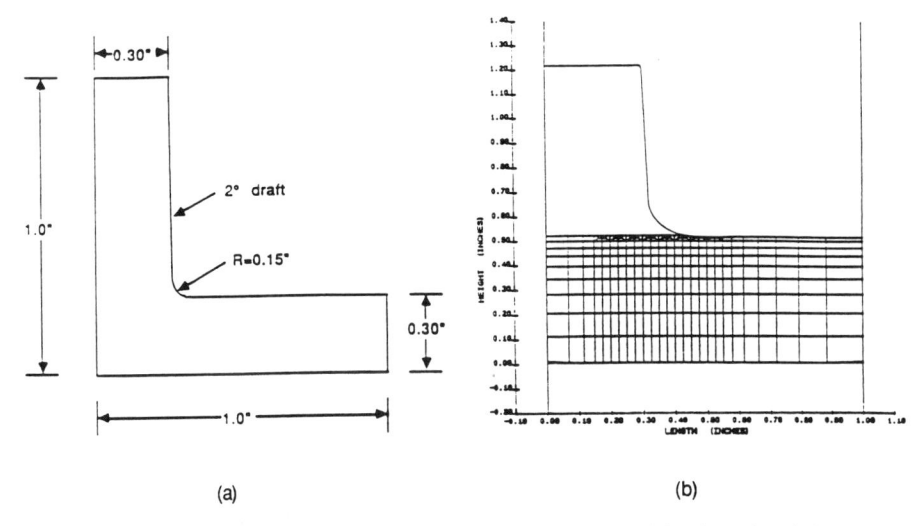

Figure 2. (a). Final dimensions of the L section (b) Initial orientation of the die and workpiece.

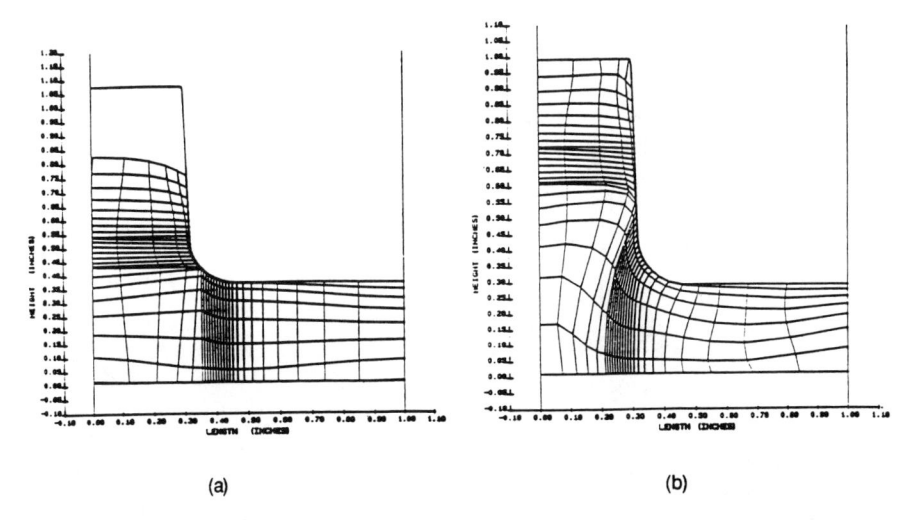

Figure 3. (a) Remeshed workpiece (b) Complete die filling of the L-section

ACKNOWLEDGEMENT

This work is supported by the DARPA Initiative in Concurrent Engineering (DICE) program at the Concurrent Engineering Research Center (CERC) at West Virginia University under contract no. MDA 972-88-C-0047.

REFERENCES

1. Lee, C.H.; Kobayashi, S.: Elastoplastic analysis of plane-strain and axisymmetric flat-punch indentation by the finite element method. Int. J. Mech. Sci., 12 (1969), 349-370.

2. Lee, C.H.; Kobayashi, S.: New solutions to rigid-plastic deformation problems using a matrix method. J. Eng. Ind., Trans. A.S.M.E., 95 (1973), 865-873.

3. Zienkiewicz, O.C.; Godbole, P.N.: A penalty function approach to problems of plastic flow of metals with large surface deformation. J. of Strain Analysis, 10 (1975) 180-183.

4. Rebelo, N.; Kobayashi, S.: A coupled analysis of viscoplastic deformation and heat transfer, I: Theoretical considerations, II: Applications. Int. J. Mech. Sci., 22 (1980), 699-705, 707-718.

5. Oh, S.I.; Lahoti, G.P.; Altan. T.: ALPID - A general purpose FEM program for metal forming. Proceedings of NAMRAC-IX, (1981) 83-88.

6. Onate, E.; Zienkiewicz, O.C.: A viscous shell formulation for the analysis of thin sheet metal forming. Int. J. of Mech. Sci., 25,5 (1983), 305-335.

7. Nagtegaal, J.C.; Veldpaus, F.E.: On the implementation of finite strain plasticity equations in a numerical model. Numer. analysis of forming processes, John Wiley & Sons (1984), 351-372.

8. Park, J.J.; Rebelo, N.; Kobayashi, S.: A new approach to preform design in metal forming with the finite element method. Int. J. Mach. Tool Design Res., 23 (1983), pp. 71-79.

9. Park, J.J.; Kobayashi, S.: Three dimensional finite element analysis of block forging. Int. J. Mech. Sci., 26, 3 (1984), 165-176.

10. Schreurs, P.J.G.; Veldpaus, F.E.; Brekelmans, W.A.M.: Simulation of forming processes using the arbitrary eulerian-lagrangian formulation. Comp. Met. App. Sci. Eng., 58 (1986),19-36.

11. Yang, H.T.Y.; Martin, H.; Shih, J.M.: Adaptive 2D finite element simulation of metal forming processes. Int. J. Num. Meth. Engr., (1989), 1409-1428.

12. Arsenault, R.J.; Beeler, J.R.; Esterling, D.M. (eds.) Computer simulation in material science. Metals Park, OH: ASM International (1985).

Strategic Value of Concurrent Product and Process Engineering

EDWIN R. BRAUN
University of North Carolina Charlotte
Charlotte, NC

JASON R. LEMON
International TechneGroup Incorporated
Milford, OH

Introduction

Concurrent Product and Process Development (CP/PD) is a disciplined, computer integrated, product and manufacturing process development methodology. Similar to simultaneous engineering, CP/PD combines marketing, finance, design, engineering, manufacturing, purchasing and suppliers in the development process from concept initiation to customer delivery using computer database technology.

CP/PD is not a magical process that instantly gives unique answers that solve all business problems. It is not a new set of buzz words and it is not a new set of software algorithms. CP/PD is a rigorous product development process that identifies and quantifies market opportunities. It determines related customer needs, does competitive assessments and converts the "voice of the customer" into measurable product specifications and requirements using Quality Function Deployment (QFD) methods and software. CP/PD defines and evaluates multiple product and process alternatives using computer simulation capabilities. The CP/PD product development team, through predictive "what-if" analyses, focus on design for manufacturing, design for assembly, design for quality and design for cost. Finally, CP/PD utilizes business models and decision tables to allow management to select market, product, manufacturing, and business strategies that are "optimized" for their global competitive environment. A recently reported example is Cincinnati Milacron's 40% reduction in cost, increased productivity, and a 50% reduction in the normal development time cycle as detailed in FORTUNE magazine.

There are no easy roads to worldclass product performance, cost structures, quality levels and competitive time to market. Any one that implies otherwise simply does not understand the complexity and difficulties involved.

Technology and Quality

Level of technology to be used and cost of total quality are key factors when assessing the manufacturing approach for achievement of total product quality. Level of technology examines the extent to which technical tools and methods are applied to the product development process. The cost of total quality evaluates costs associated with all phases of product development and manufacturing operations. A low technology producer achieves acceptable product quality mainly through a "build and test" product development approach and through a heavy reliance on inspection methods throughout manufacturing operations. A middle range producer achieves acceptable produce quality primarily through rigorous implementation of statistical methods; i.e. design of experiments, Taguchi methods, etc. for product development and statistical process control and related shop floor quality systems. Worldclass manufacturers achieve total acceptable product quality primarily through design; i.e. by implementing product and manufacturing process development methods that achieve quality by design.

Most producers fall somewhere in the middle range. They are very good at statistical methods in product/process development and manufacturing operations; but they are not advanced with regard to worldclass capabilities to achieve and control total quality via fundamental product and process design and development.

Cost to Achieve Total Product Quality

The cost to achieve total quality involves all product development and manufacturing phases and can be evaluated in four categories.

1) Product development costs; such as design of experiments, durability life tests, etc.

2) Manufacturing planning and engineering cost to develop systems for incoming inspection, in process gauging, statistical process control, final product run off testing, etc.

3) Manufacturing operations cost after the product is released to production, to conduct and control quality systems.

4) Field warranty and product recall costs.

An objective of most CP/PD programs is to reduce costs to achieve total product quality by a factor of two or three to one compared with today's methods. These reductions are achieved by designing both products and processes for total quality and production control without heavy reliance on statistical or manual inspection methods.

Cost of Engineering Changes

The cost of engineering changes to achieve and maintain total acceptable product quality is dependent upon the development phase in which the changes occurs. The cost of engineering changes, involving production, tooling and equipment modifications, after production has started, can exceed $100,000 per change (on average) in many product industries. If changes are made before production, in the build and test phase, the cost might be $20,000 per change (on average). While impossible to nail down precisely, a cost estimate if these same changes are made in the predictive simulation phase within the computer, is $1,000 to $2,000 per engineering change (on average).

The CP/PD process relies extensively on predictive methods and computer simulation. CP/PD is a cultural change for most companies and requires major revisions in the way products and processes are developed, compared with the traditional "build and test" serial approach. A major CP/PD strategy is to examine **multiple** product and manufacturing process alternatives at the earliest stages of development using computer simulations.

Total Product Cost

While impossible to identify precisely, an estimated 75% or more of a product's costs are "locked in" when the first layouts are developed during the concept design phase. Internal cost reduction programs and supplier cost reduction programs "squeeze and squeeze" on the remaining 25% or less, but very little real savings can be achieved after a product concept is selected.

Evaluating multiple product and process alternatives and comparing relative product and manufacturing costs of these alternatives (i.e. material costs, capital investment, amortization costs, etc.) and buy versus make trade-offs, CP/PD program teams attempt to attack total product costs when changes can be made easily and quickly in the computer.

In the product development processes used by most industrial manufacturers today, at the end of the concept phase an estimated 50% or less of final product and process decisions necessary to achieve total product cost and quality goals, have been made. In addition, only 5% to 7% of the total development budget has been expended at this stage. Utilizing predictive analysis and simulation capabilities, it is the objective of CP/PD programs to be 80% to 90% certain that the correct product concepts, and the correct manufacturing and assembly strategies have been selected at the concept design stage.

Product and related manufacturing process decisions must be influenced significantly by related product family mix and volumes in order to achieve worldclass costs and total product quality compared with competition...**before detail design begins**. This mandates that management be willing to allocate four to five times more budget and resources at the concept stage of product/process development; i.e., 20%-25% of the total product development budget, compared to the typical 5% to 7% today.

Time to Market

The final economic driver is time to market; from concept development to fully implemented production. Advanced technology introduced by competitors or demanded by customers, obsoletes current products and forces new product development programs for companies to remain competitive. Product life cycles are becoming shorter and in most industries will be even shorter in the future.

Currently, it can take as long as six years or more to develop a new product with total quality characteristics demanded by the market place. If it takes six years of a ten year product life cycle for product development, there simply is not way to recover product and process development and related capital investments. Reduced time to market is strategic ... regardless of cost...to remain viable in business tomorrow.

CP/PD Overview

Given the above economic drivers the strategic importance of CP/PD can be evaluated. Current processes for developing and producing new products take too long, are too expensive, are narrow in focus and do not always result in products with the precise performance and features that customers want to buy. Concurrent product and process development methods and capabilities if implemented properly, can:

A. Achieve and **improve** on worldclass standards for **product quality**.

B. **Reduce** overall **product costs** by 25% to 30% compared with product costs of leading competitors that are still using today's "build and test" serial product and process development methods.

C. **Shorten time to market** by 35% to 50%.

D. **Reduce** product and process **development costs** by 20% to 30%.

E. **Reduce capital investment** for state-of-the-art automation by 40% or more.

F. And at the same time **lower** overall product **business** risk significantly.

Specific CP/PD objectives are established by management at the beginning of each concurrent product and process development project. These objectives become the criterion by which program status and success are measured.

Similar Product Development Processes

CP/PD's activities are similar to product development processes currently being implemented by selected major manufacturers. The GM Four (4) Phase vehicle development process is similar to the CP/PD program. The Ford Concept to Customer product and process development program likewise is similar. Caterpillar's New Product Introduction (NPI) program within the Engine Division follows the CP/PD activities directly.

Organizing and Structuring for CP/PD Programs

CP/PD programs involve parallel versus serial development. The organization must be structured for team management of multiple parallel related activities. Successful implementation of CP/PD methods is guided by structured **program management** and is **continually enhanced** through application of experience and data gained in each previous CP/PD project.

Program Management

Product Control Management is responsible for selecting the pilot projects. During the project(s), product control management approves major milestones; i.e. moving the program forward from one activity to the next, recycling the program, if necessary, or killing the program if results do not warrant continuation.

Top Management is responsible for establishing objectives and providing the environment for CP/PD program managers and core teams to function. CP/DP requires established methods and procedures for managing resources and resolving conflicts among functional groups.

The first step in a CP/PD project is to organize the **Core Multifunctional Team** under a strong **Program Manager**. The team must be carefully selected to insure that proper skills are available to support the project. The core multifunctional team is the focal point of the CP/PD process but must be implemented within an overall organizational structure that supports the achievement of CP/PD objectives. The **Program Steering Committee** is made up of top functional organization managers and is chartered to work with the program manager and core team to resolve scheduling, resource, budget and technical problems judiciously and effectively when they arise in the program. The program manager and core team must identify and coordinate required resources with related functional organizations.

Often the CP/PD process is new and radically different. When this is the case, it is useful to use and outside consultant as a facilitator. The outside facilitator can take the "blame" for the perceived disruption to the normal routine until the process becomes part of the corporate culture. At that time, the consultant leaves and the perceived disruptive force is gone and the staff is left with a cohesive process. In a client follow-up study the comment was made," I never would have believed that we (product engineering) would be working together to close or with manufacturing as we do now."

CP/PD reduces time to market by permitting work to be done in parallel instead of sequentially. Multiple interrelated parallel activities introduce severe scheduling, tracking and resource planning requirements. Manual project management methods will not work effectively in this environment. The CP/PD project manager, related functional managers, team members, suppliers and customers, as appropriate, must have immediate access to the project plan and status, market models, product and process alternative data, capital and cost estimates, market and business models, etc., as needed. Electronic interactive project planning and resource allocations are essential. Current technology for data and information sharing is powerful and yet very inexpensive compared to its importance to CP/PD program success. Modern PC based interactive project management capabilities are extremely powerful, very user friendly and inexpensive. Many powerful vendor packages are available in the $1,000 to $2,000 price range.

Alternative Product and Manufacturing Concepts

Defined alternative market opportunities together with related functional specifications and target costs make it possible to formulate alternative product and manufacturing concepts accordingly. Predictive CP/PD simulation and analysis methods encourage CP/PD teams and related functional groups to identify multiple product and process concepts.

Day one on the project, the team defining current product extensions, competitor concepts and innovative concepts. Coarse simulation and analyses methods are used to evaluate each concept against specific functional specifications and engineering requirements defined using QFD capabilities and procedures. It is important that the team not get bogged down in detailed modeling and simulation at this juncture. Multiple product and process alternatives must be evaluated very quickly at the early development stages.

The combined QFD requirement capture technique for multiple potential market targets, together with multiple product and process alternative simulation, analyses and competitive evaluations, generate huge amounts of data and information. A very detailed decision support system is required to allow the team to make recommendations and for management to gain sufficient knowledge to make correct selections.

With management decisions can be made on both product and process concepts, and repeated at the subsystem design level. Combined system simulation and QFD methods are used to define detailed subsystem engineering requirements for the overall concept or concepts selected for further development. Alternative materials and processing strategies are evaluated for core components. Buy/make trade-offs are assessed for all major subsystems and components.

Again at early product and process development phases manufacturing, assembly and buy/make trade-off evaluations are done coarse and quickly. Sometimes as few as three or four alternative mix and volume assumptions are used to plot a cost/volume curve for a given manufacturing strategy; i.e., manual, semi-automated, fully automated cellular manufacturing for example.

The key to these efforts are to define rough manufacturing cost estimates and mix/volume ranges at strategy cross over points. These decisions involve major capital investment considerations in addition to fundamental manufacturing cost structures that the client must compete and survive with over and extended period of time; fifteen years or more in some cases.

The economics usually warrant significant up-front evaluation and analysis to assure that management makes correct business decisions; i.e. capital investment decisions (refurbish an existing production facility, build a new factory within an existing factory, build a greenfield facility, etc.), what should the company buy, what should it make, what costs structures are needed to compete globally in selected market segments, today, in the future, etc. These usually are very difficult decisions that end up being made by product engineering and manufacturing planning departments instead of by management. CP/PD programs attempt to provide management with the information and knowledge that they need to make such decisions intelligently, with a longer range focus but with less business risk.

Engineering and Manufacturing Analyses

Given subsystem concept selection and related manufacturing strategies by management at the end of the component design phase, overall product and manufacturing/assembly simulation models are upgraded for those choices and evaluated in an intermediate level of detail. High risk product and process assumptions are prototyped and tested to assure simulation and analysis validity. The type of automation and the level of automation is evaluated for each product concept (and related bill of material). The team identifies optimum manufacturing strategies based upon forecast mix and volume for the product family. The shop floor information integration requirements are defined as well as interface requirements for higher level computer integrated manufacturing (CIM) requirements defined by the client organization. Concurrent with product development, the team evaluates process plans, tooling and fixtures, purchase costs, work standards and scheduling rules for each manufacturing and assembly operation. Dynamic process simulation and analyses capabilities are used to identify process bottlenecks, throughput, cycle times, queue storage, outage recovery sequences, proper statistical burden rates, etc. Classical cost volume curves are recommended to identify correct manufacturing strategies for alternative product mix and volumes.

One very important lesson learned by working with clients on CP/PD programs is the importance of designing the product for the type and level of automation to be used in the production facility. Regardless of manufacturing strategy used; i.e., job shop, flexible cellular machining and assembly cells or higher volume flexible transfer line systems, **design for manufacturing (DFM) and design for assembly (DFA) pay handsome economic benefits.** A client turbocharger design improved efficiency by 300%, assembly time by 600% and reduced part count by 40%. Another client reduced part count by 25% assembly time by 30% and capital costs by over $2 million on a fuel system component.

CP/PD uses manufacturing process simulation capabilities extensively to evaluate and optimize various manufacturing and assembly lines and/or cells. Again, coarse layouts and simulations are developed from process plans and routings and are used until fundamental alternative manufacturing processes and strategies are developed and evaluated for various product and business alternatives. Similar to product simulations and analyses, manufacturing process simulations help the CP/PD team and manufacturing planning organization to identify high risk operations, estimate value added manufacturing plant costs and to estimate capital requirements for each alternative considered. These simulations become more detailed as manufacturing and assembly systems are developed for the selected processes and manufacturing strategies.

Advanced manufacturing development must be an ongoing process to identify and develop advanced technologies for incorporation into CP/PD programs. Strategic suppliers and vendors must joining the CP/PD team. The establishment of processes and strategic relationships must be piloted on the first CP/PD programs to initiate the network of suppliers needed to fully implement CP/PD. Purchasing is involved day one and strategic suppliers (major cost and quality related systems and components) are selected shortly thereafter. Strategic suppliers participate as part of the CP/PD team and are expected to achieve similar overall objectives. Major costs are attacked early but continuous improvement cost reduction programs continue after production start-up as normal. The effective use of business modeling provides the team and management with the tools and information to make good decisions. The business models perform the basic business calculations that are standard in the business community. The key issue is the data flow and integration with other CP/PD models. The business model must efficiently accept data from other models to support the many "What-If" alternatives and multiple iterations required to support the decision process. The business models must also support the process through the coarse-to-detail sequence. Examples of these business model inputs include: costs from manufacturing analysis; margin percentages from pricing models; volumes from market forecast models; capital investment estimates from manufacturing analyses; and product/process development investment from project plan.

The output of models must be easily accessible to the team and management. With many iterations and "What-If alternatives, the communication and change control requirements are more significant compared with more traditional development programs. Examples of business model outputs include: revenue streams; costs; profit stream; investments; cash flow; and any graphics and/or financial analysis (see Figure 13). Often strategic opportunities become clear. Using the output from a business model, one client moved from a quality with premium price niche to quality with price leadership.

ALTERNATIVE	REVENUE	CASH FLOW	INVESTMENT
Product	$ 25m	$(.3)m	$7.6m
All Clients	$275m	$ 59m	$33m
OEM	$225m	$ 96m	$49m
Client + OEM	$475m	$355m	$69m

These data show unit costs reduced by 47% by adding production volume thru OEM and client product integration. The results of the above example indicate the need for management involvement throughout the process. Without the broad perspective of management, many programs are not continued or are carried on without achieving maximum strategic benefits.

Summary - Key Management Issues

There is not shortage of ideas today concerning "what to do" to become more competitive.. "how to do it" and "doing it" are quite different matters. There are no "quick fixes." The common denominators for success with any new program have to include plain hard work, effective application of the basic, tenacity of purpose and attention to detail. The key issues for management are: recognize need; establish and communicate objectives; establish environment for success; participate in and support CP/PD project; establish measures of performance; and provide appropriate recognition.

The Design Process for Concurrent Engineering

NICHOLAS J. YANNOULAKIS, SANJAY B. JOSHI and RICHARD A. WYSK

Industrial and Management Systems Engineering
Pennsylvania State University
University Park, PA

Abstract
This paper lays a foundation for the construction of computerized concurrent engineering systems, by analyzing and outlining the design process of mechanical components in a way that readily lends its self to computer application. The design process is broken down to two phases that account for manufacturing and life-cycle issues. Two levels, that address component and assembly specifics are also considered.

1.0 Introduction

Recent requirements for increased manufacturing quality and production of frequently changing products have triggered an interest in concurrent engineering. This trend has been supported by the existence of mature engineering applications software and the development of more powerful computer hardware. Although limited-scope applications have been very successful, a formal description of the structure underlying such systems has yet to be presented. The use of a well-defined structure will be to provide an architecture which can be followed in the development of a concurrent engineering system. Such an architecture will provide the guidelines needed for proper definition of each subsystem's function, efficient data flow, and coordination between the design and manufacturing functions.

Older design theories [3, 5] considered design a stand-alone process with the sole purpose of generating a solution to a need. The means to realizing the solution (manufacturing) were not always factored into the design process. A recent report of the National Science Foundation (NSF) [4] points out that the design activity should be partitioned into "narrow vertical slices." These vertical slices should describe the design process from conceptual design to manufacturing and be narrow in order to be manageable. Suh et al. [6] also suggest the use of manufacturing knowledge during the design process. This knowledge is in the form of axioms which place a great importance on the function of the product and also dictate design simplifications.

The need for specific theories for classes of problems is well recognized [1, 4] and a taxonomy of mechanical design problems for the development of such theories is proposed in [2]. However, a general architecture under which specific design problems will be addressed has not been defined. This paper describes such an architecture, which will accommodate the various "vertical slices" of the domain of mechanical design. The different stages of the design process are described and the design-manufacturing link is discussed.

2.0 The Design Process

In contrast to the older, "stand-alone" philosophy, the design process must now be seen as a link between a need and the realization of the need. Mechanical design is accomplished in two phases: In phase I, the design engineer must completely describe the product. Since manufacturing requirements must be incorporated in this process, this phase is called *Integrated Design*. In phase II, prototypes are manufactured and put to operation or are tested under real-life conditions. Since this phase deals with issues that relate to the life cycle of the product, it is called *Life Cycle Design* [4].

2.1 Phase I--Integrated Design

Figure 1 depicts the various stages of the integrated design phase and their relationships. The whole process starts with a need which is translated into a set of *Functions* and *Constraints*. The functions describe what must be accomplished by the device or system that will be designed. The constraints describe the limitations on the functions. Each step of the design process is directly related to the set of functions and constraints because the designer must always make sure that the functions are satisfied and no constraints are violated.

The functions and constraints are also the first step of the design process. They are the rough guidelines that are used during the *Conceptual Design* stage to formulate the basic ideas for the final product. The result of the conceptual design stage is a number of alternatives (schemes) that must be evaluated. During the *Engineering Analysis* stage,

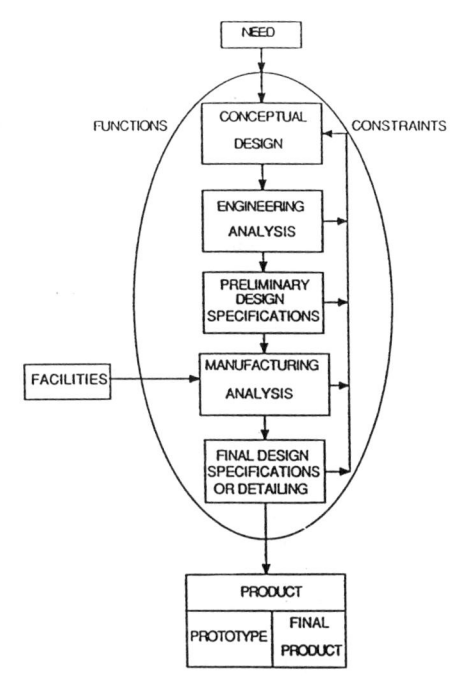

Figure 1. Phase-I of the Design Process -- Integrated Design

which follows, the conceptual designs are evaluated with technical criteria, like strength, heat transfer, vibration, and properties of materials. The criteria for evaluation are the applicable functions and constraints, while the methods vary greatly (FEM, simulation).

The calculation results, sketches, and computer printouts obtained up to this point are organized during the *Preliminary Design Specifications* stage into a drawing. Dimensions, tolerances and other technical details are specified here. The product of the three design steps described so far is a drawing that, in the classical sense of design, would be given to a draftsman for detailing and production would follow immediately. However, the principle of integrated design dictates one more step: *Manufacturing Analysis*. During this stage, the processes needed to manufacture and/or assemble the product are determined, based on the whole design and the functions and constraints. Furthermore, the ease of manufacturing and the available facilities are taken into consideration. As a last step in the integrated design phase, the detailed part drawings and specifications can be finalized for manufacturing (phase II).

Like in earlier analyses, the borders of each stage are not strictly defined, and neither are the tasks. An important realization that should be made is that several iterations of phase I must be completed in order for a product optimum for design and manufacturing purposes to be finalized. The advantage of the model presented here is that the range of each step corresponds to a distinct, well-defined knowledge base. This partitioning makes the design process less ambiguous and suitable for immediate computer implementation.

The diagram of Figure 1 essentially indicates the control structure of an expert system (particularly one with a blackboard architecture) capable of performing integrated design. The different stages can be seen as local experts that process the product of their previous stage and update the design global data base. At the same time, they indicate infeasibilities and generate suggestions for design improvements, which are fed back to conceptual design for consideration.

The challenge in completing this phase successfully is an accurate prediction of manufacturing difficulties (manufacturing analysis) without actually making or even planning the production of the designed part. For this reason, the development of comprehensive measures of manufacturability, like indices or expert systems, is of utmost importance. Such devices, applicable to specific design areas (stamping, extruding, casting) have appeared in recent literature.

However, the subject of evaluating machined parts has not been addressed adequately because of the amount of variables involved and their interdependencies. The quantification of manufacturing difficulty is also imposing a major hurdle in the process. A design evaluation method currently under investigation involves the separation of the part into manufacturing features which are evaluated with respect to a "most efficient" operation. The result of these comparisons is a *manufacturability index* that indicates the ease with which the part can be machined.

2.2 Phase II--Life-Cycle Design

Phase II of the design process is an evaluation of the product designed in the first phase, and has two branches, as depicted by Figure 2. The one branch examines the

```
        PROCESS PLANNING
               │
               ▼
    MANUFACTURING PROCESSES
               │
               ▼
     ANALYSIS ──▶ PROTOTYPE
        OF           │
     PROCESSES       ▼
        │          TESTING
        │            │
        │            ▼
        │      ANALYSIS OF
        │        RESULTS
        │            │
        ▼            ▼
      CONCEPTUAL DESIGN,
          PHASE I
```

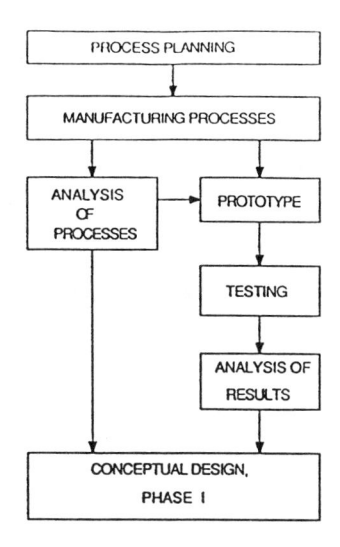

Figure 2. Phase-II of the Design Process -- Life Cycle Design

manufacturing aspect of the product and the other its functional aspect. Phase II starts by specifying the processes needed to make the product. Then, the processes are executed; if major problems occur, they are analyzed and suggestions are fed back to the conceptual design stage. If there are no manufacturing problems or they are small enough not to impede the production of a prototype, the prototype is made and tested.

This testing could be done in a laboratory or real setting or both, and indicates how the response of the final device differs from the response expected during phase I. The results indicate the areas of possible design improvements with respect life-cycle issues: fatigue strength, serviceability, and ease of use.

3.0 System and Component Design

The design process is divided in two levels: 1) design of components, and 2) design of systems. In both cases, the model of the design process that was described previously can be used. However, the line of thought and the specific tasks that must be accomplished during each stage are fundamentally different between the two categories. In the case of component design, functions like strength, size and surface finish of features are of the greatest importance. A lot of weight is placed on detail and performance of individual parts here. In the case of system design, the weight is placed on concept and overall operation of a system (mechanism). The exact dimensions of the components are not important but their relationships and interactions are.

System design, logically, comes first, especially in novel design. At this design level, the components are described only conceptually and are designed in greater detail later. In the cases of adaptive and, mostly, variant design, system design may be omitted.

The information that must be exchanged between the two levels should flow only from the one conceptual design stage to the other. This way, consistency in the structure of the design system is maintained.

4.0 Conclusion

A computer- and manufacturing-oriented definition of design is given in this paper. The model presented here gives an outline for structuring concurrent engineering system and can be used as a basis for the development of problem-specific design strategies and producibility evaluation schemes.

References

1. Dixon J. R., *On Research Methodology Towards a Scientific Theory of Engineering Design*, Artificial Intelligence for Engineering Design, Analysis, and Manufacturing, v.1, no. 3, 1988.

2. Dixon J. R., Duffey M. R., Irani R. K., Meunier K. L., Orelup M. F., *A Proposed Taxonomy of Mechanical Design Problems*, Proceedings of the ASME International Computers in Engineering Conference, v. 1, 1988.

3. French M. J., *Conceptual Design for Engineers*, Springer-Verlag, The Design Council, London, 1985.

4. National Science Foundation, *Research Priorities for Proposed NSF Strategic Manufacturing Research Initiative*, Report of a National Science Foundation Workshop Conducted by Metcut Research Associates Inc., March 11-12, 1987.

5. Pahl G., Beitz W., *Engineering Design*, Springer-Verlag, The Design Council, London, 1984.

6. Suh N. P., Bell A. C., Gossard D. C., *On an Axiomatic Approach to Manufacturing and Manufacturing Systems*, Journal of Engineering for Industry, v.100, no. 2, 1978.

Modeling Concurrent Manufacturing Systems Using Petri Nets

KELWYN A. D'SOUZA

Department of Industrial and Management Systems Engineering
University of South Florida
Tampa, FL

SUMMARY

A Computer-Integrated Manufacturing (CIM) System is complex, with distributive data, concurrent, and interacting processes. Individual processes can be carried out concurrently at each workstation under the control of a host computer. Petri Nets is a powerful tool for modeling and analysis of asynchronous, concurrent systems which exhibits synchronization and contention for shared resources, making it suitable for application to CIM systems. This paper discusses modeling, analysis, and performance evaluation of CIM systems using petri Nets. An application of Generalized Stochastic Petri Net model to a Computer-Integrated Assembly Cell is illustrated.

1. INTRODUCTION

Computer-Integrated Manufacturing (CIM) systems will play a major role in the development of advanced manufacturing techniques, and U.S. industries are expected to invest over $40 billion on CIM systems during the next few years [1]. A CIM system is used to manufacture a variety of products under the control of a host computer. It is a complex system handling distributive data, concurrent and interacting processes, which poses a major problem in studying the relationships and interactions among these structures. The concurrent and asynchronous nature of a CIM system could result in deadlocks and conflicts, thus affecting the system performance. In a manufacturing cycle, it is essential that once a process is started, it must terminate without any intermediate "dead-end" situations, and according to some set precedence relationships. Modeling techniques are required to study these characteristics off-line. Several modeling techniques have been developed for manufacturing systems [2]. Each have their own advantages and disadvantages. Petri Nets (PNs), with its rich mathematical background have gained popularity as a powerful tool for modeling automated manufacturing systems [3], [4], [5], [6]. This paper discusses the application of the Generalized Stochastic Petri Net (GSPN) for modeling, analysis, and performance evaluation of CIM systems, and is organized as follows: Section 2 introduces the theory of PN modeling. Section 3 contains a brief description of the analytical tools for the PN models. Section 4 illustrates the application of the GSPN to a Computer-Integrated Assembly Cell (CIAC). In section 5 the conclusion and proposition for future research are presented.

2. INTRODUCTION TO PETRI NET (PN) THEORY

This section follows the approach of Peterson [7].
A PN is a graphical model of information flow in a system. The structure is defined as a four-tuple, $C = (P,T,I,O)$ where :
$P = \{p_1, p_2, -- p_n\}$ is a finite set of places, $n \geq 0$, $T = \{t_1, t_2, -- t_m\}$ is a finite set of transitions, $m \geq 0$.
$P \cup T \neq 0$ and $P \cap T = 0$
$I : T \rightarrow P^\infty$ is the input function.
$O : T \rightarrow P^\infty$ is the output function.
Place p_i is an input place of transition t_j if $p_i \in I(t_j)$ and it is an output place if $P_i \in O(t_j)$. A PN structure can be represented as a PN graph, where a place is denoted by a circle - ◯, and a transition is denoted by a bar |. The input function I and output function O, relates the transition and places. This graph represents the static properties of the PN such as precedence and input-output relationships. Dynamic property is incorporated by tokens, represented by black dots •, which reside in the place P. A marking m is a function, $m : P \rightarrow N$. It is the assignment of tokens to the places P of a PN, and can be represented by an n - vector, $m = (m_1, m_2, -- m_n)$ where $n = |P|$; $m_i \in N$, $i = 1, 2, --- n$.

A PN can be executed by the firing of transitions which are enabled. Firing a transition will change the marking m of the PN to a new marking m' given by :

$$m'(p_i) = m(p_i) - \#[p_i, I(t_j)] + \#[p_i, O(t_j)] \text{-----------1)}$$

where $\#[p_i, I(t_j)]$ and $\#[p_i, O(t_j)]$ denotes the number of occurrences of the place in the input and output bag of the transition t_j. If we can obtain $m'(p_i)$ from $m(p_i)$ by using (1), then we say that m' is immediately reachable from m.

The reachability set of a PN C with marking m includes all markings reachable from m, and is defined as $R(C,m)$. The execution of the PN will continue as long as there is at least one enabled transition, resulting in a sequence of marking (m^0, m^1, m^2, ----) and a sequence of transitions which were fired (t_{j0}, t_{j1}, t_{j2}, ----). The sequence of marking and transitions are related by the function $\delta(m^k, t_{jk}) = m^{k+1}$
for $k = 0,1,2, ----$.
δ is called the next-state function.
A marked graph is defined by $M = (P,T,I,O,m)$.

3. ANALYSIS OF PN MODELS

A modeling tool is of little use, unless it can be used to analyze the system for the presence of desirable and undesirable properties. PNs can model CIM system graphically, and supports an in-depth analysis of the associated properties. Undesirable properties can be eliminated by a redesign of the system. There are several properties in CIM system which can be analyzed by a PN model. They can be classified [8] as those depending on the initial marking (marking dependent or behavioral), and those independent of the initial marking (structural). The former are the dynamic properties, while the latter are the static properties. Properties commonly addressed by researchers are reachability, boundedness, liveness, persistence, conservativeness, repetitiveness, and consistency. A detailed description of these properties can be found in [8].

Properties can be analyzed by the following techniques:
1) The Coverability (Reachability) Tree.
2) Matrix Equations.
3) Reduction or decomposition.
4) Invariant.

The Coverability (Reachability) Tree

For a bounded PN, the coverability tree is called the Reachability Tree (RT), since it covers all possible reachable markings. The RT is a finite representation of a reachability set R(C,m). The nodes represent markings of the PN and the arcs represent the possible changes in state, resulting from the firing of transitions. RT enables testing for safeness, boundedness, conservation, coverability, and reachability. If the tokens in the tree are ≤ 1, then the PN is safe. And if the number of tokens is a finite quantity 'k', then the PN is k-bounded. A PN will have a weight vector $w=(w_1,w_2,---w_n)$, where w_i is the weight for each place $p_i \in P$. If w exists such that the weighted sum of all reachable markings is consistent, then the PN is conservative. Inspection of the tree will indicate for a given marking m', if m" > m' is reachable.

Matrix Equation

Matrix equations can be used to study the conservative and reachability properties [7]. A PN is conservative if for a weight vector w, D.w = 0. If marking m' is reachable from m then there exist a firing sequence $\sigma = t_{j1} t_{j2}---t_{jk}$, which will lead from m to m', we can say, m'= m + f(σ).D --------------- (2).
f(σ) is the firing vector of σ.
D is the composite change matrix.

Reduction or Decomposition

A large problem can often be solved by reducing it into smaller subproblems. Transformations are available for performing reduction, which preserves the properties of liveness, safeness and boundedness. Research is required to develop transformation which allow hierarchical or stepwise reduction. Reduction methods applicable to CIM systems are introduced in section 4.

Invariant

An (nx1) non-negative integer vector x is called an S-invariant, or P-invariant iff:
$$x^T C = 0$$
An (mx1) non-negative integer vector y is called a T-invariant iff:
$$Cy = 0$$
C denotes the incidence matrix.
Invariant are useful for analyzing the structural properties of the PN model, and locating modeling errors [8]. A PN is structurally bounded if $x^T C \leq 0$, conservative if $x^T C = 0$, and consistent if Cy = 0.

4. APPLICATION OF GSPN TO A COMPUTER-INTEGRATED ASSEMBLY CELL

In the basic PN theory, the time for firing a transition was not considered. The PN model is asynchronous, and did not require time to control the sequence of events. Incorporation of transition time into the basic model is required for performance evaluation of the system. In most manufacturing applications, the transition firing (operation) times are assumed to be random variables. Stochastic Petri Nets (SPN) consider these times to be exponentially distributed random variables [9]. SPN can be defined as $S = (M, \lambda)$, where M is the marked PN defined earlier, and $\lambda = \{\lambda_1, \lambda_{21}, --- \lambda_m\}$, are a set of marking dependent transition rates. SPNs are isomorphic to homogeneous Markov process, due to the memoryless property of the exponential distribution. This property is useful for obtaining performance estimates. SPNs become very complex for large manufacturing systems, and the number of states of the Markov chain (MC) grows as the dimension of the model increases.

When the time durations of transition firing times differs by orders of magnitude, it is possible to associate times only with the activities having largest impact on the system performance. This extension of the SPN is termed as Generalized Stochastic Petri Nets (GSPNs). The GSPN [10] has two different classes of transitions called immediate (drawn as thin bars) and timed (drawn as rectangular boxes) transitions. By defining these two classes of transitions, GSPNs reduces the number of MC states, and are still equivalent to Markovian models. Once enabled, the immediate transitions fire in zero time (infinite firing rate), and timed transitions fire after a random, exponentially distributed time. When several transitions (Set H) are simultaneously enabled by a marking, the enabled transition fires with a probability:

$$\lambda_i / \Sigma_{k \in H} \lambda_k$$

If H consists of immediate and timed transitions, only the immediate transitions can fire with some defined probability distribution. And this subset of H forms a random switch where the associated probability distribution is called switching distribution.

The GSPN model can be applied to analyze and evaluate the performance of a CIM system. The CIM system can be said to comprises of immediate transitions (start process) and time transitions (complete process). Greater flexibility has been added to the GSPN model by introducing deterministic and exponentially distributed firing times. Such models have been termed [11] as Deterministic and Exponential SPN (DSPN). The enabled transitions in DSPN can fire after either a deterministic or exponentially distributed time. This model is applicable for CIM system having some deterministic operation times (eg. robot operations).

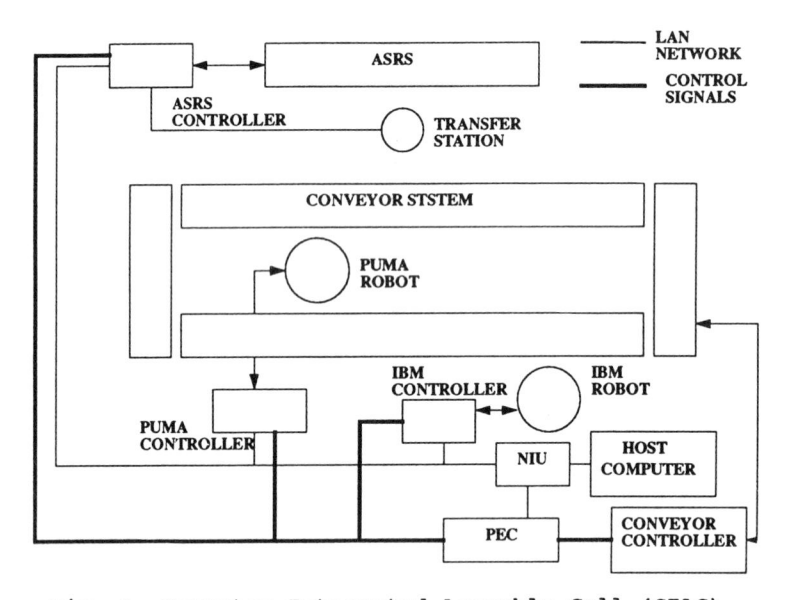

Fig. 1. Computer-Integrated Assembly Cell (CIAC)

The Advanced Automation Lab of the Industrial and Management Systems (IMSE) Department has recently setup a computer integrated assembly cell (CIAC). The cell consists of Robots, ASRS, Programmable Electronic Controller (PEC) and conveyor system, all connected by a communication networking system to a host computer. Figure 1 shows an outline of the cell. Programs for assembling a range of products are stored in the host Computer, and are downloaded to the workstations according to the job schedule. During operation of the cell, the host computer controls the input and output signals to the workstations and conveyor through the PEC. The host computer maintains statistics on work station performance and throughput rates. Assembly operations are carried out concurrently on the work station while parts are loaded and unloaded from the ASRS. Prior to commencement of production, it would be essential to analyze the assembly process plan, to avoid conflict and deadlock situations. Also, performance evaluation of the cell is to be conducted to determine the throughput and station utilizations.

Once the assembly process plan for a product is known, a GSPN model can be constructed. Equipment breakdown and repair activities can also be modeled into the GSPN. Figure 2 shows a partial view of the GSPN model (Details have been suppressed due to lack of space). The GSPN model will be analyzed for invariant, using the GreatSPN software package [12]. The package can also perform a performance evaluation for each workstation. Feedback from the analysis and performance evaluation will be used by the process designer to detect logical errors of the system and maximize throughput.

GSPN models become complex with an increase in the number of reachable states and events, giving rise to errors in modeling and analysis of the system. Hence, it is essential to reduce the net size using decomposition techniques, which retain the desirable properties of the original nets. The GSPN model developed for the CIAC will be decomposed using the Hierarchical Reduction methods suggested by Lee and Favrel [13], [14]. Due to the presence to timed transitions which differ by orders of magnitude, the method of Time Scale Decomposition (TSD) will also be applied [15]. These are parts of the current research plan of the IMSE Department.

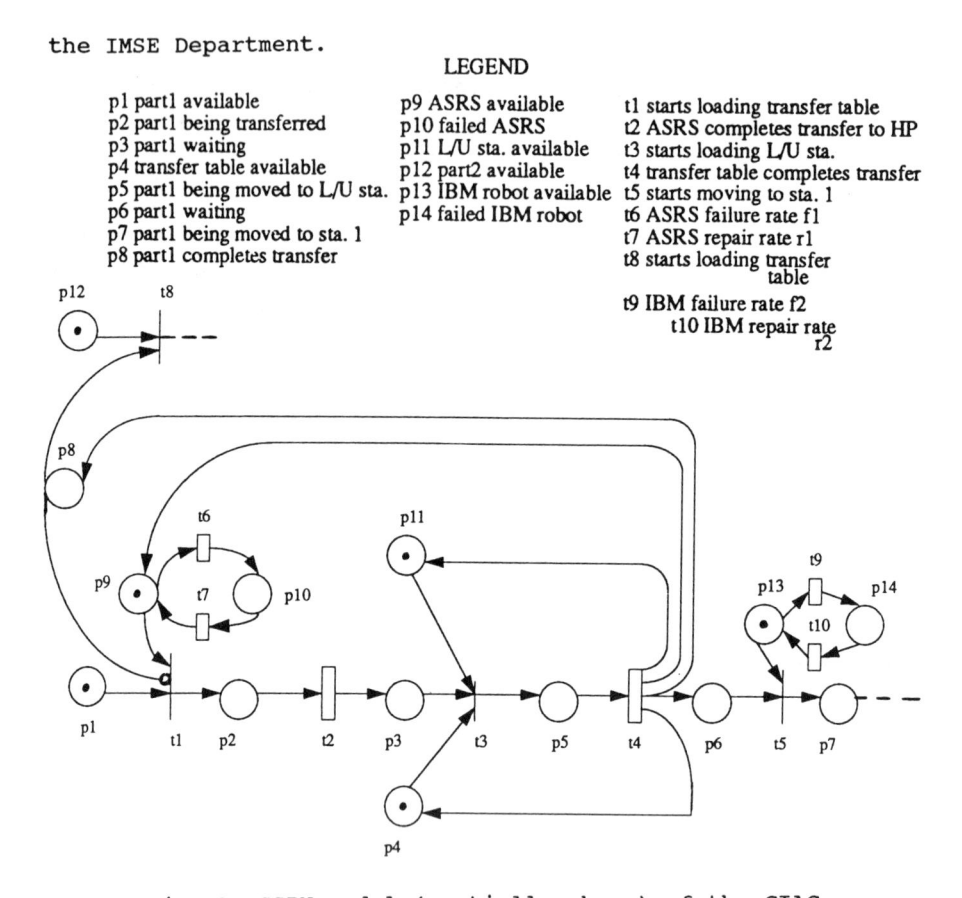

LEGEND

p1 part1 available
p2 part1 being transferred
p3 part1 waiting
p4 transfer table available
p5 part1 being moved to L/U sta.
p6 part1 waiting
p7 part1 being moved to sta. 1
p8 part1 completes transfer

p9 ASRS available
p10 failed ASRS
p11 L/U sta. available
p12 part2 available
p13 IBM robot available
p14 failed IBM robot

t1 starts loading transfer table
t2 ASRS completes transfer to HP
t3 starts loading L/U sta.
t4 transfer table completes transfer
t5 starts moving to sta. 1
t6 ASRS failure rate f1
t7 ASRS repair rate r1
t8 starts loading transfer
   table
t9 IBM failure rate f2
t10 IBM repair rate
   r2

Fig. 2. GSPN model (partially shown) of the CIAC.

5. CONCLUSIONS AND SCOPE FOR FUTURE RESEARCH

This paper has developed an approach for modeling a CIM system using the theory of PNs. An extension to the basic PN called GSPN appear to be suitable for modeling such systems. The model is being implemented for product assembly at the CIAC of the IMSE Department. This research will enable the development of a generic model which will be applicable to a more generalized CIM system. The complexity of the GSPN model can be reduced by the application of decomposition techniques. The model discussed considers a single product being assembled on the CIAC at a time.

For the cell to be truly flexible, it must be able to assemble products at random. More than one product could be loaded to the cell at a time. Such a model would be more complex than the case discussed, and the use of Colored PNs [16], or Colored Stochastic PN [17] may have to be applied. Research in software development for the analysis and performance evaluation of large size models is required.

REFERENCES

1. Branam, J.M.: JIT vs FMS - which will top management buy?. Flexible Manufacturing Systems, T.J. Drozda (ed.). SME, Dearborn, MI (1988) 30-36.

2. Suri, R.: An overview of evaluative models for Flexible Manufacturing Systems. Annals of Op. Res. **3** (1985) 61-69.

3. Al-Jaar, R.Y.; Desrochers, A.A.: A Survey of Petri Nets in Automated manufacturing systems. $12^{th.}$ IMACS World Congress, Paris (1988) 503-510.

4. Hillion, H.P.; Proth, Jeane-Marie: Using Timed Petri Nets for the scheduling of job-shop systems. Engineering Costs and Production Economics. **17** (1989) 149-154.

5. Zhang, W.: Representation of Assembly and Automatic Robot Planning by Petri Nets. IEEE Trans. Syst., Man and Cyber. **19** (1989) 418-422.

6. Teng Sheng-Hsien; Black, J.T.: Cellular Manufacturing System Modeling: The Petri Net Approach. Jr. Mfg. Syst. **9** (1990) 45-54.

7. Peterson, J.L.: Petri Net theory and the modelling of systems. New Jersey: Prentice-Hall, Inc. 1981.

8. Murata, T.: Petri Nets: Properties, Analysis and Applications. Proc. of IEEE **77** (1989) 541-580.

9. Molly, M.K.: Performance analysis using Stochastic Petri Nets. IEEE Trans. on Computers **C-31** (1982) 913-917.

10. Marsan, M.A.; Conte, G.; Balbo, G.: A class of Generalized Stochastic Petri Nets for the Performance evaluation of Multiprocessor Systems. ACM Trans. on Comp. Syst. **2** (1984) 93-122.

11. Marsan, M.A.; Chiola, G.: On Petri Nets with deterministic and exponentially distributed firing times. Advances in Petri Nets. G. Rozenberg (ed.). Berlin, New York: Springer-Verlag (1987) 132-145.

12. Chiola, G.: GreatSPN User's Manual Ver. 1.3. Dipartimento di Informatica Universita degli Studi di Torino corso Svizzera 185, 10149, Torino, Italy (1987).

13. Lee Kwang-Hyung; Favrel, J.: Baptiste, P.: Generalized Petri Net Reduction method. IEEE Trans. Syst. Man and Cyber. **SMC-17** (1987) 297-303.

14. Lee Kwang-Hyung; Favrel, J.: Hierarchical Reduction method for analysis and decomposition of Petri Nets. IEEE Trans. Syst. Man and Cyber. **SMC-15** (1985) 272-280.

15. Ammar, H.H.; Islam, S.M.R.: Time scale decomposition of a class of Generalized Stochastic Petri Net models. IEEE Trans. Software Engg. **15** (1989) 809-820.

16. Jensen, K.: Coloured Petri Nets and the Invariant-Method. Theoretical Computer Science **14** (1981) 317-336.

17. Zenie, A.: Colored Stochastic Petri Nets. IEEE Intl. Workshop on Timed Petri Nets, Torino, Italy (1985) 262-271.

Production Planning and Control in the Factory of the Future

W.H. ISKANDER and M. JARAIEDI

Department of Industrial Engineering
West Virginia University
Morgantown, WV

Abstract
In this article, the concept of Concurrent Engineering and the modules of Integrated Production Systems are presented. These are essential ingredients for the success of American companies in the future. Implementation factors for proper communication and integration at both the design and production stages are also discussed.

Introduction

To survive the fierce competition from Europe and Japan, and to secure a good share of the world market, American companies have to apply modern technology and state of the art in their production planning and control, such that high quality products can be designed and produced in a short time frame and with a minimum cost.

Two factors are essential for successful planning and control both at the design stage and at the production stage of the product. These are "Communication" and "Integration". All elements involved either at the design or at the production of an item need to be integrated together to reduce or eliminate redundancies and contradictions. An effective communication system is essential for a successful integration.

Concurrent Engineering

At the design stage, the recent concept known as simultaneous or concurrent engineering is becoming more and more important and popular. According to Winner et. al. (1), Concurrent engineering is a systematic approach to the integrated, concurrent design of products and their related processes, including manufacture and support. This approach is intended to cause the developers, from the outset, to consider all elements of the product life cycle from conception through disposal, including quality, cost, schedule, and user requirements. Concurrent engineering focuses on the decisions made in the first twenty percent of program effort which drive approximately eighty percent of product performance, producibility, reliability, maintainability, schedule, and life cycle cost (2).

The main objective of concurrent engineering is to reduce the time needed to introduce new high technology materials, products, and processes. Its strategy is to accelerate schedules by running phases of a project in parallel rather than in series. In essence, research and development, design, testing, and qualification assessment are performed simultaneously and interactively. Concurrent Engineering can result in high product quality and low cost by designing the product for quality, cost, manufacturability, and assembly.

To achieve its objectives, concurrent engineering requires the establishment of good communication channels for close and timely interaction between various members of the design and manufacturing teams. Controlled discussions and design iterations conducted by the "project leader" through electronic "blackboards" insure that input from all disciplines are incorporated into the product design.

Integrated Production System

At the production stage, a fully integrated computerized system should be used to control the flow of information and material, to coordinate scheduling and production between the different departments, and to provide management with good decision support tools. Such a system should be capable of tracking all raw material, in-process inventory, and finished goods throughout the plant. It should also be able to keep track of the availability of all personnel and facilities. The integrated system should have modules that cover all areas of production, starting from the forecasting of demands, to the process of order entries, inventory control, process planning, and scheduling of operations. A brief description of these modules and their principal features is given below.

Forecasting of Demands

Forecasting forms the basis for the development of good process planning and scheduling and for the control of inventory levels. Long term forecasting can be very helpful in setting the companies goals and strategies, while short term forecasting can play a vital role in the development of production and maintenance schedules, and in the determination of raw material and parts to be ordered. Management must forecast the future demands of its products and on this basis provide for the materials and labor required to fulfill these needs. A reliable forecasting system can provide much needed control of material inventory to minimize in-process inventory storage and inefficient process utilization. Forecasting techniques can be categorized into three groups. First is the "qualitative" group, where all information and judgement relating to individual items are used to forecast the items' demands. This technique is often used when little or no reliable history of demand history is available. Second is the "causal" group, where a cause-and-effect type of relation is sought. Here a forecaster seeks a relation between an item's demand and other factors, such as business, industrial, and national indices. Third is the "time-series analysis" group, where a statistical analysis of past demands is used to generate forecasts. A basic assumption here is that the underlying trends of the past will continue in the future.

Order Entry

It is important for manufacturing facilities to have an order entry system which receives customer requirements, translates them to the facility's language and operations system, and acknowledges the order to the customers. These orders are then entered into control books, which may be computerized, and forwarded to the scheduling and production planning systems. Coupled with a good forecasting system, the order entry system forms the basis upon which operations scheduling and production planning are performed. In addition, order entry systems can have by-products that may include, but are not limited to, booking controls, bookings by month, week, or by tons of product, order acceptability, specification acceptability, credit audit, processing information, packaging, loading and shipping information, processes and changes to be accomplished at each step of the operation, ordered quantities, and promised delivery dates.

The customer order entry system should provide a complete, timely, and accurate database of activities in all stages from forecasting through finished product inventory. It should provide the user with an automated, computerized file generation system that eliminates the need for manual key punch and thus improves both speed and accuracy. It should also provide an interface with the customer so as to provide a direct entry of orders, acknowledgments to the customer, and pricing and delivery promises. The system should allow for ways to cross check customer orders against the production records to guarantee highest levels of accuracy. It should also allow changes to the orders to be entered directly and immediately evaluated and be either accepted or rejected based on order status and conditions at the plant.

Inventory Control

Inventory can be viewed as idle resource that has some economic value. The true cost of inventory is usually not known and may not be easy to accurately estimate. Recent methodologies like MRP and JIT have been developed in order to reduce, or if possible eliminate all inventories. Such methodologies may work very well in some environments, but not in all industries. Some plants may have a wide variety of products that may be quite diverse and that may be requested in small batches. Also some of the operations may require very large set up costs and times. Large set up costs or small batches of diverse products may prohibit the application of JIT principles and may necessitate the intentional production of in-process inventory. Other inventories may also exist due to processing errors, shipping date uncertainty, or order cancellation. Inventory should be either eliminated or dealt with as a necessary evil. It should be kept under control and at minimum levels at all times. An on-line computerized system usually needs to be applied to keep track of the inventories of all raw material, and in-process and final products at all times and at all production stages. The application of such a system, together with good order entry and production planning and scheduling systems should help control inventories at acceptable levels.

To develop a strategic inventory management system one has to first develop systematic and computerized approaches to estimate the costs involved in carrying inventories at all production stages. These should cover as a minimum the following cost components: cost of capital tied up, storage cost, deterioration, insurance cost, and material handling cost. Software can then be developed to keep track of the amount and location of each type of raw material, in-process inventory, and final product. This information, together with the order entry data are vital to the success of the production planning and scheduling system, which is the proper vehicle to keep the inventory levels under control.

Process Planning and Scheduling

Scheduling is an important decision-making function in most industries. When large volumes are produced or when large number of resources are allocated over time to different jobs, the scheduling process becomes rather difficult. With proper planning and scheduling, small improvements can sometimes be translated into millions of dollars saved. These improvements may be in the form of reducing lead times, reducing in-process inventories, better utilization of resources, increasing throughputs, reducing production cost per unit, etc.

Generating manual schedules for complicated operations may not be the best answer to the scheduling problem. Using algorithms that have been well tested, and taking advantage of computerized methods that can quickly produce and evaluate several schedules can definitely help produce better schedules. A successful schedule must be well integrated with other components, namely long and short term forecasting, order entry, automatic and manual data collection, and inventory management.

The development of a computerized scheduling system start with the performance of a complete and thorough analysis of the resources available, flow of material, and current processes. An investigation of the computerized systems used for order entries, automatic data entries, and inventory control systems is essential for the proper integration of the scheduling system with all other systems. One should also analyze the scheduling practices applied at the plant, define the possible bottlenecks and the operations that may need special attention, and identify the different objectives of the schedules produced. Scheduling algorithms should then be developed to satisfy as much as possible the objectives defined. These algorithms must be flexible enough to allow the planner to tailor the schedule to meet different objectives that may be dictated by the prevailing conditions and market constraints. Simulation models can be developed to test the developed algorithms. After validating and adjusting the algorithms developed, they should be incorporated in a computerized system and integrated with the overall production control system.

References

1. Winner, R.I., et. al.: The Role of Concurrent Engineering in Weapon System Acquisition. Report R-338, Institute of Defense Analyses, Alexandria, VA. (1988)

2. DARPA: Workshops on Concurrent Design. Key West, Fl. (1987,1988).

Expert Control of Turning Process

P.S. SUBRAMANYA, V. LATINOVIC and M.O.M. OSMAN

Department of Mechanical Engineering
Concordia University
Montreal, Quebec, Canada

Summary

There has been a steady attempt to use Artificial Intelligence (AI) and Expert system techniques for real time control systems.The purpose of this paper is to illustrate the methodology for successfully adopting rule based expert system with process knowledge for the control of turning process. This is done by treating control of the machining process as a decision making problem, involving high degree of uncertainty, thereby establishing a decision making framework to be applied in tandem with multiple sensors.The system was tested for reliability under simulated turning conditions.

Introduction

Recent developments in computer networks have enabled full integration of DNC and CNC in a hierarchical manufacturing system with distributed control. Within such a system the individual machine tool should ideally be able to carryout the machining activities independently and adaptively without human intervention at the machine hardware level. This is the goal of research into adaptive control.

Adaptive control involves a) sensing the changing environment (e.g., tool wear or work material hardness variations etc.,) b). decision making, and C) taking actions (altering cutting speed and/or feed rate etc.,). The research described here is predominantly concerned with the decision making database.

Previous work in adaptive control of machine tools can be classified into two groups:

The first utilizing the well known Taylor's equation (or extended ones) for the particular tool workpiece combination and economic criteria to optimize the cutting speed and feed rate. This method is usually called Adaptive Control of Optimization (ACO) [8].

The second focussing attention on maintaining a "safe" optimization under certain physical constraints. This method avoids difficulty associated with optimization and is called Adaptive Control of Constraints (ACC) [3-5].

Many ACO and ACC systems have been investigated [4-7]. The ACO method faces the problem that a huge database need to be established before the method can be applied. Furthermore, its reliance on Taylor's tool life equation to obtain the optimum working conditions makes it unsuitable for implementing on-line real time control, in the general sense. For example, it is not clear that stabilizing cutting forces or temperature are an advisable policies for all machining processes. In the second method it is difficult to approach the global optimum working conditions with an incomplete knowledge of relationship among the machining variables.

Expert System Method

Expert systems techniques are often effective for controlling complicated problems containing a substantial degree of uncertainty, and thus offer a possible approach for controlling the machining process [2].This is done by treating the control of the machining process as a decision making problem and by establishing a decision making framework to be applied in tandem with multiple sensors technique. The objective of expert control is to encode process knowledge and decision capabilities to allow intelligent decisions and recommendations automatically, rather than to preprogram logic which treats each case explicitly.This approach attempts to model those aspects of a problem which are not naturally amenable to numerical representation or which can be more efficiently represented by heuristics. Fig.1 shows the expert control scheme and Fig.2 expert system architecture adopted for control of turning process.

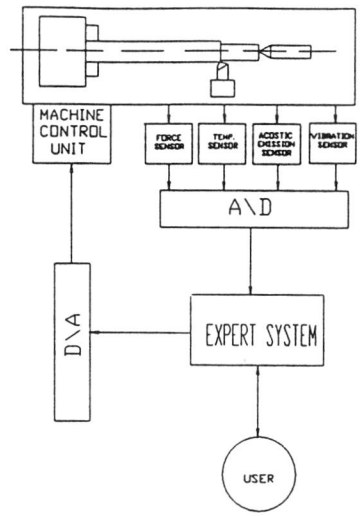

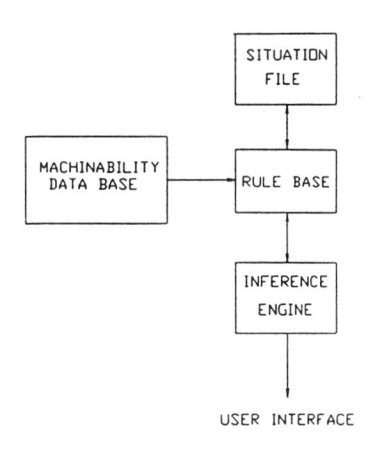

Fig.1 Expert Control Scheme   Fig.2 System Architecture

The System Data Base

In order to achieve on line prediction and identification of the machining process, it has to be represented by mathematical models. These models provide the hypothesis for selection of proper machining data. This can be achieved by suitable adaptive optimization algorithms.

There are various methods for estimating the parameters for the machinability models. In order to fulfill the present requirement of continuous monitoring of the changing dynamics of the machining process the use of sequential Maximum a posteriori(MAP) estimation procedure based on the Bayesian statististical approach was found very useful [1]. The recursive equations for this procedure are proposed as

$$A_{u,i+1} = \sum_{k=1}^{p} X_{i+1,k} P_{uk,i}$$

$$\Delta_{i+1} = \sigma_{i+1}^{2} + \sum_{k=1}^{p} X_{i+1,k} A_{k,i+1}$$

$$K_{u,i+1} = \frac{A_{u,i+1}}{\Delta}$$

$$e_{i+1} = Y_{i+1} - \sum_{k=1}^{p} X_{i+1,k} \, b_{k,i}$$

$$b_{u,i+1} = b_{u,i} + K_{u,i+1} \, e_{i+1}$$

$$P_{uv,i+1} = P_{uv,i} - K_{u,i+1} \, A_{v,i+1}$$

Where $u = 1,2,\ldots,p$, $V = 1,2,\ldots,p$, p is the number of parameter estimates, X is the independent variable, Y is the dependent variable, σ^2 is the variance of Y, P is the covariance of estimators, A is the product matrix, K is the gain matrix and b is the estimator. This adaptive estimator becomes the hypothesis part of the system database.

The Rule Base

The rule base contains the production rules which are typically described as :"if <situation> then <action>". In order to achieve a high speed decision as desired by the real time control system, metarules are provided in order to invoke only the set of rules which are required for reaching a goal. In order to obtain reflex-action like decision for unforeseen circumstances, metalevel inference is made possible by this type of grouping of rules.

The 'situation' as a part of the rule represents the actual situation in machining, namely signals from the data which re interpreted as chatter or temperature at the tool point or acting force levels etc. The 'action' can represent the activation of a controller or an estimation algorithm. The rules may be seen as functions operating on the state, since the data base is broader in concept then usual notion of state in control theory, production rules are also richer in content than common transition functions.

Since it is a common knowledge that the machining process follows some hypothesis only if the machining operation is chatter free i.e., stable, the priority will be given to chatter suppression rules over other rules. For example when rough machining steels,

```
IF <chatter>
THEN < get(hyp(stable(v,s))) >
AND  <alter(v,s)>
```

Fig.3 illustrates the stability hypothesis. The detailed discussion of rules and hypothesis are beyond the scope of this paper.

V = Cutting speed.
S = Feed rate.
+   implies 'increase'
-   implies 'decrease'
Signed numbers represent change in stability with confidence factor.

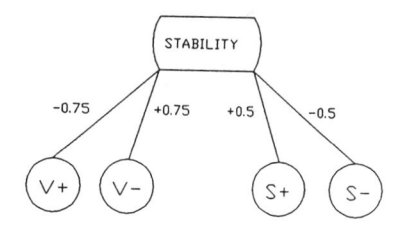

Fig.3. Stability hypothesis

Inference Engine

The purpose of the inference engine is to decide from the context (current data base of facts, evidence, hypotheses and goals) which production rules to select next. Even though various strategies have been developed and utilized in general expert systems, these techniques cannot be directly applied in expert control of process due to the long execution time that can result if the knowledge base becomes really complex. In order to overcome this problem a different kind of search technique has to be used. The graph method of inferenceing and dividing the inference process among different modules has been proved to be the best method available[6].

User Interface

The main purpose of user interface in expert control of turning process is to override the actions of the machine in case the operator decides so and also to explain the line of action being taken. It also provides the editor for rule base and the system database.

Conclusions

The expert system method for controlling the turning operation is found to be very effective when subjected to simulated turning conditions. It is observed that the system can provide a satisfactory performance even with incomplete and unreliable data and knowledge. The main aspects of expert control of turning process are:
1. Signal interpretation to infer state of machining process,
2. Multisensor function to combine information from several different signal sources in order to get a more dependable description of the process,
3. Anomaly detection,
4. Classification of anomaly into process anomaly and environmental anomaly etc.,
5. Diagnosis of the process to determine the causes of the situation under monitoring.

References

1. Beck, J.V.; and Arnold, K.J.: Parameter Estimation in Engineering and Science. John Wiley & Sons. 1977.

2. Chryssolouris,G.; Guillot, M.; and Domroese, M.: A Decision Making Approach to Machining Control. Trans. ASME Jr. of Engg. for Industry. 110 (1988) 397-398.

3. Jaeschke, J.R.; Zimmerly, R.D.; and Wu, S.M.: Automatic Cutting Tool Temperature Control. International Jr. of Machine Tool Design and Research. 7 (1967) 465-475.

4. Kaminskaya, V.V.; et al.: Trends in Development of Adaptive Control. Machines and Tooling. 45 (1974) 66-70.

5. Opitz, H.: Adaptive Control- Fundamental principles for Numerical optimization of Cutting Conditions. ASME Proc. of the International Conference on Manufacturing Technology. (1967) 25-28.

6. Subramanya, P.S.; Latinovic, V.; and Osman, M.O.M.: Expert System Architecture for Machining Process Control. Unpublished report. Concordia University. Montreal.

7. Wick, C.:Automatic Adaptive Control of Machine Tools. Manufacturing Engineering. (1977) 38-45.

8. Yonetsu, S.; Insaki, I.; and Kijima, T.: Optimization of Turning Operations. 6th North American Manufactruing Research Conference Procedings. (1978) 17-23.

Expert System for Milling Process Selection

B. GOPALAKRISHNAN and M.A. PATHAK

Department of Industrial Engineering
West Virginia University
Morgantown, WV

Abstract
The application of artificial intelligence in the field of manufacturing engineering have been quite widespread, especially in the areas of diagnostics, debugging, process evaluation, planning, design, and classification problems. Expert systems are valuable tools to be used in manufacturing engineering, especially in the selection of various parameters that contribute to the quality and cost of the product being manufactured. In machining, particularly milling, the selection of the appropriate process (es) to be used to machine the part is dependent on heuristic and domain dependent knowledge which can be represented as production rules. This paper will outline the design and development of an expert system which will enable the selection of milling processes for the manufacture of a product, describe data and knowledge acquisition methods, examine inference procedures, and identify effective expert system design methods.

Introduction - Expert Systems
Artificial Intelligence (AI) has been gaining popularity since the 1960s. Expert Systems (ES) within artificial intelligence are now beginning to be applied in industrial and manufacturing areas as a means of solving problems and as decision making aids. Expert systems are also called Knowledge Based Expert Systems (KBES). Feigenbaum [1] has defined KBES to be intelligent computer programs which use knowledge and inference procedures to solve problems that are difficult enough to require significant human expertise for their solutions. This human

---

Acknowledgement
This work has been sponsored by the Defense Advanced Research Projects Agency (DARPA), under contract No. MDA972-88-C-0047 for DARPA Initiative in Concurrent Engineering (DICE).

expertise is obtained from an expert. An expert is a person who through training and experience can perform a task with a high degree of skill. An expert system has three main components (Figure 1), namely the knowledge base, the inference engine, and the working memory.

Knowledge Base
The knowledge base stores the facts and heuristics of domain experts. This knowledge is acquired from documents and consultation with experts. Knowledge acquisition is accomplished by meetings between people developing expert system and domain experts. The knowledge may be represented as semantic networks, frames, or rule based knowledge. Each section of the knowledge base may be independent in terms of conclusions and input data.

The Inference Engine
The inference engine provides the system control. The method used to reproduce part of an expert's reasoning process is called chaining. There are two methods presently available, namely, Forward Chaining, and Backward Chaining. In forward chaining, the applicable set of known facts are used to trigger applicable rules. These facts are stored in working memory. From the total set of facts collected, additional rules are triggered till a goal condition is reached. In backward chaining, the expert system starts out with a known goal searching for rules or subgoals that satisfy this or part of this goal. Each of the subgoals are checked against already existing data or data input by the user. This process continues until the goal is verified as true or false, based on data.

The Working Memory
The working memory contains the information that the system has received about the problem at hand. In addition, any information the expert system derives about the problem is also stored in the working memory.

Milling
The milling process is one of the most important metalworking processes. Milling is a machining process which removes material by the relative motion between a workpiece and a rotating cutter with multiple cutting edges to produce flat and curved surfaces. The basic milling operations can be categorized as peripheral milling, face milling, and end milling. In peripheral milling, the cutting teeth are located on the outside periphery of the cutter body and are most often parallel to the main cutter axis. In face milling, the cutter uses two different cutting edges to

achieve the finished surface while in end milling the cutter has teeth on its periphery and on its edge. These processes can generate various kinds of slots. In this research four kinds of slots, namely, square, t-slot, v-slot and dovetail slots are considered. Two of these slots are shown in Figure 2.

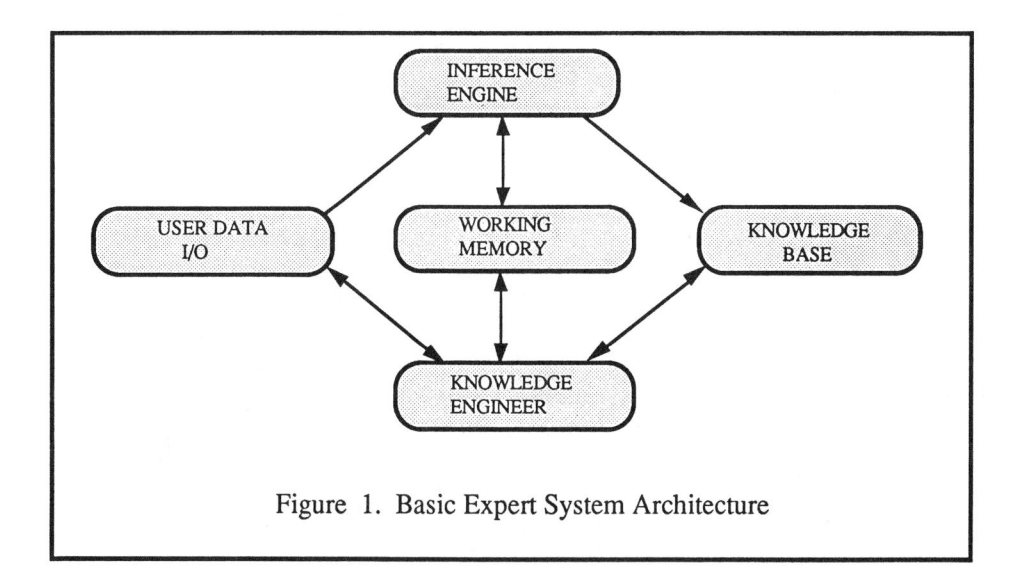

Figure 1. Basic Expert System Architecture

Milling Process Selection

The development of a process selection module is very important in the case of non-NC milling as opposed to NC milling. The geometry of the slot will dictate the milling process/es and their sequence. These geometric are used as an input to the expert system. Since the features are "text-like" having a numerical value, symbolic manipulations of them is easily done using expert system. LASER is the expert system used in this module.

LASER

LASER is an object oriented expert system [2]. Object oriented programming is a set of techniques that allows programs to be built using objects as the basic data item and using actions on objects as the active mechanism. An object here is a data structure that contains all information related to a particular entity. Each rule has a right hand and a left hand side. If the conditions specified in the left hand side (IF) of a rule are satisfied by the current state of the data, then the actions specified in the right hand side (THEN) of the rule will be accepted when the rule is selected and executed.

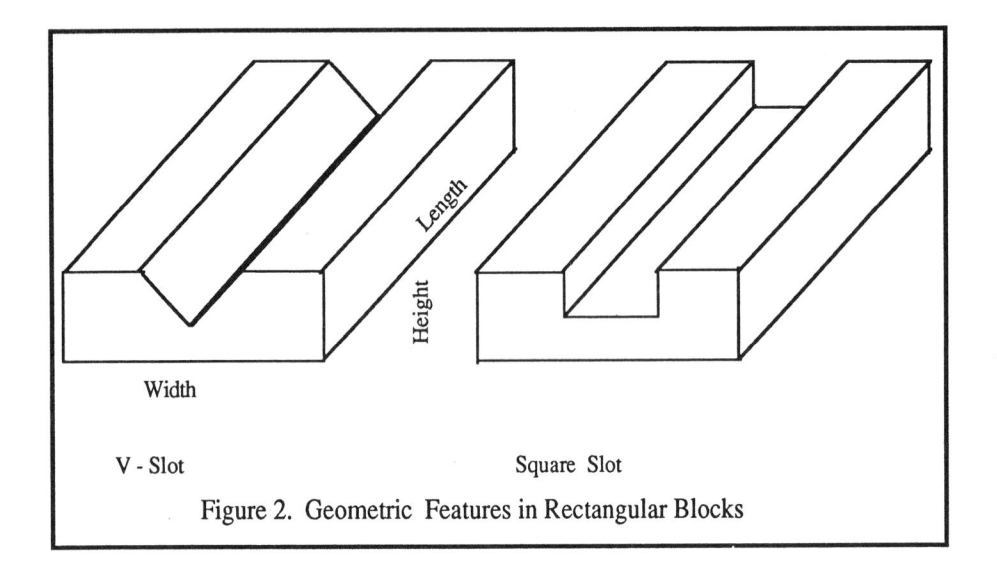

Figure 2. Geometric Features in Rectangular Blocks

This expert system contains the knowledge of milling process so that it can conclude on the operation type, number of operations, their sequence and the width of cut for each operation. This knowledge required to build expert system was effectively collected from various handbooks and manufacturing textbooks and intuitively transformed into rules.

Expert System Development
There are five variables involved in this expert system to select a milling process for four kinds of slots. These are 1.) width of the slot, 2.) depth of the slot, 3.) face mill diameter, 4.) end mill diameter and 5.) width of the peripheral mill. If a user needs to know the process to generate a square slot, then the get_info rule, shown below, will ask the user to input the type of slot, the width and depth of the slot. Next he/she is asked to input the face mill diameter, end mill diameter, peripheral mill width closest to the slot width value.
{ get_info

```
    instanceof # : LSR_GLOBAL_RULE
    LHS : "slot slot_type = 'lsr_null'"
    PRIORITY : 1
    RHS : "read read_name ?NAME "
        " makeobj ?NAME"
        "setval slot slot_type ?NAME"
        "read read_wide ?WIDE "
        "setval slot slot_width ?WIDE"
        "read read_deep ?DEEP "
        "setval slot slot_depth ?DEEP"
        "read read_fdia ?FDIA "
        "setval slot face_dia ?FDIA"
        "read read_edia ?EDIA"
        "setval slot end_dia ?EDIA"
        "read read_pwidth ?PWIDTH"
        "setval slot p_width ?PWIDE"
}
```

The values input by the user are stored in an object called slot. These data are used by the inference engine to examine and select rules.

```
{ slot
    end_dia : 0.8
    face_dia : 2.5
    face_dia1 : 0.0
    p_width : 3.2
    slot_angle : 0.0
    slot_depth : 1.1
    slot_depth1 : 0.0
    slot_type : "square"
    slot_width : 2.3
    slot_width1 : 0.0
}
```

Next, the inference engine goes to all the rules that have the same slot type in their left hand side as input by the user. As can be seen from the input that the slot width is less than the face mill diameter and peripheral mill width but it is greater than the end mill diameter. The depth is less than three times the width of the slot thus satisfying all the conditions and actions on the left & right hand sides of the rule 'square_six' respectively. The inference engine selects this rule and executes the actions specified in the right hand side of this rule. The output consists of the milling process type, process sequence, and the width of cuts for each process.

{ square_six
   instanceof # : LSR_GLOBAL_RULE
   LHS : "slot face_dia = ?FDIA"
      "slot end_dia = ?EDIA"
      "slot slot_width = ?WIDE"
      "slot p_width = ?PWIDE"
      "slot slot_type = 'square'"
      "slot slot_width < ?FDIA"
      "slot slot_width > ?EDIA"
      "slot slot_depth < (3*?WIDE)"
      "slot slot_width < ?PWIDE"
   PRIORITY : 1
   RHS : "write 'The process selected is end milling. \n'"
      "bind ?QUO (?WIDE/?EDIA)"
      "bind ?LFT ((?WIDE-?EDIA)/2)"
      "write 'The width of cuts are ?EDIA inch, taken ?QUO times'"
      "write ' (ignore the fraction), and last cuts are ?LFT inch on both sides.'"
}

The output in this case shows that the process selected is end milling, the width of cut is 0.8 inch taken two times and the last width of cuts are 0.75 inch on both sides of the slot.

Conclusion

An expert system can be a vital ingredient in the manufacturing industry. It can help improve the manufacturability of the product and reduce the total costs attached with the product. The expert system developed is able to consider four different kinds of slots and suggest the milling operation, the operation sequence and the width of cuts for each operation. This suggestions from the expert system will enable the manufacturing engineer decide on appropriate machines, tools and eventually the complete process plan for a particular product.

References

1. Barr, A., and Feigenbaum, E. A., The Handbook of Artificial Intelligence, William Kaufman, Menlo Park, California, Vol. 1, 1981.
2. LASER Reference Manual, Bell Atlantic Knowledge Systems, 1984.

Forging Die Design with Artificial Intelligence

S.K. PADHY, R. SHARAN, S.N. DWIVEDI and D.W. LYONS

Department of Mechanical and Aerospace Engineering
West Virginia University
Morgantown, WV

Abstract

Forging has been a critical manufacturing process for production of components with superior tensile strength, toughness and grain flow. Rib and web type forgings have been of significant importance and have been the most difficult to forge because they require several design considerations. Design algorithms for rib and web forging formulated based on several practical data. A geometric approach is presented to evaluate the shape complexity factor. To determine the tolerance for the different forging designs, an expert system is developed.

Introduction

Forging is a forming process involving plastic deformation of the metal. This process is capable of producing components of high quality at moderate cost. The advantages of forging include: high ultimate strength, controlled grain flow, uniform density of grain structure, low machining cost and wide range of shapes, sizes and materials. Among the different shapes of forgings the most difficult and most significant ones are the Rib and Web forging. Webs are the thin section on the part that are parallel to the parting plane whereas ribs are thin sections perpendicular to the parting plane. Ribs are designed to provide rigidity whereas the webs are designed to supply metal to fill the die cavities, to provide necessary connections for integrating the strengthening elements of forging, to provide surfaces for fastener and holes etc.

Compared to thicker sections both the rib and web are difficult to forge because of the rapid cooling of metal and the force and energy requirement in these sections are different from other sections. In addition, the rib is formed in a deep cavity of the die and high pressure must be generated by the resistance of the flash in order to secure the material flow required for complete die filling. The rib and web design is analyzed in this paper. Forging tolerance varies with the dimension, weight and many other parameters of the forging design. An expert system is developed to provide the forging tolerance and this system provides the tolerance for a given set of data that include: dimension, shape complexity factor and weight of the forging.

Shape Complexity Factor

In forging the metal flow is greatly influenced by the part or die geometry. Spherical and block-like shapes are the easiest to forge in impression or closed die forging. Parts with long thin sections or projections like ribs and webs are more difficult to forge as they have more surface area per unit volume. Again, the variations in shape increases the effect of friction and temperature changes and influence the pressure required to fill the die cavities. To quantify the shape complexity factor, Teterin et. al. [2] defined the shape complexity factor for round forgings (having one axis of symmetry) as :

$$\alpha = \frac{X_f}{X_c}$$

with $\quad X_f = P^2 / A \quad$ and $\quad X_c = P_c^2 / A_c$

where P is the perimeter of the axial cross section of the forging (surface that includes entire axis of symmetry), P_c is the Perimeter of the circle circumscribe the forging and A and A_c are respective cross sectional area.

However, on round forgings, bosses and rims placed farther from the center and increases the forging difficulty. A lateral shape factor β can be introduces [2] as:

$$\beta = \frac{2R_g}{R_c}$$

where $\quad R_g$ = radial distance from the symmetry axis to the center of gravity of half of the cross section
$\quad R_c$ = Radius of the circumscribing circle.

With these factors, the shape difficulty factor becomes:

$$S = \alpha \beta$$
$$= \frac{2R_g \, P^2 \, A_c}{A \, P_c^2 R_c}$$

The factor S describes the complexity of a half cross section of a round forging with respect to that of circumscribing cylinder.

Since all the forgings are not round a factor χ is introduced to take care of the non-round ones. The value of the χ need to be evaluated from practical data. Incorporating this modification the general shape complexity factor can be given by:

$$S = \chi \frac{2 R_g P^2 A_c}{A P^2 R_c}$$

The calculation of the perimeter of a forging is time consuming though not difficult However if the perimeter is divided in to some set of standardized perimeters then the calculation will be easier. The plan area (and perimeter) of forgings in general can be expressed as the combination of the areas (and perimeter) of square, rectangle, triangle and circle.

Perimeter Evaluation

The importance of the perimeter of the forging cross section is of significant importance as it influences the load, the stresses and cost of the forging. Consider the forging plane with a circle attached to a rectangle (Figure 1). Now due to the combination of the two geometries, some portion of both the rectangle and the circle are not to be accounted for in calculation of the total perimeter of the cross section. To calculate the effective perimeter (P_c) the correction factor ζ is defined as:

$$\zeta = P_c / P = \frac{\pi - [\sin^{-1}(b/2r)]}{\pi}$$

Generalizing, the equation becomes:

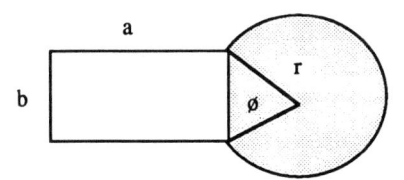

Figure 1. A Plan Area of Forging

$$\zeta = \frac{\pi - [\Sigma \sin^{-1}(b_i / 2r_i)]}{\pi}$$

Similarly, the correction factors can be found for other geometrical combinations. Hence, the total perimeter is given by:

$$P_t = \Sigma P_c$$
$$= \Sigma (P_i \zeta_i); \qquad i = 1, 2, 3, \ldots\ldots\ldots n.$$

Area Evaluation

The effective area calculation is as follows. For sections with no circular shape, the effective area becomes the summation of individual areas. For a plan area with a circular shape, the area contribution of the circular shape can be calculated to be (Figure 1):

$$A_c = \pi r^2 - \left\{ r^2 \frac{\sin^{-1}(b/2r)}{2} - \frac{b\sqrt{(4r^2 - b^2)}}{4} \right\}$$

So the correction factor becomes

$$\wp = A_c / A$$

$$= 1 - \frac{\left\{ r^2 \frac{\sin^{-1}(b/2r)}{2} - \frac{b\sqrt{(4r^2 - b^2)}}{4} \right\}}{\pi r^2}$$

In general, the total plan area is given by:

$$A_t = \Sigma A_c$$
$$= \Sigma (A_i \wp_i); \quad i = 1,2,3,\ldots\ldots\ldots n.$$

Once the product is available, the cross sectional area and the perimeter of the cross section are decomposed to simple predefined features for which the the correction factors are known to calculate the shape complexity factor.

Rib and Web Design

In this section, various relationships between different design parameters are presented and formulations are developed analyzing empirical data.

a. Web Design

The permissible web thickness (T_w) varies with the height (H) of the web linearly. The relation is given by:

$$T_w = 0.064 + 0.1434\, H \quad \text{for W/H} = 4$$
$$ 0.0717 + 0.1586\, H \quad \text{for W/H} = 5$$

The transverse distance has an influence on the web thickness for the transverse distance increases the web thickness has to increase. The forging will be difficult to achieve with a higher transverse distance and lower web thickness. The relation is given by the following equations for Titanium, Steel and Aluminium respectively:

For Ti :: $T_w = 0.0567 + 0.0912W - 0.0137W^2 + 0.0011W^3$

For Steel :: $T_w = 0.0552 + 0.0768W - 0.0118W^2 + 9.722 \otimes 10^{-4} W^3$

For Al :: $T_w = 0.0542 + 0.0264W - 0.0032W^2 + 2.778 \otimes 10^{-4} W^3$

where T_w is the Minimum web thickness and W is the transverse distance.

The web thickness should increase with the increase of area (A_w) of the web. For rapid completed forging the equation is:

$$T_w = 0.0992 + 0.0551 A_w - 0.0047 A_w^2 + 5.419 \otimes 10^{-6} A_w^3 + 2.096 \otimes 10^{-5} A_w^4 - 8.151 \otimes 10^{-7} A_w^5$$

where $\otimes$ is the multiplication symbol.

The minimum web thickness is also affected by the forging plan area. It is found that the minimum web thickness increases with plan area, but the rate of increase in minimum thickness decreased with the increased plan area. The relation between the minimum web thickness and plan area is as follows:

$$T_w = 0.0754 \, A^{0.2527}$$

where T_w is the minimum web thickness and A is the plan area.

b. Rib Design

The minimum thickness of a rib is determined by its height and the metallurgical properties of the material being forged. The location of the parting line also influences the rib geometry, and the rib thickness also varies with the rib height in the forging. The equations governing the design for the rib thickness with the rib height are as follows:

For Ti ::	$t_r =$	0.38	for H < 1 inch
		$0.1027 + 0.2343\,H$	for H ≥ 1.0
For Steel ::	$t_r =$	0.15	for H < 0.5
		$0.0441 + 0.1774\,H$	for H ≥ 1.0
For Al ::	$t_r =$	0.19	for H < 0.5
		$0.0595 + 0.2092\,H$	for H ≥ 1.0

where t_r is the rib thickness and H is the rib height.

Fillet Radii and Corner Radii

Confined webs are difficult to forge as the metal flow is more difficult; therefore, web thickness and radii have to be greater. The fillet (R_f) and corner (R_c) radii are calculated as follows:

$$R_f = 0.5157\,H + 0.1666\,W.$$
$$R_c = 0.1747\,H^{0.4054}$$

Tolerance Advisor

The forging tolerance advisor is an expert system designed to find the tolerance required for a forging for a given weight, dimension and shape complexity factor. The output of the system is the tolerance, the upper limit, the lower limit, residual flash and mismatch allowed.

More than 200 rules are incorporated into the knowledge base and the system is built using the VP-Expert Expert System Shell which is easy to use and user friendly. This expert system can run any personal computer with 640K ram and IBM compatible. A backward chaining is employed to solve the goal, in this expert system shell. The rules are in the form of IF ..... THEN. A sample rule is as below:

Rule 12
If dimension >160
and dimension <=250
and weight >0
and weight <=0.4
and scf >0.16
and scf <=0.32

Then tol = 2.0
Display " tol_uplimit =1.3
and tol_lolimit =-0.6
and residual_flash = 0.5
mismatch =0.4";

Figure 2 illustratres a sample screen of the consultation with the expert system.

```
┌─────────────────────────────────────────────────────┐
│  ** TOLERANCE ADVISOR FOR FORGING **                │
│                                                     │
│  THIS EXPERT SYSTEM DETERMINES THE TOLERANCE,       │
│  MISMATCH AND RESIDUAL FLASH FOR A GIVEN            │
│  DIMENSION, SHAPE COMPLEXITY FACTOR AND WEIGHT      │
│  OF THE FORGING.                                    │
│           tol_uplimit = 0.7                         │
│           and tol_lolimit = -0.4                    │
│           and residual_flash = 0.5                  │
│           and mismatch = 0.4                        │
│                                                     │
│                              tol = 1.1              │
└─────────────────────────────────────────────────────┘
```

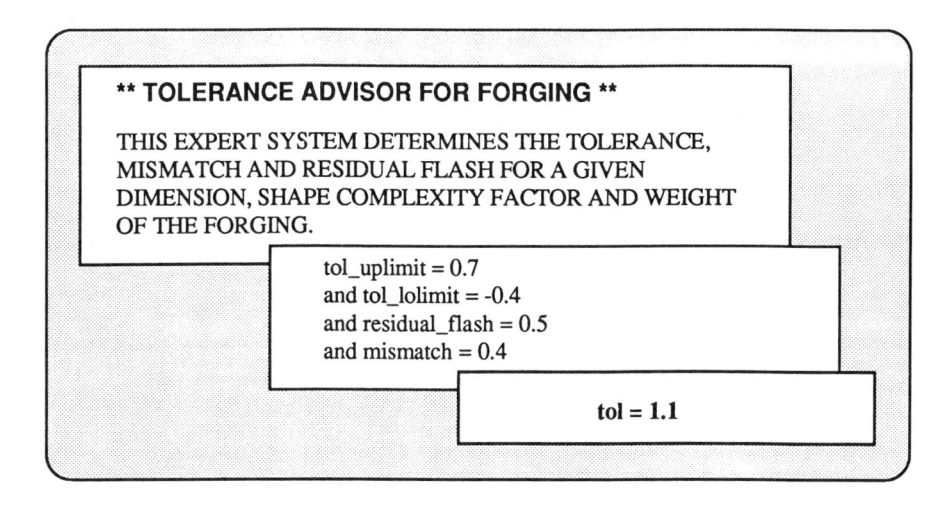

Figure 2. A Sample Screen of Tolerance Advisor

Conclusion

In this paper we describe a geometric approach to determine the shape complexity factor. The importance of this approach can be appreciated if computers are to be used for designing forgings. Since various correction factors can be stored as data in a computer program, and given a combination of predefined geometric features for the plan area of the forging and connection between them, the program can evaluate the effective area and perimeter and thus the shape complexity factor. Emperical data for rib and web forgings are analyzed and formulations pertaining to the rib and web design are also established. Furthermore, an expert system is developed to determine the optimum tolerance for a product design.

References

1. Sharan, R., Prasad, S.N. & Saksena, N.P.:" Forging Die Design and Practice", S. Chand & Co., India, 1982.
2. Altan, T., Oh, Soo-lk and Gegel, H.:" Metal Forming: Fundamentals and Applications", American Society for Metals, 1983.
3. Padhy, Sisir K.,"On Some Considerations for Forging Design", Unpublished Work, West Virginia University, 1990.
4. VP-Expert , Paper Back Software , CA, 1989.

Chapter VIII

Process Simulation and Automation

Introduction

Computer aided modeling and simulation, group technology and knowledge-based systems are the keys of today's advancements in manufacturing technology. The first paper of this chapter addresses the application of animation to enhance the simulation model for increased productivity. A framework for integration of simulation/animation with other facility planning techniques is proposed to produce a more powerful CIM system design. An expert system called EXSEMA is presented in the next paper which provides advice on the most appropriate simulation software environment for a particular manufacturing application. The system has been developed using a level five knowledge engineering shell with more than 100 rules. The third paper describes the development of a system called CL^2AUDIA. A method suitable for automatic clustering of any kind of manufacturing data and incomplete ordinal data is proposed and incorporated in the system. The system has been practically proven. The next paper explores the complexities of issues involved in dispatching the automated guided vehicles (AGV). Modeling of these AGVs and their behavior and some heuristic solutions for some associated problems are also discussed. It is advocated that the combined use of neural networks and artificial intelligence is promising for modeling and solving non-deterministic problems and to extract relevant information from fuzzy data. The next paper presents a review on the current practices of integration of design, fabrication, and testing using PC class computers. Planned integration of automation equipment into the Raspet Flight Research Laboratory is described. The implication of such an automation for aerospace engineering education is also presented. The next paper discusses an automated knowledge-based system which helps managers to plan the organizational change needed to implement computerized manufacturing processes. The system has been tested and found very successful. The final paper in this chapter discusses the planning and realization of skill based flexible automation for developing countries.

Simulation Modeling in CIM Systems Design

COLIN O. BENJAMIN, MELINDA L. SMITH and DEBRA A. HUNKE

Department of Engineering Management
University of Missouri-Rolla
Rolla, MO

Abstract
Simulation modeling can be enhanced by appropriate animation to increase the productivity of CIM system designers. This paper proposes a framework for integrating simulation/animation with other facility planning techniques, in order to provide a powerful methodology for CIM system design.

Introduction
Computer Integrated Manufacturing (CIM) attempts to control all of the phases of manufacturing, including operational functions and information processing [1]. A CIM system thus incorporates computer-based methods to increase productivity through better integration of the design, planning, scheduling, and control functions. Simulation, a management tool capable of modeling an entire manufacturing system, can provide the means for a system designer to evaluate the effectiveness of different system components, assess their effect on the entire system, and make appropriate design modifications to improve system performance. This paper proposes a framework for integration of simulation modeling and manufacturing planning tools into a methodology for CIM system design. The proposed methodology will be illustrated with a case study describing the design and analysis of a flexible manufacturing cell.

Simulation Modeling
Surveys have confirmed the popularity of simulation modeling among operations research practitioners [2] and underscored its ever-increasing relevance as a tool for manufacturing systems engineers in facilities planning and operation [3]. Modeling techniques available to designers and planners range in decreasing levels of abstraction from mathematical models

through computer simulation to a pilot plant. The modeling techniques become increasingly visual as they move closer to reality. Computer simulation with animation falls in the middle of the spectrum, and enables the designer to have an improved understanding of the system's operational performance without incurring the additional costs associated with a physical model or pilot plant.

An important decision in simulation modeling is the choice of the simulation software to be used in developing the computer model. A software survey can provide a good starting point for simulation software selection. The survey by Law and Haider [4] provides a good summary of 23 current simulation products. An expert system EXSEMA [5], can assist in identifying the simulation software which best matches the user's requirements and environment.

Application in CIM System Design
General Methodology
Simulation modeling can be effectively integrated into a CIM system design project. This integration, shown in Figure 1, will typically proceed in the following four phases:

1. Orientation
This involves the definition of the CIM project, its objectives, the external constraints or outside influences, the scope and form of the study's output and development of a plan to conduct the study.

2. General or Overall CIM Plan
Simulation modeling could be used to evaluate alternative CIM plans which may embody a variety of design concepts. Highly aggregated models using relatively simple simulation/animation software e.g. XCELL+ [6], would be appropriate in this phase.

3. Detailed CIM Plan
Here, a more detailed simulation model of the CIM system plan could be constructed and used iteratively to analyze, develop, adjust, work-out, refine and evaluate the details of the

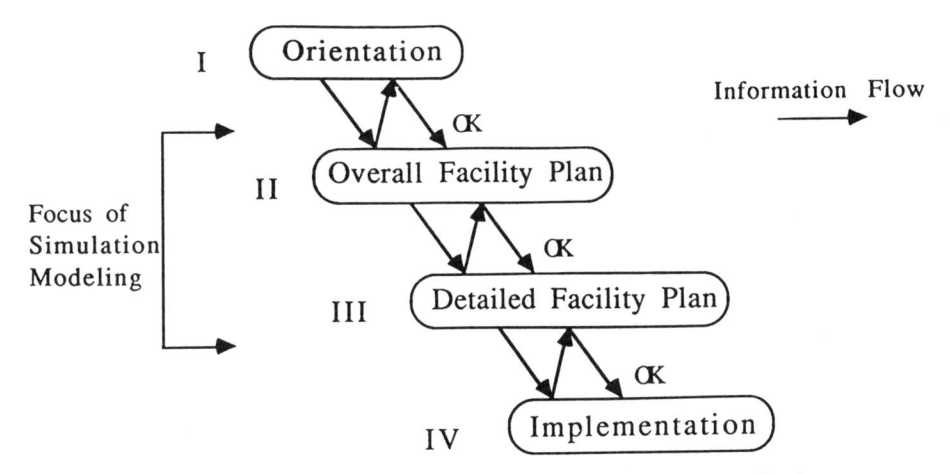

Figure 1 - Simulation Modeling in CIM System Design

system. The model's results can be interpreted and used to guide final decisions. A more sophisticated simulation environment e.g. SIMAN/CINEMA [7, 8] would be more applicable at this stage.

4. Implementation
Action is now required to get funding for the actual installation, pilot plant, or next-step refinements. Here the simulation/animation model developed can provide a very effective means of "selling" the project to upper management.

CIM Design Methodology
The first phase of developing a CIM system includes finalizing the design of the products to be manufactured. The use of Computer-Aided-Design and Drafting (CADD) packages for conceptual product design, development and modification can significantly enhance design productivity. This first phase also includes studies of the feasibility of the product, the market potential, production methods and costs, and the production equipment and ancillaries required in the manufacturing cell.

Next, the operations required to manufacture the product will need to be identified and finalized. Process charts incorporating standard ASME symbols can be used to record all of the information pertaining to part routings, operations sequence and times. An integrated facility plan then needs to be established for the manufacturing system required. Systematic approaches generally adopted by industry, such as Muther's SPIF (Systematic Planning of Industrial Facilities) [9] can be used to facilitate integration of the important components of the manufacturing system viz. Layout or Machine Arrangement, Handling and Storing, Communications and Controls, Utilities and Auxiliaries, Buildings and Structures. Finally a plan for implementation should be established.

Simulation modeling can extend the scope of the traditional facility planning methodologies by allowing the critical evaluation and analysis of alternative system designs. When enhanced with animation, it can serve as a powerful visual tool in "selling" the proposed CIM system to various interest groups.

A Case Study
The Flexible Manufacturing Cell shown in Figure 2, has been developed to produce a range of toys and souvenir items. A GE model P60 robot and a Mercury robot are utilized for material handling within the cell. The AS/RS stores finished products as well as parts needed for production. The parts are delivered to the GE robot from the AS/RS by a conveyor. The GE robot moves the parts between the conveyor and the CNC milling machine or CNC lathe as required. After inspection, the Mercury robot loads and unloads the packaging machine and returns the finished product to the conveyor for transport to the AS/RS for storage.

In this case a simulation model constructed using the XCELL+ simulation environment aided the evaluation of the performance of the cell and helped identify changes in the cell design and layout to improve operating performance. The CNC mill was identified as the major obstacle to increasing system throughput. An alternative cell design proposed the addition of a vision system to provide automated inspection capability.

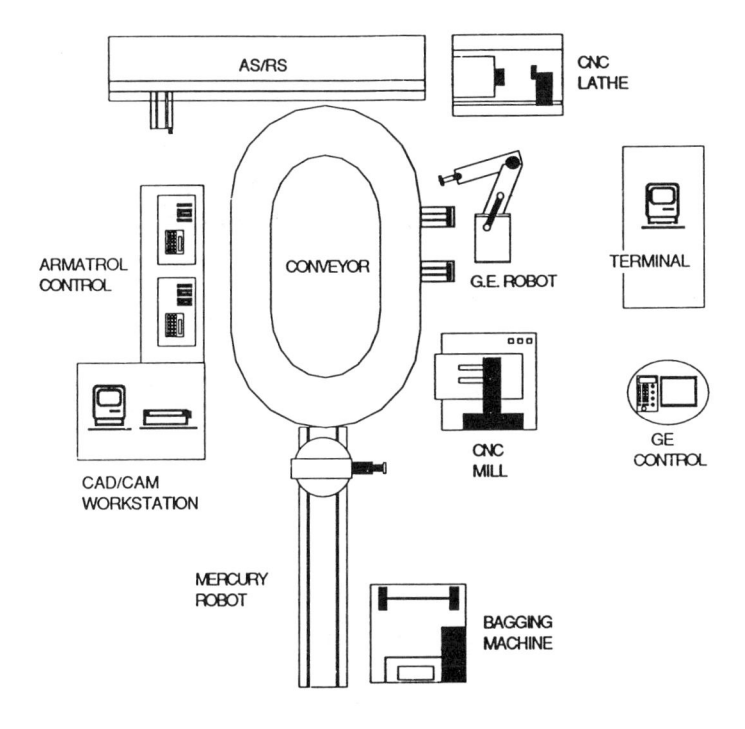

Figure 2 - Flexible Manufacturing Cell

More detailed models have been developed using the SIMAN/CINEMA simulation/animation software to permit in-depth analysis of alternative system designs.

Conclusion
A review of the literature on simulation modeling, [10], confirms its emergence as a powerful tool in manufacturing system design. It allows general "what-if" analyses which can lead to a reduction in risk and unveil unforeseen problems in the CIM system at the design stage.

The scope of a simulation model can be limited by the software environment chosen. For example, XCELL+ is limited in model size and flexibility. The outputs are all standard and cannot be customized. However its minimal cost, ease of learning, user friendliness, and menu driven features. make it quite attractive for simple modeling applications.

More complex simulation languages require considerable investment of time and effort on the part of the user in order to establish a valid working model. Never-the-less, the advantages of being able to evaluate system performance before incurring extensive capital investment, make simulation modeling an invaluable tool in CIM system design.

Acknowledgements

This paper is based on research work supported under NSF Grant #8954338.

References

[1] Nazametz, J.W.; Hammer, Jr., W.E.; Sadowski R.P. (Ed.), "Computer Integrated Manufacturing Systems: Selected Readings", Industrial Engineering and Management Press, Norcross, Georgia, 1985.

[2] Shannon, R.E.; Lone, S.S.; Buckles, B.D., "Operations Research Methodologies in Industrial Engineering: A Survey", AIIE Transactions, Vol. 12, No-A, pp. 364-367, 1980.

[3] Kosda, D.F.; Romano, J.D., Countdown to the Future: The Manufacturing Engineer in the 21st Century, A.T. Kearney Research Study, Profile 21, Executive Summary, SME, Fall 1988.

[4] Law, A.M.; Haider, S.W., "Selecting Simulation Software for Manufacturing Applications: Practical Guidelines & Software Survey", Industrial Engineering, pp. 33-46, May 1989.

[5] Benjamin, C.O.; Hosny, O., "EXSEMA - An Expert System for sElecting Software for Manufacturing Applications", Working Paper #90-12-37, Department of Engineering Management, University of Missouri-Rolla, Rolla, MO 65401, May 1990.

[6] Conway, R., "XCELL: A Cellular, Graphical Factory Modeling System", 1986 Winter Simulation Conference, Washington D.C., December 1986.

[7] Pegden, C.D., "Introduction to SIMAN", 1986 Winter Simulation Conference, Washington D.C., December 8-10, 1986.

[8] Healy, K., "CINEMA Tutorial", 1986 Winter Simulation Conference, Washington D.C., December 8-10, 1986.

[9] Muther R.; Hales, L., Systematic Planning of Industrial Facilities, Management and Industrial Research Publications, Kansas City, MO, April 1988.

[10] Wortman, D; Miner, R., "The Role of Simulation in Designing Manufacturing Systems", Autofact '85 Conference Proceedings, Detroit, Michigan, November 1985.

EXSEMA-An EXpert System for SElecting Simulation Software for Manufacturing Applications

COLIN O. BENJAMIN and OSSAMA A. HOSNY

Department of Engineering Management
University of Missouri-Rolla
Rolla, MO

ABSTRACT
This paper describes the development of EXSEMA, an expert system (ES) prototype for providing advice on the most appropriate simulation software environment for a particular manufacturing application. During a typical consultation session the user is prompted for information which describes the proposed application and the existing development and organization constraints. EXSEMA then identifies the simulation system which provides the best fit with the user's needs and provides appropriate explanations.

INTRODUCTION
Simulation, one of the more powerful and popular Operations Research (OR) techniques, [1] involves the construction of a computer model of a problem on which we experiment and test alternative courses of action. In recent years, increased interest in simulation has led to an explosion in the number of simulation packages with a strong orientation toward manufacturing problems [2]. The engineer/analyst trying to select simulation software for a particular application is now faced with a bewildering variety of choices in terms of technical capabilities, ease of use and cost. In this paper, we discuss the development of a knowledge-based expert system for providing the simulation analyst/engineer with expert advice on the simulation software environment most suitable for his particular manufacturing application.

LITERATURE REVIEW
Simulation Software
An important decision in simulation modeling is the choice of the language to be used in developing the computer model. Although simulation models have been constructed using high-level languages (Fortran, Pascal, Basic, etc), in the

manufacturing environment, the realistic alternatives are:

(1) *A General Purpose Simulation Language* [3], (eg. SIMAN, SIMSCRIPT, SLAM II), which is general in nature, but may have special features for manufacturing such as workstations or material handling modules.

(2) *A Manufacturing Simulator* [4], (eg. SIMFACTORY, PROMOD), which allows one to model a specific class of manufacturing systems with little or no programming.

Expert Systems
Expert systems (ES) are computer programs that provide "expert quality" solutions to problems in a specific domain by mimicing the decision-making process of a human expert. These systems are finding increasing acceptance in industry [5,6].

ES prototyping requires collaboration among a domain expert, a knowledge engineer and potential users of the system [7]. The domain expert is a knowledgeable person in a particular field. The expert's knowledge is utilized in developing rules that employ facts and heuristics to arrive at an acceptable solution in a timely manner. The knowledge engineer is the artificial intelligence (AI) language and knowledge representation expert. The end-user should also participate in specifying the purpose of the system and defining its operational features.

SYSTEM DEVELOPMENT
System Architecture
In the proposed ES, the user interacts with the system through the user interface that makes access more comfortable for him and hides much of the system's complexity [8]. The system is menu-driven with the user being prompted for data inputs during a consultation session. The program keeps track of case-specific data, the facts, conclusions and other relevant information on the required features in the simulation software. This information is separated from the general knowledge base. The explanatory subsystem allows the program

to explain to the user the reasons for selecting a software for a certain application. The general knowledge base is represented in the form of "IF-THEN" rules and contains the problem-solving knowledge for the selection of the simulation software. The inference engine applies the knowledge to the selection of the appropriate software. It is the interpreter of the knowledge base.

System Prototyping

System development passed through the following stages:

#1 *Knowledge Acquisition:* A general literature review was supplemented by interviews with experienced practitioners.

#2 *Knowledge Representation:* The knowledge was represented by decision tables and production rules.

#3 *Knowledge Base Design:* The Level Five knowledge engineering shell [9] which uses a backward search strategy and runs on the IBM personal computer XT or AT under the DOS operating system was selected for this application. The overall knowledge base is made up of over 100 rules coded in the Level Five syntax.

#4 *Prototype Development:* A demonstration prototype model was developed prior to establishing the detailed system in order to test the system, refine its knowledge and correct its shortcomings.

#5 *System Development:* The demonstration prototype was tested and corresponding corrections and additions to its knowledge base were made until the system was considered ready for implementation.

System Verification and Validation

Preliminary testing using different sets of data inputs confirmed that the rules were being executed where appropriate. For this demonstration prototype, this verification phase was vitally important as it aided the timely rectification of errors in the system's logic and identified areas for refining the prototype. The test case scenarios used in validating the system are described below.

Scenario #1:
A small producer of high-end, contract office and institutional furniture wants to undertake a significant expansion to increase its production capacity and increase its competitiveness in international markets. The company president, as part of the planning process for the envisaged expansion, would like a simulation model to be built of the proposed manufacturing system. This new production system will utilize CIM concepts to achieve greater flexibility and efficiency and would represent a radical departure from the current labor-intensive system. In this case, EXSEMA recommended PC Model and Simple-1.

Scenario #2:
The Director of the CIM laboratory in a technological university is considering acquisition of simulation software to assist in planning and designing an integrated, flexibly automated manufacturing and assembly cell. This cell would be used to produce a range of souvenir items and would be incorporated into the Department's teaching, research and extension programs. Here, EXSEMA recommended SIMAN/CINEMA.

Scenario #3:
A large manufacturing company is planning a multi-million dollar project to improve its material handling system by acquiring additional automatically guided vehicles (AGVs) and expanding its guide-path system. The company would like to construct a simulation model to evaluate alternate designs for its complex guide-path system. In this case, EXSEMA recommended AUTOMOD II.

Scenario #4
A medium-sized consulting firm specializing in manufacturing systems integration has just secured a contract to develop a $250,000 automated material handling system for the paint shop

of a motor vehicle assembly plant. To minimize the project risks, the client requires a PC-based computer simulation and animation model to be developed to facilitate the evaluation of alternative system designs. The consulting firm's design engineer assigned to the project has a very good knowledge of material handling systems but possessed limited programming expertise. Here, PROMOD was selected as the most appropriate simulation environment.

SYSTEM EVALUATION

Usefulness

Using information available on twenty three popular software currently available on the market, EXSEMA will provide the user with the simulation software that best meets the desired features required for his particular application while taking into consideration the organizational environment. This will be done through asking the user some simple questions from which the system will be able to determine the application orientation, the availability of programming expertise, the available development time, and the importance of cost, animation, and so forth in the required software.

Limitations and Further Development

The selection is limited only to those twenty three software packages tabulated in the 1989 survey by Law and Haider [2]. The recommendation will therefore not consider changes or software enhancements introduced after May 1989. Neither will it consider new simulation software launched on the market after that date. Good decision-making would require regular development and updating of the software database to incorporate new simulation software appearing on the market and significant enhancements made to existing software.

CONCLUSION

The EXSEMA knowledge based system can assist the simulation analyst/engineer in selecting the simulation software that best fits the needs of his particular manufacturing application. Through a simple consultation process, the user is prompted for information about his needs and work environment. The system then recommends the software that best meets these conditions from a database of twenty three software packages currently available on the market.

REFERENCES

1. Shannon, R.E.; Long, S.S. and Buckles, B.D., "Operations Research Methodologies in Industrial Engineering: A Survey", AIIE Transactions, Vol. 12, No. 4, pp. 364-367.

2. Law, A.M. and Haider, S.W., "Selecting Simulation Software for Manufacturing Applications: Practical guidelines and Software Survey", Industrial Engineering, May 1989.

3. Law, A.M., "Introduction to Simulation: A Powerful Tool for Analyzing Complex Manufacturing Systems", Industrial Engineering, May 1986.

4. Haider, S.W. and Banks, J., "Simulation Software Products for Analyzing Manufacturing Systems", Industrial Engineering, July 1986.

5. Uzel, A.R., "Guidelines for Expert Systems Application", Chartered Mechanical Engineer, February 1987, pp. 40-45.

6. Leonard-Barton, D. and Sviokla, J.J., "Putting Expert Systems to Work", Harvard Business Review, March-April, 1988, pp. 91-98.

7. SME, Expert Systems, SME Blue Book Series, CASA/SME Technical Council, Society of Manufacturing Engineers, Dearborn, Michigan 48121, 1990.

8. Luger, G.F. and Stubblefield, W.A., "Artificial Intelligence and the Design of Expert Systems", Benjamin/Cummings Publishing, California, 1988.

9. Information Builders, Inc., Level Five Reference Manual (PC Version), 1985.

Group Technology Analysis for Manufacturing Data

ABDELLAH NADIF
Laboratoire MECATRONIQUE
Ecole Nationaler d'Imgenieurs de Metz
Metz, France

RENE-PIERRE BALLOT
Agence nationale pour le Developpement de la Productique Appliquee a l'Industrie
France

BERNARD MUTEL
Ecole Nationale Superleure d'Art et Industrie de Strasbourg
Strasbourg, France

SUMMARY

The Group Technology (G.T.) is a concept for industrial data analysis which helps to rationalization of production It consists in clustering economic and technical manufacturing data into homogeneous families. The G.T. facilitates the development of solutions using the manufacturing know-how and using analogies relative to shapes, dimensions, estimates,...., and manufacturing processes.

The present paper proposes a methodology in order to attain to rational use of data analysis methods. This methodology has been applied to the development of a software called **CL²AUDIA** (CLuster analysis and AUtomatic CLassing of Industrial Analytical Data). An industrial application of this software to production data in the field of manufacturing has been developed. Satisfactory parts families have been obtained. The main difficulties against which G.T. traditional clustering systems came up have been smoothed away.

I-INTRODUCTION

In order to organize production while making the best of manufacturing knowhow, manufacturers are facing a large amount of data. Group Technology is a tool which can help them to rationalize the design and the manufacturing process using analogies between parts (Designing Department) or between process plans (Process planning Department, Workshop).

More generally, it is necessary to divide the production system which is the result of complex interactions, into sub-systems (families) of similar components. Taking into account the difficulty of this situation, this paper presents a general methodology for group technology implementation, based on the concept of similarity used in Data Analysis. The methodology has been applied to the development of a software called **CL²AUDIA**.

In the production system, parts are described using a code containing their shape and dimension characteristics (dimensions, function, tolerances,...), as for example the Multi-M code [1]. A process plan is defined as an ordinal series of machining operations, the number of which differs from one part to the other. Process plan data are therefore considered as incomplete ordinal data.

When compared to other clustering systems available nowadays, this methodology offers the possibility of analyzing simultaneously several data sets of distinct natures, which has never been studied before to our knowledge. For example, it is possible to analyse separately codes ($C=\{c_1, c_2,...,c_i,...c_n\}$), or process plans ($G=\{g_1, g_2,...,g_i,...,g_n\}$), or the combination of them (References: $r_i = (c_i, g_i)$).

Examples used in this paper are the result of the application of CL²AUDIA to a real set of 52 tridimensional mechanical parts (references consisting in Multi-M codes + process plans)

II-DESCRIPTION OF THE PROPOSED METHODOLOGY

This methodology is organized into several stages, which are the following:

1- setting (recoding) of different type of data to be treated simultaneously (codes and process plans)
2- choice for suitable similarity measures (distance) and proximity calculation between data type by type (d_c:distance between codes, d_g:distance between process plans)
3- proximity calculation between references from those relative to codes and to process plans ($[dr]=f(\alpha_c,[d_c],\alpha_g,[d_g])$:distance between references, α_c and α_g weighing factors for codes and process plans respectively)
4- data plotting
5- automatic clustering (search for families)
6- description families
7- result validation

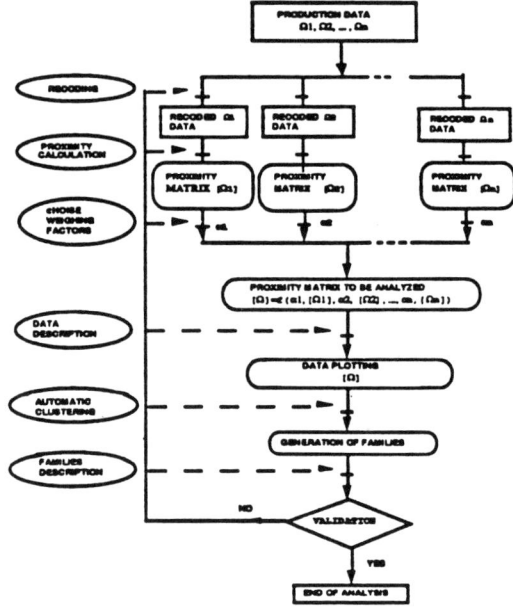

figure-1: Flowchart of CL²AUDIA methodology

II-1 Data description

Group Technology involves 2 types of set: data and variables describing them. The data set Ω represents for example codes, process plans or references. The variable set is **V**. Each variable v_j is defined by symbols called modalities, for identification of the different states of the variable. For example, if variable v_j represents a part materials, this material can be steel or copper... The number of modalities can change from one variable to the other.

One process plan is considered as a sole variable. The set of modalities representing it (here the work centers necessary for parts manufacturing) is provided with a total order relationship, corresponding to the succession of the various work centers.

II-2 Setting of data

Recoding is performed in order to obtain binary independant data suitable for treatment

II-2-1 Case of process plans

Recoding of a process plan [2] consists in building a square matrix, the dimension of which is total number of machines involved in the process plans set. The general term of this table is t_{ijk} with:

- $t_{ijk}=0$ if machine m_j does not appear in process plan g_i or if it appears after machine m_k

- $t_{ijk}=1$ if machine m_j appears in process plan g_i before machine m_k or at the same time.

II-2-2 Case of codes

Codes are recoded into a line table. Its general term β_{ijk} is **1** if variable v_j is represented by its modality m_{jk} in code c_i, and $\beta_{ijk}=0$ if not

II-3 Proximity calculation

Evaluation of the proximity between parts from the set Ω is performed using a distance defined as an application from Ω^2 into $\mathbf{R}^+$:

$$\forall (\omega_i,\omega_l) \in \Omega^2 \qquad \Omega^2 \longrightarrow \mathbf{R}^+$$
$$(\omega_i,\omega_l) \longrightarrow d(\omega_i,\omega_l)$$

having the 3 characteristics of a distance (reflexivity, symmetry and transitivity).

Owing to the strong difference between process plans (which are incomplete ordinal data) and codes, it was necessary to choose distinct similarity measures, suitable for each set of data.

II-3-1 Case of process plans (dg)

The best way to identify a suitable distance for process plans has been to define several industrial criteria to test the various similarity measures available. As a result of this, the index of **Dice-Kzekanowski** [3] has been selected and transformed in to a distance [4].

$$d_g(\omega_i,\omega_l) = \sqrt{\frac{\sum_{j=1}^{p}\sum_{k=1}^{p}(t_{ijk}-t_{ljk})^2}{\sum_{j=1}^{p}\sum_{k=1}^{p}(t_{ijk}^2+t_{ljk}^2)}} \qquad \text{with p=total number of work centers}$$

II-3-2 Case of codes (d_c)

Contrary to incomplete ordinal data (process instructions or assembly plans), any distance is well adapted. In this methodology **Euclid distance** has been choosen:

$$d_c(\omega_i,\omega_l) = \sqrt{\sum_{j=1}^{p}\sum_{k=1}^{n_j}(\beta_{ijk}-\beta_{ljk})^2} \qquad \text{with p=total number of variables and } n_j\text{=number of modalities of variable } v_j$$

For each data set Ω_k, it is now possible to calculate the distance matrix $[d_{\Omega k}]$.

II-4 Building of a similarity matrix for references

Contrary to other automatic clustering algorithms, this work offers the possibility of simultaneous treatment of several data sets of different types Ωk (k=1..m). For this purpose, a distance matrix is deduced from the distance matrix relative to each data set:

$$[d_\Omega] = \frac{\sum\limits_{k=1}^{q} \alpha_k [d_{\Omega_k}]}{\sum\limits_{k=1}^{q} \alpha_k} \quad \text{with} \quad \begin{cases} \text{-q: number of set taken into account} \\ \text{-}\alpha_k\text{: weighing factor for each data set } \Omega_k. \\ \text{-}[d_\Omega]\text{: global distance table} \\ \text{-}[d_{\Omega_k}]\text{: distance table between data of } \Omega_k \end{cases}$$

α_k values are choosen by the user. In particular, it is possible to perform the treatment:

a) on the sole codes $[d_\Omega] = [d_C]$
b) on the sole process plans $[d_\Omega] = [d_G]$

c) on the references (combination of codes and process plans) $[d_\Omega] = \dfrac{\alpha_C [d_C] + \alpha_G [d_G]}{\alpha_C + \alpha_G}$

II-5 Data plotting

The method selected for data plotting is called ACPTD (Principal Component Analysis on a Distance Table) [5]. Using this descriptive method, it is possible to represent a set of points defined in a p-dimensional space in a q-dimensional space with p>q while minimizing the information loss, in particular concerning distances between points.

The ACPTD method can treat any distance table. The result of this treatment is the projection of the data set on the best plane as regards information loss (first factorial plane). Factorial axes are classed by decreasing inertia. The first factorial plane, defined by the first and the second factorial axes, is the plane of maximal inertia. The factorial representation consists in projecting data in this plane. The quality of plotting is characterized by its explanation rate τ defined as the sum of inertiae of the 2 first factorial axes. The higher τ is, the better the factorial representation. Figure-2 shows the results of ACPTD treatment applied to the real set of 52 references given as an example.

Although this plotting is not an exact copy of reality, it offers an apparent repartition of data as reliable as possible

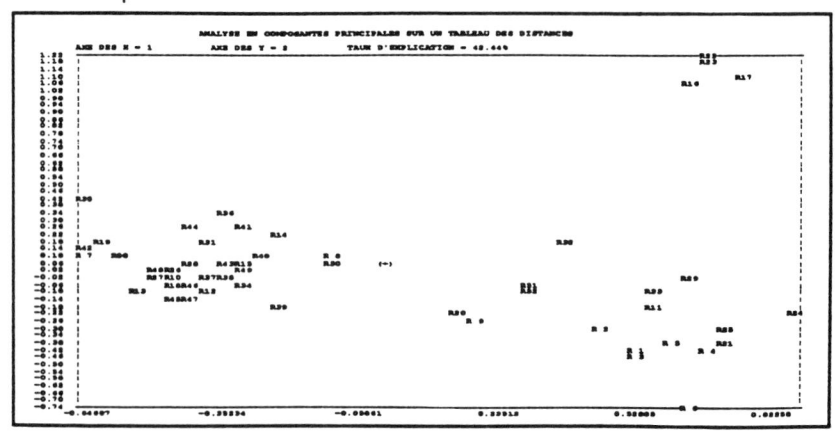

figure-2: Factorial representation of the set of 52 real references

II-6 Automatic clustering

As a result of a review of the various clustering methods available in Data Analysis, the Dynamic Clustering Algorithm has been selected (DCA) [5].

The purpose of the DCA method is to cluster data (codes or process plans,...) into a number of homogeneous families. For a given data set Ω containing n objects, it means searching for a partition P_k of Ω into k classes such as objects from a same class be very similar. To this end, it is necessary to define:
-a representation function g, in order to be able to connect to each class p_j an element called its nucleus λ_j, representative of it. Λ_k is the nuclei set corresponding to the partition P_k
-an assignment function f, in order to allocate each object of Ω to the nearest nucleus, so that objects from different classes be strongly dissimilar.

From a first nuclei set $\Lambda_k^{(0)}$ (initialization) the DC Algorithm consists in calling in turn the assignment function f and the representation function g. It is necessary to define a mathematical criteria W, in order to stop the process.

Unfortunately, the DC Algorithm gives results depending on the initial conditions. This is not satisfactory from the industrial point of view. Nevertheless, it is possible to counterbalance this effect using the strong pattern Algorithm. When performing several DCA treatments with distinct initial conditions, objects which have always been associated build very homogeneous families called strong patterns. Figure-3a shows the results relative to the real set of 52 references choosen as an example. After 3 treatments based on 4 nuclei, 9 strong patterns are obtained.

II-7 Aggregation of families

The number of strong patterns is generally rather large and, aggregation of families can be necessary. This is possible using an Ascendant Hierarchical Analysis [6] (CAH): each class being represented by its nucleus, aggregation of nearest neighbours provides partitions of the data set which are less and less fine.

Referring to the previous application, 9 strong patterns had been obtained after DCA treatment. Figure-3b shows the tree resulting from Ascendant Hierarchical Analysis. If needed, manufacturers will be able to choose the suitable number of families, in this case 8, 7 or 5 classes.

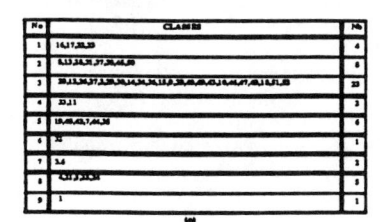

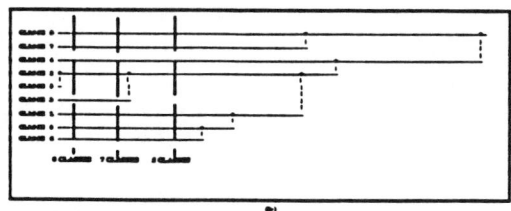

figure-3: (a) Strong patterns resulting from DCA treatment (real sample studied)
(b) Ascendant Hierarchical Analysis of the 9 strong patterns

II-8 Description of families

Description of families must adjust to each specific conditions. It will be detailed here for manufacturing data such as codes and process plans.

Each identified family of process plans is reprented by its an automatic generated mother process plan as a graph. This graph aggregates the work centers involved for manufacturing the family parts. It is deduced from a matrix **U**, the general term of which u_{ij} is the number of times when machine j follows machine i for all the process plans of the family.

Each identified code family is represented by its spectrum **S**. The general term s_{ij} is the number of times when modality j is present at position i.

III-CONCLUSION

Given the complexity of the manufacturing system and the large number of Data Analysis algorithms available, it is generally difficult for manufacturers to choose the one fitting in with their specific data and background.

To meet these requirements, the present paper proposes a methodology suitable for automatic clustering of any kind of manufacturing data, even incomplete ordinal data. Thanks to its flexibility, it can adjust to any manufacturing background.

Applications of this methodology to production data in the field of wood as well as shoe or mechanical industries have been developed with excellent results.

Even very large files (>1000 references) have been successfully treated at **ADEPA**.

IV-REFERENCES

[1]-A.HOUTZEEL, Computer and Industrial Engineering, vol.6, 2, 1982, pp.159-168

[2]-A.NADIF, C.COSTANTINI & B. MUTEL, Mesures de ressemblance de gammes de fabrication. RAIRO, APII, vol 19, 5, 1985, pp 455-470

[3]-B.FICHET & G.LE CALVE, structure géométrique des principaux indices de dissimilarité sur signes de présence-absence, STATISTIQUE ET ANALYSE DE DONNEES, vol. 9, 3, 1984, pp 11-44

[4]-A.NADIF, Contribution à la classification automatique de données de production en Technologie de Groupe.Thèse de Docteur de l'Université de Metz, Sept. 1987

[5]-E.DIDAY, J.LEMAIRE, J.POUGET & F.TESTU, Eléments d'Analyse des Données, BORDAS, 1982

[6]-J.P.BENZECRI & F.BENZECRI, Pratique de l'Analyse des Données, DUNOD, 1984

Dispatching Mobile Robots in Flexible Manufacturing Systems: The Issues and Problems

HIMANSHU BHATNAGAR and PATRICK D. KROLAK

Center for Productivity Enhancement
University of Lowell
Lowell, MA

ABSTRACT

The problem of dispatching automated guided vehicles (AGVs) and autonomous mobile robots (AMRs) in flexible manufacturing systems (FMSs) has, so far, not been addressed in depth. This research is an attempt to explore and present the complexities of issues involved in dispatching these vehicles, the techniques for modeling them and their behavior, and some heuristic solutions for some of the associated problems.

1. INTRODUCTION

On-line real time dispatching of mobile robots for material and tool transportation in a FMS factory is a problem which is receiving attention only recently. In conventional job shop environments the transportation time is considerably less than the set up and production time for a part, hence the issue of material handling did not receive the importance that has been associated to the problem of job-shop scheduling and dispatching. To stay competitive, an increasing number of factories are now adopting the just in time (JIT) inventory and flexible manufacturing concepts. These factories have multi-purpose, possibly automated, workstations which keep the bare minimum inventory in stock, process orders in stages and request for parts when they need. These, workcell characteristics, require frequent and punctual pick ups and deliveries which, combined with small set up times and comparable processing times makes the transportation time and loading/unloading time significant factors in overall production planning and operation control [28][9]. Rapid advances in FMS technology have shifted the mode of transport from bulky and inflexible conveyer belt systems to more flexible automated guided vehicle systems (AGVS)[20]which are either paint or wire guided. Lately more versatile versions of AGVs, the autonomous mobile robots (AMR), are undergoing industry testing [9]. These mobile robots have sophisticated on-board sensing and navigation system which obviates the need for chemical (paint) or wire guides. This makes them more flexible and hence more difficult to control, especially in real time.

In FMS factories, a large number of automated workcells are constantly supplied (on demand) the tools and input material by AGVs and AMRs. Dispatching these vehicles, on-line, to perform single or multiple pickup and delivery tasks is a complex operation[21] and needs a very careful understanding. Thus far the research on dispatching AGVs [10][11] and AMRs [22] has focused only on the issue of scheduling and routing of vehicles. However, the scope of real time dispatching of mobile robots extends well beyond that. For example it should include the effects of cancellations and no shows. Cancellations of pick up and delivery requests occur due to several independent and randomly occurring events like manufacturing delays, extended lead times, and tool jamming and failures. A no show may occur if the cancellations are not communicated well in advance or if there is a delay in manufacturing at the workcell. Since the cancellations and no shows can adversely affect and limit the system performance, it is imperative to include them while modeling a realistic dispatching problem. Whether the scheduling is done a priori (the day before or the shift before) or on demand, additional problems arise. These are: maintaining a status of all the previously routed robots; keeping their utilization profile; estimating and updating the approximate location of each robot; the quantity and part mix that they are carrying; and staying informed of their general mechanical state[32]. In certain situations the workcells do not indicate the pick up time for the sub-assembly, but call whenever it is ready. Assigning a robot to handle that request can be computationally cumbersome and is of concern to the dispatcher. However, if the dispatcher could predict such calls in advance, robots could be dispatched before time to the areas where calls are expected to originate. Continuity of material flow is an even bigger concern in dispatching. Since workcells have finite storage or buffering capacity, stock piling of material at the out-bay could slow down the workcell operation considerably and reduce its throughput.

A lot of these problems are unforeseen and occur when the dispatcher is unprepared to handle them, causing a considerable loss of manufacturing resources. In this paper we are trying to enumerate such problems under several classes and investigate the various techniques for modeling these problems, and also the heuristics to solve them. This paper is divided into four sections: 1) a survey of research done in dispatching, 2) analysis of the issues involved and problems faced in dispatching mobile robots in a FMS environment, 3) modeling techniques, 4) heuristic solutions for them.

2. DISPATCHING IN FMS - A SURVEY

Most of the research done so far, concerning dispatching in FMS has been in the area of job flow scheduling and dispatching without a separate treatment of material handling system[8][9][26][29]. A survey of 113 dispatching rules for job flow can be found in [26]. They classify the rules under three categories, a) Priority rules, b) Heuristic Rules and c) Other rules. Sabuncuoglu [29] provides a comparison of FMS performance under different job-shop and AGV scheduling rules. Denzler and Boe [9], experimentally analyzed the performance of scheduling rules for FMS. A few have addressed the issues in dispatching the autonomous guided vehicles (AGVs) [10][11]. Egbelu and Tanchoco [11] discuss five dispatching rules for AGVs which Egbelu in [10] describes as *push* (source driven) type of rules and shows that in FMS-JIT kind of environment *pull* (demand driven) type of rules are more suitable, though in certain situations the system fall backs on the source driven rules. Only a very few references can be found on dispatching in FMSs which use autonomous mobile robots [34][36][22]. Taghaboni [34], addresses the problem of immediate scheduling by using a branch and bound heuristic, and also deals with the problem of idle vehicle dispatch. Yeung et al [36] propose a connectionist solution to the problem of load balancing but do not deal with the topic of on line dispatching or scheduling. A study of the various dispatching rules for the AGVs and AMRs are under preparation and can be found in [5]. The dispatching rules mentioned above are implemented either as heuristic algorithms [10] or as expert system rules [21] or as a combination of both [19]. A survey of 20 expert systems used in FMS can be found in [18]. A study done on real time scheduling in FMS can be found in [6].

3. DISPATCHING MOBILE ROBOTS : ISSUES AND PROBLEMS

In this section we shall identify the issues and problems faced in dispatching the mobile robots in a FMS. In flexible manufacturing environment a production plan dictates the flow of material (raw material, tools, sub-assembly or assembly) from workcell to workcell. This plan can be easily mapped into a *list* of paired pick up and delivery requests for each material during a work shift. The material handling system (MHS) processes this list and assigns mobile robots to handle the requests. The issues which the dispatcher has to deal with and the problems it has to face before and after the robots are on the floor, have been categorized as follows :

3.1 SCHEDULING AND ROUTING

The scheduling of robots involves four steps, a) selecting the most suitable robot for handling the task, b) inquiring from the robot, its willingness to perform that task, c) actual assignment of the task to the robot, and d) handling of residual tasks, if any, of the selected robot, which result from the new assignment. Routing, on the other hand, involves actual path selection. Collision free path selection has thus far been touted as the biggest problem in routing. Attempts have been made to include collision avoidance in the routing/scheduling algorithms, however, a collision free pre-scheduled route may not succeed because of various delay factors which arise both periodically and stochastically. Besides collision, there are several other detracting factors, like cancellations, no shows, global and local traffic congestion, blocked paths and workcell breakdowns, robot's incapability of handling specialized materials, robot storing and load capacities, robot speeds, capability of traversing congested areas, task (material and tool) priorities, operational costs, and the intelligence of the robot.

3.2 ROBOT AND WORKCELL STATUS

Whether, a long list covering the entire shift, is transmitted to the material handling system (MHS) of FMS at the beginning of the shift or several short lists are handed over during the shift, the dispatcher has to keep track of the state of all the MHS entities on the floor. It has to know the status of all the previously routed **robots**; their utilization profile, their current loads and part mix, their traffic and specialized load handling capabilities, the level of intelligence of the control software on-board the robot, and their general mechanical state. Besides this, it has to continuously estimate and update the location of each robot. At the same time the dispatcher has to store **workcell** related data like their processing speeds, level of automation, past performance statistics vis-a-vis on time job completion, off-loading/on-loading materials from/to mobile robots, wait queue lengths, number of no shows and cancellations. The statistics helps the dispatcher to heuristically over-ride a program generated scheduling decision. Other workcell related issues, it has to deal with, are input/output buffer sizes, repeated presence of debris (obstacles) outside workcells, lateness in readying the material, failure-to-communicate-cancellations frequency, locational problems, alternate workcell availability, clustering of multiple but identical workcells, workcell idling, workcell deadlocks, loading/unloading capabilities, and their tool failure rates.

3.3 IDLE VEHICLE DISPATCH

The dispatcher also has to handle situations where robots are idle. In some situations a robot could be dispatched to handle a request in the far corner of the factory just before a cluster of workcells on the other end make an immediate pick up request.

3.4 REQUEST ANTICIPATION AND TOUR PROJECTIONS

If the dispatcher has the capability of projecting future requests, based on the requests made in the previous few days, then idle vehicles can be handled efficiently. However, this would give rise to another problem, future request projection. If the dispatcher could project the tours of already dispatched robots for a future time segment, it would immensely help the dispatcher in assigning new trips to robots.

3.5 MATERIAL FLOW

Yet another class of problems which the dispatcher faces is that of maintaining a smooth flow of material. Restricted by finite capacities, the workcell buffers, as a result of some missed pick ups, could very quickly overflow, resulting in workcell slow down and even closure, queue build up at the input end, and most importantly missing of deadlines. An example where the stock-piling could really hurt the continuity is when an important tool is lying on the out-bay of a workcell, waiting to be picked up, while another workcell is accumulating the material which has to be processed by this tool. Even if the tool is in extended use at a particular workcell and is badly needed at another, it can cause back logging of manufacturing tasks.

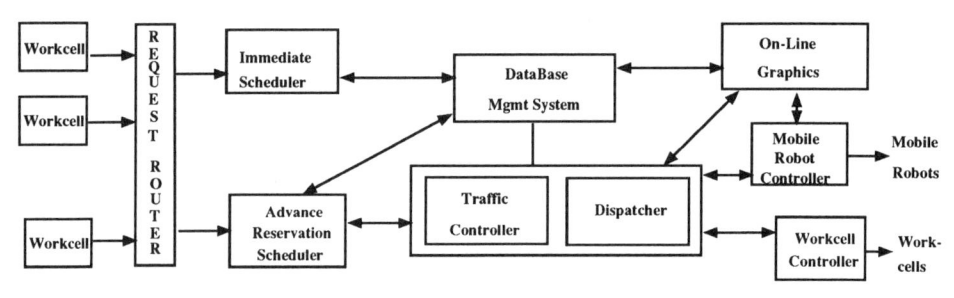

Information and Control Flow in a Typical Material Handling System

3.6 TOOL MANAGEMENT

Efficiency in tool management within FMS is greatly enhanced if the dispatcher can handle situations where the tools jam, break, fail, arrive late, are idle or are in use beyond the prescribed time limits while being in need at a different location. Note that the dispatcher as in the case of other materials shall not be involved in the decision making for replacing tools. It will be only responsible for detecting these problems and reporting to the higher level tool manager, and to subsequently re-assign robots to handle the modifications requested by the tool manager.

3.7 INFORMATION FLOW AND DATABASE MANAGEMENT

Another issue which the dispatcher has to deal with is that of broadcasting globally any status information which would be of interest to the objects of the FMS. Besides global broadcasting the dispatcher has to periodically check for the state of different entities in order to make sure that things are proceeding in the planned way. It also has to continuously update its data base with changes in the environment like link/workcell breakdowns, and traffic congestion and with changes in the state of various objects of FMS. Any robot or workcell can, at any time demand any information stored in the database. It is the responsibility of the dispatcher to be able to provide latest information to the robots and workcells.

3.8 TRAFFIC CONTROLLER INTERACTION

If the material handling system has a large number of mobile robots, there will be a need for traffic controller which will control all traffic related problems. The dispatcher's functioning and decision making will depend on traffic controller's capabilities and also will effect its performance. Thus the dispatcher and the traffic controller will have to act in tandem and have fluent communications between them.

4. MODELING

In [24] Noi, stresses the importance of a systematic representation of the problems in order to extract information relevant to it, from the environment. Several mathematical models for different aspects of FMS have been suggested. However, Lin and Chung [21] and Stecke [33], have indicated the ineffectiveness of mathematical programming techniques in handling the very unpredictable dynamics of FMSs. Lin and Chung have used *event graphs* to model the production events including the machine deadlock. Raman et al [28], have proposed that modeling of the entire production process including the material handling should be modeled as a *resource constrained project management* problem. In it, each machine and cart is modeled as a resource and each manufacturing job is considered as a series of machine and cart operations. Shaw and Whinston [32] model the workcell events as *expert system rules* and their interaction by *petri-nets*. From above we note that modeling through expert systems is being preferred over mathematical modeling.We, however, think that mathematical modeling should not be ignored. In fact it should be tied in with the different AI techniques and other heuristics, as in [19].

Research in detailed modeling of AGVs, mobile robots and the events related to their operation and control (dispatching), has been rather sparse. Taghaboni and Tanchoco [34], have used frames (artificial intelligence) to model the events and mobile robots. Mathematical models that have been suggested in [13] and [7] are at best theoretical in nature and need more work to be of any practical use. The use of event graphs and petri nets in conjunction with artificial intelligence, seems like an attractive modeling tool for AGV and mobile robot dispatching. However, because of the unpredictability of events, fuzziness of knowledge in certain situations, and the requirement of adaptivity to new situations, neural networks together with expert systems seem to be a better alternative in those situations. Neural networks have been used for modeling constrained optimization problems [27], multiple traveling salesman problems [35], however, with these solutions the settling time for the networks is always in question. Neural networks are particularly suited for the 'look ahead' (future) projections, as they can learn from previous data. Their learning capabilities can also be utilized to i) store and update performance factors of various hardware entities of FMSs, ii) topological information and iii) environmental knowledge. Yeung et al [36] have represented mobile robots,workcells, and their interaction alongwith the load balancing issue as a connectionist (neural network) model. This discussion indicates that an integrated neural network and AI approach is very attractive for modeling the dispatching problems. Recently, we have developed an integrated environment for neural networks and expert systems, called N-CLIPS [2]. In short , we recommend the use of

1. Event graphs, petri nets and other graph theoretical modeling techniques, in conjunction with expert system rules and facts, for modeling deterministic situations,

2. Integrated neural network and AI techniques (N-CLIPS) for modeling non-deterministic and fuzzy situations, and

3. Use of distributed expert systems [25], [23] in case of independent but parallel events.

5. SOME HEURISTIC SOLUTIONS

As noted above, heuristic solutions have been strongly favored over the mathematical programming, for solving real time dispatching and scheduling problems. Dispatching rules have been suggested for optimally selecting a vehicle for handling a task. These and other dispatching rules meant for job flow and workcell operation have been extended in [5] to provide dispatching rules for assigning vehicles to complete tasks on demand. These rules follow the 'pull' strategy [10] rather than the 'push' strategy as recommended in [11]. The issues like material re-routing and re-assignment after a collision or after a robot skips a workcell because of a blocked path or a long queue or because a robot is running very late, can also be dealt with by the extensions of the same dispatching rules [5]. However, since the decision making has to be faster, and preferably parallel, we use a parallel expert system shell PCLIPS(Parallel C Language Integrated Production System)[23]. For the rules PCLIPS globally asserts a fact over a network of heterogeneous computers and hence is also useful for global information broadcasting. The solution provided by [34] for idle vehicle dispatch involves computation over all vehicles and takes no consideration of future requirements. A N-CLIPS based projection of requests in future provides an ideal handling of idle vehicles. For dynamic scheduling of mobile robots in FMS factory, we have extended a heuristic TAXI [14] which schedules rides for the elderly and handicapped on a day's notice. In this extension [15] the starting location and time of already assigned tours of the vehicles are updated to last stop and time reported. We are also experimenting with a parallel version of this extension. Another heuristic called SELECT uses a method of ellipses [17] to select the best three or four vehicles for a particular task. The algorithm tries to fit the pick up and delivery ellipses in a tour's ellipse and assigns a score to it, and the tours with first three scores win. We are investigating the use of competitive learning [16] in conjunction with various expert system rules for dispatching to parallelize and enhance the performance of SELECT. As far as dynamic collision avoidance is concerned, it should be left for a traffic controller[3] to handle. The scheduling should be done with enough slack so that the traffic related delays can be accounted for. We have also developed a simulating environment for the robots[4].

6. CONCLUSION

Detailed discussions on the heuristic solutions, and a list of dispatching rules have been left out of this paper, but can be found in [5],[15] and [16]. Some of the problems enlisted here are implementation related, while the others are planning related, however for efficient operation of the mobile robots, none of these can be treated lightly. The combined use of neural networks and AI (N-CLIPS) is very promising for modeling and solving non-determinsitic problems and to extract relevant information from fuzzy data. TCP/IP or NCS could be used for networking, while for database management, a data base package which has networking capability will be required.

7.REFERENCES

[1] Ammons J.C. Govindraj,T.,Mitchel,C.M.,"Human Aided Scheduling for FMS: A Paradigm for Human Computer Interaction in Real Time Scheduling and Control",1986, Proc. of the Sec. ORSA/TIMS Conf. on FMS,Elsevier Sc. Publ. B.V., Amsterdam.

[2] Bhatnagar H., Krolak P., "A Neural Network Simulation Package in CLIPS", First CLIPS users group conference", Aug, 1990, Houston, TX.

[3] Bhatnagar H., Krolak P., McGee B. "An Intelligent Controller for Automated Material Handling Systems", Fourth Adv Tech. Conf.", Nov, 1990,Washington D.C.

[19] Kusiak A., "Scheduling Automated manuf. Sys. : A Knowledge Based Approach," Proc. of the third ORSA/TIMS Conf. on Flex. Manufac. Sys.",Aug, 1989, Cambridge, MA, 379-382.

[20] Lasecki R.R. " AGVs:the latest in material handling technology",CIM Tech., vol. 5, no.4, 90-94.

[21] Lin L.-E.S.,Chung S.-L.,"A Systematic FMS Model for Real -Time On-Line Control and Question Answer Simulation Using Artificial Intelligence.",Proc. of the Sec. ORSA/TIMS Conf. on FMS,1986.567-79.

References(contd).

[4] Bhatnagar H., Krolak P., "A Simulation System for Prototyping Advanced Material Handling Systems", Fourth Adv Tech. Conf.", Nov, 1990,Washington D.C.

[5] Bhatnagar H., "Dispatching Rules for Mobile Robots in FMS",#CPE-TRANSP-90-5,Working Paper, Ctr. for Prod. Ehnc.,Lowell,MA.

[6] Chang Y.-L., Sullivan R.S., Bagchi., "Experimental Investigation of real-time Scheduling in flexible manufacturing Systems",Proc. of the first ORSA/TIMS Spec. int. Conf.on FMS,1984, Michigan.307-312.

[7] Chen C.C., Lee C.S.G., McGillen C.D., "Task Assingnment and Load Balancing of Autonomous Vehicles in a Flexible Manufacturing System", Dec 1987, IEEE Journal of Robotics and Automation.

[8] Co H.C., Jaw T.J.,Chen S.K, "Sequencing in Flexible Manufacturing Systems and other Short Queue-Length Systems", 1988, J. of Manuf. Sys. 7(1) 33-45.

[9] Denzler D.R.,Boe W.J.,"Experimental investigation of Flex. Man. Sys. Scheduling Decsion Rules", 1987, Int'l.J.Prod.Res.,Vol.25,No. 7, 979-994.

[10] Egbelu J.,"Pull Versus Push Strategy for Automated Guided Vehicle load Movement in a Batch Manufacturing System",Joural of manufacturing systems, Volume 6/no.3,1987, pp -209-221..

[11] Egbelu P.J.,Tanchoco J.M.,"Characterization of Automated Guided Vehicle Dispatching Rules, 1984, Int. Journal of Production Res. 22(3), 359-374.

[12] Gould L.,"Is Off-Wire AGV Guidance Alive or Dead ? ", May 1990 -39 ManagingAutomation.

[13] Grossman D.D., "Traffic Control of Multiple Robot Vehicles ", Oct 1988 , IEEE Journal of Robotics and Automation ,Vol 4 no. 5.

[14] Krolak P., "A Realistic Advanced Reservation Dial-A-Ride Algorithm," DOT/TSC Report.1982.

[15] Krolak P., Bhatnagar H. "A Real Time Scheduling Heuristic : RealTaxi", #CPE-TRANSP-90-4, Working Paper, Center for Prod. Enhnc., U. Lowell, Lowell,MA.

[16] Krolak P., Bhatnagar H., "Selecting Vehicles for Immediate Scheduling of Mobile Robots in FMS: A Neural Solution", #CPE-TRANSP-90-6,Working Paper, Center for Prod. Enhnc., Lowell,MA.

[17] Krolak P., "The Method of Ellipses(MOE):A Vehicle Selection Heurisric for the GRASP System," DOT/TSC Report, Dec. 1980.

[18] Kusiak A., Chen M., "Expert Systems For Planning a18nd Scheduling Manufacturing Systems", 1988, Euro. Journal of Operational Research 34,113-130.

[22] Milberg I.J., Lutz I.P., "Integration of Autonomous Robots into Industrial Production environment." , 1987, IEEE.

[23] Miller R., Krolak P., "PCLIPS : A Distributed Expert System, First CLIPS users group conference", Aug, 1990, Houston, TX.

[24] Nof S.Y.,Whinston A.B., Bullers W.I.,"Control and Decesion Support in Automatic Manufacturing Systems", 1980(12), AIIE Transc.,150-69.

[25] Parunak H.V.D., "Distributed AI Systems,1987, A I : Computer Integrated Manuf. IFS Kempston, Bedford,U.K. and Springer, New York.

[26] Panwalker S.S., Iskander W.,"A Survey of Scheduling Rules", Jan-feb 1987, Operations Research,25(1).

[27] Poliac M.O., Lee E.B., SlagleJ.R., WickM.R., "A Crew Scheduing Problem".779-86. ------.

[28] Raman N.,Talbot F.B.,Racamadugu R.V., "Simultenous Scheduling of Machines and Material Handling Devices in Automated Manufacturing",Proc. of the sec. ORSA/TIMS Conf. on FMSystems,1986, Elsevier Science Publishers B.V., Amsterdam.454-65.

[29] Sabuncuoglu I.,Hommertzheim D.L.,"An Investigation of Machine & AGV Scheduling Rules in an FMS", Proc. of the Third ORSA/TIMS Conference on FMS,1989,Elsevier Sc. Publ. B.V., Amsterdam,261-66.

[30] Seidman A., Tenenbaum A.,"Optimal Stochastic Scheduling of Flexible Manufacturing Systems with Finite Buffers", August 1986,Proc. of the second ORSA/TIMS Conf. on FMS.

[31] Shaw M.J.P.,Whinston A.B., "Task Bidding and Distributed Planning in Flexible Manufacturing", Dec. 1985, The Second Conf. on Artificial Intellegence Applications,Miami Beach,11-13,184-89.

[32] Shaw M., "A Pattern-Directed Approach to FMS Scheduling",1986,Flexible Manufacturing Systems: Operations Research Models and Applications, Elsevier, NewYork 545 -554.

[33] Stecke K.,"Design, Planning, Scheduling and Control Problems of FMSs," Proc. First ORSA/TIMS Conf. on FMS, 1984, p1-7.

[34] Taghaboni,F., Tanchoco J.M.A."A LISP-based controller for free-ranging autmoated guided vehicle systems,"Int. J. Prod. vol. 26, no. 2, 173-88, Feb 1988.

[35] Wacholder E., Han J.,Mann R.C.,"A Neural Network Algorithm for the Multiple Travelling Salesmen Problem", 1989, Biol. Cybern.61,11-19.

[36] Yeung D.Y., Bekey G.A.,"Coordinating Multiple Robots for Materials Handling",AAAI,Seatle, 1987.

Automation of Prototype General Aviation Aircraft Development

GEORGE BENNETT

Raspet Flight Research Laboratory
Aerospace Engineering
Mississippi State University
MS

Summary

Developments in low cost computing and multi-axis automation are making an impact on the development process for prototype composite general aviation aircraft. The integration of design, fabrication, and testing using PC class computers makes possible reductions in the cost and time required. The methods currently used for the prototype process are reviewed and the activities which can be automated are identified. Planned integration of automation equipment into the Raspet Flight Research Laboratory aircraft development projects is outlined. Implications of automation for aerospace engineering education and the general aviation aircraft industry are briefly discussed.

Composite Aircraft Concepts

Composite materials have been of interest to the aircraft designer for many years because the fiber can be aligned in the direction of the load path and three-dimensional, smooth surfaces can be easily fabricated. Graphite fibers holds the additional promise of significant structural weight reduction due to high strength and stiffness. The fabrication process for composite structures is fundamentally different from metal structures which affects the automation required. The fabrication of composite structures requires a mold to define the shape of the fibrous materials. The designer must define the direction of each woven or unidirectional layer. The resin matrix is preimpregnated (prepreg) into the fiber layer (ply) to improve productibility and material properties. The prepreg plies are stacked in the mold in such a way that after curing, very little machining is required. The material properties and the structural integrity of the part are affected by prepreg, the layup design, the bagging of the layup, and the cure cycle. The geometry of the finished part is dependent upon the thermal expansion characteristics of the mold during the cure cycle.

RFRL Prototype Aircraft Development Activities

The Raspet Flight Research Laboratory (RFRL) has engaged in aircraft modification and development activities since August Raspet began the Lab in 1948. There was a long series of sailplanes and light aircraft which were highly modified to support the STOL research program under U. S. Army support during the 50's. The RFRL has developed a series of prototype fiberglass aircraft beginning with the Marvel II constructed in the early 60's. The Marvel was modified in 1982, (Marvel II), (Ref. 1) as part of a project to demonstrate a capability to operate in the deep sand environment of the Saudi Arabian Empty Quarter. The new wing constructed for the Marvel II used Kevlar prepreg in a vacuum bag 250 degrees F oven cure process. The latest prototype project was the construction of composite wing and tail surfaces for a Beech A-36 Bonanza. The graphite structure had autoclave cured spars and vacuum bag 250F oven cured surfaces. Other composite vehicles developed were a 50 lb. towed RPV and a 450 lb. RPV.

Figure 1. Marvel II Aircraft

RFRL Composite Aircraft Prototype Development Methodology

The experience gained from the long history of prototype development in a university environment has led to a structured approach. The development of a prototype aircraft by a small group requires a close relationship between the engineer and the fabricator. At the RFRL, most of the faculty have hands-on fabrication experience. The RFRL has experienced technicians who oversee the fabrication tasks, but the bulk of the work is done by young engineers, thus the RFRL has a workforce which can easily move between engineering and fabrication tasks. Considerable thought has to be given to the management of all aspects of the development process to minimize the possibility of an accident. The ground rules for all activities are:

1. Safety first. The RFRL is very reluctant to operate any of the prototype aircraft near the design conditions.
2. Make sure there is a low probability of unsafe flight characteristics.
3. Weight reduction is lower priority than structural integrity.
4. Analysis and critical verification tests lead fabrication and flight test.
5. An engineer is responsible for a structural assembly throughout the entire development cycle.

The fabrication aspects of the configuration and structure are considered during the design activity. Usually the moldmaking and fabrication capabilities control the design. The analysis aspects are secondary for the general aviation class aircraft. Documentation requirements for prototype aircraft can be minimized if the design engineer works closely with the fabrication group. The primary requirement is the definition of the geometry of the aircraft and the structure. It has become apparent that the procedures and capabilities are in place to prepare for the automation for many aspects of the prototype aircraft development process.

Composite Structural Molds

The molds used to fabricate the composite parts drive the development process. The accuracy, durability, heat transfer characteristics, and thermal stability are important aspects of a mold. At this time, the process is complex, time consuming and costly. Automation of the mold making process is most important to improve the economics of composite structures. It has become apparent that a large 5-axis gantry robot would be a valuable tool in the fabrication of molds and assembly jigs. The gantry robot provides a fixed coordinate system to define all geometrical aspects of the structures. The designer has defined the geometry using a software package such as AUTOCAD which is directly accessible by the robot toolpath software package.

Figure 2. Mold Fabrication

Composite Part Layup

It has been found that the layup of a thick part such as a spar is very labor intensive. First a master mylar template must be drawn and the mold marked for important points. Then a large number of patterns must be drawn to be used to cut all of the many plies for the layup. It has become clear the 5-axis robot could be used to draw master layout templates and the cut the ply shapes direct from the drawing database.

Composite Cure Process

The cure of prepreg structures either using the oven or the autoclave requires control of the temperature cycle and monitoring of the temperature and pressure. An "expert" process control computer system is needed to manage the cure of composite parts. It is expensive to develop the cure process for each large part thus the automation of the cure process control is of great interest.

Composite Inspection and Trimming

The inspection and trimming of parts after curing is a difficult and tedious task. The 3-dimensional shape of the parts makes alignment and marking of the part a hard task. The 5-axis robot is ideally suited for this task if 3-D setup alignment points are cured into the part.

Automation in the Prototype Development Process

The RFRL plans to expand its capabilities in the areas of design, analysis, fabrication, and testing. It has recently become feasible to consider the automation of many tasks in the prototype composite general aviation aircraft development. There have been significant improvements in the capabilities per unit cost of computers, software, and multi-axis gantry type robots. Integration of these elements is required to change the methods used in the aircraft prototyping process. Automation is suitable for the university environment since it has computer literate faculty and students with an intense interest in automation. The design and analysis will be enhanced with more powerful computers and software, and with the integration of the design database with the fabrication automation. The economics of prototype automation will be compared with a study of production automation (Ref. 2). The fabrication and ultrasonic automation inspection will be new capabilities for the RFRL.

Aircraft Design Automation

The external geometry of an aircraft is composed of wings and fuselage with transition fillets at the intersections. The wing cross-sections is defined by the airfoil section chosen by the designer and is a complex smooth shape which cannot be discontinuous or wavy. Spanwise, the wing is a ruled surface. The fuselage is analytically defined by conic sections with a smooth variation in the control parameters prescribed by the designer. AUTOCAD 10 is capable of generating a grid between defined sections, but the treatment of discontinuous transitions and intersections such as windshields is not clear.

RFRL Fabrication Automation

The automation of the fabrication processes in the modern aerospace company has been in place for many years. Heavy duty, precision, very expensive equipment was required for metal working because of the high tool loads and the tight tolerances for fasteners. Recently, the furniture industry began to utilize automation. The tolerances are relaxed and the tool loads are lower. This technology is sufficient for the prototype aircraft tasks such as moldmaking, cutting and inspection.

The most useful robot for aircraft fabrication is the gantry configuration. The parts of an aircraft are relatively slender about one axis. The X-Y-Z coordinate system used to define the geometry of the gantry configuration maximizes the accuracy and simplifies the calculation of the tool position. Since the RFRL was inexperienced in robot operations, it was decided to begin with a small unit and move up to a larger unit as soon as possible. There are

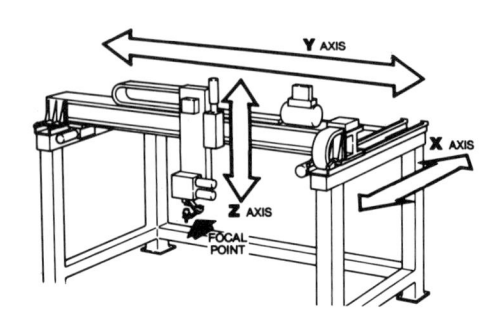

Figure 3. 5-Axis Gantry Robot

many small operations such as rib molds and prepreg cutting which can be accomplished on the smaller machine.

The RFRL has on order a small prototype gantry unit with a 5-axis working section of 4 x 4 x 1.5 feet and 2.5 hp router cutter. This unit was chosen because it was completely turn-key with PC and CAD toolpath software supplied. Another attractive feature was that the cost was approximately $85,000. This unit can be used for other tasks such as cutting and ultrasonic inspection by replacing the router unit. The RFRL is considering the relative merits of purchasing a 6 x 10x 3' 5-axis woodworking gantry robot priced in the $150,000 range, or to retrofit, inhouse, a CNC system on a 6 x 20 x 5' planermill which can be purchased used in the $30-40,000 range.

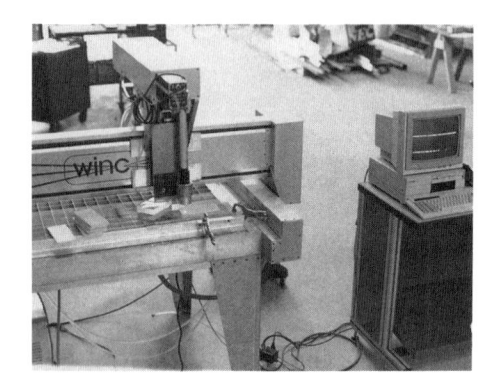

Figure 4. Low Cost Gantry Robot

The larger gantry robot is needed for fuselage and wing molds. For the ultrasonic imaging research and for "C" scanning inspection tasks, there has been developed small light duty 3-axis stepper motor drive gantry units. The working dimensions are 1.5 x 3 x 1' with an 8-axis controller which allows for expansion. These units cost approximately $9,000.

Composite Cure Automation

The RFRL is installing a PC control and data acquisition computer to conduct the cure of the composite structures in the ovens, press, and autoclave. Integrated software that can set up arbitrary cure cycles and adjust the control based upon input from thermocouples imbedded in the part will increase the quality of the part. Failsafe aspects of the computer control is being investigated. The total cost of computer, interface hardware, and software is $7,000. A 5' dia x 20' length autoclave with thermoplastic (300psi-850F), capabilities is being installed at a cost of around $180,000.

Low Cost Automation in the Aerospace Engineering Curriculum

There has been interest for many years in the integrated approach to aerospace education (Ref. 3), but the facility costs were too high. Now the integration of design and fabrication hardware and software could be used to change the curriculum at a reasonable cost. The student can be introduced to all aspects of the product development process in a realistic way. The faculty and students could make contributions to enhance the productivity of aircraft development.

Automation in General Aviation Aircraft Development

Integrated design and fabrication are going to significantly change the general aviation aircraft development process. There is the potential for cost reductions or enhanced performance through the increased flexibility of the automation equipment. It will be possible to make changes in the configuration shape without having a major impact upon the development schedule. The automation should make possible a delay in the commitment to mold fabrication and other items which will permit a more refined development of fabrication processes. Integrated design and automation should permit increased productivity even for low production rates for small aircraft. Perhaps this will allow the industry to remain competitive in the world markets.

References.

1. Bennett, G., Bryant, G., "Design and Fabrication of Marvel II Wings," SAE Paper 850891, 1985.
2. Foley, M., Bernardon, R., "Thermoplastic Composite Manufacturing Cost Analysis for the Design of Cost Effective Automated Systems," SAMPE Journal, Vol 26, No. 4, July/Aug 1990.
3. Bennett, G., "Design-Build-Fly, An Effective Method to Teach Undergraduate Aerospace Vehicle Design", AIAA Paper 73-785, August 1973.

Determining Organizational Readiness for Advanced Manufacturing Technology: Development of a Knowledge-Based System to Aid Implementation

DONALD D. DAVIS
Old Dominion University
Norfolk, VA

ANN MAJCHRZAK and LES GASSER
University of Southern California
University Park, Los Angeles, CA

MURRAY SINCLAIR and CARYS SIEMIENIUCH
University of Loughborough
U.K.

Summary

We describe a portion of an automated knowledge-based system designed to help managers plan the organizational change needed to implement computerized manufacturing processes.

Introduction

Computerized integrated forms of manufacturing (CIM) technology promise to revolutionize the production process. Unfortunately, the productive promise of CIM is seldom fulfilled in American companies. U.S. managers buy and install the hardware and software but they use it inadequately, particularly when compared to their Japanese counterparts [1].

Recent studies have revealed that from fifty to seventy-five percent of the attempts to implement advanced manufacturing technologies result in failure to achieve predicted benefits [2]. This failure stems largely from mistakes managers make when preparing their organizations for these new technologies. Some managers we have worked with report losses of more than one million dollars per day due to faulty implementation of advanced manufacturing technologies. Successful implementation of CIM requires new forms of organization and management [3].

This paper describes a portion of HITOP-A (Highly Integrated Technology Organization People--Automated), a knowledge-based system (KBS) for use by managers to guide them when planning the implementation of advanced manufacturing technologies such as

CIM. This KBS is based on a model developed by Majchrzak [4]. HITOP-A currently focuses at the level of the flexible manufacturing cell, which is defined as a set of information and machining technologies which produce a family of parts.

Written in Knowledgecraft and Common Lisp and employing coordinated problem-solving techniques [5], the KBS allows users to provide inputs which represent important technical, strategic, philosophical and organizational assumptions of the organization's management team. Given these inputs, the system uses several hundred decision rules to generate a detailed prescriptive model of the human infrastructure for the new technology.

One of the purposes of HITOP-A is to direct managers' attention to characteristics of their organization requiring change and to assess the degree of change needed. The KBS is used prior to technology implementation so that managers may view the consequences of their technical and organizational choices and may determine how best to reconfigure their organization in preparation for the new technology. The KBS is based on the sociotechnical principle of "joint optimization" of the social and technical systems of the organization [6].

HITOP-A focuses on the four major areas of the human infrastructure which are most relevant to the implementation and use of technologies such as CIM [4]. Based on users' technical choices, the KBS's output suggests to users the type of (1) work design, (2) skills/selection/training, (3) performance management and (4) organization structure/culture which should be in place prior to implementation of the new technology. The KBS also tells users how ready their organization is to adopt the necessary elements of human infrastructure. This paper describes one portion of the KBS--determination of the organization's readiness to change in response to the new technical system.

Organizational Readiness to Change

It is axiomatic that the implementation of advanced technologies such as CIM requires organizational change. Unfortunately, managers too seldom recognize the extent of change required by their organization until after problems emerge. The KBS is intended to help managers plan organizational change.

The organizational readiness to change component of HITOP-A focuses the attention of managers on more than one-hundred characteristics of the human infrastructure of their organization representing the four areas described above (work design, skills/selection/training, performance management, organization structure/culture). Some examples of these human infrastructure characteristics include:

- Work Design — mix of job specialties, reliance on resources outside the flexible manufacturing cell, team organization of work

- Skills/ Selection/ Training — operator knowledge of: cell equipment, procedures, process scheduling and control, quality monitoring

- Performance Management — type of reward system, degree to which rewards are contingent on individual/team performance

- Organization Structure/ Culture — differentiation, formalization, integration, span of control, professionalism, thirteen cultural norms, e.g., individual initiative

Some characteristics of the human infrastructure serve as preconditions for technology implementation. For example, Adler and Helleloid [7] point out that the pace with which new technology can be implemented will be dictated by the preexisting level of skills and implementation procedures. Other important factors which must be in place include the motivation to change, an effective plan to reduce resistance to change, and a clear technology plan which specifies the role of the new technology [8]. These preconditions become more important as the complexity of the technology increases [9]. The organization should delay implementation until preconditions such as these are in place.

HTOP-A allocates the remaining characteristics of the human infrastructure across four levels of analysis: individual, intragroup, intergroup and unit/organizational. These characteristics are used to compute organizational readiness for change. This calculation is done by creating two organizational profiles. The first profile represents the current state of the organization (AS-IS profile). The second profile represents the future state of the organization subsequent to technology implementation (TO-BE profile). The AS-IS profile represents the existing human infrastructure. The TO-BE profile, on the other hand, represents the human infrastructure which will exist after the technological and organizational changes are complete. In a sense, the TO-BE profile is an ideal type or goal toward which the organization must move.

The AS-IS profile is calculated from user inputs. Users of HITOP-A describe their organization in terms of the features which we have chosen. For example, they describe the existing mix of job specialties, reward system, cultural norms and so forth.

The TO-BE profile characteristics are generated by the KBS itself based upon rules written into the program. These rules represent predictions concerning the future state of the human infrastructure given organizational constraints, the AS-IS profile, and the technology to be adopted.

Readiness to change is calculated by comparing the AS-IS and TO-BE organizational profiles on all of the human infrastructure characteristics. Gaps between the profiles of the human infrastructure characteristics depict the difference between what the organization is and what it must become in order to implement the new technology successfully.

Readiness to change is determined by making two comparisons. The first contrast is the size of the gap between the AS-IS and TO-BE profiles on each human infrastructure characteristic. Large gaps denote that a greater degree of change is needed; the new practice is radically different from current practice. Large gaps must be narrowed in order to successfully implement the new technologies.

The second contrast used to indicate readiness to change examines the number of gaps between human infrastructure characteristics on the AS-IS and TO-BE profiles. Fewer gaps denote a greater readiness for change. A large number of small gaps between the AS-IS and TO-BE profiles indicates the need for many incremental changes.

Comparison of the AS-IS and TO-BE organizational profiles to calculate organizational readiness for change in the KBS yields several benefits. First, it gives to users a detailed picture of what the human infrastructure of their organization will look like after implementation of the planned technical and organizational changes. This is important because managers frequently underestimate the extent to which the impacts of new technologies will ripple throughout the organization and produce unintended consequences.

A second benefit is the ability to isolate where the organization is least ready to change. That is, we can show managers specific features of their organization which must be changed before the new technology should be implemented. For example, we may discover from the AS-IS profile that management allows only limited decision making authority among operators, but the TO-BE organizational profile prescribes self-managing teams (which by definition require autonomy), resulting in a large gap for this human infrastructure characteristic. The organization might take several actions to close this gap and move in the direction of greater autonomy. The organization should first select workers who prefer having greater authority, it should provide training in group process and decision making, it should tie rewards to group performance, and so forth.

Finally, users can run the KBS before committing resources to the organizational and technical change. Users can judge whether making the organizational changes required for the new technology is worth the cost. Costs may be compared for different organizational/technical configurations prior to making a major financial commitment. The configuration with the fewest and narrowest gaps is preferred. Managers who have tested the KBS report that their planning for technological implementation has been reduced from several months to a few days.

References

1. Jaikumar, R. Postindustrial manufacturing. Harvard Business Review, 86 (1986) 69-76.

2. Works, M.T. Cost justification and new technology addressing management's No! to the funding of CIM. In L. Bertain & L. Hales (eds.), A program guide for CIM implementation (2nd Ed.). Dearborn, MI: Society of Manufacturing Engineers 1987

3. Manufacturing Studies Board. Human resource practices for implementing advanced manufacturing technology. Washington, D.C.: National Academy Press 1986.

4. Majchrzak, A. The Human side of manufacturing: Managerial and human resource strategies for making automation succeed. San Francisco, CA: Jossey-Bass 1986.

5. Gasser, L., & Hill, R. Engineering coordinated problem solvers. Annual reviews of computer science 4, (1990).

6. Cherns, A. The principles of sociotechnical design. Human Relations, 29 (1976) 783-792.

7. Adler, P.S., & Helleloid, D. A. Effective implementation of CAD/CAM: A model. IEEE Transactions on Engineering Management, 34 (1987) 101-107.

8. Davis, D. D. Integrating technological, manufacturing, marketing, and human resource strategies. In D. D. Davis (ed.), Managing technological innovation: Organizational strategies for implementing advanced manufacturing strategies. San Francisco, CA: Jossey-Bass 1986.

9. Gerwin, D. A theory of innovation processes for computer-aided-manufacturing technology. IEEE Transactions on Engineering Management, 35 (1988) 90-100.

Planning and Realization of Skill Based Flexible Automation for Developing Countries

S. KUMAR
Department of Production Engineering
Birla Institute of Technology
Mesra, Ranchi, India

A.K. JHA
Foundry Forge Plant
Heavy Engineering Corporation
Ranchi, India

Abstract - After giving a brief about Skill Based Flexible Automation (SBFA) concept and its linkage with the competitive strategy of the industry the paper describes the importance of SBFA and related matters relevant to future course of action. The ideas predicted in this paper will provide strategies for design - planning and building skill based flexibly automated industries for developing countries.

1. INTRODUCTION

During last few years there have been several concepts in automation not only in automated machines but also in the process of decision making regarding planning, design and operation of these machines. The developed countries with high per capita income do realize the importance of increasing productivity through use of automated manufacturing technologies. This is because the high technical development will promote their standard of living. Their socio-economic set up is totally integrated with technology as it ranks among the real wealth producing activities. On the other hand in developing countries, the sensitivity of socio-economic structure is much different. The per capita income is low and have surplus manpower. Manpower is looking for new jobs and employment opportunities. They fear that increasing use of automated manufacturing technologies and techniques will take awaytheir jobs, so the society as a whole do not much encourage complete automation of existing and future upcoming industries. Society is a were that new technology will improve their standard of living and create more job opportunities for the future generation, but at present economic considerations and the race for earning more money are their immediate necessity of life.

They insist on incentive and immediate jobs and believe in increased productivity through mass participation. The quantum of sensitivity of human behaviour in a particular socio-economic environment is a critical factor which should be considered before automated manufacturing is planned in developing countries. In view of these considerations, presently developing countries are in a situation of choice with strategic options for future production systems regarding organisation, technology, skills etc. In order to generate efficient production structures, appropriate decision on work organisation, on the division of functions and on the forms of skills and interaction between man and machine have to be made.

The concept of the 'Unmanned Automated Manufacturing', where skilled work on the shop floor and human expertise in the design office is assumed to be replaced by machine artefacts to the optimum level, seems to turnout being illusionary. In recent years many attempts of bringing man back in rather than replacing him by computers and robots have become visible. Today situation is to look for different partition of functions between man and machine by reintegrating tasks and using human skill and judgement in manufacturing more comprehensively in order to make it more flexible and productive. Further more, existing skills would be rejected, while skills which do not existed would be required.

2. FLEXIBLE SKILL PATTERN FOR FLEXIBLE AUTOMATION

The objective of the productivity is to use wisely all the available human, scientific and technological resources in order to obtain the most competitive and highest quality results. The whole world is in the race of tremendous market competition and in order to survive one has to be ready. In order to obtain full benefits offered by new technologies, it is necessary to re-organise the human skill pattern and implement it in an appropriate manner. In any advanced automated manufacturing system there are manual tasks as well as supervisory tasks. Technically and economically it is not possible to automise complete manual tasks. All the tasks that are needed to make the whole system work could be put into three categories, primary, secondary and tertiary tasks. The primary tasks are regular, anticipated and planned. They are mostly routine tasks of physical nature. The secondary tasks are foreseen. They are regular and irregular and can be planned. Their content is both physical and mental. The tertiary tasks are not anticipated and irregular. The workman here deals with unexpected and uncommon events. The mental content of these tasks is a characteristics feature. At a fully automated system the primary, secondary andtertiary tasks would be mechnized. For flexible automation (to accomplish these tasks) the following three major changes in the human skill patterns are identified :

i) There is clearly a need for certain kinds of new skills-programming, electronic equipment operation and maintenance, diagnostics, and so on.

ii) There is also a need for multiple skills-combining new and existing skills into a broad package rather than the traditional pattern of single specific skills.

iii) The third area of importance is in the ability to deploy these above skills flexibly and this has implications for factors like work organisation (where the trend is towards team based work) and working practices.

Table 1 presents few categories associated with human skill. To be able to meet the requirements of the proposed SBFA concept, thinking in system terms and inter disciplinary team work are urgently needed.

3. DESIGN-PLANNING AND IMPLEMENTATION STRATEGIES OF SBFA CONCEPT

It is quite obvious that flexiblity, especially in manufacture, will be a decisive strategic competitive factor in the future. The mere implementation of advanced computing technology is by far not the only issue to be taken care of. As far as computer aided work (CAD/CAM/CIM etc.) is concerned, it is extremely important to acquire profound knowledge/skill of the functioning of the computer system used in order to enable the workers towse it as a tool. It is not sensible trying to develop and to apply turnkey compputerised systems in automated manufacturing. These are profound differences between industrial cultures (production concepts, skill pro-

Table - 1

Different Technological Categories Versus Human Involvement

Categories	Human Association
1. Technological	Human as a tool
2. Ergonomical	Human as a biological organisation
3. Psychological	Human as an active, thinking and evaluating individual.
4. Sociological	Human in the social context.
5. Educational	Human as a learned.

files etc.) and industrial relations in different countries, which do imply adaptations of work organisation and production technology. Further, production concepts even differ between single firms raising diverse functional requirements for the computer system applied. For successful designplanning of SBFA concept for a particular industrial set up (in a particular country) the manufacturing structures, work organisation and skill profiles have to be considered and, if necessary, adapted to future market conditions. Only from that the functional requirements for an appropriate technology can be derived.

A) Planning and Work Organisation Skills
i) Dicide which products should be manufactured and how much.
ii) Decide on product priorities.
iii) Allocate human and material resources
iv) Where to automate
v) Co-ordinate production operations between facilities.
vi) Co-ordinate information from supervisory controllers.

B) Supervisory Control and Programming Skills
i) Computer aided process planning/design
ii) Part programmers
iii) Human supervisory controllers of different subsystems such as FMS, robotic systems etc.

C) Assembly, Machine Operation and Maintenance Skills
i) Maintenance operators
ii) Human NC operators
iii) Human robot tasks
iv) Human assembly operators.

A correct system design procedure determining work organisation and task allocation prior to the development andimplementation of the technicalsystem makes it possible to start with skill formation in an early phase. Thus skill formation (efforts to quality the work force) can become and should be made an essential part of system design-planning and the development task.

In implementing automation, specially in developing countries where the cost of labour is much cheaper than developed countries, the critical issue is how to make the transition from traditional system to automated system (may be "manned" or "unmanned") so that thetechnological change will be jointly supported by all levels. The management of implementation should not only give consideration to technical and economic factors but also to ergonomic factors too.

The economical factors that will have effect on the employee are : Employment, Wages, Job status and job security, Inter plant transfer. The social factors that will be affectedby the introductionof advanced manufacturing techniques are : Organisational relationships, worker adaptations in new environment and family and communityassociations. The advanced manufacturing techniques will also have an impact onthe psychological factors like apprehension and resistance to change, worker attitudes and satisfaction and worker insecurity. For successful implementation of SBFA concepts, finely management action is needed with regard to : advance planning, training and workestation design.

Advance planning for the implementation of SBFA concepts should be made with the objective in mindthat it is not merely planning for new machinery and equipment, it is planning (re-orientation) forhuman skills. The possible problems must be anticipated and dealt with frank atmosphere. All concerned people (affectedby the technological change) should participate in the advance planning including management personnel, production and maintenance workers and their supervisors, manufacturing engineers and union representatives. All shall develop a "consensus" towards a common objective. The SBFA concept will require high skills to program machine instructions but once this is accompalished, production is almost smoothlyautomatic and little labour is required. Fewer skilled and semi-skilled workers will be needed for production, whereas, a highly specialised or trained maintenance groupwill be required. Undoubtly, the work and responsibilityof the production and maintenance substantial change as a consequence of the automated manufacturing technologies. In view of this a continuous training program should be developed to update workers skills, so that they are able to cope with the anticipated technological change. To derive optimum benefit from the SBFA concept, attention must be given to workstation design. The workstation should be compatible not only with systems performance requirements but also the user. Work-tation dimensions should be compatible with anthropometric characteristics of the anticipated user.

4. CONCLUSION AND RECOMMENDATIONS

The technological change based on SBFA concept will bring following changes (positive and negative) in the work related to traditional automated manufacturing.

a) Positive Changes -
- The work is more interesting and there will be an increase in the use of human skills and abilities.
- Increase of status of own work.
- Much more scope to learn new things and increased qualifications needed at work.
- Better opportunities for career growth.

b) Negative Changes -
- Routine work tasks
- Decrease in freedom to 'control own work'
- Confinement to the workplace
- Decrease in own decision making and planning of work.

Since it is impossible to wait for a new generation of workers to emerge, it is necessary to develop new and adequate ways to learn considering the large amount of experience being at the disposal of skilled workers with several years of practice. Presently the factory of future is at a cross roads between "Unmanned" and "Skill Based Flexible Automation" and that we are in a situation of choice. In order to

develop an appropriate production concept for a company's specific needs following question gain strategic importance
* How to find a good production concept ?
* How to design the system ?
* How to acquire and integrate the suitable skill ?

A firm's organisational structure linked with its social interests and priveleges, existing polarized skill profile and traditional production concepts (Tayloristic and fordistic) all prove to be strong forces of inertia and hard to overcome. Unless the existing idealogythat cannot imagine any other improvement in production than replacing human capabilities by machine artefacts is surmounted only little changes will occur. The industrial relations also tend to stabilize tratitional production systems and to resist to change. All such problem,have to be put asidefor successful design planning and implementation of skill based flexible automation system. The concept is likely to be the superior choice towards automation in developing nation which have rigid socio-economic boundries and commitment for employ - ment.

REFRENCES

1. Weatherall, Alan, "Computer Integrated Manufacturing" Butterworths & Co Ltd., Great Britain, 1988.
2. Midha, P.S. and Trmal, G.T., "Human Element in the Factory of the Future", Proc. Int. Conf. on CAD,CAM, Robotics, and FOF, TMH, New Delhi,1989.
3. Besant, J and Haywood, B, "Flexible Skills for Flexible Manufacturing" IFAC Skill Based Automated Manufacturing Kerisruhe, FRG, 1986.
4. Brodner, P, "Skill Based Production the Superior Concept to the "Unmanned Factory," Toward the Factory of Future, spring-verlag, Berlin, 1985, pp. 500-505.

Chapter IX

PCB Manufacturability and Assembly

Introduction

Expert systems have been found very useful in today's industry. This chapter deals with some recently developed expert systems. The first paper describes a framework for an expert system based concurrent engineering system for Printed Circuit Board (PCB) manufacturing using surface mounted technology (SMT). The importance of such a system for PCB manufacturing is also discussed. The second paper proposes a relative piloting policy for flexible assembly lines (FAL). Different heuristics for dynamic piloting have been tested on a simulated FAL. The third paper presents a knowledge-based system for printed circuit board manufacturability evaluation. The system is built using an object-oriented paradigm with a Smalltalk-80 language interface. Hybrid technology has been considered in creating the knowledge base. The system employs a rating method to rank various designs. A hierarchical rule representation scheme is developed and described. The next paper discusses an IGES post processor and its integration with a workcell for printed circuit board assembly. The post processor operates on a CAD file to produce the Cartesian locations of the electronics component as specified by the designer and produces a formatted output specific to the workcell. The information is then downloaded to the workcell and the assembly sequences are executed. A discrete optimization assembly method for an automated workcell is described in the fifth paper. The method of implicit enumeration is employed and it provides a good basis for selecting an optimum assembly sequence. A module called TPS (trajectory planner and simulator) for printed circuit board assembly is presented in the next paper. In the final paper of this chapter, a routine is presented for an optimal lead to pad matching technique for surface mounted component placement in PCB assembly. The problem is formulated as a non-linear optimization problem with linear objective function and quadratic constraints. The routine takes into consideration the errors associated with the shape of components. Using this routine, elimination of fiducials or centering devices has been reported.

An Expert System Based Concurrent Engineering Approach to PCB Assembly

K. SRIHARI

Department of Mechanical and Industrial Engineering
Thomas J. Watson School of Engineering
State University of New York
Binghamton, NY

Summary

A framework for an Expert System based concurrent engineering system is described. The domain under consideration is Printed Circuit Board manufacture using surface mount technology. The working of the system's components are discussed along with its advantages. The system will enhance PCB design quality while considering manufacturing needs.

Introduction

The design and manufacture of Printed Circuit Boards (PCBs) is becoming increasingly complex [5]. The advent of factors such as fine pitch components, new manufacturing processes, and the segregation of design and manufacturing knowledge into a large number of specialized areas of expertise have resulted in the need for tools to help reduce the time from PCB design to actual manufacture. A major problem is that designers who have an excellent view of technology from a design standpoint often have a limited view of the needs of manufacturing [7,8].

This paper presents a framework that will use Expert System (ES) techniques to promote concurrent engineering in the PCB manufacturing domain. This research describes a set of interrelated systems that are designed to assist the engineering design function, improve and support human interactions, promote integrated decision making, and create an user friendly system. The benefits of this effort include:

* Improving the information levels of all the key players in the PCB design and manufacture process.
* Providing a structured interface between design and marketing.
* Allowing for the implementation of concurrent engineering systems.
* Reducing expensive and time consuming product redesigns.
* Reducing product and life cycle cost.
* Reducing time to market.
* Reducing design cost.

PCB Assembly Using Surface Mount Technology (SMT)

Electronic Packaging can be divided into three levels: Integrated Circuit (IC) chip packaging, Printed Wiring Board (PWB) assembly, and electronic system assembly [1]. The printed circuit (or printed wiring) process continues to be the basic interconnection technique for electronic devices. Virtually every electronic packaging system is based on this process, and will continue to be in the foreseeable future. Since printed wiring technology was invented more than half a century ago, several methods and processes have been developed for manufacturing PCBs of various types [5]. Modern electronics packaging has become very complex. The choice of which packaging technology to use is governed by many factors including electrical, thermal, and density requirements, cost, and material used [1,5].

Manufacturing that falls under the electronics packaging umbrella has often been neglected in manufacturing systems research [3]. ES techniques have been used in the development of Computer Aided Process Planning (CAPP) systems for electronics packaging domains such as PCB assembly. These include systems such as COPES and EPPSEA [6], and PWA1 [3]. ES based systems have been used in trouble shooting the wave soldering process [4]. Recent trends however indicate that research efforts in this field are gradually increasing [2,3].

The use of SMT has gained popularity quite recently and is fast replacing insertion mount (through hole) technology [1,4]. Through hole components are mounted to the printed wiring board by means of leads inserted through the board. Surface Mount Components (SMCs) are soldered directly to the copper conductors on its face [4]. Through hole printed wiring technology is no longer adequate to meet the needs of high performance electronic assemblies [1,4]. The processes involved in PCB assembly using SMT are strictly ordered. They generally include: board loading, solder paste application, adhesive dispensing, component placement, solder reflow, final cleaning, inspection, and in-circuit testing (Figure 1). This research considers boards which have only SMCs mounted on one side or on both sides.

Surface mount assemblies are classified on the basis of how they combine surface mount and through hole components [1]. They are commonly referred to as:
1. Type 1: SMCs placed exclusively on one or both sides of a planar board. The components are attached via the reflow soldering process.
2. Type 2: Combination of both Insertion Mounted Components (IMC) and SMC on

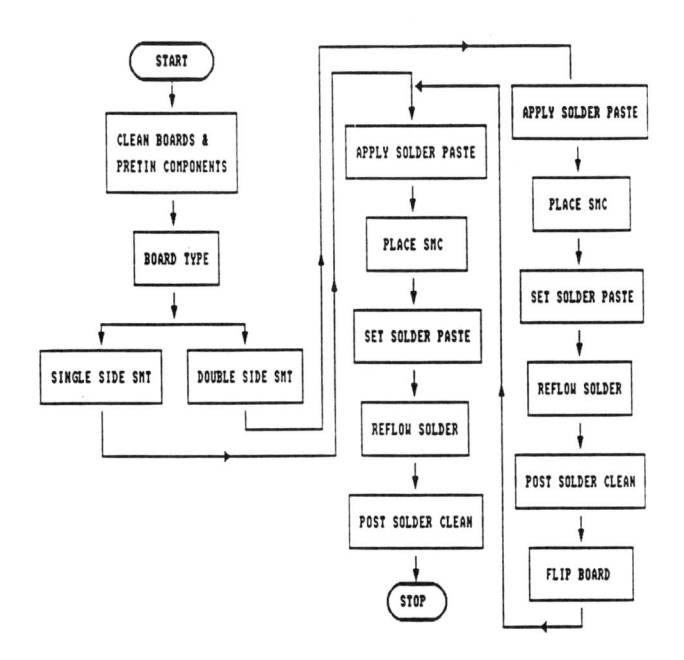

FIGURE 1 - A TYPICAL SMT PCB MANUFACTURING PROCESS

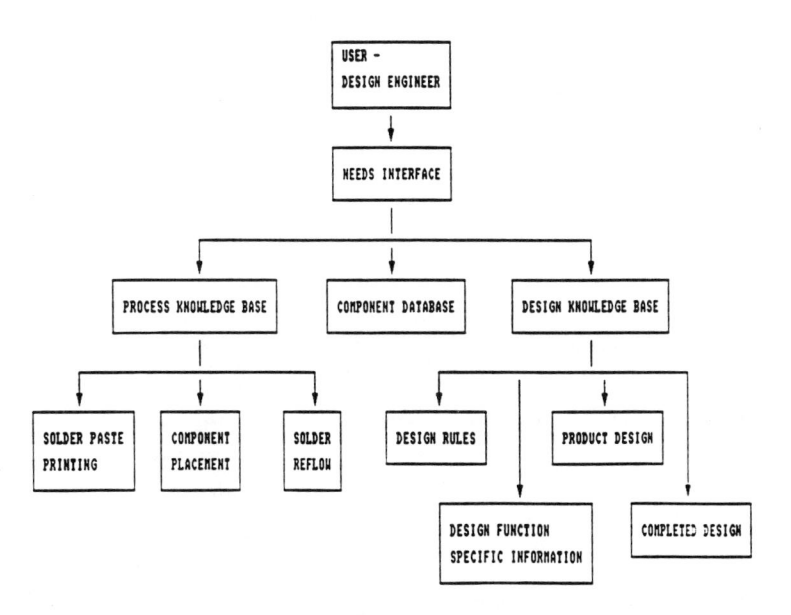

FIGURE 2 - AN EXPERT SYSTEM BASED CONCURRENT ENGINNERING SYSTEM ARCHITECTURE

a planar board. Both SMC and IMC are assembled on the top side of the board and SMC on the bottom side of the board. Top side components are reflow soldered, while the bottom side components are wave soldered.
3. Type 3: It consists of IMC on top and SMC on bottom, soldered in place via the wave solder process.

SMT offers many benefits over conventional through hole technology. The advantages of SMT over through hole are described in Coombs [1]. Some of these benefits are reduced size, better interconnectivity, lower cost, better performance, reduced electrical noise, improved shock and vibration performance, a controlled manufacturing process, reduced materials handling, and better quality. SMT is also more amenable to automation. The two principal surface mount soldering technologies in use are [4]: adhesive attach - wave soldering and reflow soldering

Expert System Based Concurrent Engineering System Design
The ES based concurrent engineering system presented in Figure 2 will be used primarily by the PCB design engineer. It will contain both design and manufacturing knowledge. The system is designed for a facility that designs and manufactures PCBs for use in-house. It is under development with some components currently being functionally operative. The short term goal is to develop the individual Knowledge Bases (KBs) such that they can be used as independent, stand alone systems. Over the longer term, the system will function in conjunction with a CAPP system. The concurrent engineering system described in this paper will be an advancement to presently employed systems in the completeness of automation, integration of human and automated components, and in the standardization of material, processes, and manufacturing methods.

The system will assist the design engineer in designing for testability, manufacturability, and reliability. It will provide the designer with information that would assist in identifying the consequences of design decisions early in the design process. The system will allow for the consideration of a multitude of factors, resulting in a higher quality initial design which requires fewer design cycles to eliminate problems and sub-optimal decisions discovered downstream, thus shortening design cycle time and lowering production cost.

The ES based concurrent engineering system (Figure 2) would increase the productivity and efficiency of the design engineer by providing a supportive and intelligent design environment. The inference mechanisms coupled with the KBs would provide the user (the designer) with expert knowledge that would otherwise be available only by meeting with experts or by the designer becoming an expert. It is intended that the system would be proactive in nature. For example, when a component is placed at a specific location on the PCB or a specific board type chosen, the system would invoke the appropriate inference mechanisms, fire all the appropriate rules, and provide the designer with the implications of the PCB design from a design and manufacturing stand point.

The design engineer, upon receipt of the need to design a new PCB or redesign an old one, would access the needs interface, and through it access the product design KB. This KB would provide the user with information on previous designs that might be relevant in the current scenario. The designer would then access the function specific KB. This KB would provide the user with specific information on the parameters that are critical to the current design effort. The design rules KB would provide the user (designer) with guidelines on good design practice for PCBs.

The process knowledge KB (Figure 2) would provide the designer with manufacturing information relevant to the design process for PCBs. This could be in the realm of solder paste screen printing, solder paste choice and limitations, the component placement process, and the solder reflow process. The designer, through the ES based system, would use these KBs to ensure that the final design is developed considering design for manufacture principles. An example would be in the design of PCBs with regard to the location of components on the PCB. If the PCB is densely populated with components in one segment and sparsely populated in another, it could result in problems in reflow soldering the PCB. Similar problems could arise due to the location of large, black components that could act as heat sinks and prevent other nearby components from being reflow soldered satisfactorily. Analogous manufacturing related design issues can be identified for every segment of the PCB manufacturing process. The system would identify the problems associated with various factors of a PCB design, and proactively inform the user about them.

The component database helps the PCB designer choose components for use in specific PCBs. It helps the user assess the needs and identify specific choices. The component database helps the designer readily access information on components available, their sources, and their functional specifications.

The role of the component database is to help reduce duplicity in component use, reduce the number of components on a PCB by combining multiple components into one if possible, reduce the types (number) of PCBs manufactured, facilitate the design of new PCBs, increase the inventory turns on PCBs, reduce component inventory, reduce the PCB design cycle, and reduce the yearly manufacturing cost of finished PCBs.

Conclusion

This paper describes the structure of a supportive and intelligent system that assists in the concurrent engineering function for PCB assembly. It is a valuable tool to the PCB designer and manufacturer. It will help improve the designer's productivity and efficiency. The system will enhance the quality of the PCB design, consider manufacturing needs, and assist in providing global optimization to the design process. The system will reduce redesign, manufacturing heartaches, design cycle time, and overall cost. It is evident that within automated PCB operations, there is considerable gain to be made by designing, researching, and developing ES based systems.

REFERENCES

1. Coombs, C.F., 'Printed Circuit Board Handbook', Third Edition, McGraw Hill, New York, 1988.

2. Ghosh, B., & Gupta, T., 'A Survey Of Expert Systems In Manufacturing And Process Planning', Computers In Industry, No.11, 1988, pp.195-204.

3. Imerese, P., & Wang, H.P., 'PWA1 - Process Planning Advisor', Proceedings -IIE Integrated Systems Conference, Atlanta, Georgia, November 1989, pp.463-467.

4. Komm, R., & Warner, D., 'An IBM Case Study Implementation Of Expert Systems For Wave Solder', Proceedings - NEPCON West '90, Anaheim, California, March 1990, pp. 677-687.

5. Lea, C., 'A Scientific Guide To Surface Mount Technology', Electrochemical Publications, Ayr, Scotland, 1988.

6. Sanii, E.T., Liau, J.S. & Srinivasan, K., 'Computer Aided Process Planning for Electronics Manufacturing', Proceedings - 1989 IIE Integrated Systems Conference & Society For Integrated Manufacturing Conference, Atlanta, Georgia, 1989, pp.450-454.

7. Solberg, V., 'Design Guidelines For Surface Mount Technology', TAB Professional And Reference Books, Blue Ridge Summit, Pennsylvania, 1990.

8. Wilson, D.R., & Hutchison, K.K., 'Integrating Design With Manufacturing', Proceedings - NEPCON West '90, Anaheim, California, 1990, pp.446-455.

Real Time Production Scheduling and Dynamic Parts Routing for Flexible Assembly Lines

J.P. BOURRIERES, O.K. SHIN and F. LHOTE

Laboratoire d'Automatique de Besancon / CNRS
Institut de Productique
Besancon, France

Abstract A reactive piloting policy for Flexible Assembly Lines (FAL) is proposed, where the sequencing of the operations as well as the assignment of tasks to workstations are not pre-determined but driven by the actual state of the FAL. For each work-in-process coming out from a workstation, the next destination is determined by minimizing a temporal criterion taking in account:

> the time needed to reach the destination,
> the load of the workstation to reach,
> the duration of the operation to be completed in the destination station,
> the disponibility of product components in this station.

The purpose of our piloting policy is to manufacture a given quantity of products as rapidly as possible by balancing the amount of work allocated to workstations and to reduce the efforts required for scheduling the production of short series of diversified products. Two strategies of dynamic assignment are described and evaluated on a simulated FAL.

Keywords Assembly Automation, Reactive Scheduling, Dynamic Process Control

1 Introduction

Controlling manufacturing workshops is traditionally based on a two-step decision process in which sequences of operations and task assignment to machines are determined by short time production scheduling and thereafter transmitted to the piloting system. This predictive approach seems today to be unsuited to respond to quality and productivity new requirements in FMS, because of:
- optimal scheduling of multi-product series on a flexible system can be very complex,
- scheduling optimization effort is based on an idealized production model that does not coïncide with the real system,
- one cannot face up easily the perturbations of the system during production performing,
- since sequences of operations and parts routing are predetermined, the flexibility of the system cannot be used 'on line' to its full potentiality.

In this context, an advanced piloting system must be dynamic i.e. capable of running the manufacturing system on the basis of its actual state, in order to react to the environment changes such as production flow variations, new products series, equipment breakdowns, etc.. [1],[5].

We propose here a new approach of dynamic piloting for Flexible Assembly Lines (FAL) where traditional control is unefficient as it is faced with complex part flows and frequent production aleas. The purpose of our piloting policy is to get rid of the combinatorial scheduling problem and to take full advantage of the flexibility of the production system: the flexibility of the FAL and, given a product or a family of products, the multiplicity of possible sequences of operations.

The piloting method presented in this paper is based on the following suppositions:
- the transport network of the FAL uses free transfer technology; any pallet can be transferred to any workstation by at least one route.
- the manipulators are more or less flexible.
- the operations performed in the FAL are mainly assembly operations, but can also be complementary operations like drilling, marking, etc..

We first formalize the assembly process, then present the dynamic piloting strategy and finally discuss the results of evaluation tests on a simulated FAL.

2 Assembly process

The first concern is to decompose the product(s) into elementary parts (Fig.1), then to study the various ways to build it (them) in a FAL. Here, our goal is not to identify exhaustively all possible orders of operations [2],[6], but to take into account a limited number of them so as to introduce a degree of freedom in the piloting strategy.

The different ways to build a product are clearly described by a Petri net (Fig. 2) on which the transitions represent the tasks and the places denote the material objects that result from the tasks. In this product-oriented approach, a task is an abstract operation (assembly operation or physical transformation) defined only by its contribution to the final product without any consideration on the involved equipment. Loading a part on a pallet and removing the final product from a pallet (Task X in Fig. 2) are seen as complementary tasks. Note that the number of assembly tasks required to get a product is invariant for all sequences of tasks [8], what can be verified on Fig. 2.

In the following, we call 'manipulation' the performing of a task by a definite manipulator.

3 Resources modelling

Regarding real time piloting of a FAL, flexibility is to be considered in manipulators capabilities as well as in transfer network potentiality.

We suppose that each workstation is composed of one input buffer, an unique manipulator and one ouput buffer. Each of the N manipulators of the FAL is capable of a specific set of operations. Let:

t_{ij} be the execution time of manipulation j performed by manipulator i

Since in assembly systems transfer times of pallets between workstations are often much longer than manipulation times, the piloting policy of a FAL has to take quite precisely into account the transfer dynamics. Topology and temporal performances of the transfer network are modelized by a distance matrix:

$$D = [d_{ik}] \quad d_{ik} \geq 0 \quad (i,k = 1,..N)$$

whose entry d_{ik} denotes the minimal distance

-expressed in time units- from manipulator i to manipulator k. When there is no route -permanently or temporarily- from a manipulator to another, the corresponding entry of matrix D is infinite.

4 Dynamic piloting policy

In an intuitive approach, task assignment may be done in order to perform tasks as soon as possible. Another strategy is to balance workstations loading. Those two approaches seem to be similar, but in the second case transfer times are not taken into consideration. We then retain the first approach and formalize it in the following:

In our piloting policy, neither the order of tasks -for a definite product- nor the task assignment are predetermined. When a pallet is leaving a station, i.e a task has just been completed, the next task to be performed is defined as a function of both the state of the product and the state of the FAL.

The state of the product is determined by the set of tasks previously completed, in other words by the -unique- marked place of the Petri net describing the different sequences of tasks. For example, the state ACE/P on Fig.2 results from the tasks A,C and E whatever the order and can evolve to ACDE/P or ABCE/P depending on the decision.

The state of the FAL is characterized on the one hand by workstations load:

q_{ij} actual number of pallets in the queue of manipulator i, expecting manipulation j

and on the other hand by the availability of manipulators:

w_{ij} working/breakdown state of manipulator i regarding manipulation j

($w_{ij} = 1$ if the manipulation is possible, $w_{ij} = 0$ otherwise)

Furthermore, we have to consider the availability of the components required by assembly tasks in stations. To be capable of assigning any task to any station, it is necessary to regulate the number of components in the stations. We then introduce complementary state variables of the FAL:

τ_{ci} c-type components ratio in station i

T_c global ratio of c-type components in the FAL

Finally, each time a pallet is leaving station i, the choice of the next task and of the destination station k -expecting to manipulation j- must take into account the following factors:

a) transfer time d_{ik} to reach destination k

b) load of station k: $L_k = \sum_j q_{kj}$

c) working time t_{kj} in station k

d) components availability in station k

Model

The amount of work in station k is modelized by the iterative equation:

$$L_k(h) = L_k(h-1)$$
$$+ \sum_{r,j} v_{rk}(h-1).t_{kj}$$
$$- \sum_{s,j} v_{ks}(h).t_{sj} \qquad \text{(eq. 1)}$$

(initial load $L_k(0)$ is given)

where

h is the abbreviation of h.δt

(δt sampling period)

$v_{ik}(h) = 1$: a pallet out of station i is attributed to station k for the next manipulation

$v_{ik}(h) = 0$ otherwise

The first term in the right member of eq.1 denotes the amount of work of station k at time hδt, the second one the amount of work assigned to station k at time (h-1)δt and the third one the duration of a task which has just been achieved by manipulator k.

Taking into account the factors a) to d), dynamic assignment of tasks can be formalized as follows:

For any pallet leaving station i at moment h, find destination k such as criterion

$$C_{ik}(h) = \beta_1.\text{Max}(L_k(h), d_{ik}) \qquad \text{(eq. 2)}$$
$$+ \beta_2.(v_{ik}(h). t_{kj})$$
$$+ \beta_3. |T_c - \tau_{ci}(h)|$$

(β_1, β_2, β_3 : weighting factors)

is minimal.

Algorithm

0. FOR each pallet which has been finished a manipulation DO
1. Search for feasible tasks at the next step
2. FOR each feasible task DO
   2.1 Search for manipulations corresponding to the task
   2.2 FOR each manipulation j found, search for a suitable manipulator k
   2.3 FOR each manipulator k -if not broken down- DO
     2.3.1 Get execution time t_{kj}
     2.3.2 Calculate amount of work L_k in the input buffer of station k
     2.3.3 Get distance d_{ik} from last to next station
   2.4 FOR each manipulator DO
     2.4.1 Calculate cost C_{ik}
     2.4.2 Keep destination k whose cost is minimal
3 Attribute the pallet to the station found at 2.4.2
4 IF the required quantity of products is not attained THEN GOTO 0 ELSE Stop.

Note that this algorithm leads to a non optimal production because each decision is made without taking into account its effect on the following tasks ('greedy searching'). We tested a more sophisticated algorithm consisting in minimizing the cost of two consecutive tasks assignments (look-ahead searching)[7].

5 Simulation

Those algorithms were tested on a simulated FAL composed of a free transfer ring and 6 off-line robotized workstations. The simulation software -made in our laboratory [4]- allows fine analysis of pallet traffic with dynamic display (Fig.3).

Given two types of products, different cases have been examined:

case n°1: single sequence, fixed assignment

Here, only one sequence of tasks is programmed for each product and each task is assigned to one single workstation. Furthermore, the physical order of stations along the transfer ring corresponds to the order of tasks. Though dynamic scheduling is not justified in that case, the goal is here to test our strategy on a conventional FAL before introducing flexibility.

Simulation results in Fig.4 show the production duration vs the number of pallets. It can be seen that mixed production (type 1 and type 2 together) is more efficient - with about 9 % advantage- that sequential production of type 1 then type 2. Moreover, saturation point of the system can be reached by introducing a sufficient number of pallets. The corresponding production time is then minimal and is here considered as a reference for dynamic piloting evaluation.

case n°2: greedy searching, without component constraint.

Simulation is necessary to optimize the weighting factors β_1 to β_3, depending of the configuration of the FAL. Fig.5 shows the effect of ratio β_1/β_2 without taking the disponibility of components into consideration ($\beta_3 = 0$). Here, it can be seen that high values of the ratio deteriorate the performance.

case n°3: look-ahead searching, with component constraint.

When taking into account the disponibility of components in the stations, the efficiency of our piloting strategy is lower (Fig.6). In the best situation (type 1 and 2 mixed), the production time is about 112% higher than the reference.

6 Conclusion

Different heuristics for dynamic piloting have been tested on a simulated FAL. The conclusions are the following:

We have observed that mixed production always leads to the best results (especially in case n°3).

Since task assignment is not predetermined, it is necessary to regulate efficiently the level of components stocks in the stations. For that reason, the 'kitting' technique is suitable when using dynamic routing strategy.

When all is said and done, the dynamic piloting concept described in this paper makes the FAL capable of reacting to operations failures in the workstations, what is a strong argument in assembly systems. Moreover, scheduling is practically eliminated.

Simulation results have shown that the algorithms proposed in this paper do not lead to notable drop of efficiency in comparison with static piloting strategy.

ooo

Bibliography

[1] Stecke, K.E 'Design, planning, scheduling and control problems of flexible manufacturing systems', Annals of operational Research, 3, pp.3-12, 1985.

[2] A. Bourjault 'Contribution à une approche méthodologique de l'assemblage automatisé: élaboration automatique des séquences opératoires' Thèse d'Etat, Univ. de franche-Comté, Besançon, Nov. 1984.

[3] S.B. Gershwin and al. 'Short term production scheduling of an automated manufacturing facility', Internal report of the Lab. for Inf. and Dec. Syst., MIT, LIDS-FR-1356, Cambridge, Jan 1984

[4] F. Lhote and al. 'Modelization of transfer systems for flexible assembly lines', Proc.of the Int. Conf. IASTED pp. 398-402, Paris, June 1987.

[5] O.Z.Maimon, S.B.Gershwin 'Dynamic scheduling and routing for flexible manufacturing systems that have unreliable machines', Internal report of the Lab. for Inf. and Dec. Syst., MIT, LIDS-P-1610 Cambridge, July 1987.

[6] T.L. De Fazio, D.E. Whitney, 'Simplified generation of all assembly sequences', IEEE Robotics and automation Vol. RA-3, n°6, pp. 640-658, Dec. 1987.

[7] O.K. Shin 'Contribution au pilotage d'îlots flexibles d'assemblage' Thèse de l'Univ. de Franche-Comté, Besançon, Nov. 1989.

[8] J.P.Bourrières 'Contribution à la modélisation intégrée des systèmes flexibles d'assemblage', Thèse d'Etat, Univ. de Franche-Comté, Besançon , Janv. 1990.

Ink (E)   Cartridge (A)

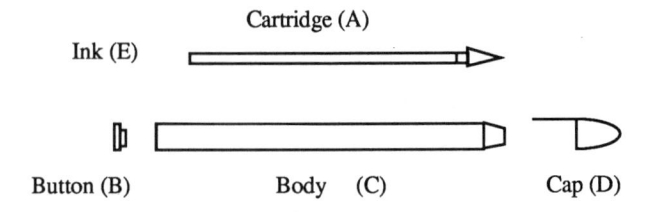

Button (B)   Body (C)   Cap (D)

Fig. 1 Product and components

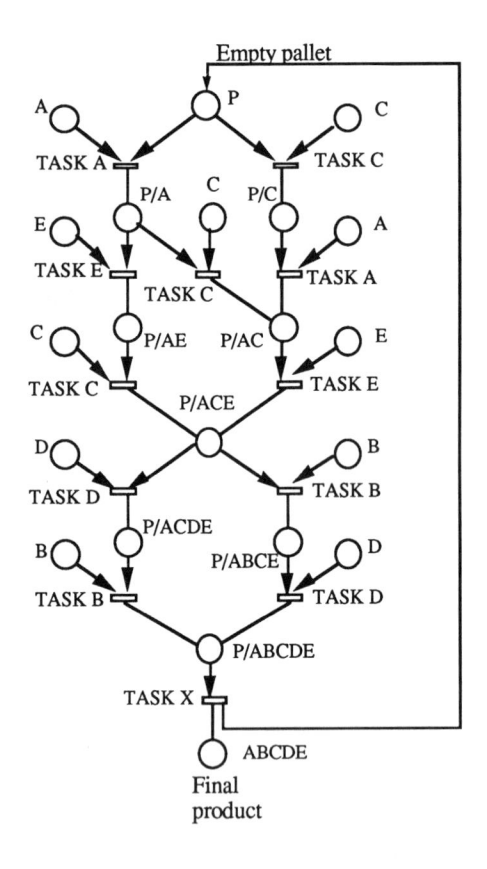

Fig.2 Feasible sequences of tasks
(representation by Petri net)

```
                    station 6    station 5    station 4
┌──────────────────────────────────────────────────────┐
│            ↓       ↑    ↓       ↑    ↓       ↑       │
│  22                    20←                            │
│                 in work    queue        on way        │
│         station 1  [ 5]  [         ]  [16 22          │
│         station 2  [29]  [ 7 31 15 ]  [26 8 20 27 32  │
│         station 3  [14]  [         ]  [               │
│         station 4  [12]  [         ]  [ 1 11       27 │
│         station 5  [23]  [         ]  [            11 │
│         station 6  [  ]  [ 4       ]  [               │
│              8     32                         26      │
│  16     ↓       ↑    ↓       ↑    ↓       ↑           │
│                 5     15 31 7 29          14          │
└──────────────────────────────────────────────────────┘
                  station 1    station 2    station 3
```

type of product	goal	started	achieved	number of pallets
1	200	39	25	16
2	100	35	20	16

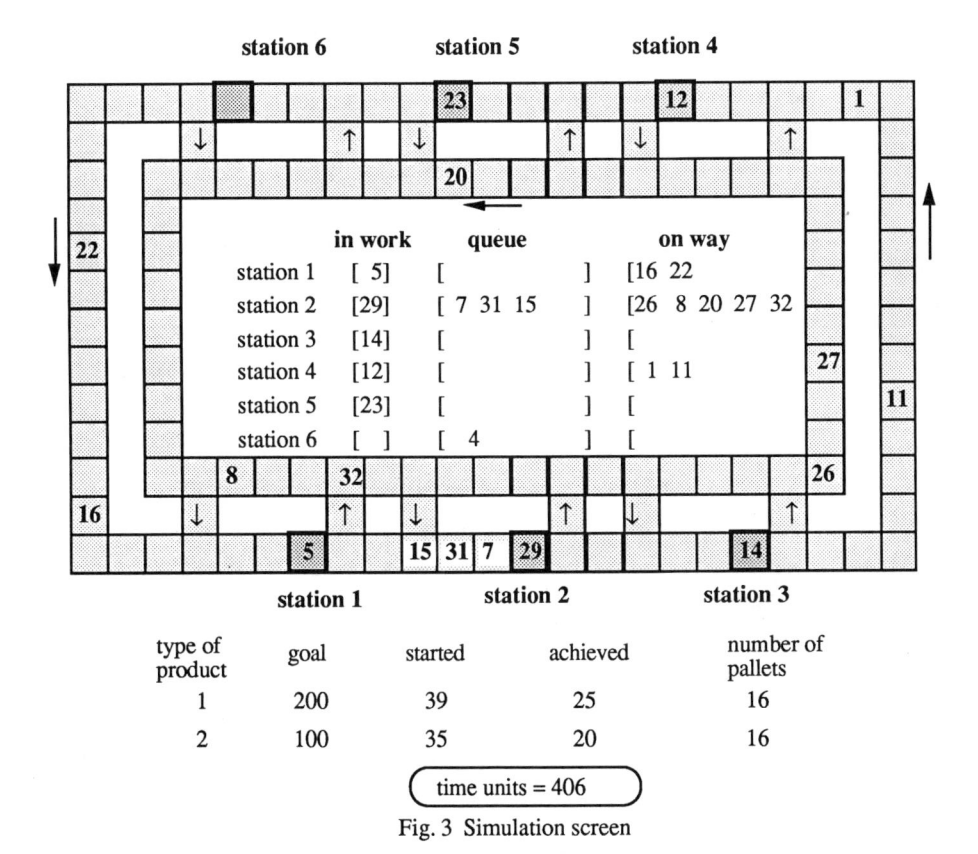

(time units = 406)

Fig. 3 Simulation screen

Fig. 4 Simulation results: (case n° 1)

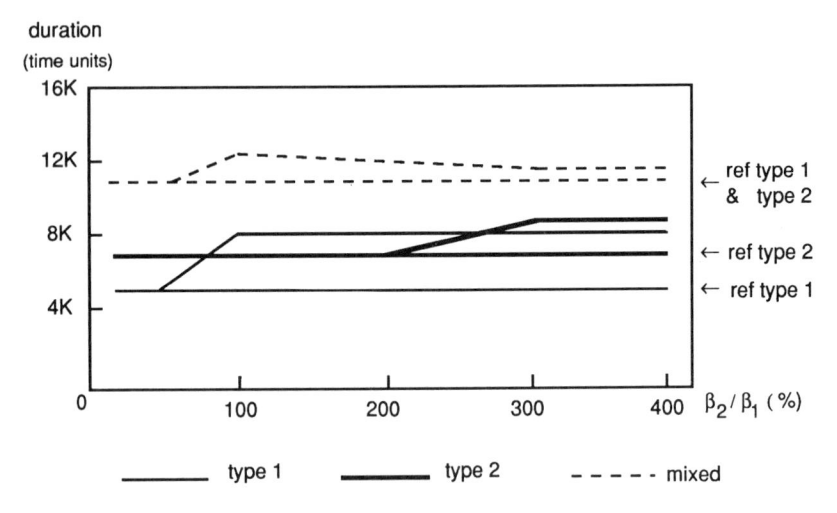

Fig. 5 Simulation results: 'greedy searching' without constraint of components (case n° 2)

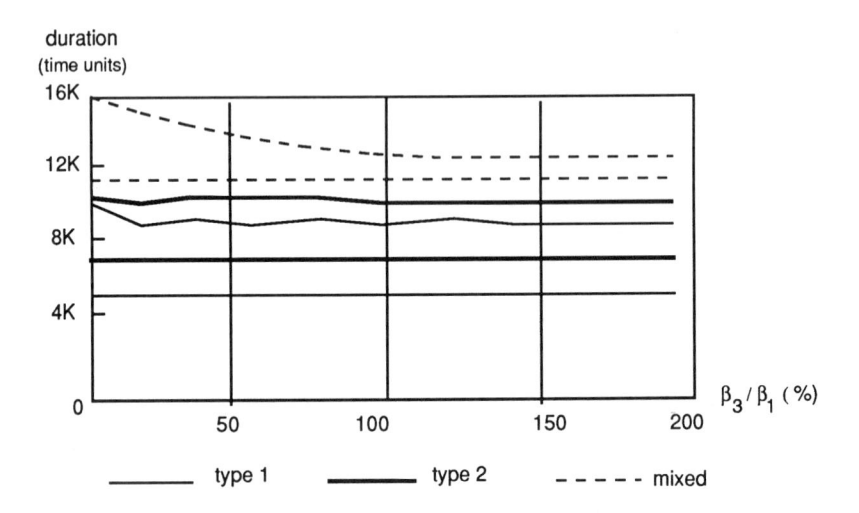

Fig. 6 Simulation results: 'look-ahead seraching' with component constraint (case n° 3)

A Knowledge-Based Approach for Manufacturability of Printed Wiring Boards

SISIR K. PADHY and S.N. DWIVEDI

Department of Mechanical and Aerospace Engineering
West Virginia University
Morgantown, WV

Abstract

Cooperative or concurrent product development is gaining popularity and will become norms of the Industry of the Future. To avoid problem created by designs not amenable to manufacturing, the designer should be aware of manufacturing constraints. Electronics manufacturing is one of the fastest growing segment of manufacturing domain today. The rapid change in printed wiring board manufacturing with the induction of surface mounted components and robotic assembly makes the system design more complex. A knowledge base is developed in an expert system environment to capture the manufacturability constraints of the PWB manufacturing. Rules and guidelines pertaining to manufacturability are formulated and incorporated into the knowledge base which is developed on an object-oriented paradigm. The system is capable of advising whether to go for manufacturing or not with a particular design and suggests design modification to the designer at the early stage of design to eliminate the manufacturing problems in the later phase of board production.

Introduction

Artificial Intelligence (AI) is playing a vital role in design and manufacturing automation, and AI systems have been applied in various domains starting from material selection to production, packaging and testing. We describe the application of an advanced AI system to the electronics manufacturing domain which is the fastest growing segment of the manufacturing world.

The reliability and quality of the electronics devices are primarily governed by its design and manufacturing of the printed wiring board (PWB) and the components. With the dramatic change in technology in the electronics industry, design and manufacturing methods are taking new shape. To address the manufacturability of the design, a knowledge-based system has been developed using object-oriented paradigm and applied to hybrid technology (combination of surface mounted technology and through-hole mounted technology).

---

* This work is sponsored by Defense Advanced Research Projects Agency (DARPA), under the contract no. MDA 972-88-C-0047 for DARPA Initiative in Concurrent Engineering (DICE).

The General Electric Company has developed the MRS (Manufacturing Rating System) knowledge-based system [7], for through-hole component assembly. Struttmann [9] applied Design for Manufacturability (DFM) principles to PWB producibility. His software, written in ANSI C, consists of 47 production rules and is limited mainly to the assembly of through-hole components. McKirchan and Bao [4] described the DMRS (Design Merit Rating System) for assembly of the printed wiring boards using surface mounted technology. Questions and design parameters are weighted to calculate an overall rating of the design. The system was able to rate the design from the assembly point of view. In another work, Bao [2] described the PC-DFM expert system for assessing the assemblability of the printed wiring boards with the rating scheme based on his previous scheme. However, the software is limited to assembly considerations and focuses primarily on surface mounted technology. In an earlier effort Padhy and Dwivedi [5] built a knowledge-based system using VP-Expert shell with forty-five rules. The system has very limited capability and a poor user interface. In a recent paper, Wilson, et. al. [10] from Texas Instruments presented an integrated tool set for printed wiring board design with emphasis on circuit design.

System Development

The present system is developed using an object-oriented knowledge-based system development environment DICEtalk [8,9]. It provides a high-level graphical interface and is able to solve complex problems. All the rules are built upon the vocabulary of a dictionary created a priori. Objects that are used in rule building are defined in this dictionary. The object values are also specified. Using antecedent and consequent relations the objects are related to each other and form a hierarchical relationship among each other. This object structure is shown in Figure 1. The rules are created in the declarative knowledge base using the surface percept description language (SPDL) and Smalltalk-80 object-oriented language. For example, rules in the system have the following structure:

>   *rulename kb#*
>   *IF*
>   >   *subtask 1*
>   >   >   *and subtask2*
>   >   >   .   .   .
>   >   >   *and subtask n*
>
>   *THEN*
>   >   *conclusion.*

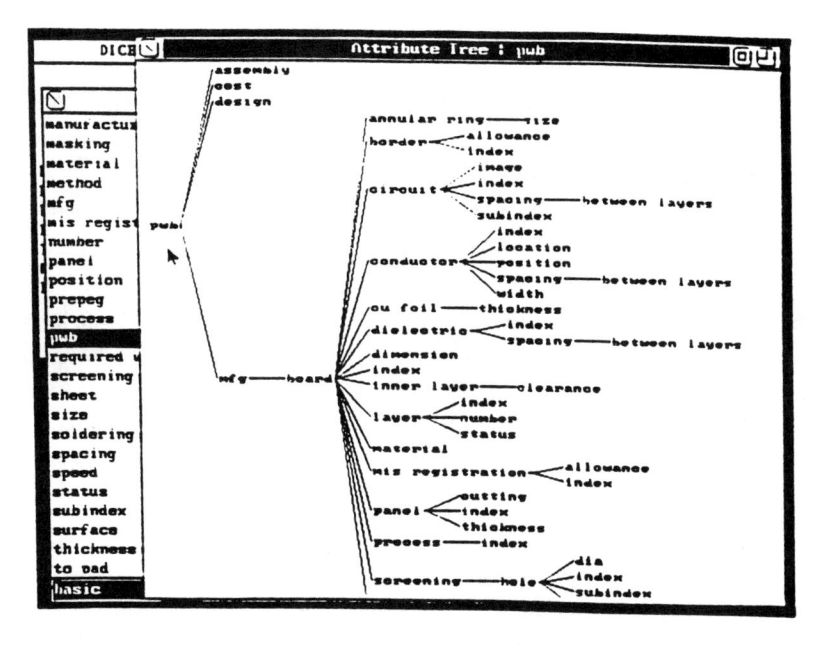

Figure 1. Object Structure in Dictionary of the System

A name is given to the rule to organize the rules into rule groups and thus it becomes easy to access them. The *kb* (knowledge base) number is automatically added to the rule after it is created or defined the number can be the user. A sort method is used in the system to sort all the rules by rule name or by *kb* number. The last subtask of the rule structure contains a message to the procedural knowledge base about the grade or comment or both. This subtask is a predicate in the form:

[kb grade: $letter Comment: aString]

The printed wiring board involves a number of independent parameters that affect the board's overall manufacturability. To compare different designs, a rating system based on a *university grading* pattern, which uses the letter grades of *A, B, C, D* with numerical values of 4, 3, 2, and 1 respectively. A grade is associated with each design parameter. A design parameter gets a grade of *A* if the design parameter satisfies the manufacturability criteria; *B* if it is acceptable, although not the best, i.e. some other value or method is preferred than this parameter value or method; *C* if the design parameter is not acceptable, but does not influence the manufacturability seriously; *D* if the design parameter is not acceptable, but influences overall production.

The goals in the knowledge base are organized into a hierarchical structure. The main goal of the system is represented as *evaluate mfg index* with four subtasks; of these three are subgoals and the fourth one is the predicate. The subgoals are:

        1. *evaluate board index*
        2. *evaluate manufacturing index*
        3. *evaluate process index*

These subgoals in turn consists of other subgoals . For instance, to evaluate the board index the inference engine has to solve for the border index, mis-registration index, conductor index, layer index and surface index. The benefits from this hierarchy of rule structure are as follows:

    i. The consultation can be done only part by part. If a subgoal produces a poor rating then the design parameters of that particular subgoal are studied thoroughly for improvement. The main goal is improved by improving the subgoals.

    ii. Understanding the inference problem solving becomes easier since the conflict set contains a small number of rules.

The rules are edited using the Smalltalk text editor in the DICEtalk environment. Because of a large number of queries asked by the system, the user input is provided in the initial data format to the knowledge base. A design file is created that can be loaded for problem solving. Procedural knowledge base contains procedures written using conventional or OOP language and these procedures are called upon by the inference engine at the time of problem solving. They can be considered as external programs interlinked to the main program. In the system, these procedures are Smalltalk methods. Eight methods have been written to accomplish the tasks of initializing the knowledge base after each task or subtask is solved, assigning numerical values to the literal grades in declarative knowledge base, calculating the average grading, providing a means of checking the validity of the calculation with intermediate results and displaying design modifications on the conclusion screen of the system.

    The system contains more than 100 rules and the average problem solving time for these knowledge bases without meta-level is 10 minutes 39 seconds. To reduce the problem solving time the meta-level knowledge is implemented. In addition knowledge-sources are created for the rules in the declarative knowledge base. The knowledge sources names are incorporated in the control advices for the rules in the declarative knowledge base. This reduces the number of rules in the conflict set for a particular problem solving. Instead of all possible rules being in the conflict set, only the rules which are advised by the meta-level knowledge are in the conflict set. Thus the problem solving time is reduced to 3 minutes with same initial data (design). Following example illustrates the meta-level implementation. *BOARDINDEX* is a psuedo-rule in the declarative knowledge base which has three subtasks; of these three, the first two subtasks are normative. Control advices are described for these subtasks as follows:

Rule: BOARDINDEX kb136:
          IF
                    boarder index is evaluated
                              and mis registration index is evaluated
                    and [kb mfgIndex]
          THEN
                    board index is evaluated
Control Advices:
                    ast1: source is borind, normative
                    ast2: source is mrind, normative

The knowledge-sources in the meta-knowledge level are created as follows:
          borind mkb234:
                    source borind: rules are border01-border02
          mrind mkb251:
                    source mrind: rules are mreg01-mreg04

Execution

The execution sequence for the system starts by loading the knowledge base and the goal is selected in the Problem Solving Browser. The execution of system is carried out in a batch mode consultation with an apriori-created design file. The user creates a design file by providing initial data to the declarative-knowledge base browser interactively for the first time and then saving the design. Once the design is in a file, design parameters can be easily changed, to create new designs and to compare different result for different designs; consequently, the best design is selected. For the batch mode consultation, the *solve again* option is chosen from the goal pane menu. To be able to capture the inference engine activities, the *tracing* option in the problem solving menu is kept at *yes*. During problem solving, the inference engine findings are displayed in the problem solving browser. After the problem is solved, the conclusions are displayed in the problem solving browser. Figure 2 illustrates the problem solving browser with conclusions. By saving the design, the findings and conclusions are written to the design file.

Conclusion

In this paper, an object-oriented knowledge-based system is presented. This is the first knowledge-based system to incorporate the hybrid technology in creating a system. Since the present trend has been toward hybrid technology, the system is capable of fitting into the real world production.

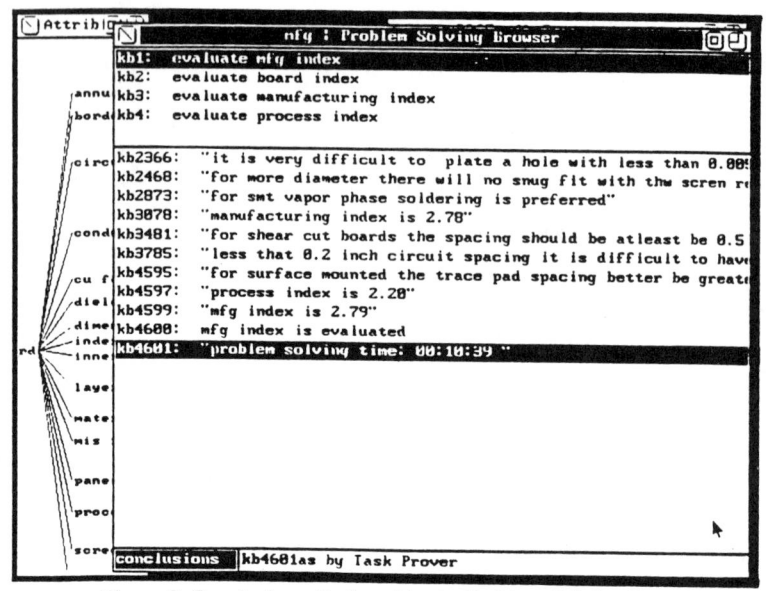

Figure 2. Conclusions displayed in the Problem Solving Browser

The declarative knowledge base in the system has been built using production rules in the form If ... Then. The percept knowledge representation scheme is used to create the rules, and the procedural knowledge base is developed in Smalltalk-80. A hierarchical paradigm is introduced to solve the goal in conjunction with several subgoals, and each subgoal in turn contains further subgoals. The incorporation of meta-level knowledge has resulted in the reduction of execution time by two-thirds. A rating method based on university grading pattern is utilized to rank different designs. The system is the first knowledge-based system to be built using Smalltalk-80 object-oriented language and DICEtalk knowledge base environment. The feasibility of using an object-oriented paradigm in developing knowledge-based systems for engineering applications has been demonstrated in this paper. The system is capable of analyzing designs and suggesting design modifications for better manufacturing and assembly of the printed wiring board.

References

1. Bao, H. and Reodecha, M.,"An Approach to Appraising the Manufacturability of a Printed Circuit Board", Proc. of First International Conf. on Product Design for Assembly, Newport, RI, April 15-17, 1986.

2. Bao, H.,"An Expert System for SMT Printed Board Design for Assembly ", Manufacturing Review, Vol 1, No 4, pp 275 -280, Dec. 1988.

3. Luger, G.F. and Stubblefield, W.A., Artificial Intelligence and the Design of Expert Systems, Benjamin/Cummings, 1989.

4. McKirachan Jr., J.F. and Bao, H.P.,"Producibility Consideration for SMT Printed Circuit Card Assembly", Proc. Third International Conf. on Product Design for Assembly, Newport, RI, June 6-8, 1988.

5. Padhy, S. K. and Dwivedi, S.N.," A Rule Based Expert System for Design and Assembly of Printed Circuit Board", 4th International Conf. on CAD, CAM, Robotics and Factories of the Future, New Delhi, Dec 19-22, 1989.

6. Padhy, S.K., A Knowledge-Based System for PWB Manufacturability in Concurrent Engineering Environment, Master's Thesis, West Virginia University, Morgantown, 1990.

7. Skaggs, C.W.,"Design for Electronics Assembly", Proc. of First International Conf. on Product Design for Assembly, Newport, RI, April 15-17, 1986.

8. Sobolewski, M.,"DICE Percept Knowledge and Concurrency", Proc. 2nd National Symposium on Concurrent Engineering, Morgantown, WV, February 7-9, 1990.

9. Sobolewski, M., "DICEtalk: An Object-Oriented Knowledge-Based Engineering Environment", 5th International Conf. on CAD/CAM, Robotics and Factories of the Future, Norfolk, VA, Dec 2-5, 1990.

10. Struttmann, J. D.," Design for Manufacturability P.C. Cards PERC, A Post Processor for Scicards", Proc. of Second International Conference for Manufacturability: Building in Quality, Orlando, FL., Nov. 13-15, 1988.

11. Wilson, D.R., Martin, Cinthia C. and Hutchinson, K.K.,"An Integrated Tool Set for Printed Circuit Board Design", Unpublished Work. Texas Instruments, Inc.1990.

Design of an IGES Post Processor and Integration with a Robotic Workcell

R. H. WILLISON and G. M. PALMER

Department of Mechanical and Aerospace Engineering
West Virginia University
Morgantown, WV

Summary
Computer aided design (CAD) packages offer the electronic designer the ability to translate complicated circuit ideas into printed circuit board layouts. When given the relative X,Y,Z and theta coordinates of the components to be placed, it can accurately position the fine pitch components on the printed circuit board.

Concurrency in the design, engineering and manufacturing of printed circuit boards would link the design process with the manufacturing operations. A standard for CAD data exchange is the Initial Graphics Exchange format, or IGES. This structured format readily allows graphical and textual information to be disseminated. Utilizing the IGES format, the IGES Post Processor (IPP) operates on the CAD file to produce the cartesian locations of electronic components as specified by the designer and produce a formatted output specific to the workcell. It then has the ability to download the information to the workcell and execute the assembly sequences.

The combination of CAD/CAE systems and the latest chip technologies (Surface Mount Devices, SMDs) can produce PCB designs that are difficult to populate. The ability to place the components on the board is a prime concern. The closer land specifications of fine pitch SMDs demands a high level of automation (2). In placing the newer fine pitch chips onto boards, highly accurate pick and place robotic machines are needed. The prominent configuration is the gantry style robot. This approach allows the robot to move in the "X" and "Y" directions (parallel to the PCB's plane) and the end effector to move in the "Z" direction (perpendicular to the PCB's plane) and the ability to rotate the IC in a "theta" direction.

The concept of concurrent engineering relies on integration. If the operations of the CAD/CAE workstation and workcell can access a common database of information, the requirement to program the robot by hand, or "teach", the robot would be eliminated. It would then be desirable to have the capability to download information from a CAD/CAE workstation to a robotic workcell.

The Initial Graphic Exchange Specification (IGES) is based on the concept of entities. In the design of a post processor, it is important to determine what information is required to be passed from the CAD/CAE work station to the workcell. The minimum information that the workcell requires to place components are the board relative coordinates in the "X", "Y" and "Z" direction. Furthermore, the angle of the component's rotation relative to the local coordinates of the PCB is required. This angle is "theta". In order for the workcell to properly determine the orientation of the board, a fiducial reference is required.

The occurrences of each SMD is encoded the Data Entry section of the IGES file. The positional information for each component are encoded in the Parameter Data section. If more than one of the same type of component is placed on a PCB, it will have multiple entries in the PD section. The translational software itself will here after be referred to as the "IGES Post Processor", or IPP.

The following is a sample IGES file. Certain lines in each sections have been omitted because of its length. The actual file is 2196 lines long and describes a single sided PCB with two Flat Pack 100 pin SMDs.

The first two lines are the START section of the IGES file. The line numbers are determined by the right justified numbers in columns 73-79. The next three lines are the GLOBAL section. This section describes global parameters used throughout.

Following the GLOBAL section is the DATA ENTRY (DE) section. The "110" in the first column (Line D0000001), indicates that this entity is a line. The second field in the DE section indicates where the associated data is in the PD section. In this case it is located on the first line in the PD section.

In order for the IPP to locate the entity known as a SMD, the first column must contain a "308". The "308" is how IGES organizes its entities/sub-entities. The "308" is a label for a grouping of lines (or any other entitiy). The IPP then searches the first fields in the DE section for the "308" marker. This can be seen in line "00000200D0001413. The "308" does not gaurantee that the entity detected is actually a component. Furthermore, if it is determined to be a component, the IPP must compare it to the workcell list to see if it is a part that the workcell can place. There are several other paramters that are required to be extracted from the "308" entity line to make that determination.

The line number in the PD section is given by the second column next to the "308". Here, it is the number "707"; meaning the specifics about where the entitiy is located can be found in the "1413P00000707" (the next section: PD). The "1413" references the PD section back to the DE section.

Knowing that the "308" is correct entity and that the specifics can be found in the PD section at the line "707", the IPP advances one line to "D0001414". The last field in this row has the reference: "FP100". Assuming that the "FP100" is found in the workcell list, the IPP will advance to the PD section.

The concept of entities allows for the component to be described only once; in terms of lines. If the component is located in several places on the PCB, the re-definition of every line is not necessary. This example IGES file describes a PCB with two "FP100" components. It is therefore necessary to distinguish the occurrences of these two components. This is done with reference designators. In this example these are arbitrarily labeled "U1" and "U2". The reference designation must be attached to the "FP100" entity. That is to say, when the designer does something to change one part of the component (rotate it, erase it, move it), the reference designation must also change. This is done by using three pairs of DE lines beginning with a "212", "406", "408". The "212" is the reference to the printed text. The descriptions of each are contained in the PD section described by fields to the right, namely "752","753", and "754" (from lines D0001423, D0001425, D0001427 repectively).

Having confirmed that the entitiy is a component, that the component is one that the workcell can handle, and that it has a reference designation, the IPP will look for the line that describes where the component is located on the PCB. This is done by using the sub-entity connection between "308" and "408". The IPP now has the information required to determine where to find the component's X,Y,Z and Theta information in the PD section. The first occurrence that needs to be recognized is located in line "1413P0000707". The "707" was the reference specified in the DE section. The line begins with a "308", the component name "5HFP100", and the number of sub-entities that make up the "308". Here, there are 605. This means the 605 numbers that follow are references to the "lines" that make up the entity called a "FP100", or SMD. The 1413 is the reference back to the DE section. The "408" in line "1427P0000754" is where the first of the two SMDs coordinates reside. The line again references "1413" (the DE section line number), followed by the X,Y,Theta and Z coordinates. These are listed as "60.0,60.0,0.0,1.0".

Since there are two components on this particular PCB of the same type, a second occurrence of the "408" can be seen in the next to last line "1433P0000757". This references the DE section by the number "1433" preceding the section and line number. The coordinates "90.0.60.0,0.0,1.0" are read from this line.

Sample IGES File:

```
IGES file generated from an AutoCAD drawing by the IGES              S0000001
translator from Autodesk, Inc., translator version IGESOUT-2.0.      S0000002
,,3HWVU,7HWVU.IGS,13HAutoCAD-10,11HIGESOUT-2.0,16,38,6,99,15,3HWVU,  G0000001
1.0,1,4HINCH,32767,3.2767D1,13H890906.103524,1.0D-8,150.0,6HThroop,  G0000002
14HAutodesk, Inc.,4,0;                                               G0000003
      110       1       1       1       5              00010200D0000001
      110               5       1                              D0000002
      110       2       1       1       5              00010200D0000003
      110               5       1                              D0000004
      110       3       1       1       5              00010200D0000005
              Portions Omitted for Illustrative Purposes (1405 lines)
      110     704       1       1       2              00010200D0001407
      110               2       1                              D0001408
      110     705       1       1       2              00010200D0001409
      110               2       1                              D0001410
      110     706       1       1       2              00010200D0001411
      110               2       1                              D0001412
      308     707       1                              00000200D0001413
      308              41                          FP100       D0001414
      110     748       1       1       4              00000000D0001415
      110               4       1                              D0001416
              Portions Omitted for Illustrative Purposes (7 lines)
      212     752       1       1       5       0      00020100D0001423
      212               5       1                              D0001424
      406     753       1       1       5              00020000D0001425
      406               5       1    7901                      D0001426
      408     754       1       1       4              00000000D0001427
      408               4       1                              D0001428
      212     755       1       1       5       0      00020100D0001429
      212               5       1                              D0001430
      406     756       1       1       5              00020000D0001431
      406               5       1    7901                      D0001432
      408     757       1       1       4              00000000D0001433
      408               4       1                              D0001434
110,-7.25,9.5,0.0,-7.25,-1.025D1,0.0;                          1P0000001
110,-6.5,1.025D1,0.0,-7.25,9.5,0.0;                            3P0000002
110,7.25,1.025D1,0.0,-6.5,1.025D1,0.0;                         5P0000003
110,7.25,-1.025D1,0.0,7.25,1.025D1,0.0;                        7P0000004
              Portions Omitted for Illustrative Purposes (695 lines)
102,1,1395;                                                    1397P0000699
110,-10.0,-9.425,0.0,-7.6,-9.425,0.0;                          1399P0000700
102,1,1399;                                                    1401P0000701
110,-10.0,-9.25,0.0,-10.0,-9.425,0.0;                          1403P0000702
110,-7.6,-9.25,0.0,-10.0,-9.25,0.0;                            1405P0000703
110,-7.6,-9.6,0.0,-7.6,-9.25,0.0;                              1407P0000704
110,-10.0,-9.6,0.0,-7.6,-9.6,0.0;                              1409P0000705
110,-10.0,-9.425,0.0,-10.0,-9.6,0.0;                           1411P0000706
308,0,5HFP100,605,1,3,5,7,9,15,17,19,21,23,25,27,29,31,33,35,39,  1413P0000707
41,43,45,47,49,53,55,57,59,61,63,67,69,71,73,75,77,81,83,85,87,   1413P0000708
              Portions Omitted for Illustrative Purposes (38 lines)
1357,1359,1361,1363,1365,1369,1371,1373,1375,1377,1379,1383,   1413P0000746
1385,1387,1389,1391,1393,1397,1401,1403,1405,1407,1409,1411;   1413P0000747
110,0.0,0.0,0.0,150.0,0.0,0.0;                                 1415P0000748
110,150.0,0.0,0.0,150.0,125.0,0.0;                             1417P0000749
110,150.0,125.0,0.0,0.0,125.0,0.0;                             1419P0000750
110,0.0,125.0,0.0,0.0,0.0,0.0;                                 1421P0000751
212,1,2,4.0,3.0,1,,0.0,0,0,58.0,43.0,0.0,2HU1;                 1423P0000752
406,2,2HU1,2HU1;                                               1425P0000753
```

```
        408,1413,60.0,60.0,0.0,1.0,1,1423,1,1425;                    1427P0000754
        212,1,2,5.0,3.0,1,,0.0,0,0,8.75D1,43.0,0.0,2HU2;              1429P0000755
        406,2,2HU2,2HU2;                                              1431P0000756
        408,1413,90.0,60.0,0.0,1.0,1,1429,1,1431;                     1433P0000757
        S0000002G0000003D0001434P0000757                                  T0000001
```

(This is the end of the sample IGES file.)

The IPP will translate a file created by any CAD/CAE package that supports IGES (any version) and represents its ICs graphically by utilizing the entity/sub-entity function. It will decode the IGES file for the fiducial locations and the component's local PCB coordinates and place the data in a format that will be used by other application programs requiring the positional data.

References

1   ((Jerry Lyman, Fine-pitch Assembly Affects Every Phase of Surface Mounting, Electronic_Design, August 25, 1988, pgs.60-66))

2   ((Designing with CAD, Electronic and Wireless World, July 1989, Vol 95, Num. 1641 pg 694))

Discrete Optimum Assembly Methods for Automated Workcells

KENNETH H. MEANS and JIE JIANG

Department of Mechanical and Aerospace Engineering
West Virginia University
Morgantown, WV

Introduction

The assembly of components is a fundamental part of most manufactured products. Where high volume production is required or where the components are numerous the order of component assembly becomes important in reducing the assembly time. This is true for both manual and automated assembly schemes. The optimum assembly sequence of a group of components to minimize assembly time is the goal of this study. In the 1990's, automated assembly workcells will become prevelant in the manufacturing area. These workcells must be programmed efficiently to minimize the assembly time and reduce production costs.

Review of Assembly Methods

Since World War II operations research has been modeling manufacturing systems using mathematical techniques. In the early 1960's simulation software such as GPSS and CIM emerged but had little real effect on manufacturing methods. There was little validation of these methods until computing systems became more advanced and larger more detailed problems were examined using linear programming and numerical techniques.

As far as assembly methods are concerned, some important works are now given as a background for this study.

In 1981 Wilhelm [1] presented a methodology to describe operating characteristics of assembly systems. A numerical approximation technique was applied using a strategy of setting due dates for parts. This was shown to have a significant influence on the assembly sequence.

Sarin and Das [2] also considered the problem of optimal part delivery dates on a small lot assembly line used to manufacture large and costly assemblies. A dynamic programming algorithm is developed to minimize the cost in setting delivery dates.

The optimal solution of an order picking problem in a warehouse with aisles was acheived by Goetschalckx and Ratliff [3]. Their optimal algorithm resulted in 30% savings in travel time.

Heuristic rules have also been incorporated in assembly optimization schemes. In fact, most of life involves a sequence of decisions to carry on the daily tasks of our lives. These sequences are normally determined through some heuristic procedure. Das and Sarin [4] also used heuristic rules with an analytical programming technique to determine optimal delivery dates for parts.

Various mathematical tools have been used to study the optimum sequence problem. These include many optimization techniques. For example, a directed network method is used to find the longest and shortest paths in a complicated assembly scheme. In robotic assembly much effort has been spent on the path and circle time of a single task. This is to be distinguished from network optimization which is concerned only with the origin and destination of the parts and not the branches connecting two nodes [5,6].

The complete assembly optimization must consider both the path and cycle time of the robot as well as the network of parts to be assembled.

Method of Optimization

In any single assembly task, the robot motion may be made up of several segments of varying lengths and velocities. The path length and motion may be optimized. However, once the assembly task has been decided, the robot motion and cycle time become fixed quantities. (The optimization of the robot motion and cycle time is a non-linear problem that is not the focus of this paper).

In this study the robot motion and cycle time for any one given task is considered to be a constant. Thus, for a series of separate assembly tasks the decision to be made is which sequence of assembly is best. Each controllable variable can have a solution value of 0 or 1 only.

In a n-variable 0,1 problem, there will be 2^n possible solutions. Most of these solutions are infeasible. This problem could be solved by brute force, that is, by enumerating all possible solutions rejecting those which violate one or more constraints and selecting the one sequence that minimizes the objective function. Generating the 2^n possible solutions is easy with a few variables but becomes a formidable task when the number of components to be assembled is above 7.

However, it is not necessary to enumerate every possible solution to find the optimum assembly sequence. Given the objective function to be minimized as:

$$\min \sum_{i=1}^{m} \sum_{j=1}^{n} C_{ij} X_{ij}$$

Where

$$X_{ij} = \begin{cases} 1, & \text{If the part is assigned from feeder i to position j} \\ 0, & \text{otherwise.} \end{cases}$$

and

C_{ij} = The trip distance from feeder i to position j.

The problem is subject to constraints of the form:

$$g_{ij} \geq d$$

where d is a constant

The constraints state simply that only one component may occupy a place on the feeder tray and on the final assembly location. In addition, there can be no more trips than the number of parts to be assembled.

The method is similar to an implicit enumeration method [7]. Any sequence of assembly steps is compared to the objective function. If the objective function is not reduced the sequence is discarded. The advantage of this method lies in its ability to discard whole branches of sequences that do not lower the objective function without actually enumerating each individual case.

Figure 1 shows the flow chart of the optimization method. We begin this program by assigning to $\bar{Z}$ (the objective function) a value for all $X_{ij} = 1$ and calling it f. We also assign to S_o the null set indicating that at this point no variables have been assigned specific values. Next the constraint equations are checked for violations. If the constraint set V is empty, we complete the partial solution set S by setting to zero all variables not in

S. This completed solution becomes the incumbent solution $\bar{X}$ and the new value of the objective function at $\bar{X}$ becomes the new value of $\bar{Z}$.

If the constraint set V is not empty we find the value of f_p, which is the value of f obtained by setting to 0 all variables not in S and calculating the objective function coefficient limit $\bar{Z} - f_p$. We store in the set T the variables not in set S but have an objective function coefficient less than the limit $\bar{Z} - f_p$ and a positive coefficient in some constraint in the set V.

In each violated constraint we raise the variables to 1 that are included in set T. If any constraint is still violated under the previous condition, then no feasible solution is possible.

If no constraint is violated, the next step is to select the most helpful variable in set T and add it to set S. This becomes the partial solution. At this point the backtracking process begins. Every element in set S which has not yet been replaced with its complement in that partial solution will be referred to as positive; every element in set S which has been complemented will be referred to as negative since a minus sign precedes the variable number in the listing in S.

The general procedure is to complement the rightmost element in S which has not yet been complemented and drop any elements in S to the right of the variable currently being complemented and begin again to fathom the new partial solution.

An additional rule is never to add to set T a variable with an objective function coefficient that would result in an objective function value greater than or equal to the current value of $\bar{Z}$.

Sample Solution

The objective is to minimize the assembly time of a printed circuit board (PCB) with 15 components in an automated workcell with three tray-type part feeders as shown in Figure 2. Feeder 1 contains 4 parts, Feeder 2 is assigned 4 parts and Feeder 3 is assigned 7 parts. This results in 18 constraint equations. The initial objective function is $\bar{Z}$ = 330. The optimum solution is reached after 10,000 iterations giving a $\bar{Z}$ min = 145.11. When compared to the default sequence used by the workcell which was the arbitrary parts list as input by the operator, the reduction in assembly time was over 25%. The total number of possible solutions is over 32,000. This method did not require the enumeration of all solutions but arrived at the optimum assembly sequence after eliminating two thirds of the possible solutions.

Conclusions

The method of implicit enumeration provides a good basis for selecting an optimum assembly sequence in automated assembly workcells. The method is being extended to large numbers of components with guides to reduce computer time. Future work will include the optimization of the entire workcell arrangement as well as the assembly sequence.

Acknowledgements

This research was made possible by support under the DICE Program at the Concurrent Engineering Research Center at West Virginia Unviersity (DARPA Contract No. MDA 972-88-C-0047).

Bibliography

1. W. E. WILHELM, "A Methodology to Describe Operating Characteristics of Assembly systems", IIE Transactions, September, 1982.

2. SUBHASH C. SARIN and SANCHOY K. DAS, "Determination of Optimal Part Delivery Dates in a Stockastic Assembly Line", INT. J. PROD. RES., 1987, Vol. 25, No. 7, 1013-1028.

3. MARC GOETSCHALCKX and H. DONALD RATLIFF, "Order Picking in an Aisle", IIE Transactions, March, 1988, Vol. 20, November 1.

4. SANCHOY K. DAS and SUBHASH C. SARIN, "Selection of a Set of Part Delivery Dates in a Multi-Job Stochastic Assembly System", IIE Transactions, March, 1988, pp. 3-11.

5. G. V. REKLAITIS, A. RAVINDRAN and K. M. RAGSDELL, "Engineering Optimization", A Wiley-Interscience Publication, 1983, pp. 292-343.

6. K. HITOMI and M. YOKOYAMA, "Optimization Analysis of Automated Assembly Systems", Journal of Engineering for Industry, May 1981, Vol. 103, pp. 224-232.

7. D. R. PLANE and C. MCMILLAN, Discrete Optimization, 1971, Prentice-Hall.

387

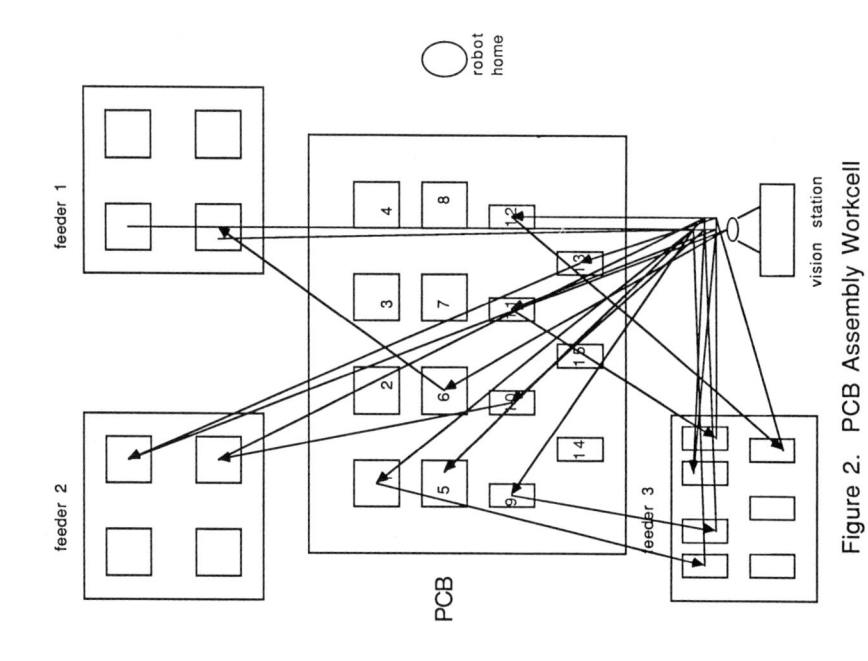

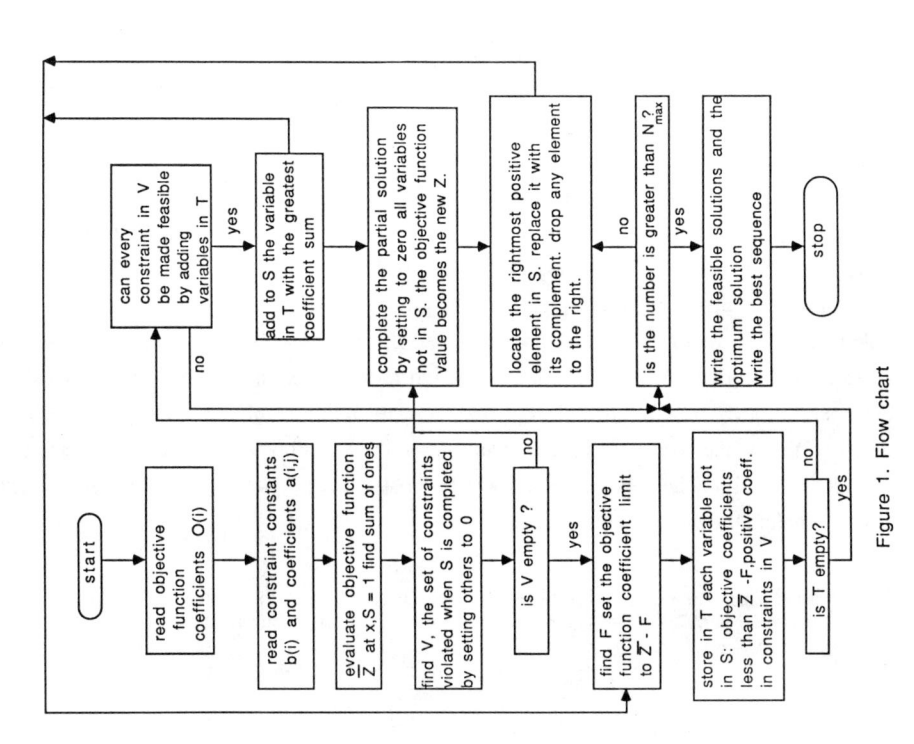

Figure 2. PCB Assembly Workcell

Figure 1. Flow chart

Trajectory Planning for Obstacle-Avoided Assembly of Planar Printed Circuit Boards

TAK-LAI LUK and JOHN E. SNECKENBERGER

Concurrent Engineering Research Center
West Virginia University
Morgantown, WV

Abstract

The use of robots to assemble Printed Circuit Boards represents a cost-effective, efficient approach to the electronics manufacturing industry. The ability to plan and simulate the assembly robot workcell operations early in the PCB design stages is essential. This concurrent engineering concept leads to the goal of reduction of design time by using an assembly planner and simulator to predict assembly workcell problems as early as possible in the product definitions stage. An assembly robot workcell trajectory planner and simulator were implemented to approach the goal. This paper addresses the issues of the development of such robot trajectory planner and simulator. The trajectory planner and simulator (TPS) can plan assembly paths for the assembly for a given PCB layout and assembly sequence with a collision avoidance algorithm implemented. TPS also provides the capability to graphically visualize workcell mechanics and to provide advisory information to the user for a PCB High Density Electronics assembly.

I. Introduction

One of the main research efforts devoted to the DICE (DARPA Initiative in Concurrent Engineering) program at WVU is to implement a Design For Assembly Advisor in the DICE computer network. The overall objective of this research task is to develop a computer-based Design For Manufacturability/Assemblability (DFM/A) module for the DICE Design For X workstation. This DFM/A module will, for example, enable the lead designer to perform concept-level evaluations between the physical design specifications and the assembly workcell operations early in the concurrent engineering product and process development cycle. The ability of being able to *plan* and *simulate* the assembly workcell operations at the early part design stages is essential and productive. It will shorten the design-to-manufacture cycle by giving advice to the designer, through an assembly planner, to optimize the assembly operations; and helps the designer, through an assembly simulator, to foresee the assembly workcell problems, and hence provides the designer an opportunity to correct or make necessary changes to a design as early as possible, before the manufacturing stages begin.

[†]Acknowledgements - This works has been sponsored by the Defence Advanced Research Project Agency (DARPA), under contract No. MDA972-88-C-0047 for DARPA Initiative in Concurrent Engineering (DICE).

The primary focus of the DFM/A module development has been on enhancing the manufacturability and assemblability of electronic printed circuit board (PCB). Among the several modules implemented in the DFM/A module, the one that currently emphasizes the assemblability of the electronic PCB is called the Assembly Planner and Simulator (APS).

II. Assembly Planner and Simulator

APS is an integrated module for the robotic assembly of printed circuit boards. It consists of three modules with networked exchanges of files and messages to provide the most productive workcell assembly of a given CAD-based board. Figure 1 shows the three integrated modules associated with the APS tool as well as the functional data exchanges between the modules. The three modules are:

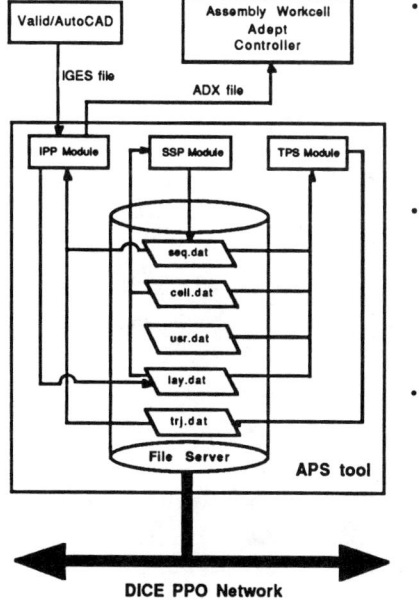

- **IGES Pre/Postprocessor (IPP) module**: an advisor/tool for the assembly of PCBs that uses a file transfer mechanism to generate a board component location and orientation file and a board assembly workcell file from a CAD-based board input file which is in the IGES format.

- **Spatial Sequence Planner (SSP) module**: an advisor/tool for the assembly of PCBs that uses an optimization method to provide the best feasible assembly sequence for the board components in terms of spatial workcell considerations.

- **Trajectory Planner and Simulator (TPS) module**: an advisor/tool for the assembly of PCBs that uses collision detection and trajectory planning algorithms to compute robot assembly paths, and provides computer graphic simulated manipulator path and motion mechanics for board assembly by a robot assembly workcell.

Figure 1. Internal structure of APS tool which consists of the IPP module, SSP module and TPS module.

This APS tool provides a highly user-friendly graphic interface environment in which a CAD-based design of a PC board layout can be evaluated and an optimized workcell plan for assembling the board can be generated. A simulation of the board being assembled can also be provided for visual verification of the workcell operations. The APS tool produces a command file to drive the electronics assembly workcell controller.

The advisory aspects of the APS module will permit the user of the DFM/A module to address and/or resolve design problems such as (1) enabling the translation of a CAD-based layout file for a printed circuit board into an efficient workcell controller file to be quickly achieved, (2) enabling the placement sequence of components on a printed circuit board for a specified assembly workcell configuration to be spatially optimized, (3) enabling the robot motions for printed circuit board assembly considering collision detection, minimum paths and trajectory speeds to be graphically simulated, as well as to achieve process solutions such as (4) enabling the programming time for achieving assembly workcell production of printed circuit boards to be drastically reduced. The current TPS module is particularly developed to address the third issue indicated above.

III. Trajectory Planner and Simulator

The TPS module in the APS tool is a graphic interactive simulation program which runs on the Silicon Graphics Personal IRIS-4D workstation. As shown in Figure 1, TPS takes the following four data files as input files:

(1) *cell.dat* : contains assembly workcell layout information, such as feeders and trays locations, parts allocation in workcell, assembly robot home position and inspection camera location, etc. ;
(2) *lay.dat* : contains the layout description of a PC board as well as the assemble parts information such as part names, part types, and part dimensions;
(3) *seq.dat* : contains the optimum assembly sequence information;
(4) *usr.dat* : contains user-specified parameters such as assembly requirements like minimum total assembly time, or simulation speed, and sampled rate for the trajectory planner.

TPS processes these files by verifying the possibility of the assembly operations for each step in the optimum assembly sequence given in *seq.dat* generated by the SSP module, checking any violation of the user-specified assembly requirements, determinating the obstacle-free assembly path for the robot gripper for each chip, and graphically simulating the assembly robot motion with displays of assembly time, travel distance, robot speed and acceleration, etc. TPS plans the time trajectory for each collision-free assembly path and saves the trajectory information into an output file called *trj.dat*. The IPP module will use the path and trajectory information for the generation of the "path database" in the workcell command file for the robot workcell controller.

IV. Functional Description of TPS

The primary functions of TPS are (1) verification of the possibility of assembly operations; (2) confirmation of the user-specified assembly requirement; (3) determination of assembly paths with collision avoidance; (4) generation of time trajectories for assembly paths; (5) simulation of the assembly operations with graphic user interface.

After the TPS module is initiated, TPS will process the four input data files described in the previous section and display a graphic window titled as "DICE.APS.tool" as shown in Figure 2. TPS starts to determine the collision-free assembly path for each chip based on the assembly sequence given in *seq.dat*. The verification of the possibility of the assembly operations is also performed in this phase. The collision-free path for each chip is determined by a path finder which considers all the on-board assembled chips and the cameras as workcell obstacles. The path finder uses a collision detection algorithm derived from the polygon clipping algorithm [2] in computer graphics theory, and an A* searching routine [3,5] to compute the collision-free path for each chip from the part feeder to the inspection camera and then to the PCB. It takes less than 2 seconds wall-clock time to determine all the assembly paths for a 20-chip PCB.

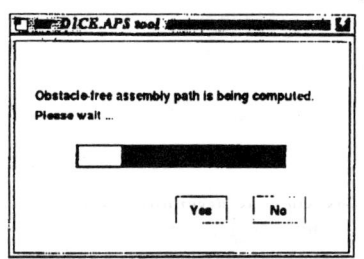

Figure 2. TPS will display a window showing that it is computing the collision-free path for each of the chips.

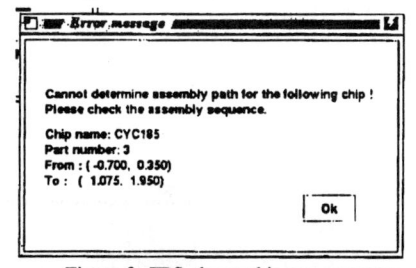

Figure 3. TPS shows this error message window when it detects a chip cannot be successfully assembled on the PCB.

In the case that the TPS module predicts that the assembly sequence cannot be successfully executed in the workcell, another window titled as "Error message" will come on the screen shown in Figure 3. This window will provide information referring to the particular chip which causes the error. On the other hand, if the computation of assembly paths succeeds, the user has an option to continue the assembly path planning and simulating by selecting a desired trajectory planner. There are two different trajectory planning methods available in TPS: a 3-4-3 spline function trajectory planning method and a parabolic blend trajectory planning method [1,4]. The first method works better for curve trajectories while the second method is more suitable for straight line trajectories. Both methods are capable of determining time optimum trajectory.

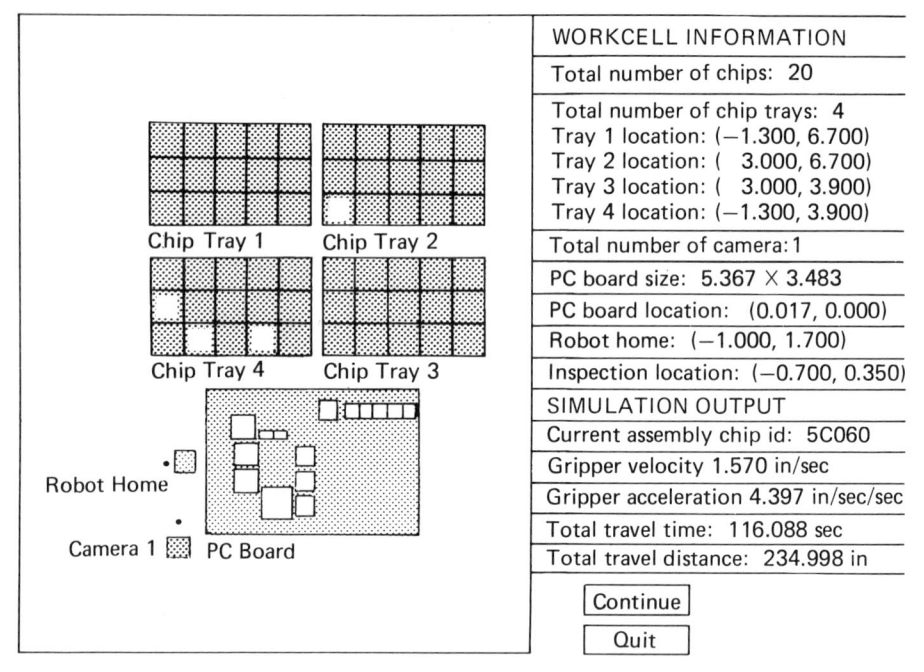

WORKCELL INFORMATION
Total number of chips: 20
Total number of chip trays: 4 Tray 1 location: (−1.300, 6.700) Tray 2 location: (3.000, 6.700) Tray 3 location: (3.000, 3.900) Tray 4 location: (−1.300, 3.900)
Total number of camera: 1
PC board size: 5.367 × 3.483
PC board location: (0.017, 0.000)
Robot home: (−1.000, 1.700)
Inspection location: (−0.700, 0.350)
SIMULATION OUTPUT
Current assembly chip id: 5C060
Gripper velocity 1.570 in/sec
Gripper acceleration 4.397 in/sec/sec
Total travel time: 116.088 sec
Total travel distance: 234.998 in

Figure 4. Simulation window shows the simulated robot motion and workcell information.

The graphic simulating program is automatically activated after a trajectory planning method is selected. TPS will open a full-screen simulation window with the workcell layout displayed on the left-hand side of the window and the workcell information listed on the right-hand side of the window as shown in Figure 4. The workcell information consists of two portions. The upper portion contains static workcell information such as number of trays, location of inspection camera, etc. The lower portion contains dynamic workcell information, such as total assembly time or robot gripper velocity, which will continuously be updated while the simulation is executing.

In the simulation, TPS will graphically animate the assembly motion of the robot gripper in the left-hand side of the simulation window and will keep monitoring any violation of the user-specified assembly requirement in the data file *usr.dat*. Figure 5 shows the message window that comes up during the simulation when TPS has detected that the simulated assembly time exceeds the user-specified time. In Figure 5 the user-specified assembly time is 30.0 seconds but the total travel time is already 30.148 seconds. When the simulation is done, an output file containing all the path information will be generated and will be used to be down-loaded as the path database to the robot controller.

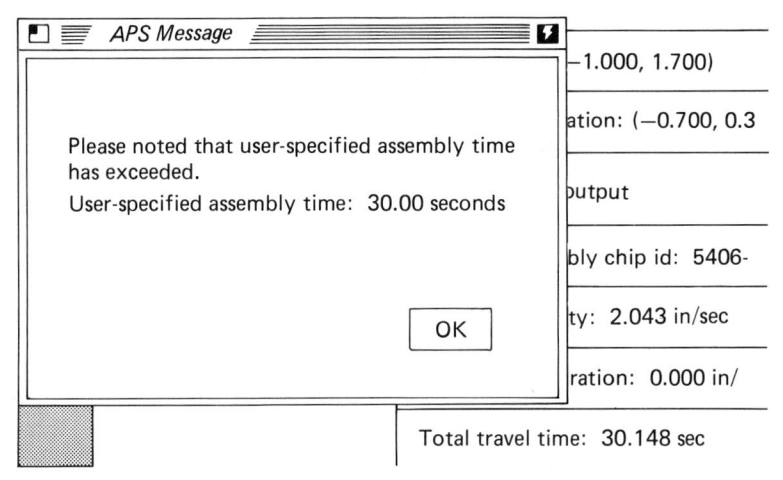

Figure 5. APS displays a message window to notify the user that the user-specified assembly time has exceeded.

V. Conclusion

TPS is the initial effort for the development of a generic assembly planner and simulator which is known to be beneficial and helpful in a concurrent engineering environment. The current implementation of the TPS is able to handle a PCB of at least 60 chips of Surface Mounted Devices (SMD). It is envisioned that the TPS module will be enhanced to be able to plan paths for more than one robot in the same assembly workcell, and also to be able to manage mechanical assembly planning.

VI. References

1. P.G. Ránky and C.Y. Ho, *Robot Modelling: Control and Applications with Software*, IFS (Publications) LTD., 1985.
2. D.F. Rogers, *Procedural Elements for Computer Graphics*, McGraw-Hill, 1985.
3. R. Sedgewick, *Algorithms*, Addison-Wesley, 1983.
4. J.J. Craig, *Introduction to Robotics: Mechanics and Control*, Addison-Wesley, 1986.
5. E. Rich, *Artificial Intelligence*, McGraw-Hill, 1983.

Development of a Vision Assisted Optimal Part-To-Pad Placement Technique for Printed Circuit Board Assembly

S.H. CHERAGHI, E.A. LEHTIHET and P.J. EGBELU

Department of Industrial Engineering
Pennsylvania State University
University Park, PA

ABSTRACT

In this paper, an optimal lead to pad matching technique for surface mount component placement in PCB assembly is developed. The problem is expressed as a non-linear optimization problem with linear objective function and quadratic constraints. A solution procedure is developed and an example is given.

INTRODUCTION

A substantial portion of the electronic industry is replacing traditional through hole connections with surface mount assembly technology. Within this technology, the trend is towards smaller components with very closely spaced leads (25 mils). Together with fine pitch requirements, global and local errors inherent in boards and components create difficulties for automated component placement. Board and component position errors at assembly time are typical of global errors; board pads and component leads position and shape errors on the other hand are typical of local errors. Board pad position and shape errors may result from imperfect etching while component leads are easily bent during manufacturing and handling (McW 1989). The large magnitude of the resultant error has made the use of vision systems unavoidable for placement of fine pitch components (AMI 1986).

A typical system for low volume high mix applications usually consists of a SCARA type manipulator with appropriate vacuum type or mechanical gripper end effectors, a vision system and appropriate board fixture and part presentation devices. The vision system is typically a two camera system with a downward looking camera for board feature acquisition and an upward looking camera for component feature acquisition.

A number of current vision based placement systems acquire board and component errors and then use a weighted average of lead position errors to guide placement. Under this method, a couple of bent leads will not affect the weighted average significantly; on the other hand, a large number of slightly bent leads might heavily affect the weighted average (McW 1989). These cases may lead to placement difficulties.

With increased likelihood of local errors on fine pitch parts and demand for higher placement accuracies at lower cost, a vision based procedure able to perform optimal matching of each lead and corresponding pad is highly desirable. This paper describes development of such a procedure.

MATHEMATICAL FORMULATION OF THE PLACEMENT PROBLEM:

Figure 1 shows portion of a component on its pad. This component has positional error.

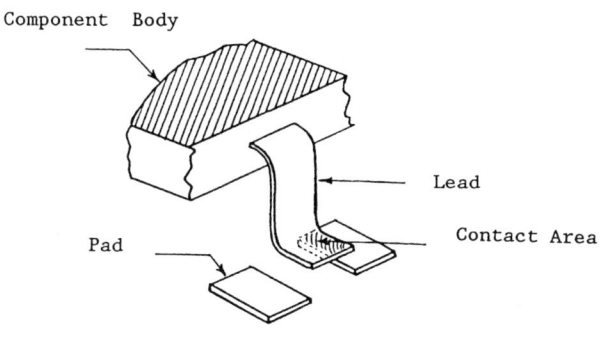

Figure 1 Portion of a component on its pad

The objective of placement is to maximize the area of contact between each lead and pad and thus promote conditions for a good solder joint. This problem can be stated as follows:

$$\text{Maximize } A_{ii} \quad \forall i,j \tag{3.1}$$

subject to: $A_{ii} \geq \alpha A_i$,   i=1,..,n. (3.2)

where:

A_{ii} = area of contact between lead i and pad i.

A_i = area of lead i.

α = constant, $0 < \alpha \leq 1.0$. $\alpha = 0.75$ is often recommended (FIE 1986).

n = number of leads/pads.

This statement is equivalent to maximizing the minimum area of contact between each lead and pad. This goal can be achieved by way of the following formulation:

Minimize (max(d_{ij})), i=1,..,n,   j=1,..,4. (3.3)

where:

d_{ij} = Euclidian distance between corner j of lead i and corner j of pad i.

The following notation is used in the remainder of the text. The notation refers to positions of leads and pads after the component is moved to its placement position.

PX_{ij}, PY_{ij} = x and y coordinates of actual position of corner j of pad i.

$\qquad$ i = 1,...,n and j=1,..,4.

LX_{ij}, LY_{ij} = x and y coordinates of actual position of corner j of lead i.

$\qquad$ i = 1,...,n and j=1,..,4.

Once a component is aligned with its pad, suppose it is translated by dx, dy and rotated by an angle of θ radian. Lead corners in their new positions would have the following coordinates (i = 1,n and j = 1,4):

$$LX'_{ij} = \cos\theta \, LX_{ij} - \sin\theta \, LY_{ij} + dx \qquad (3.4)$$

$$LY'_{ij} = \sin\theta \, LX_{ij} + \cos\theta \, LY_{ij} + dy \qquad (3.5)$$

Using a small angle approximation for sinθ and cosθ leads to the following expressions:

$$LX'_{ij} = LX_{ij} - \theta \, LY_{ij} + dx$$

$$LY'_{ij} = \theta \, LX_{ij} + LY_{ij} + dy$$

Expressing d_{ij} in function of pad and lead corner coordinates leads to the following formulation:

P1: Minimize $f(x,y,\Theta)$ $\qquad\qquad\qquad\qquad\qquad\qquad\qquad$ (3.6)

where;

$$f(x,y,\theta) = \underset{\forall i,j}{Max} [(PX_{ij} - LX'_{ij})^2 + (PY_{ij} - LY'_{ij})^2] \qquad (3.7)$$

Consider the function $f(x,y,\theta)$. Substituting for LX'_{ij} and LY'_{ij} in 3.7 leads to:

$$f(x,y,\theta) = \underset{\forall i,j}{Max} \; [(PX_{ij} - LX_{ij} + \theta \, LY_{ij} - dx)^2 + (PY_{ij} - LY_{ij} - \theta \, LX_{ij} - dy)^2] \qquad (3.8)$$

Let $\quad a_{ij} = PX_{ij} - LX_{ij}$

$\qquad b_{ij} = PY_{ij} - LY_{ij}$

After substitution, the function f(x,y, θ) becomes:

$$f(x,y,\theta) = \underset{\forall i,j}{Max} \; [(a_{ij} + \theta \, LY_{ij} - dx)^2 + (b_{ij} - \theta \, LX_{ij} - dy)^2]$$

An equivalent problem to P1 can now be stated as follows; find dx, dy, dθ to:

P2: Minimize z (3.9)

subject to:

$$(a_{ij} + \theta LY_{ij} - dx)^2 + (b_{ij} - \theta LX_{ij} - dy)^2 \leq z \quad (3.10)$$

$$i = 1,..,n \quad \text{and } j = 1,..,4.$$

There are 4n constraints, one for each corner of a lead and pad.

SOLUTION METHODOLOGY

For a given θ value; let

$$CX_{ij} = a_{ij} + \theta LY_{ij},$$
$$CY_{ij} = b_{ij} - \theta LX_{ij}.$$

Problem P2 simplifies to:

P3: Minimize z

subject to:

$$(CX_{ij} - dx)^2 + (CY_{ij} - dy)^2 \leq z \quad (3.11)$$

$$i = 1,..,n \quad \text{and } j = 1,..,4.$$

Each constraint in (3.11) represents a circle of radius z centered at (CX_{ij}, CY_{ij}). Mathematically, problem P3 is equivalent to finding the smallest circle which contains all points with coordinates (CX_{ij}, CY_{ij}). Among all the algorithms that solve problem P3, the one due to Elzinga and Hearn is used here because of its efficiency (FRA 1974). Further information regarding this algorithm can be found in reference (ELZ 1971).

A procedure is developed to solve problem P2 using Elzinga's algorithm as a subroutine; the procedure is given below:

1- Select an initial value θ^* for θ. And consider a step size β in θ.

2- Using Elzinga's algorithm, evaluate the objective function in P3 at

$\theta^*-\beta, \theta^*, \theta^*+\beta$. Let the corresponding objective values be

$z(\theta^*-\beta), z(\theta^*), z(\theta^*+\beta)$ respectively.

3- Check which of the following conditions holds:

(a) If $z(\theta^*) \leq \min\{z(\theta^*+\beta), z(\theta^*-\beta)\}$ then;

    if β is very small, stop. {dx, dy, and θ } give the optimal solution.

    otherwise set $\beta = \frac{\beta}{2}$ and go to 2.

(b) if $z(\theta^*+\beta) \leq z(\theta^*) < z(\theta^*-\beta)$, set $\theta^* = \theta^*+\beta$ and go to step 2.

(c) if $z(\theta^*-\beta) \leq z(\theta^*) < z(\theta^*+\beta)$, set $\theta^* = \theta^*-\beta$ and go to step 2.

(d) if $z(\theta^*) > \max\{z^*+\theta, z^*-\theta\}$ then;

if $z(\theta^*-\beta) < z(\theta^*+\beta)$,set $\theta^* = \theta^*-\beta$, go to step 2.

else set $\theta^* = \theta^*+\beta$, go to step 2.

EXAMPLE APPLICATIONS

The algorithm described above was implemented in Fortran 77. Two different components are tested; a two terminal chip resistor and an 8 lead IC. Local errors are intentionally added to leads and pads. For the chip resistor, three versions are considered. In versions 2 and 3, widths of the land patterns are purposely made smaller and smaller to check the sensitivity of the technique to cases where sizes of leads and pads are very close to each other.

Figure 2 shows an SOIC with 8 leads on its land pattern both before and after the technique was applied. All dimensions are nominal dimensions taken from different sources (HIN 1988, PRA 1989).

Results for all cases are given in table 1. As can be seen from table 1 and figure 2, lead to pad area coverage is more uniformly distributed after the technique is applied; this is true in all cases. Note also that variances of areas of contact between leads and pads are smaller after application of the optimization procedure.

Execution time for the first case was about 0.03 seconds; for the second case about 0.1 seconds. As the number of leads or pads increases; the processing time will increase but at a smaller rate. This is because only a limited number of points will be considered.

CONCLUSION

A lead to pad matching optimization routine has been developed. This routine receives information on position and shape errors associated with component leads and board pads from a vision system and tries to optimally match each individual lead and pad. Unlike existing methods, this routine takes into consideration errors associated with the shape of components. It eliminates any need for centering devices or fiducial marks.

REFERENCES

Amick, C.G., "Close Doesn't Count," Circuits Manufacturing, September 1986, 35-43.

Elzinga, J., Hearn,D.W., "Geometrical Solutions for Some Minimax Location Problems," Transportation Science, Vol.6, No.4, 1971,pp.379-394.

Englander, A.C., "Assuring Precision Placement of Surface Mount Devices by Means of Machine Vision Feedback Control," Vision 87, Chapter 11, 29-41.

Field, J., Payne, J., Cullen, C., "SMD Placement Using Machine Vision," Electronic Packaging and Production, January 1986, 128- 129.

Francis, R., White, J., Facility Layout and Location, Prentice-Hall, inc., Englewood Califfs, New Jersey, 1974.

Hinch, S.W., Handbook of Surface Mount Technology, Longman Group UK Limited., 1988.

McWalter, K., "Future Trends for Vision Systems in Surface Mount Placement," Surface Mount Technology, April 1989, 37-39.

Prasad, R.P., Surface Mount Technology, Principles and practice, Van Nortrand Reinhold, New York, 1989.

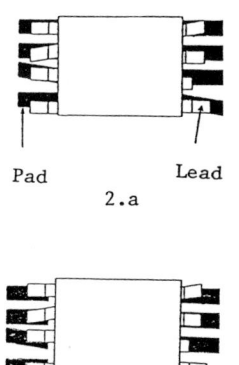

Case	i	Solution			% Area of Contact		Variance	
		dx	dy	θ	Before	After	Before	After
1.1	1	2.38	-3.53	0.3	81	100	0.095	0.000
	2				100	100		
1.2	1	1.74	-2.98	0.25	75	100	0.125	0.025
	2				100	95		
1.3	1	1.35	-2.15	0.178	65	100	0.14	0.025
	2				93	95		
2	1	0.75	-1.5	0.00	95	75	21.08	8.67
	2				50	100		
	3				75	100		
	4				100	83		
	5				50	87.5		
	6				95	87.5		
	7				87.5	96		
	8				100	87.5		

Table 1 Approximate lead i to pad i areas of contact for different cases before and after applying the optimal matching technique.

2.a   Pad   Lead

2.b

Figure 2 SOIC on the placement pad
(a) before applying the optimal matching technique
(b) after applying the optimal matching technique

Chapter X

Quality Control Techniques

Introduction

Quality function deployment (QFD) represents a broader concept than design for manufacturing (DFM) since the design specifications are driven by the voice of the customer. QFD is concerned with manufacturing productivity and quality improvement. The first paper analyzes the QFD and DFM. The next paper presents a study of consumer evaluation and assessment of product quality. Quality value function is determined by consumer preference for product quality in a deterministic manner. The expected quality value is derived. The third paper analyzes implementation of computer aided quality (CAQ) systems in a CIM environment involving a mechanical component production. Requirements, risks, possible disadvantages, profit and advantages justifying the CAQ are presented. The next paper describes the benefits achieved by the introduction of CAQ systems in a light engineering plant. A software program has been developed in-house and results are presented. The fifth paper proposes an enhancement of existing design for assembly tools with the introduction of part quality into assembly. The effect of fraction defectiveness on average assembly time is considered. A valve assembly analysis has been done with this new enhancement and results are presented. The final paper of this chapter discusses the selection of acceptance sampling plans through knowledge based approach.

Quality Function Deployment, a Technique of Design for Quality

CHIA-HAO CHANG

Department of Industrial and Systems Engineering
University of Michigan-Dearborn
Dearborn, MI

Summary

Most Design for Manufacture (DFM) techniques do not involve the consideration of customer requirements, and hence are not sufficient enough to make a company stay competitive. Quality Function Deployment (QFD) method drives the product design and process planning by customer requirements. It is a technique of Design for Quality (DFQ).

Design for Manufacture

Design for manufacture (DFM), as stated by Stoll [12], recognizes that a company cannot meet quality and cost objectives with isolated design and manufacturing engineering operations. Therefore, the essence of the DFM approach is the integration of product design and process planning into one common activity. The simultaneous engineering approach is used so that all relevant components of the manufacturing system can participate in the design from the very start. DFM tries to run the design and process planning concurrently, and improve the quality of early design decisions. Design from the manufacturing engineering's point of view reduces the number of iterations in sending the design back and forth from the manufacturing engineers to the designers. No doubt, the effective use of the design for manufacture techniques will lead to cost and design time reduction, and ease of manufacture and assembly process; yet those benefits brought by DFM approaches are not enough to guarantee customer satisfaction and make a company stay competitive.

Today more and more customers are looking for products that are tailored to their needs and preference. The definition of quality changes from defect-free to the fulfillment of the "voice of customers" (customer requirements). Marketing, the customer connection, becomes the crucial component of all factories of the future that wish to stay alive and be competitive. To assure that a product be well accepted by customers, industries have to identify the customer requirements, convert them into finished product characteristics, and deploy them into process operations and production planning. A process like this is called quality function deployment (QFD).

Quality Function Deployment

Quality Function Deployment (QFD) was originated in the early seventies at the Kobe Shipyards in Japan. There the technique was used successfully in the design and manufacture of ships, and was subsequently adopted by other Japanese industries. At present, QFD is widely used by companies that recognize and emphasize on the importance of customer satisfaction. As defined by Fortuna [4]: "QFD is a systematic means of ensuring that customer or marketplace demands are accurately translated into relevant technical requirements and actions throughout each stage of product development."

QFD also manages the design process by a team of persons from marketing, product engineering, manufacturing engineering and quality assurance. It builds up a framework for the integration of customer requirements with product design and process planning. In a sense, QFD represents a broader DFM concept. It concerns not only the manufacturing and assembly process, but also assures that the ultimate goal of all processes is towards customer satisfaction. Quality, the fulfillment of the "voice of customers", is emphasized throughout the design process.

The QFD process starts with a marketing study to identify customer needs for product design. This is the most familiar and widely used QFD phase. Marketing study is tasked with collecting the "voice of customers" in the following forms:

primary customer wants and expectations, customer complaints, claims, murmurs, actual customer specifications, and user opinions. Competitor information from market surveys, industry publications, etc. are also gathered in this phase.

To tie up those collected customer requirements with the languages spoken by the designers and manufacturing engineers, secondary or tertiary requirements, specific customer specifications, and competitor products' performance are then matched and combined in order to identify specific product quality features which are controllable in the manufacturer's viewpoint. In other words, the customer "wants" are transformed into engineering "how." For example, customers want the door of an automobile "easy to close." The design engineers, working with marketing personnel, will interpret it into an engineering quantity called "closing effort". Through some further studies, they will then decide this so-called "easy to close" can be considered as the closing effort that should be no more than certain foot-pounds.

In as much as sales function is in direct communication with the customers, product complaints, customer claims, design-betterment suggestions are more easily transmitted into the desired engineering changes to improve product design. Failure- related complaints and returns are forwarded to the reliability laboratory for analysis and evaluation. Competitor products' information is also analyzed. This way, quality features that satisfy customer requirements and at the same time avoiding weaknesses and poor design features found in competitors' products are achieved. Furthermore, competitive benchmarks are analyzed in order to assist the deployment of specific target values later in the product design in order to stay ahead of the competition, and be the best of the class.

Besides marketing study and product planning, phase II of QFD processes, part/mechanism deployment, designs the product with the critical part/mechanism characteristics. Phase III, process planning, identifies the key processes in operations and develops a quality control plan. Most DFM approaches are applied during phase II and III where the product is designed with the concern of

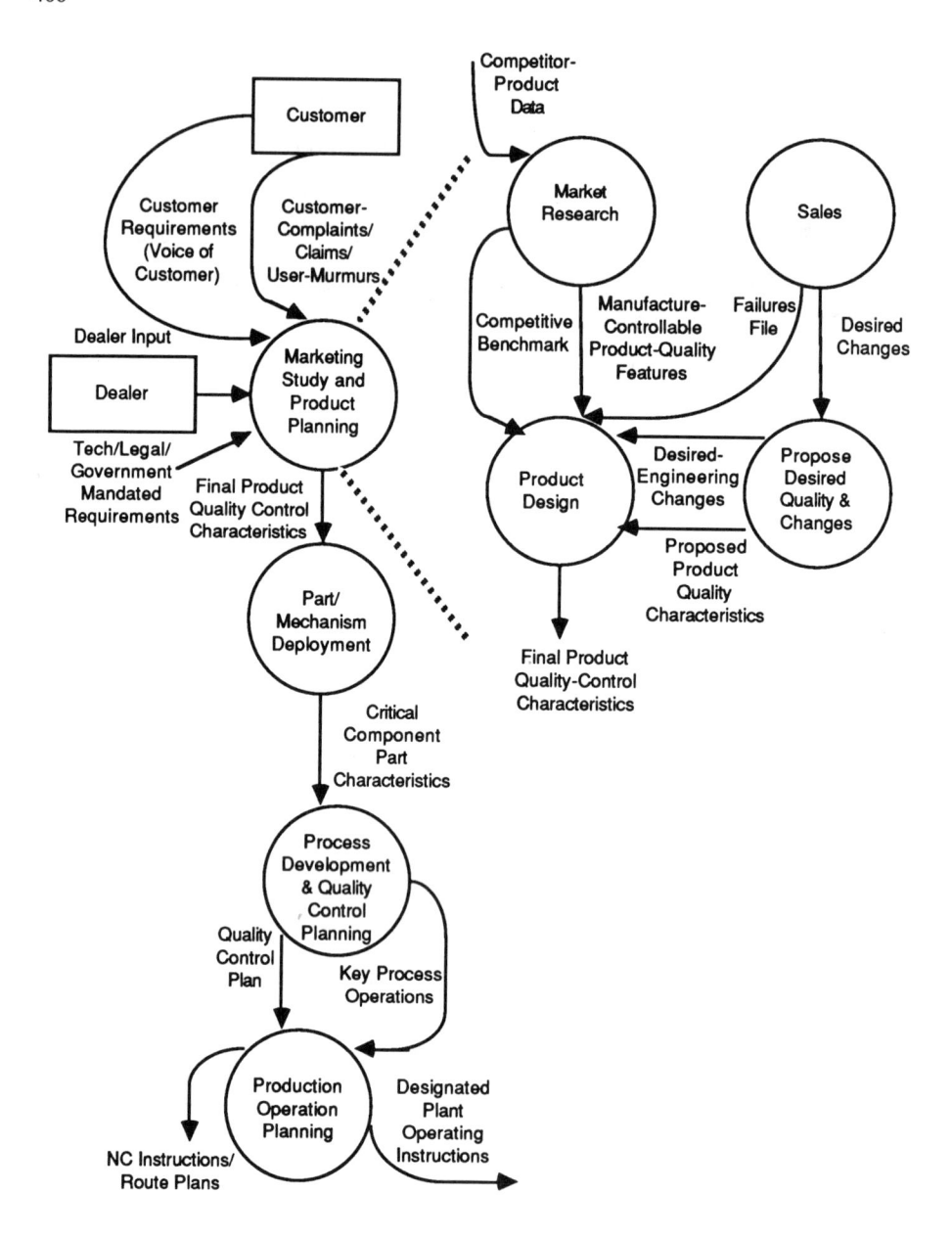

Fig. 1. Data flow diagram of quality function deployment process

manufacturing quality and productivity. For example, failure mode and effects analysis (FMEA), an important DFM method mentioned by Stoll [12], is considered a critical element of QFD operation in both phase II and III. Phase IV, production operation planning, identifies quality check and control procedures and major production requirements including equipments and manpower to support the production operations. A brief data flow diagram representing the key operations of QFD can be found in figure 1. It also includes the functions covered in phase I of QFD. Detail data flow diagrams of QFD processes can be found in Chang's work [1]. Figure 2 compares the coverage of QFD versus the operations of a typical DFM process.

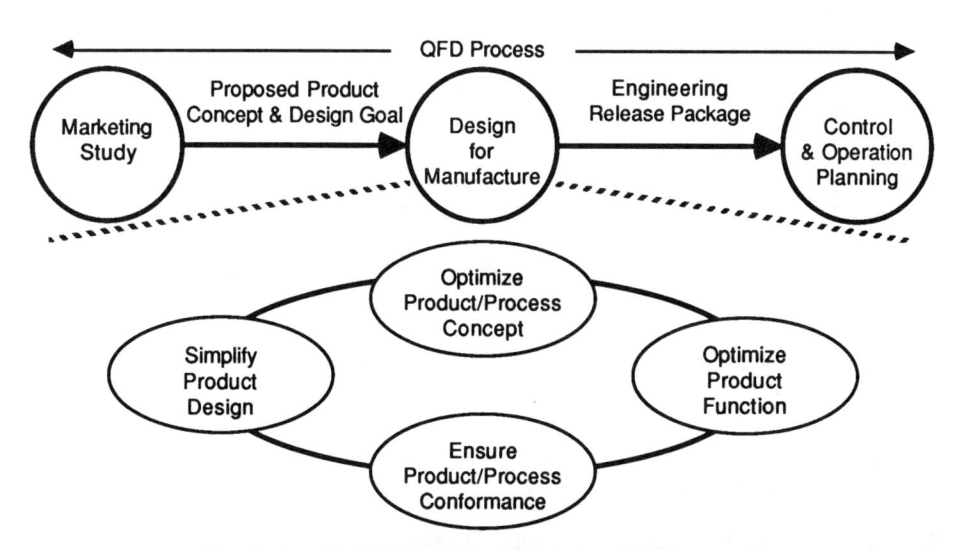

Fig. 2. A typical DFM process verses QFD process

Conclusion

The views of product design and planning have undergone dramatic changes in recent years. Like most DFM methods, QFD concerns about manufacturing productivity and quality improvement. Since the design is driven by the voice of customers, QFD therefore represents a broader concept than just another DFM method. Marketing, product design, manufacturing process planning and control activities are incorporated into one integrated design process for customer satisfaction, which is a new definition of quality. Thus QFD should be called a technique of Design for Quality (DFQ).

References

1. Chang, C. H.: Quality function deployment (QFD) processes in an integrated quality information system. *International Journal of Computers & Industrial Engineering*, 17(1-4), (1989).

2. Chang, C. H.: The structure of quality information system in a computer integrated manufacturing environment. *International Journal of Computers and Industrial Engineering*, 15(1-4), (1988).

3. Chang, C. H. and L. Tsui: Marketing and management aspects of today's quality control function. included in Hamza, M. H. (ed.) *Advances in reliability and quality control*, Anaheim, Acta Press 1988.

4. Fortuna, R. M.: Beyond quality: taking SPC upstream. *Quality Progress*, June (1988).

5. Hauser, J. R. and D. Clausing: The house of quality. *Harvard Business Review*, May-June (1988).

6. Ishikawa, K. *What is total quality control? The Japanese way,* Englewood Cliffs, Prentice-Hall Inc. 1985.

7. Joyner, J. M.: Marketing's role in quality improvement. *Quality Progress*, June (1986).

8. Kogure, M. and Y. Akao: Quality function deployment and CWQC in Japan. *Quality Progress*, October (1983).

9. Kukla, B.: Meeting customer needs. *Quality Progress*, June (1986).

10. McHugh, J. E. *Quality function deployment.*. Dearborn, American Supplier Institute, 1986.

11. Plsek, P. E.: Defining quality at the marketing/development interface. *Quality Progress*, June (1987).

12. Stoll, H. W.: Design for manufacture. *Manufacturing Engineering,* January (1988).

13. Sullivan, L. P.: Policy management through quality function deployment. *Quality Progress*, June (1988).

14. Sullivan, L. P.: Quality function deployment. *Quality Progress*, June (1986).

Quality Value Function and Consumer Quality Loss

FU QIANG YANG, MAJID JARAIEDI, and WAFIK ISKANDER

Department of Industrial Engineering
West Virginia University
Morgantown, WV

Abstract
This paper presents a study of consumer evaluation and assessment of product quality. Consumers experience a loss due to quality discrimination because they pay same price for a product but get different quality. Quality value function is determined by consumer preference for product quality in a deterministic manner. This paper derives expected quality value and consumer-based quality loss which depend on the form of the quality value function employed.

Introduction
One of the basic assumptions in classic microeconomics is that products are homogeneous, which implies that there is no difference in product quality to affect consumer satisfaction. Chamberlin [1] and Dorfman and Steiner [2] viewed quality difference as shifts in a product's demand curve to avoid violation of the assumption of homogeneous product, but this approach faced a difficulty to explain the effect of quality variation on consumer decision making. Taguchi and Wu [3] provide a meaningful quality loss function to illustrate social and producer's quality loss due to product quality variation. Since their loss function is derived mathematically based only on producer's cost function, development of a more comprehensive loss function based on consumer quality assessment is required.

Quality Discrimination and Quality Value Functions
Quality discrimination for consumer is based on the fact that consumers pay the same price for a certain product but could get different product quality. Comparing with the value of the highest quality in the product, consumers experience a loss under quality discrimination. The larger the product quality variation, the more serious the quality discrimination. If the product price is determined by its quality,

no quality discrimination will occur. However, it is very difficult to establish a price based on a product quality before it is used.

Phadke [4] gave four types of quality characteristics. The quality that consumer attaches to the price, T, is the mean value (μ) in the nominal-the-best and the asymmetric types of quality characteristics; while it is the largest value in the larger-the-better type of quality characteristics, or the smallest value in the smaller-the-better type of quality characteristics. We define consumer quality loss, CQL, as the difference between the highest quality value, w_H, and the actual quality value, w, realized by the consumer.

$$CQL = w_H - w \tag{1}$$

Quality value function for a product is derived from consumer assessment for deterministic quality. The consumer is willing to pay higher price for higher product quality. No uncertainty and no risk are involved in the determination of quality value function. Consumer quality value function can be rationaly assumed to be a monotonic increasing and differentiable function over a range of product quality. Consumer marginal quality value (first derivative of the value function) can be decreasing, constant, or increasing in general, which correspond to concave, linear and convex curves over a range of product quality. Three typical quality value functions, the linear, the quadratic function with decreasing marginal value and the quadratic function with increasing marginal quality value have the following forms in the nominal-the-best type of quality characteristics:

$$w = w_H - k_1 |z - \mu| \tag{2}$$

$$w = w_H - k_2 (z - \mu)^2 \tag{3}$$

$$w = w_L + k_3 (|z - \mu| - \Delta)^2 \tag{4}$$

where z - actual value of quality characteristics, unit, $|z - \mu| \le \Delta$;

μ - mean value of quality distribution, unit;

w_L - quality value corresponding to the lowest quality, $;

k_1, k_2, k_3 - loss coefficients, $/unit for k_1, and $/unit² for k_2 and k_3;

Δ - limit for quality variation, unit.

The loss coefficients k_1, k_2 and k_3 in the above equations can be determined by setting $|z - \mu| = \Delta$ in equations (2) and (3) and $z = \mu$ in equation (4), as shown below

$$k_1 = (w_H - w_L)/\Delta \tag{5}$$

$$k_2, k_3 = (w_H - w_L)/\Delta^2 \tag{6}$$

Expected Quality Value and Consumer Quality Loss

Expected quality value (EQV) depends on the shape of consumer's quality value function, $f(w)$, and the quality distribution for a given type of product. We assume that there is no substantial difference in quality preferences and beliefs among groups where consumers are relatively homogeneous. For instance, in the nominal-the-best type of quality characteristics the expected quality values for the normaly and uniformly distributed product qualities with linear quality value functions, EQV_n and EQV_u, are respectively

$$EQV_n = w_H - k_1 2 \int_0^\infty (z-\mu) \frac{1}{\sqrt{2\pi}\,\sigma} \operatorname{Exp}\left[-\frac{(z-\mu)^2}{2\sigma}\right] dz$$

$$= w_H - \sqrt{\frac{2}{\pi}}\, k_1 \sigma = w_H - 0.8 k_1 \sigma \tag{7}$$

$$EQV_u = w_H - 0.866 k_1 \sigma \tag{8}$$

where σ is the standard deviation of quality distribution.

The average consumer quality loss, ACQL, is defined as:

$$ACQL = E(CQL) = w_H - EQV \tag{9}$$

The absolute ratio of average consumer quality loss, r, for any probability distribution i to the corresponding uniform distribution for a given product is defined as:

$$r = (w_H - EQV_i)/(w_H - EQV_u) \tag{10}$$

$$EQV_i = (1 - r/2)w_H + (r/2)w_L$$

Let $\quad r/2 = p$

$$EQV_i = (1 - p)w_H + pw_L \tag{11}$$

Equation (11) is not related to parameters of any product quality distribution and is very convenient for computing the expected quality value for any shape of quality distribution if r can be derived mathematically or computed approximately.

The relative ratio of average consumer quality loss, r_{nm}, for any two products, n and m, with similar quality distribution pattern is defined as:

$$r_{nm} = (w_{Hn} - EQV_n)/(w_{Hm} - EQV_m) \tag{12}$$

$$= (w_{Hn} - w_{Ln})/(w_{Hm} - w_{Lm})$$

The difference between average consumer quality losses, d_{nm}, for the two products, n and m, is

$$d_{nm} = ACQL_n - ACQL_m = -d_{mn} \tag{13}$$

The quality value functions (QVF), expected quality value (EQV) and average consumer quality loss (ACQL) for normal quality distribution for three types of quality characteristics are shown in Table 1. Assuming that the quality specification range, $\Delta(\Delta=3\sigma$ in the nominal-the-best type of quality characteristics and $\Delta=6\sigma$ in the larger-the-better and the smaller-the-better types of quality characteristics), and the quality value difference between the highest quality value and the lowest quality value, $w_H - w_L$, are the same for linear, concave and convex quality value functions, it is obvious that

$$EQV_{convex} < EQV_{linear} < EQV_{concave} \qquad (14)$$

$$ACQL_{convex} > ACQL_{linear} > ACQL_{concave} \qquad (15)$$

As shown in Table 1, the form of consumer quality value function affects the values of EQV and ACQL significantly. The indexes of absolute ratio of consumer quality loss, r, relative ratio of average consumer quality loss, r_{nm}, and quality loss difference, d_{nm}, are shown in Table 2.

Table 1: QVF, EQV and ACQL for A Normaly Distributed Quality Characteristics

Type of quality characteristics		linear	concave	convex
The nominal -the-best	QVF	$w_H - k_1\|z - \mu\|$	$w_H - k_2(z - \mu)^2$	$w_L + k_3(\|z - \mu\| - \Delta)^2$
	EQV	$w_H - 0.8k_1\sigma$	$w_H - k_2\sigma^2$	$w_L + k_3(\sigma^2 - 1.6\Delta\sigma + \Delta^2)$
	ACQL	$0.8k_1\sigma$	$k_2\sigma^2$	$w_H - w_L - k_3(\sigma^2 - 1.6\Delta\sigma + \Delta^2)$
The larger -the-better	QVF	$w_H - k_1(T-z)$	$w_H - k_2(T-z)^2$	$w_L + k_3(T-z-\Delta)^2$
	EQV	$w_H - k_1(T-\mu)$	$w_H - k_2(\sigma^2 + (T-\mu)^2)$	$w_L + k_3(\sigma^2 + (T-\mu-\Delta)^2)$
	ACQL	$k_1(T-\mu)$	$k_2(\sigma^2 + (T-\mu)^2)$	$w_H - w_L - k_3(\sigma^2 + (T-\mu-\Delta)^2)$
The smaller -the-better	QVF	$w_H - k_1(z-T)$	$w_H - k_2(z-T)^2$	$w_L + k_3(z - T - \Delta)^2$
	EQV	$w_H - k_1(\mu-T)$	$w_H - k_2(\sigma^2 + (\mu-T)^2)$	$w_L + k_3(\sigma^2 + (\mu-T-\Delta)^2)$
	ACQL	$k_1(\mu-T)$	$k_2(\sigma^2 + (\mu-T)^2)$	$w_H - w_L - k_3(\sigma^2 + (\mu-T-\Delta)^2)$

Table 2: Indexes for Three Types of Quality Characteristics
($\sigma^2_n \geq \sigma^2_m$, $\mu_n = \mu_m$)

Type of quality Characteristics		linear	concave	convex
The nominal -the-best	r	0.533	0.333	0.633
	r_{nm}	≥ 1	≥ 1	≥ 1
	d_{nm}	$0.8k_1(\sigma_n - \sigma_m)$	$k_2(\sigma^2_n - \sigma^2_m)$	$3.8k_3(\sigma^2_n - \sigma^2_m)$
The larger -the-better	r	1	0.833	1.083
	r_{nm}	1	≥ 1	≤ 1
	d_{nm}	0	$k_2(\sigma^2_n - \sigma^2_m)$	$-k_3(\sigma^2_n - \sigma^2_m)$
The smaller -the-better	r	1	0.833	1.083
	r_{nm}	1	≥ 1	≤ 1
	d_{nm}	0	$k_2(\sigma^2_n - \sigma^2_m)$	$-k_3(\sigma^2_n - \sigma^2_m)$

Conclusions

Based on the above analysis and the assumption of quality distributions with the same mean value of quality characteristics, following conclusions can be made.

1. Consumer quality value function should be carefully evaluated. EQV and ACQL resulting from quality variation will be overestimated or underestimated if quality value function is not correctly determined.

2. Consumer quality discrimination always exists if there is variation in product quality. Average consumer quality loss, ACQL, is the difference between the highest quality value and the expected quality value.

3. In the nominal-the-best and asymmetric types of quality characteristics, consumers always favor the product with small variance regardless of pattern of the quality value function employed.

4. In the larger-the-better and the smaller-the-better types of quality characteristics, consumers with linear quality function are only concerned with the mean value of product quality distribution. Concave quality value function possesses the tendency for consumer to make decision in favor of the product with smaller quality variance. In contrast, consumer with convex quality value function will choose the product with higher quality value and ignore its larger variance.

5. The consumer attitude toward quality risk which affects the consumer quality loss and decision making was not considered in this paper. Further research for a comprehensive decision model under quality risk is necessary.

References

1. Chamberlin, E. H.: The Product as an Economic Variable. Quarterly Journal of Economics February (1953) 1-29.

2. Dorfman, R.; Steiner, P. O.: Optimal Advertising and Optimal Quality. American Economic Review December (1954) 822-836.

3. Taguchi, G.; Wu, Yuin: Introduction to Off-line Quality Control. Central Japan Quality Control Association February 1985.

4. Phadke, Madhav S.: Quality Engineering Using Robust Design. AT&T Bell Laboratories 1989.

Implementation of a Computer Aided Quality System (CAQ) in CIM Environment: Advantages and Disadvantages

M. DOMINGUEZ, M.M. ESPINOSA, J.I. PEDRERO and J.M. PEREZ

E.T.S. Ingenieros Industriales, U.N.E.D.
Madrid, Spain

Summary

In the present environment of the computer aided manufacturing, and in order to penetrate in the computer integrated manufacturing (CIM), it is very interesting to think over the computer assisted quality control or, more precisely, to enter in the computer aided quality system (CAQ). This paper analyses, from a CIM environment perspective and in a general way, the implementation of computer aided quality systems in mechanical components manufacturing environments, which are based in the flexible manufacturing system (FMS) philosophy.

Introduction

The starting point of the computer aided quality philosophy can be situated in the well known graphic control system issued by W.A. Shewhart at the beginning of this century. Indeed, the Shewhart idea was, independently of the lack of data processing machines, lead to machine foot the knowledge of the moment in the area of the statistical process control. This idea is held in the present situation but now, in the computer aided manufacturing environment, the graphic control system and the effortless formulas are substituted by automatic controls and powerful CPUs.

Quality, like all the universal concepts, changes its content with the time course and adjusts itself to each time and place progressively. Peter F. Drucker defines quality like the customer is disposed to pay according to what he gets and values. The American Society for Quality Control (ASQC), defines it like the characteristics of a product, process or service that confer it the aptitude to satisfy the user's necessities.

As can be seen, the concept of quality is able to be defined in different valid ways. Perhaps the definition best adapted to the concept of quality used in this work is that of Karou Ishikawa, who defines working with quality like design, produce and delivery a product or service useful, as economic as possible and always satisfactory for the user.

Not only the concept of quality has changed along the time but also its environment has changed. Inside this environment is to be considered the "organisation for quality", that is, the system, people and media of a company dedicated to the quality preferentialy. This organisation for quality has also developed along the time.

Figure 1 shows a possible functional diagram for a mechanical manufacturing company. In the top, the objectives of the company are placed. Subsequently are located the departments, according to the steps for a manufacturing project: design, production management, production, packing, storing and despatching.

At one of the sides, it is placed the quality area. Its task is well-defined, and it includes, first, controlling that the technical specifications and the manufacturing processes, built up by the design area, are performed according to the standards of the company, and on the other hand, it is

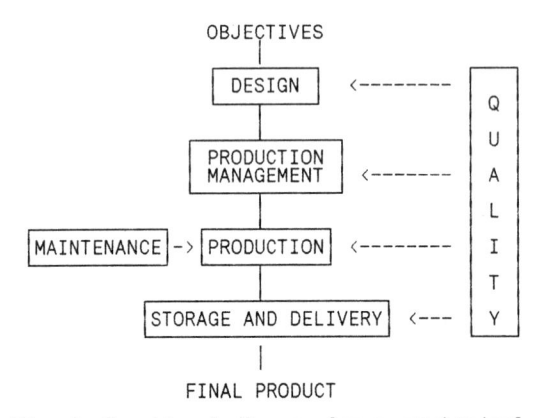

Fig. 1. Functional diagram for a mechanical manufacturing company.

responsible that the execution of the project, by the production area, is performed according to the technical specifications built up by the design area.

In a CIM environment, the objectives of the decision centre must be well-defined:

"Getting the products requested by the market with the appropriate level of quality and with the cost as low as possible, regarding the productive possibilities of the company".

This objective will have to define a master planning of manufacturing, in which the products to manufacture, it quantity and its finishing date will be specified.

Proposed model: CIM + CAQ

Figure 2 shows the basic structure of the proposed model. As it can be seen, CAQ techniques are to be implemented in each productive department of the company. In each department a terminal to access to the CAQ system is to be located, whiles the system manager is to be placed in the CAQ coordination centre. This method gets not only that each element of the organisation is responsible of the quality of its assignment but also that each department is responsible of the information about its area, which is stored in the integral data base of the company.

```
              OBJECTIVES
                  |
         ┌────────────────┐              ┌──────────────────┐
         │  DESIGN + CAQ  │<─────────────│  CAQ MANAGEMENT  │
         └────────────────┘              └──────────────────┘
                  |
     ┌───────────────────────┐     
     │  PRODUCTION MANAGEMENT│<──────────────────┐
     │         + CAQ         │                   │
     └───────────────────────┘                   │
                  |
            ┌──────────┐    ┌──────────────────┐ │
            │PRODUCTION│<───│ MAINTENANCE + CAQ│<──
            │    +     │    └──────────────────┘ │
            │   CAQ    │<────────────────────────┤
            └──────────┘                         │
                  |                              │
         ┌───────────────────┐                   │
         │STORAGE AND DELIVERY│<─────────────────┘
         │       + CAQ        │
         └───────────────────┘
                  |
            FINAL PRODUCT
```

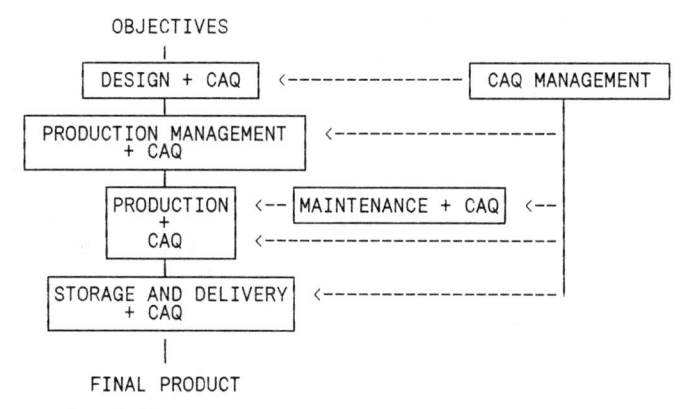

Fig. 2. Functional diagram proposed, CIM + CAQ.

The objective is to implement the philosophy of a total quality project in a project of integration of information about a manufacturing process.

Different definitions of quality were given in the introduction. Now, the concept of quality from a point of view of the integration will be analysed.

The objective of a CIM + CAQ project is "zero defects". But no defects implies no inventories in the store, no unnecessary papers, no failures, no surprises, delayed deliveries, or what is the same, no losses of time, money, resources, ...

Once set the project and the objective to reach, a medium must be set: "rejection of the big changes and tendency to the small and controlled ones".

That is, doing the things well and improving them constantly.

Such a project as the proposed in this work cannot be planed without a mention to the automation.

In the automation of the design tasks, there is an opposition between the freedom (inherent to all creative processes) and the standardisation and systematisation (needed for an automatic process). But in the mechanical manufacturing area this opposition does not exist: planning is essential. And, in order to get the automation, it is necessary a systematised organisation, once a simplification and optimisation of the process has been carried out.

The objective of the automation of the production process is eliminating the workers as much as possible, and today there are some plants which performs with few workers for watching and controlling tasks.

In the automation of the production process, some aspects must be taken into account. Tasks which does not give added value to the product must be eliminated or minimised. Tasks for preparing machines must be simplified and automatisated, in order to minimise the no-productive times. Maintenance tasks must be optimised and simplified. Manufacturing tasks must be automatisated by replacing the direct intervention of the workers by an indirect supervision.

In the beginning, the automation was centred in isolated machines. The objective was making the equipments with some automatic functions, but the worker remained being a fundamental factor. The first numerical control lathes was designed in such a way, and the worker had to type the NC program in the machine directly, oversee the process and perform some tasks, like changing the pieces, tools and broken components.

After an evolution process, the automatisated manufacturing systems have arrived at the current flexible manufacturing systems.

A typical flexible cell is composed of one or several automatic manufacturing machines, one or several robots or manipulation systems, and one or several real time-working computers, which check all the operations in the cell.

To automatise the productive process, it must be taken into account that the implantation of a FMS system does not finish at the work yard, but it requires a reorganisation of all the factory. And this is one of the factors which are preventing the automation in conventional factories.

Advantages and disadvantages

Such a project brings big advantages in a little time. But also some requirements, which can be real difficulties in many cases, must be taken into account and some risks must be assumed.

Two of the possible requirements have special importance. The first are the expenses needed for the equipments (computers and automatic machines), which may be very large. The second makes reference to the organisation. In many cases the organisational philosophy must be changed, and it is not easy to make. So this kind of projects implies a very big risk.

Other requirements and risks which can rise in such a project are the following:

- Qualified worker in continuous training statement is required.
- Staff reorganisation is required in the first stages of the project.
- Clearness in objectives is necessary. Improvisations must be avoided.
- Planning and usage of documentation is required. All must be written.
- High levels of motivation are required. Salaries must be high.

But a lot of advantages will bright few weeks before starting a CIM + CAQ project:

- Bigger productivity of the worker and higher efficiency of the productive systems are obtained.
- The best service for the customer is offered with a minimum stock and the maximum utilisation of the available capacity.
- Control over all the productive operations is got from any point of the factory.
- Tasks which do not give added value to the product are eliminated or minimised.
- Machine preparation tasks are simplified and automatisated, so that the machines are ready for production as many time as possible.
- Maintenance tasks are optimised and simplified.

- Manufacturing tasks are automatisated and the direct operation of the worker is replaced by an indirect supervision.
- Less direct workers are required.
- No-productive times are eliminated and quality is improved.
- Utilisation of computer systems are improved by eliminating duplicities and redundancies.
- More personal satisfaction is got by the workers.
- Routine tasks are eliminated.
- Obsoleted inventories, unforeseen failures, etc., are also avoided.

Conclusions

In the actual environment of advanced technology and modern, competitive automatisated-fabrication systems, it is necessary a deep knowledge of the modern techniques in the field of the manufacturing engineering.

Moreover, in the automation race, it is not possible considering the automation of one area or department of a manufacturing company and forgetting the other departments and the entire company. For this reason, the production area must not be treated independently, but it must be treated as one more piece of the engine, which is the company.

Integration consists in coupling the different automation systems in order to get a computer control of all the manufacturing tasks.

In this paper, the implementation of a computer aided quality system CAQ in an integrated manufacturing environment CIM, has been presented. Requirements, risks and possible disadvantages have been enunciated, but profits and advantages justifying the implementation, have been also presented.

Bibliography

1. Hax, A.C. "Dirección de operaciones en la empresa" Ed. Hispano-europea, 1988.
2. Buffa, E.S.; Sarin, R.K. "Modern Production / Operations Management" John Wiley & Sons, 1987.
3. Mompín, J. "Sistemas CAD/CAM/CAE. Diseño y fabricación por computador" Ed. Marcombo, 1989.
4. Schomberger, R.J. "Técnicas japonesas de fabricación" Ed. Limusa, 1987.

Computer Aided Quality Assurance Systems

V.K. GUPTA
General Manager
VXL India Ltd.

R. SAGAR
Department of Mechanical Engineering
Indian Institute of Technology
New Delhi, India

ABSTRACT

Quality has always been a major human concern and so has been the need to measure quality. Ever since the introduction of the concept of mass production, quality control has grown into an important field and has made rapid progress in technology and its application.

Traditionally quality control function is performed by a team of inspectors using conventional tools and sampling plans with the sole aim of stopping a defective item passing through the inspection stage. This process, though designed to be foolproof, lacks in implementation due to the sheer volume of work involved and dependence on human judgement at each stage. Factors such as fatigue, perception, bias etc. affect the overall outcome of this system.

Introduction of computers, tend to eliminate some of these problems, especially consistancy of judgement, and in addition provides quick storage, retrieval and analysis much needed for decision making at all levels.

This paper brings out benefits achieved by introduction of computer aided QA systems in a light engineering plant engaged in manufacture of precision engineering components and products.

BACKGROUND

History of quality is as old as the evolution of mankind. Quality is embeded in basic human nature hence the concern for quality has always remained a major concern for all individuals and societies. Productivity has been only a recent phenomenon beginning in early 20th Century. Traditionally quality control function is performed by a team of inspectors using conventional tools and sampling plans. The basic concept behind all quality control systems has been to detect a defective at inspection stage. This process of detection is cumbersome due to the volume of inspection activity needed to detect a defective and its dependance on human judgement at each stage. Factors such as fatigue, perception, bias etc. affect outcome of inspection systems.

Computer revolution has influenced quality management function. Now with use of computers and computer aided QA systems, inspection activity has been automated, reducing dependence on human judgement and also providing quick storage, retrieval and analysis of quality data needed for taking steps to prevent occurance of defectives even before these are produced. Use of statistical and forecasting techniques can prevent defects at manufacturing stage itself.

ORGANISATION

The organisation studied has been engaged in the development and manufacture of high precision engineering components and products. Most of the components manufactured are high precision and complex in nature. Design tolerances are generally tight. To meet these close tolerances, elaborate gauging and multistage inspection is required. With addition of several new products over last few years, quality control department was under tremendous pressure and needed a workable solution.

THE NEED

A detailed study was carried out to identify the causes leading to a defective product. Alternatives were evaluated to meet the organisational needs without resorting to conventional method of increasing corresponding number of inspectors to meet the increased load. Following factors emerged :

a. High rejection levels, rendering all sampling plans virtually ineffective resulting in 100% inspection at most of the stages.

b. Higher time required to inspect a component using mechanical gauges.

c. Frequent errors in measurement, needing a check & cross-check by other inspectors.

d. Bias in interpreting data.

e. Time consuming storage and retrieval system.

f. Rigid inspection schedules.

g. Use of existing conventional measuring systems for mass production as well as for small batch jobs.

IDENTIFICATION OF AREAS

Following areas were identified for introduction of computerised systems and computer aided QA systems :

1. Quality data and information

2. Standards-room operations

3. Patrol and stage inspection

4. Process capability studies

5. Receipt inspection & testing laboratories

6. Product testing

7. Components requiring 100% inspection

8. Special components

SEARCH FOR SOLUTIONS

Detailed study of quality control needs of the organisation was done. This was followed with a market research to find suitable solutions for the organisation. Following products/services were studied.

1. Measuring devices for lineer measurements

2. Measuring systems for 2D measurements

3. Measuring systems for 3D measurements

4. Transmitters/interface and receiver systems

5. Software packages

6. Retrofit systems for existing inspection equipments

7. Special computer aided QA systems

8. Integration aspects

CRITERIA FOR SELECTION

Following criteria were adopted for selection of a suitable solution for this organisation. These criteria were finalised after discussions with users and suppliers of various systems and equipments.

1. Degree of compliance in measuring QA needs
2. Ease of operation and maintenance
3. Availability of spares and services
4. Standardisation
5. Connectivity and upgradation
6. Integration
7. Overall system cost

TECHNOLOGIES STUDIED

In view of a wide range of technologies available for QA systems, the study included the following technologies :

1. 2D/3D Measurement Devices : Optical

   Laser

   Video

   Mechanical

2. Sensing Devices : Mechanical

   Pneumatic

   Electrical

   Electronic

   Optical

   Magnetic

   Laser

3. Interface : Built-in interface

   External interface :
   - Software
   - Hardware

4. Data Processors   :   Dedicated

                                  IBM compatible PC based

5. Software   :   Dedicated

                                  Standard packages

                                  In-house development

IMPLEMENTATION

Based on above criteria, solutions of computer aided QA systems were selected and procured.

Human Factor

Major attention was paid to keep the concerned people informed at each stage of the above project. A three tier programme was conceived to inform the people at all levels and seek their involvement in implementation. A half day programme was held to appraise top management on all aspects of the programme to implement computer aided QA systems. This was followed by a detailed programme for middle level managers and then for junior level managers of quality assurance department. Selection of persons to be trained on new systems was done from a list of volunteers from the various departments. They were provided comprehensive training in using these systems. A repeat training was organised after a gap of six months.

A few systems developed & implemented have been described below:

1. Computerised Quality Management System

   A comprehensive fully integrated quality management system was developed and implemented. This system covered entire spectrum of quality assurance activity and was aimed at providing uptodate information on quality at all levels through-out the organisation (Fig 1).

2. Computer Controlled Non-Contact Optical Measuring System

   This system consist of a high resolution profile projector with manual stage (optional motorised stage) with a resolution of 0.001 mm, dedicated data processor, built in dedicated software, optical edge sensor, thermal printer and interface with IBM compatible PC. The system can be programmed to perform series of calculations to compute various quality measurement routines. Upto 100 programmes can be stored and recalled. The edge sensor eliminates human error in measurement and speeds up the inspection process with the aid of motorised stage. Data once captured and stored with the help of edge sensor is automatically analysed and reports generated as per programme sequence. An inspection job requiring measurement of holes in a component and computing centre distences took thirty minutes on Univeral Measuring Microscope. In addition, it required ten minutes per component to compute various distances. On new system this job including computing distances is completed in less than one minute. IBM compatible PC with SPC software prepares SPC control charts and prepares reports after completion of each set of samples (Fig 2).

3. Retrofit Computer Aided Non-contact Optical Measuring System

A similar system as described above was fitted retrofit on exiting profile projector. This system works on the principle of detecting edge of an image projected on screen of the profile projector. An edge can be defined as the parting line between bright part of an image and dark part of the image of an object. Fibre optic sensor is calibrated to detect intensity of light, separately on bright part of the screen and also on dark part of the screen. This enables the sensor to pass on a signal to the data processor whenever the intensity of light of the part of the screen in front of the sensor changes from bright to dark or vice versa as a result of movement of the stage of projector. Rotary encoders with a resolution of 0.001 mm and fibre optic edge sensor converted existing 2 D profile projector into a semi automatic optical measuring system. This system too has been interfaced with an IBM compatible PC and provides similar advantages as given above.

4. Portable Data Processor

A portable data processor with built in thermal printer and interface with a digital micrometer provided on the shop floor capability to determine process capabilities of various machines. This system prints out XR and X charts in addition to computing process capability. Process found out of control or not capable were analysed and corrected (Fig 3).

5. SPC Software

A SPC software package was installed. This proved to be a powerful tool to build up quality data base of manufacturing operations and print out $\overline{X}R$ and $\overline{X}\sigma$ charts to allow manufacturing and QA engineers to analyse processes and take corrective action before a process went out of control, hence preventing generation of defectives. The software is capable of providing following output as standard features.

$\overline{X}R$ Chart

$\overline{X}\sigma$ Chart

Histogram

Scatter diagram

Process capability

Other statistical tests

Since data base accepts input from measuring instruments, keyboard or any ASCII File, any special requirement can be met by using conventional programming (Fig.4 - i to iv).

6. Gauge Management System

   A gauge management system was developed and implemented. This system keeps full track of each individual gauge, its life, history card, location, status, calibration due and also status of indent/order placed for a new gauge (Fig. 5).

BENEFIT ACHIEVED

Introduction of computerised systems and computer aided QA systems have benefited the organisation in following areas.

1. Reduction in inspection time, leaving adequate time to existing staff for planning and problem solving.

2. Increased accuracy and reliability by elimination of human error in measurement.

3. Speedy storage, retrieval and analysis of information.

4. Reduction in time lag between error detection and corrective action resulting in preventing a defective item being produced.

5. Reduction in rejection level from 10% to below 5%.

6. Integration with other systems

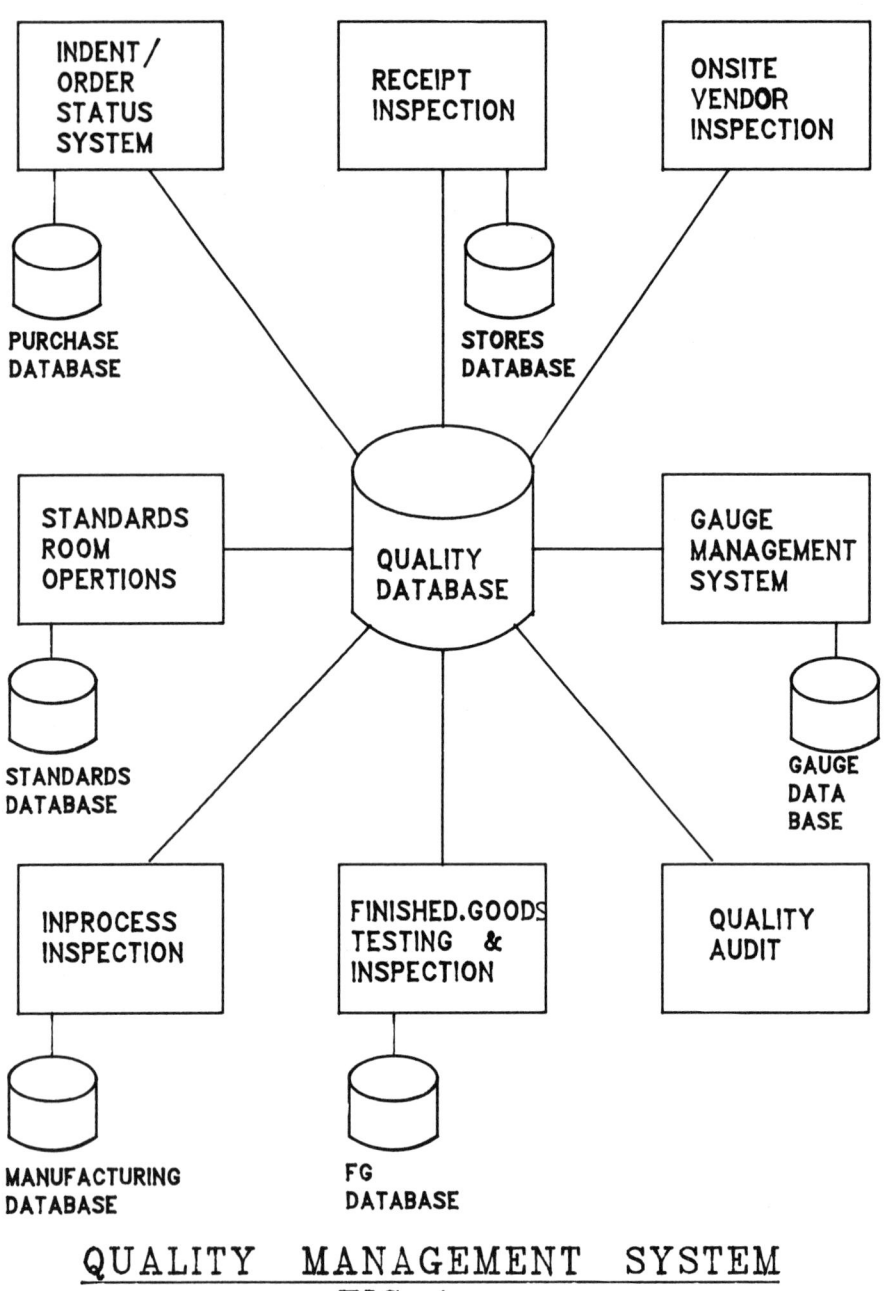

QUALITY MANAGEMENT SYSTEM
FIG-1

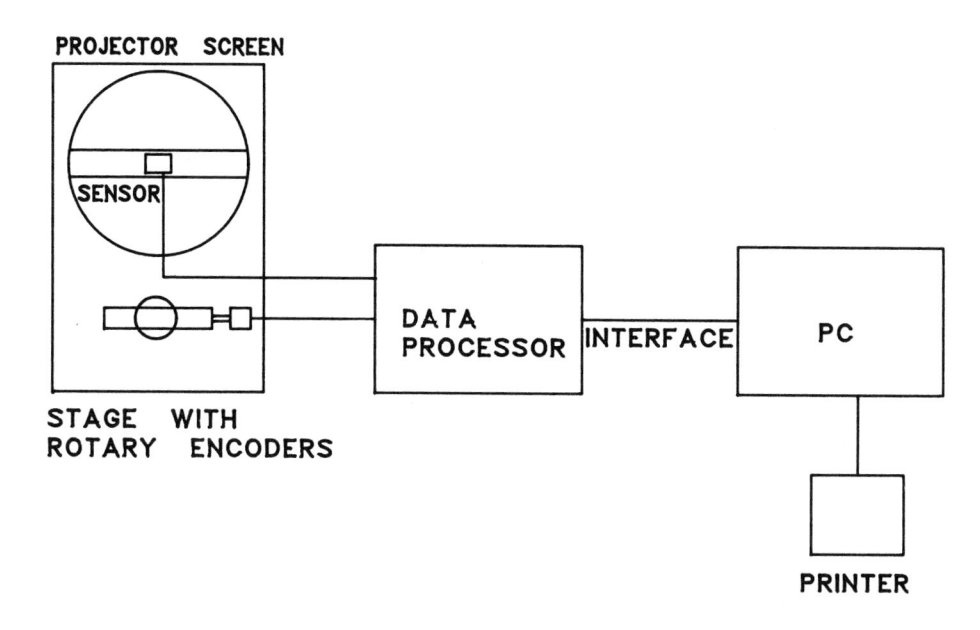

COMPUTER AIDED NON CONTACT
OPTICAL MEASURING SYSTEM
FIG-2

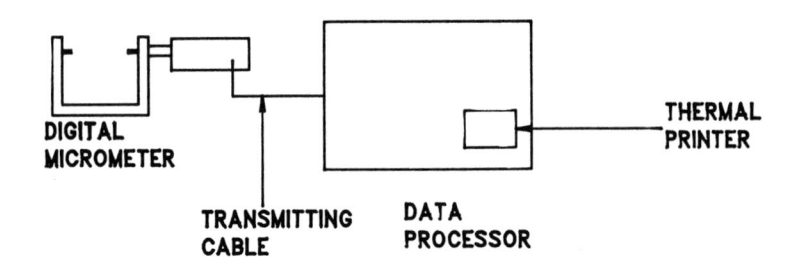

PORTABLE DATA PROCESSOR

FIG-3

FIG. 4(i)

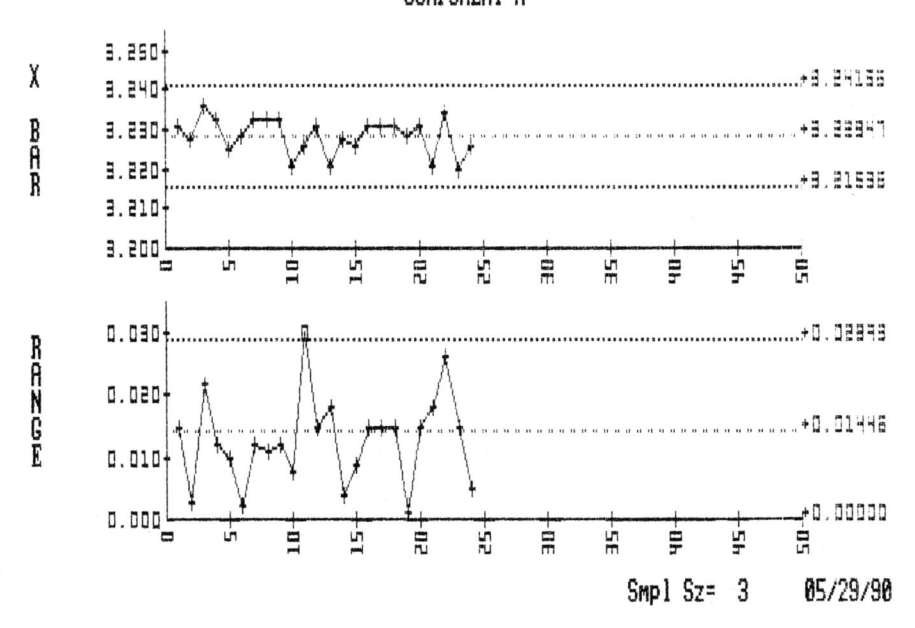

FIG. 4(ii)

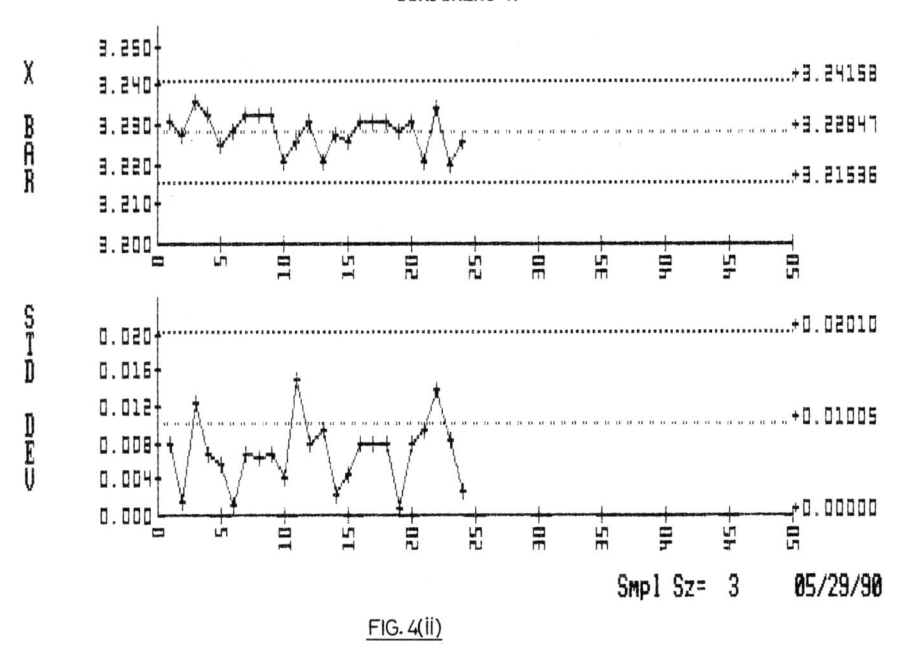

COMPONENT A

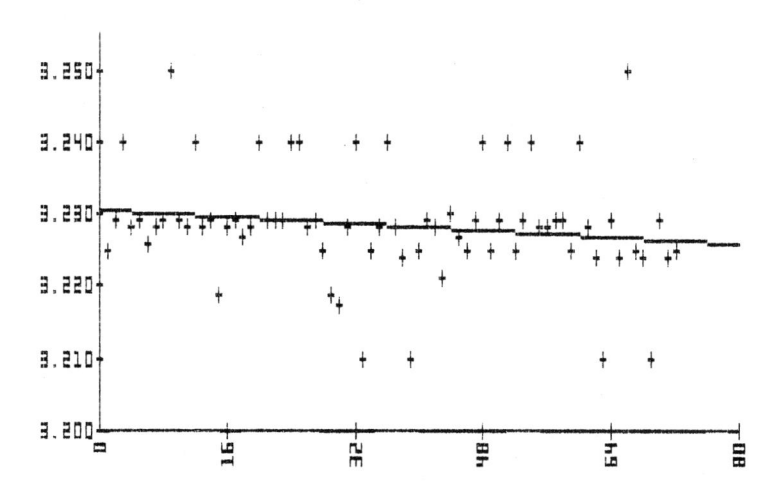

05/29/90   AVGY=   3.228478   CORR=- 0.160777

FIG.4(iii)

component A

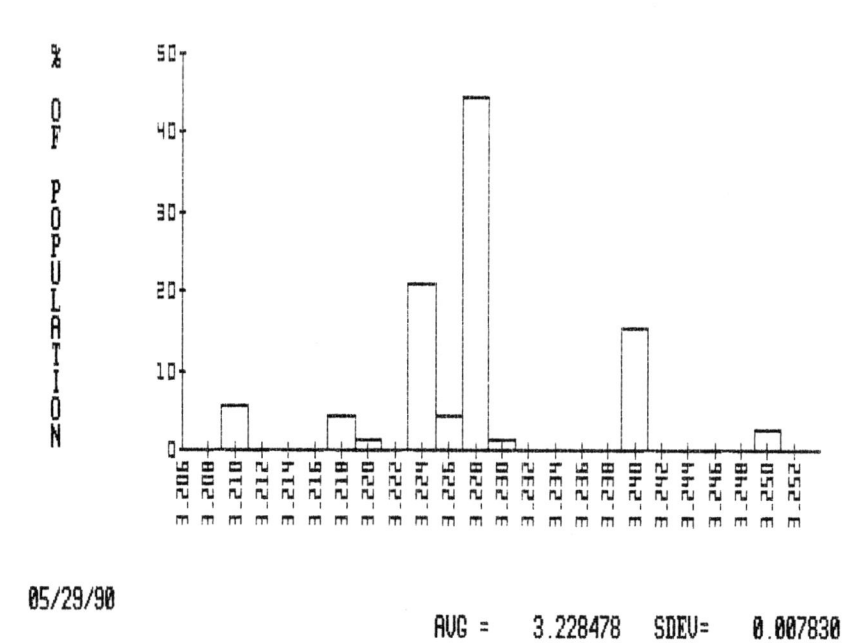

05/29/90

AVG = 3.228478 SDEV= 0.007830

FIG.4(IV)

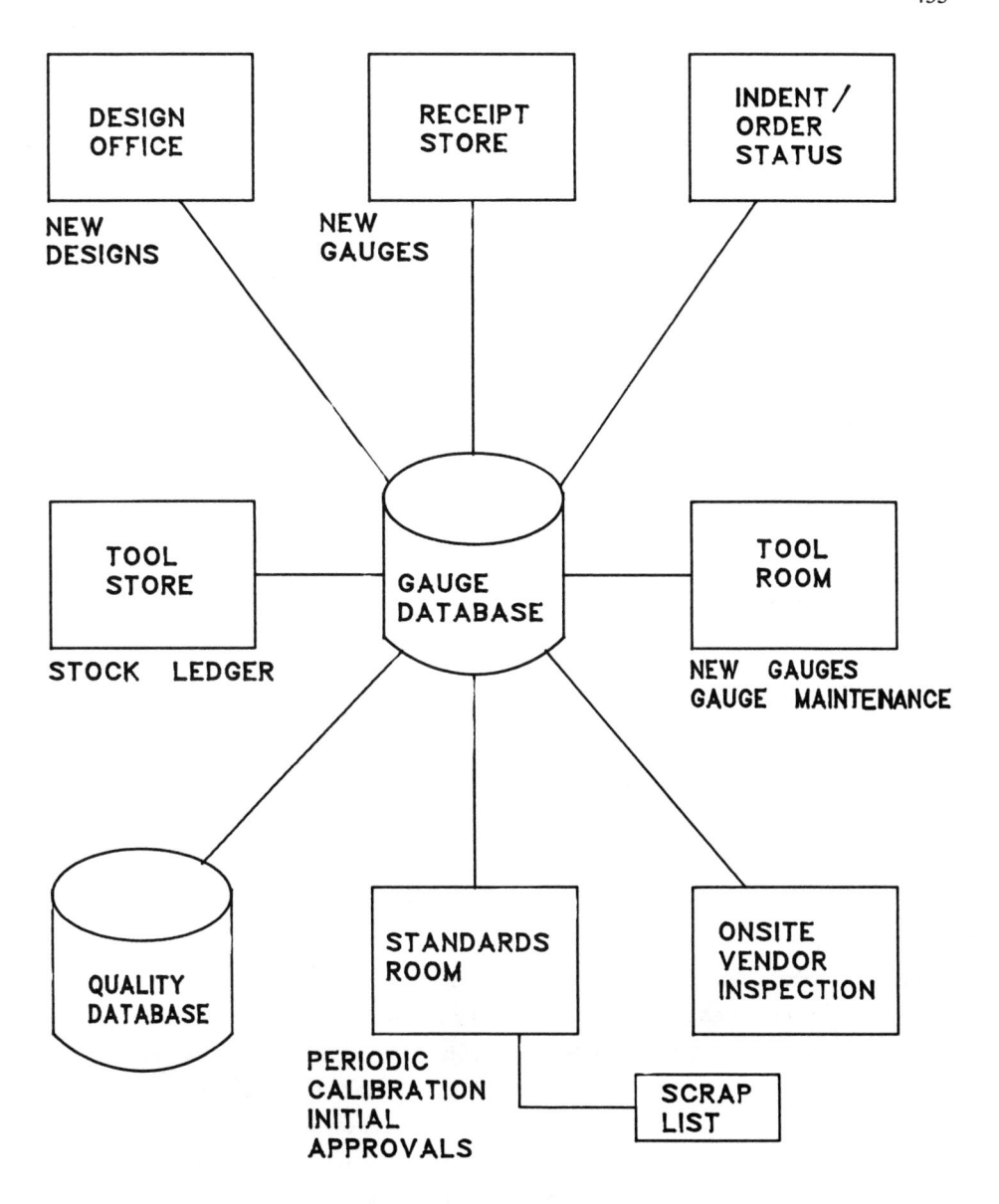

GAUGE MANAGEMENT SYSTEM
FIG-5

Quality Consideration During DFA Analysis

SUDERSHAN L. CHHABRA and RASHPAL S. AHLUWALIA

Department of Industrial Engineering
West Virginia University
Morgantown, WV

Abstract

Design for assembly (DFA) tools have the potential to assist the product designers in designing better assemblies and for improving the designs of existing assemblies. This paper proposes an enhancement of existing design for assembly assessment tools. The enhancement considers part quality when determining assemblability. The system also provides a prioritized list of parts for quality improvement. The improved design for assembly tool could be utilized in simultaneous engineering environment, where designing better products is the joint responsibility of design and manufacturing engineers.

Background

Engineers are becoming aware that manufacturing cost of the product is determined at the earliest stages of design. Once the designer decides and fixes assembly, it is too late to make the type of changes that could reduce the manufacturing costs. An interface between design and manufacturing results in the enhancement of product design [3]. Design for assembly (DFA) tools focus on product simplification. DFA tries to reduce assembly time and assembly cost [1]. Minimization of number of parts results in product simplification. It might however result in more complex part features. The resulting design with the reduced part count minimizes total manufacturing cost. Most of the design for assembly tools are stand alone in nature, i.e., not integrated with the design process. DFA analysis is typically performed "Off-line" to the basic design process. To achieve full benefit from DFA, we need to integrate it with design. There are several DFA systems. Boothroyd and Dewhrust UMass DFA Program being the most widely reported. The UMass program attacks assembly costs, which are often a major component of total manufacturing costs. It emphasizes to reduce part count in assembly. This results in reduced production cost. Using this method, different designs, with different number of components, can be compared before a detailed design commitment is made. UMass system examines end to end symmetry, rotational symmetry, ease of parts handling, parts orientation, assembly and minimum number of parts in the product [2].

Process variability is an integral part of any manufacturing process. Variability may be in raw material, in machine setting, in operator's working, or in inspection. Such variabilities contributes to the overall variability, which may produce out of specification products. The out of specifications parts are called non-conforming or defective parts, and this non-conforming fraction (also called fraction defective) should be considered while evaluating assembly time and cost. Fraction defective is a function of type of manufacturing process involved, process capability, and type of inspection associated with the components. Because the non-conforming components assembly operators need additional handling time, which affects average assembly time. Present DFA assessment tools address part count, symmetry, orientation, and handling. An underlying assumption being that assembly operator always gets good quality components; which meet all of the design and drawing specifications and ultimately assembly specifications. Such an assumption may not always be true. Quality of components is an important consideration. Presently, it is not being addressed by most DFA systems.

Enhancement of DFA Assessment Tools

In order to estimate the average assembly time, we need to estimate the time spent to assemble good quality (conforming) parts as well as time taken to handle non-conforming (fraction defective) parts. Time spent to assemble good quality parts can be estimated by an existing DFA system, such as the UMass system. A software module is developed to estimate handling time of fraction defective parts. The additional handling time is distributed over good quality assemblies in order to get an accurate estimate of average assembly time.

The software module described here deals with considering part quality during assembly evaluation. This software module is referred to as Design-Quality(DQ). It uses the output of UMass system and fraction defective data to determine the assembly time. Assembly time takes into consideration the effect of fraction defective. User also has to describe the statistical distribution type for each part. DQ software module estimates additional assembly time for parts as well as for the entire assembly. This program also determines the effect of small change in fraction defective of assembly part on the additional handling time for the part. This is called marginal additional handling time and can be used to determine the priority of assembly parts for their quality improvement. An higher value of marginal additional handling time indicates that part quality need to be improved. The relationships used to determine the additional handling time and marginal rate of additional handling time are described below.

The additional handling time (D_i) for Part # i is determined by a simulation. This software module simulates the assembly process. A Flow chart describing the simulation module is shown in Figure 1. The program generates a random number and compares it with part's fraction defective. If the random number is greater than fraction defective then it is a good part and additional handling time for this particular part is zero, and if the random number is equal or smaller than the fraction defective, the part belongs to non-conforming fraction. The program then estimates additional handling time for the part. The program generates a second random number and refers to particular type of statistical distribution, specified by the user, and estimates additional handling time for the part. Presently, the program has three options for the type of distribution, i.e., uniform, normal, and triangular. The iteration continues until the stopping criteria is satisfied. The stopping criteria is based on the relative error of 0.01. This criteria is applied to the fraction defective as well as on the average additional handling time. The generation of first random number simulates picking of a part by the assembly operator, and the second random number simulates the assembly operation. The average additional handling time (D_i) for Part# i is given by

$$= \frac{\text{Total Additional Handling Time (estimated in NS}_i \text{ iterations)}}{\text{Fraction Of Conforming Parts}}$$

$$= \frac{W_i}{NS_i * (1- F_i)}$$

Where W_i is total additional handling time estimated in NS_i iteration of simulation, and F_i is fraction defective for the part i.

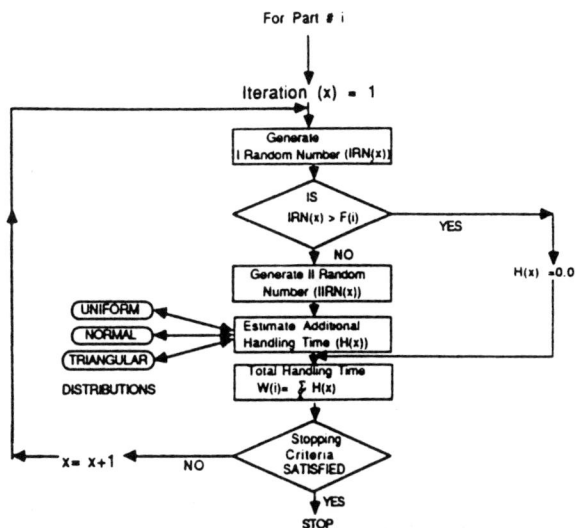

Figure 1. Flowchart for SIMULATION

The marginal additional handling time (M_i) for Part i indicates the effect of small change in part's fraction defective on additional handling time of the part. The expression is derived by partially differentiating additional handling time (D_i) with respect to fraction defective F_i, i.e.;

$$M_i = \frac{\delta D_i}{\delta F_i} = \frac{W_i}{(1-F_i)^2}$$

The average assembly time considering parts quality (B_i) for part i is determined by adding assembly time (A_i) estimated by UMass system without considering the effect of quality of part, and average additional handling time (D_i) estimated by DQ system; i.e;

$$B_i = A_i + D_i$$

Where A_i is assembly time estimated by UMass system, and D_i is estimated average handling time, which is a result of fraction defective.

Application Study

An assembly shown in Figure 2 was analyzed using this approach. The information gathered for this example is shown in the exploded 3-dimensional view of assembly. Assembly cost for assembly part estimated by UMass can be converted into assembly time. This assembly time is referred as assembly time without considering the quality of part. Partial results of UMass for this assembly analysis are shown in first three columns of Tables 1. Assembly efficiency and estimated total assembly time is shown is Table 2. DQ program modified the assembly times for the parts, for the entire assembly and estimated new assembly efficiency considering new total assembly time. The results are compared in Table 1 and 2. The difference in two values of assembly efficiencies, and total assembly times proved the importance of parts quality. The program also estimated effect of small change in fraction defective (for each part) on additional handling time, i.e. marginal additional handling time (Table 3). This information helps in deciding the priority of parts for improving parts quality (table 3).

Conclusion

This paper presented an enhancement of an existing DFA tool. The enhancement considered the effect of fraction defective on average assembly time. This DQ system provided accurate results and recommendations. More accurate results consist of more accurate assembly time for the parts as well as for the entire assembly. The program estimate the additional handling time for fraction defective parts. The additional recommendations consists of the prioritized parts list for the quality improvement. This is done by estimating and then comparing the effect of small change in fraction defective on their additional handling time.

Part Number	Part Name	Assembly Cost (in CENTS)	Fraction Defective	Assembly Time (Sec.) estimated from UMass output	Assembly Time (Sec.) Estimated by DQ system	Additional Handling Time (sec.)
1	Nut	2.80	0.05	7.00	7.46	0.46
2	Washer	2.65	0.06	6.62	7.04	0.42
3	Catcher Plate	1.02	0.04	3.00	3.15	0.15
4	Spring Plate	1.90	0.09	4.75	5.19	0.44
5	Support Plate	1.05	0.06	2.62	2.79	0.16
6	Valve Plate	1.38	0.03	3.45	3.55	0.10
7	Valve Body	2.20	0.07	5.50	5.88	0.38
8	Bolt	1.20	0.03	3.00	3.06	0.06

Table 1. Partial results of Umass and DQ systems

Total assembly time (in sec.) by UMass	Total assembly time (in sec.) by DQ	Total additional handling time (in Sec.)	Assembly Efficiency (in %) by Umass	Assembly Efficiency (in %) by DQ
45.45	48.51	3.06	13.20	12.37

Table 2. Summary of assembly analysis

Part #	Part Name
1	Nut
2	Washer
3	Catcher Plate
4.	Spring Plates (3)
5	Support Plate
6	Valve Plate
7	Valve Body
8	Bolt

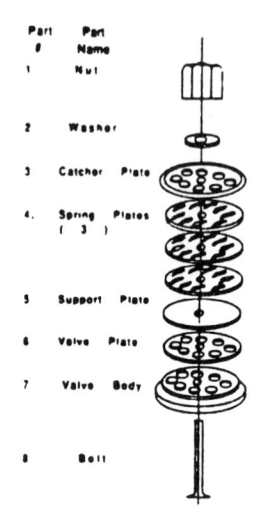

Figure 2. Exploded view of valve assembly

Marginal additional handling time	Priority	Part Number
1.4672	1	4
0.4848	2	1
0.4438	3	2
0.4108	4	7
0.1712	5	5
0.1542	6	3
0.1046	7	6
0.0581	8	8

Table 3. Prioritized parts list

References

1. Boothroyd G.,Dewhrust, P.,"Product Design For Manufacture and Assembly",Proc.3rd International Conference:Product Design For Manufacture & Assembly, June 6-8,1988, Newport.

2. Boothroyd,G., Dewhrust, P., *Design For Assembly Handbook*, Department of Mechanical Engineering, University of Massachusetts, Amhrest (1983).

3. Tanner, John, P., "Product Manufacturability", Automation, May 1989.

Selection of Acceptance Sampling Plans Through Knowledge Based Approach

S.S.N. MURTY and D. CHANDRA REDDY
Department of Mechanical Engineering
Indian Institute of Technology
Kharagpur, India

ABSTRACT

Quality is the preferred route to productivity. As an integral part of the quality system, the acceptance sampling plan is an important element in the overall approach to achieve desired quality. Expertise in quality control procedures and their applicability is needed in arriving at the best sampling scheme among the sampling schemes like AQL, AOQL, LTPD, α and β, and cost based sampling schemes, for the given manufacturing and inspection objectives. The development of the knowledge based approach for the selection of acceptance sampling scheme is described here. After a proper scheme is selected, the exact sampling plan is designed.

INTRODUCTION

The use of Statistical Quality Control (SQC) has grown rapidly in recent years. In majority of SQC software packages, interpretation of the results and reasoning is to be done by the user. Today manufacturing organisations have to meet customer needs and should make necessary changes easily and quickly by integrating the manufacturing process with total quality management. One of the recent research areas in quality control is the application of Artificial Intelligence (AI) techniques, especially Expert Systems, to computerize the process of reasoning and analysis.

It is believed that the quality of the purchased components and the training, ability and performance of the operators are the key variables influencing quality. Expert systems can preserve the expertise regarding quality control procedures and can be used to train operators. There are different application areas of expert systems which are useful in different problem domains. The application areas of expert systems relevant to quality control are monitoring, interpretation, diagnosis, prediction, planning and design. In acceptance sampling domain, application of expert systems it is yet to be explored. Acceptance sampling plans are used for taking a decision on a lot, either to accept or to reject on the basis of inspection results. As an integral part of the quality system, the acceptance sampling plan, applied on a lot by lot basis, becomes an important element in the overall approach to achieve the desired quality. There are different types of sampling schemes available (AOQL, LTPD, AQL, etc.,). In the present day manufacturing, the continuing strategy of selection, application, and modification of acceptance sampling schemes should comply with the changing environments of production and inspection. This situation necessitates a good system for selection of best sampling scheme and subsequently a sampling plan. Expertise in quality control procedures and their applicability is needed in arriving at the best sampling scheme. It is logical to anticipate that knowledge based expert systems will be of much help in assessing the situation and suggesting the proper sampling scheme[1]. The development and application of a rule based expert system is described for the selection of a suitable sampling scheme and plan.

The primary consideration of optimality in sampling plans is to minimize sample size. It is the absolute size of the sample, much more than its size relative to the lot, that governs discriminating capability of the sampling plan. Despite the existence of standards, and perhaps because of a perceived complexity of the standards, there has been a continuing interest in the design of sampling schemes to meet specific needs. **An important element in the selection of a sampling scheme should be the probable contribution of the scheme to quality improvement.** Sampling is preferred over 100 percent inspection if certain specific circumstances or conditions are present. Several criteria to be met for evaluation of a sampling scheme include the following:

a. Probability of acceptance of a lot of an acceptable quality, α
b. Probability of acceptance of a lot of rejectable quality, β
c. Average Outgoing Quality Limit, AOQL
d. Average Total Inspection, ATI
e. Discriminating power between lots of different qualities

Average Outgoing Quality Limit

In many instances, the rejection of a lot on the basis of sampling inspection results in 100% inspection of that particular lot. The accepted lots will contain certain fraction defective, although it will be slightly improved by the elimination of defectives found in the samples. For any plan, it is possible to compute the maximum possible value of the average per cent defective in the outgoing product for various incoming qualities. This maximum AOQ is referred to as the *average outgoing quality limit*, the AOQL.

Average Total Inspection

In analyzing and evaluating various sampling plans it is convenient to state the problem in terms of *Average Total Inspection* (ATI), and *Average Fraction Inspected* (AFI). The criterion that inspection effort be minimized assumes that the acceptance sampling scheme requires screening or rectifying of rejected lots. For an acceptance sampling plan ATI is a function of sample size and the number of rejected lots. The number of rejected lots is in turn is a function of the quality of the lots submitted for inspection. The formulae for computing ATI and AFI for a sampling plan are:

$$ATI = n + (1 - P_a)(N - n)$$
and $AFI = ATI/N$

All the sampling plans in the Dodge-Romig tables aim at minimizing ATI considering both sampling inspection and screening of rejected lots. One underlying assumption in computing the ATI for a particular acceptance sampling plan is that incoming quality does not vary. A second assumption regarding the above formula is that nonconforming items found in samples and rejected lots are replaced with good items.

SAMPLING SCHEMES CONSIDERED IN THE SYSTEM

Seven sampling schemes are considered in the present system. These are the following, which are discussed in detail.

α, β Sampling Scheme

The α, β risks are known as producer's risk and consumer's risk respectively. α, β risks can be satisfied at specified fraction defectives p_1 and p_2 respectively. Assuming that binomial distribution is appropriate, the sample size n and acceptance number c are solution for the equations

$$1 - \alpha = \sum_{d=0}^{c} \frac{n!}{d!(n-d)!} p_1 (1 - p_1)$$

$$\beta = \sum_{d=0}^{c} \frac{n!}{d!(n-d)!} p_2 (1 - p_2)$$

The two simultaneous equations are nonlinear, and there is no simple direct solution. Nomographs, tabular forms, and manual search procedures using statistical distribution tables are common techniques used for the solution. Each such method had limitations of accuracy, and were difficult and time consuming to use. Increasing in availability of computers led to the determination of exact sampling plan. If the decision is to favour a plan that minimizes the risk of accepting a false hypothesis (Type II error), the plan having the smaller acceptance number will be selected. If it is desired to minimize the risk of rejecting lots of good quality, the choice would be the plan having the larger acceptance number.

Sampling Schemes meeting LTPD and ATI Criteria

It is assumed that the lot is finite and the rejected lots are screened. It is also assumed that the minimization of ATI will be at some quality level such as AQL. Since it is not known that a sampling plan actually minimizes inspection costs, alternative plans having acceptance numbers of 0,1,2,3,4,5, and so on, will be evaluated. The optimal plan is the plan having minimum ATI. The procedural module for obtaining exact plan is developed.

Sampling Schemes meeting AOQL and ATI Criteria

All lots rejected by the plan will be screened. Lots being inspected are finite, and the ATI is minimized for a given apecified value of AOQL and quality p'. Formula for computing the AOQ for an acceptance sampling plan is

$$AOQ = P_a p' (N - n)/N$$

The process by which the desired sampling plan, having at the same time the specified AOQL and a minimum ATI, is to compute a series of possible plans by indexing the acceptance numbers. The sampling plan, meeting the dual criteria of AOQL and minimum ATI, should be selected. AOQL plans are used for bought out products and manufactured products. They can also be used in in-process inspection. The procedural module for arriving at the exact plan is developed.

AQL Sampling Scheme

The primary focus of this system is AQL (MIL-STD-105D) of the lots. The standard has a built-in system to encourage producers to submit products at specified AQL or better quality. The selection of sampling plan for given Inspection Level, AQL and batch size is computerized.

α-Optimal Sampling Scheme

The decision on each lot could be made based on 100% inspection, sampling inspection or no inspection. In sampling, it is reasonable to reject a lot when rejection is cheaper or to accept it when acceptance is cheaper. To achieve this goal, costs must be considered. The sampling plan in α-optimal sampling scheme is based on a linear cost function. A sampling plan (n,c) is determined, subject to a reasonable constraint, using the minimax principle on the cost function. A further assumption is that one knows the probability that the number of nonconforming items in a lot does not exceed a given number, i.e., the reliance parameter[2].

Least Cost Plans for Destructive Testing

In cases where the acceptance sampling procedure is used to control the strength of a material, a variable sampling plan will require destructive, time consuming and expensive testing of all the units in the sample. On the other hand, an attribute sampling plan designed to assure the same specification of minimal strength and providing equal risks, causes failure of nonconforming units only. Thus the remaining units in the sample are saved from destruction. This characteristic is common to equivalent compressed-limit gauging sampling plans.

In these cases, sampling plan with the lowest total cost should be selected. The total cost is comprised of the cost of sampling, cost of destructed units, cost of lot acceptance, and cost of lot rejection. The latter two components are determined by the selection of α, β risks. For given α, β risks, the least cost sampling will be selected which minimizes the cost of sampling and the cost of destructed units. As a logical consequence, sampling with relaxed-limit plans is to be considered. Here the increased cost of sampling might be compensated by the reduced number of units destructed[3].

Optimal Acceptance Plan using Prior Distribution

Because of the volume of receiving inspection decisions that are made over a period, it is important to develop and implement acceptance sampling plans which reduce the costs that are influenced by the quality control function. Such a determination requires the explicit incorporation of the costs of sampling and inspection, the costs of error in the decision and prior information regarding distribution of lot quality.

This situation is concerned with the determination of minimum cost sampling plans, when the prior distribution of lot quality and the sampling distribution are discrete. A two-stage optimization algorithm for selecting economic acceptance sampling plan, which converges on an optimal solution, is reported[4]. Three cost factors namely, inspection cost, repair cost and damage cost, are considered.

KNOWLEDGE BASED APPROACH FOR THE SAMPLING SYSTEM SELECTION

The selection of a suitable acceptance sampling scheme, for different operating and production environments and for different objectives, is accomplished. There are many conditions that affect the choice of an acceptance sampling scheme. The present system employs different object oriented modules which are activated through data driven control strategy. The system contains the modules for sampling schemes like AQL, AOQL, LTPD, α and β risks. Procedural modules for the determination of α and β risk sampling plans, AOQL plans and LTPD plans are developed in order to obtain exact sampling plans and schemes. The system also contains modules for minimum cost sampling plans with different criteria such as prior distribution of defectives, destructive testing and reliance parameter. Some of the factors considered in selecting and comparing the plans are: objectives of the inspection, desired sampling risks, destructive or nondestructive testing, availability of suppliers, costs of inspection and damage, type of production.

The system operates in three stages:

1. Selection of category of sampling plan: In this stage, it is determined using metarules whether variable acceptance sampling plan or attribute sampling plan is suitable for the given situation of process, supplier and inspection.

Example IF the inspection purpose is process standardization
         AND the supplier history is available
         AND the supplier history is good
         AND the distribution of defectives is known
         THEN variable acceptance sampling should be implemented

2. Selection of Samplig System: In this stage, the type of sampling system to be implemented is selected, depending on several parameters and objectives. Some of the objectives are minimum average total inspection, minimum cost inspection and sampling risks.

Example IF the available lot size is moderate and almost fixed
         AND the desired protection is on AOQ
         AND the inspection resources are limited
         THEN AOQL sampling scheme is applicable

         IF Acceptable Quality Level is less than 1%
         AND Rejectable Quality Level is greater than 5%
         AND Cost factors are not considered important
         THEN LTPD sampling scheme is applicable

3. Determination of exact sampling plan: At this stage, the exact sampling plan is determined either through an algorithm or through rules (depending on the sampling plan type). This stage determines the exact sample size, acceptance number and characteristics of the plan.

Features of the System

The system is written using 'C' programming language, using its own inference mechanism and implemented on Apollo DOAMIN 3500 Series. The knowledge is represented in the form of rules, as shown in the examples. A schematic representation of the system is shown in Fig. 1. A sample run of the system is shown in the appendix. The system has followig features:

- Calculation of AOQL, ATI, α, β and cost values for the given sample size and acceptance number
- Choice of distribution (i.e., Hypergeometric, Binomial and Poisson) for probability calculation
- User interface for Question-Answer session
- Graphical Output facility to display various characteristics of the scheme/plan

REFERENCES

1. Murty, S.S.N.; Reddy, D.C.: Expert Systems in Industrial Quality Control. (to appear in EQC, West Germany, 1990).
2. Elart von Collani: The α-Optimal Acceptance Sampling Scheme. J of Quality Tech. 18 (1986) 63-66.
3. Ladany, S.P.: Least Cost Acceptance Sampling Plans For Destructive Testing. J of Quality Tech. 7 (1975) 123-126.
4. Moskowitz, H.; Ravindran, A.; Patton, J.M.: An algorithm for selecting an optimal acceptance plan in quality control and auditing. Int. J of Production Res. 17 (1979) 581-594.

APPENDIX: SAMPLE RUN OF THE SYSTEM
(User response is in Italics and System response is in bold)

what is the purpose of inspection ?
   1. Process standardization   (prsd)
   2. Process verification   (prvr)
   3. To make decision about the product   (prde)
   4. To estimate the process quality of the supplier (prsp)
                           input your choice   : *prde*

what is the type of the production ?
   1. continuous   (cont)   2. continuous lots   (conl)
   3. isolated lots   (isol)   4. individual items   (indi)
                           input your choice   : *conl*

what is the category of the supplier ?
   1. new   (new)   2. existing   (old)
                           input your choice   : *new*

skipping the supplier history queries !

do you have any idea about the stability of the
        production process ? ('yes' or 'no')   : *no*

skipping the process stability queries !
what is the type of inspection [off-line(ofl)--on-line(onl)] ?
                                                                 ofl

Applicable category of sampling inspection : attribute

```
what is the type of the lot size available ?
    1. large              (lar)    2. moderate          (mod)
    3. small              (sma)    4. varying           (var)
                                   input your choice    : var

what is the type of the component ?
    1. standard           (std)    2. non-standard      (nst)
                                   input your choice    : std

what is the method of testing the component  ?
    1. Destructive testing      (des)
    2. Non-Destructive testing  (ndt)
                                   input your choice    : ndt

Any difficulty in obtaining the sample ? ('yes' or 'no')
                                   input your choice    : no

Are the cost factors important  ?  ('yes' or 'no')
                                   input your choice    : no

Is the consumer interested in certain sampling characteristics ?
('yes' or 'no')                    input your choice    : no

conditions satisfied in  rule number : 3
        implement MIL-STD-105D sampling system

what is the lot size available, AQL and RQL ?   10000,.025,.05

sample size = 200    acceptance number = 10
Probability of acceptance at AQL = 0.9874
LTPD value = 0.07600
AOQL value without replacement is 0.03241 at p = 0.041
AOQL value with    replacement is 0.03214 at p = 0.040
ATI  value without replacement is 323.210 at p = 0.025
ATI  value with    replacement is 331.500 at p = 0.025
```

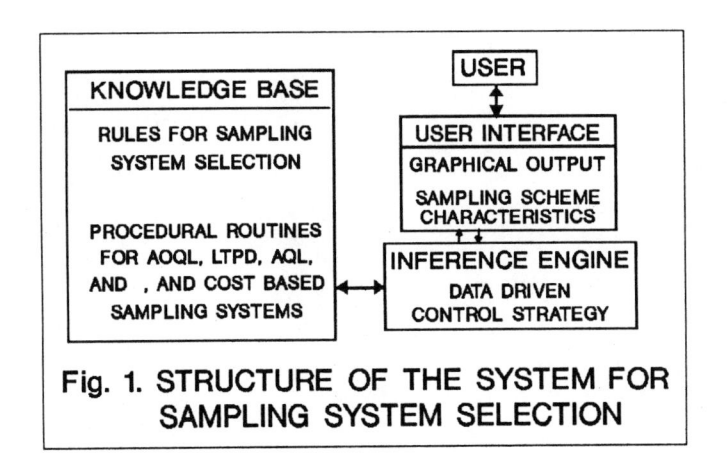

Fig. 1. STRUCTURE OF THE SYSTEM FOR SAMPLING SYSTEM SELECTION

Chapter XI

Cost Analysis Concept

Introduction

Improved quality and reduced costs are among the major objectives in implementing computer integrated manufacturing (CIM). In this chapter, different methods to improve quality as well as to estimate and reduce costs are described.

The first paper presents the quality-cost concept as an approach to identify, measure and reduce quality costs while improving quality within the CIM environment. The next paper describes the importance of cost modeling to concurrent engineering, which has resulted in the development of general forms of models for predicting cost using the improvement curve thoery. The final paper develops a time and cost estimation algorithm in BASIC which supplements the commercially available CAD/CAM software system called Pathtrace, using its database as a further step toward CAD/CAM intergration in the PC enviroment.

Analysis of Quality Costs: A Critical Element in CIM

RESIT UNAL

Engineering Management Department
Old Dominion University
Norfolk, VA

EDWIN B. DEAN

Cost Estimating Office
NASA
Langley Research Center
Hampton, VA

Abstract: This paper presents the quality cost concept as an approach to identify, measure and reduce quality costs while improving quality within the CIM environment. The effect of advanced failure prevention methodologies, such as continuous process improvement and the quality engineering methods of Taguchi, on quality and cost is discussed. Results indicate that continuously pursuing variability reduction and robust process design is the key to achieve high quality and reduce cost. CIM links robust design methods with computer aided design and provides the ideal environment for continuously improving quality and cost.

INTRODUCTION

Improved quality and reduced costs are among the major objectives in implementing computer integrated manufacturing (CIM). CIM requires the integration of design, materials, manufacturing and support for the purposes of reduced cost, reduced product development time, reduced product cycle time and improved quality throughout the life-cycle of the product.
Since improved quality and reduced costs are the major objectives, the development and implementation of quality cost measures, quality cost reports and their analysis for timely corrective action is an integral part of CIM. Once a quality cost program is implemented, it can be used by management and engineering to rapidly identify major problem areas, justify and support improvement, to guide work on reducing quality costs and to measure progress in that direction.

QUALITY COSTS

Quality costs include all costs incurred to assure and assess conformance with customer requirements and those costs associated with consequences of failure to meet the requirements. Quality costs are commonly categorized as prevention costs, appraisal costs and failure costs [7]. Prevention costs are those costs expended in an effort to prevent defects from occurring in the first place, such as the costs of improving process capability, quality engineering and quality training. Appraisal costs represent the cost of maintaining quality, such as the costs of inspection and testing. Failure costs are those costs expended due to a defective product throughout the life-cycle. During the design and development phase, cost of failure includes the cost of redesign and retest. During manufacturing, failure costs include the costs of scrap, waste and rework. In the field, the cost of failure includes the cost of warranties, costs expended by the customer when the product fails, and loss of customer goodwill. Total quality cost is the sum of these costs. Quality costs can be significant. It is estimated that about 25 percent of our nation's costs represent cost of failure [8]. Knowing the general relationship and interaction of quality costs can open many opportunity areas for cost and quality improvement.
The objective in quality cost analysis is to determine the basic relationships between quality costs to help identify major problem areas and continuously improve quality while reducing quality costs.

CLASSICAL COST OF QUALITY RELATIONSHIP

Figure-1 presents the classical relationship between quality costs and quality levels [3].

Figure-1; Classical Cost of Quality Relationship

The classical relationship between quality costs shows that prevention and failure costs rise asymptotically in opposite directions as defect free levels are achieved. An increase in prevention costs results in a larger decrease in failure costs, thereby reducing the total cost of quality until a minimum point is reached. Further increasing prevention costs beyond the optimum point can reduce defect levels but results in an increase in total cost of quality. Classical theory suggests that at the optimum point, prevention costs become saturated and no further dollars should be invested in prevention until a technological breakthrough is achieved through innovation that can shift the prevention cost curve [9].

The figure shows that the optimum lies below 100% conformance and it takes infinite investment to reach zero defects. In other words, Figure-1 suggests that high quality can only be achieved at much higher cost. However, this can be a misleading conclusion which suggests the acceptance of a certain level of defects which prevents quality and cost improvements beyond the optimum point. This is because the cost curves represent static relationships that are applicable to the existing approach to achieving high quality and the prevention methodologies being used. In reality, the costs curves are dynamic and can shift as knowledge in prevention methods change.

The approach commonly used to achieving high quality is to solve problems by building prototypes, testing them, and fixing the problems that are found by requiring expensive materials, costly components with tight tolerances and complex production processes. The result is a high quality but expensive product that may not be competitive.

However, many U.S and Japanese companies which are successfully using advanced prevention tools and new approaches, such as continuous process improvement and quality engineering methods of Taguchi, have reported that their quality costs have continued to fall with increasing quality levels [4]. In other words, attainment of increasing quality at lower total quality cost appears to be not only desirable but feasible [5].

Quality cost reductions due to the introduction of advanced prevention methods are quite drastic. Therefore, the classical cost of quality relationship needs to be modified to capture this change in prevention methods.

TAGUCHI ON QUALITY AND THE LOSS FUNCTION

The traditional approach to quality has been "Make it to specifications". A characteristic of a part which lies between the lower and upper blueprint specifications is considered to be good [1]. With this approach, manufacturing will have as its objective merely "zero defects", or producing parts that are within specifications. However, products that meet specifications also inflict a loss, a loss that is visible to the customer. A product that barely meets the specifications is, from the customer's viewpoint, as good or as bad as the product that is barely outside the specifications [11].

Taguchi proposes a different, more holistic, view of quality that relates quality to cost and loss in dollars, not just to the manufacturer at the time of production, but to the customer, and in varying degrees, to the society as a whole [11]. Taguchi defines quality as, "The quality of a product is the (minimum) loss imparted by the product to the society from the time product is shipped" [2].

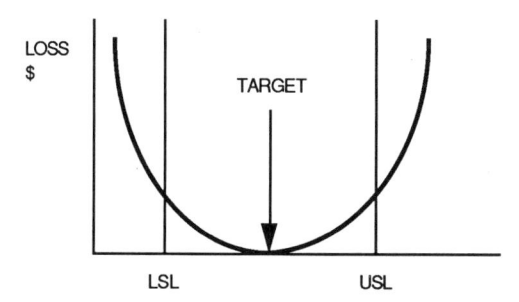

Figure-2; The Quadratic Loss Function

According to Taguchi [2], the quadratic function models this loss in most situations. The quadratic loss function can be written in the form;

$$L = K(X-T)^2 \quad (1)$$

where L represents loss in dollars, X is the quality characteristic value, T is the target value and K is the parabolic constant whose value can be determined if L is known for any particular X value.

Figure-2 illustrates the loss function and how it relates to the specification limits. When a critical quality characteristic deviates from the target value, it causes a loss. Quality, then, means no variability or very little on target performance. The loss function clearly shows that variability reduction or quality improvement, drives costs down. Lowest cost can only be achieved at zero variability. Therefore, continuously pursuing variability reduction from the target value in critical quality characteristics is the key to achieve high quality and reduce cost.

ACHIEVING VARIABILITY REDUCTION - PARAMETER DESIGN

Taguchi states that product design has the greatest impact on quality [12]. His quality engineering methods seek to design a product or process which is insensitive or robust to causes of quality problems. The three steps of quality by design are, system design, parameter design and tolerance design [11]. System design combines the innovation and specialist knowledge of the design engineer to develop a product design that meets functional requirements.

After the system architecture is decided on, the next step is to select the optimum levels for the controllable system parameters such that the product is functional, exhibits a high level of performance under a wide range of conditions, and is insensitive to "noise" factors. Noise factors are those that can not be controlled or are too expensive to control. Control factors are those that can be set and maintained. Studying these variables one at a time or by trial and error is the common approach to design optimization. However, this leads to either very long and expensive time span for completing the design or premature termination of the design process such that the product design is far from optimal.

Parameter design, sometimes called robust design, is Taguchi's approach for design optimization [11]. Parameter design methods provide the design engineer a systematic and efficient approach for determining the optimum configuration of design parameters for performance and cost. The objective is to select the best combination of controllable design parameters such that the product is most robust against noise factors in manufacturing and customer operation. Parameter design methodology uses orthogonal arrays from design of experiments theory to study a large number of decision variables with a significantly small number of experiments. It also uses a new statistical measure of performance called signal-to-noise ratio (S/N) from electrical control theory to evaluate the quality of the product. The S/N ratio measures the level of performance and the effect of noise factors on performance.

Thus, the most economical product and process design can be accomplished at the smallest development cost. Parameter design methodology reduces R&D costs by improving the efficiency of generating information needed to design systems insensitive to usage conditions, manufacturing variation and deterioration of parts. Furthermore, the optimum choice of parameters can result in wider tolerances so that low cost components and production processes can be used. As a result, manufacturing and failure costs are also greatly reduced. Typically, no manufacturing cost increase is associated with parameter design [11].

The third step, tolerance design, is only required if parameter design can not produce the required performance without special components or high process accuracy. It involves tightening tolerances on parameters where their variability could have a large negative effect on the final product. Typically tightening tolerances leads to higher cost [11].

MODERN COST OF QUALITY RELATIONSHIP

We had seen that adopting Taguchi's view of quality and using parameter design in pursuit of continuous variability reduction can lead to significant improvements in quality and cost. Most American and European engineers focus on system and tolerance design to achieve performance, overlooking parameter design which finds ways of modifying the design to gain high quality at low cost [4]. As a result the opportunity to improve quality without increasing cost is missed. Recently, however, the use of Taguchi's quality engineering methods have been increasing in the U.S. It is expected that the application of the methods will become widespread in the coming decade [11]. U.S. companies that have used continuous improvement and quality engineering methods of Taguchi in reducing variability, have achieved significant improvements in quality and cost [10].

The experiences of many companies adopting this new approach to quality suggest a new form of the cost of quality relationship (Figure-3).

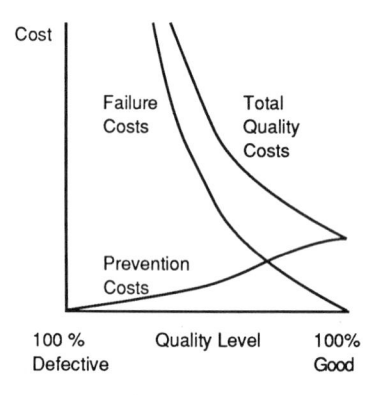

Figure-3; Modern Cost of Quality Relationship

Figure-3 suggests that prevention costs may level off as defect free level are reached. Unlike technological breakthroughs and innovation, implementation of advanced prevention tools, such as parameter design and continuous process improvement, does not introduce increased prevention costs as the quality level improves. Thus, total cost of quality may decrease as higher quality levels are reached.

USING THE LOSS FUNCTION FOR QUALITY AND COST IMPROVEMENT IN CIM

To be competitive in a global marketplace, designs must be optimized to improve both quality and cost. Implementing the Taguchi loss function within CIM links quality and cost as it transforms the quality approach from "zero defects" to "zero variability".

Taguchi's quadratic loss function can be effectively used to assist management, engineering, and production, to evaluate cost and efficiency, to prioritize those quality characteristics which need the most control during manufacturing [6], and to justify investment in quality [12].

CIM provides the ideal environment for application of the loss function. Linking control parameters and parameter design methods with computer aided design capability resident within CIM permits true concurrent engineering and leads to continuous product improvement. Quality cost data and quality cost trend reports will be immediately available to management from CIM databases, so that remedial action can be taken earlier.

REFERENCES

1. American Supplier Institute Inc, 1989, "Taguchi Methods: Implementation Manual", ASI, Dearborn, Michigan

2. Bryne, D., M. and Taguchi, S., 1986, "The Taguchi Approach to Parameter Design", ASQC Quality Congress Transactions, Anaheim, CA, p 168.

3. Campanella, J. and Corcoran, F. J. 1983. "Principals of Quality Costs", ASQC Quality Progress, XVI (4), pp17-22.

4. Cullen J. and Hollingum, J. 1987. Implementing Total Quality, Springer-Verlag, New York, N.Y.

5. Dawes, E. W., 1989, "Quality Costs- New Concepts and Methods", Quality Costs: Ideas and Applications, Campanella, J., Editor, ASQC Quality Press, Milwaukee, Wisconsin, pp 440-448.

6. Di Lorenzo, R. D. 1990. "The Monetary Loss Function - or Why We need TQM", Proceedings of the International Society of Parametric Analysts,12th Annual Conference, San Diego, CA, pp 16-24.

7. Feigenbaum, A. V. 1983. Total Quality Control, third edition, McGraw-Hill, New York, N. Y.

8. House Republican Research Committee Task Force on High Technology and Competitiveness, 1988. "Quality as a Means to Improving Our Nation's Competitiveness".

9. Morse, W. J., and Poston, K., 1989, "Accounting for Quality Costs- A Critical Element in CIM", Quality Costs: Ideas and Applications, Campanella, J., Editor, ASQC Quality Press, Milwaukee, Wisconsin, pp 400-408.

10. Sullivan, L. P. 1987. "The Power of Taguchi Methods", Quality Progress, June, pp 76-79.

11. Phadke, S. M. 1989. Quality Engineering Using Robust Design, Prentice Hall, Englewood Cliffs, N.J.

12. Taguchi, G., Elsayed, E. and Hsiang, T. 1989. Quality Engineering in Production Systems, McGraw Hill, New York, N.Y.

A Databased Time and Cost Estimation Algorithm for Piece Part Design and Manufacturing

K.W.-N. LAU and M. RAMULU

Department of Mechanical Engineering
University of Washington
Seattle, WA

SUMMARY

To get a closer step for CAD/CAM integration in PC-based workstation, a timecost algorithm is developed to supplement commercially available pathtrace system, utilizing its existing database. The part-design-timing routine is just a simple program working together with DOS batch files that mark the starting and stopping times of using pathtrace CAD/CAM system. It shows and stores the time a user spent on the computer using the pathtrace software for designing part geometries and planning the machining processes up to the final accomplishment. The stored designing time can be used in the cost estimation algorithm. Time-cost estimation algorithm determine the specific machining time and production cost for each machining process of part manufacture by retrieving appropriate database files. Utility and usefulness of this algorithm demonstrated with practical examples.

INTRODUCTION

The CAD/CAM integration efforts developing today, is to use the generic databases (e.g., geometry and machining etc.) for to the subsequent production process like process planning, inventory control, and cost estimation [1-3]. The significant advantages of PC-based CAD/CAM systems over the time-shared approach are: users are not affected by each other; faster system response time; simpler softwares used; much lower hardware and software costs; higher security, reliability and controllability as each user gets separate workstations; efficient data transfer through communication networks. These low-cost systems are generally developed on multitasking microprocessors or PCs using standard operating systems. They now make CAD/CAM more affordable to a broader base of small companies. Networking is often used to allow integration and provide for a centralized archive and backup system. In this paper, a Time and Cost Estimation algorithm in BASIC is developed to supplement the commercially available CAD/CAM software system called Pathtrace using its database for a further step towards CAD/CAM integration in the PC environment. Moreover, a Part-Design-Timing program is also written to provide additional data information for the Time and Cost Estimating program.

BRIEF DESCRIPTION OF PATHTRACE CAD/CAM SYSTEM

Pathtrace is a commercially available CAD/CAM software package for machining processes using CNC machine tools [4,5]. The control of the system is achieved through a dialogue language. With a dual screen system the CPU monitor is used solely for text, and the graphics monitor reserved for the drawing being created. A cursor, which is used to identify locations, is also displayed on the drawing area and is moved around by use of the graphics input device, a mouse. Figure 1 shows the structure of the PC-based CAD/CAM system. Functions of the system are: geometric part design; machining processes planning and design, plotting and modification of the design; CNC part programming generation using its accompanied NC postprocessor software; and direct interface with CNC machine through an external communication port, RS232.

Database Structure

Each type of data file in the Pathtrace system is identified by a prefix which is automatically added by the system as shown in Table 1. MS-DOS allows only eight characters for each file name. Each file created within Pathtrace will have a Pathtrace name and an MS-DOS name. All the Pathtrace data files are stored under the CAMDATA MS-DOS directory. The first four letters of a Pathtrace working directory name will be stored with extension .SEQ in the CAMDATA DOS directory as a file. Under the Pathtrace working directory, a number of data files with typical file types can be created with any name (as the Pathtrace data file name). The created data file will then be stored under the name beginning with the first four letters of its working directory name, together with a counter number according to the sequence the file is created. Similarly, an extension of .SEQ is also added to the data file name as distinguished as a DOS file in the CAMDATA DOS directory. Figure 2A and Figure 2B highlight an example of the pathtrace working directory files. Hence, by examining the Pathtrace working directory file in the CAMDATA MS-DOS directory, one could find out an appropriate Pathtrace data file name, and the DOS file name that the Pathtrace database is stored.

TIME AND COST ESTIMATION IN MACHINING A PIECE PART

The Part-Design-Timing Routine is just a simple program written in Basic working together with a DOS batch file that marks the starting and stopping times of using the Pathtrace CAD/CAM system. Figure 3 shows the flow chart of design-time algorithm. It shows and stores the time a user spent on the computer using the Pathtrace software for designing part geometries and planning the machining processes up to the final accomplishment. The stored designing time can be used in the time and cost estimation routine when this designing time is taken into account. The time-cost program is to determine the specific machining time and production cost for each machining process of a piece part manufacture by retrieving appropriate database files as discussed in the previous section. Figure 4 shows the flow chart of the time-cost estimation algorithm. First of all, it will show a table of operational costs. Any change could be made where applicable before proceeding to the main function. After then, the user will be prompted to provide the correct Pathtrace working directory name and machining file name. It will then search for the correct data files and parameter values. If there is any error in locating any file, the program will prompt an error message and it can be redone over again or simply stopped. If no or inappropriate parameter value is encountered, the program will ask the user to enter the required information. Once data files are located and parameter values are determined, the program will calculate the time needed for a specific machining operation (e.g., profiling cycle and drilling cycle, etc.) simply by dividing the tool path length with the feedrate. The cost of this specific operation will be calculated accordingly. Then the time and cost computations for the next machining operation will be carried out, and the process will continue until there is no more machining operation detected. Then the program will display the whole process planning with the machining time and cost computed. Finally, a total time (which may include the part designing time) and a total cost will be summarized. Nonetheless, the time and cost estimation summary together with the process planning could be stored in the computer, and a hardcopy could also be generated.

EXAMPLES

Three typical parts, namely, fluidity spiral, rod yoke and gear are chosen because of its geometric complexity and the machining operations involved in its final production. A PC-based CAD/CAM system, which includes the Pathtrace software and the supplemented time and cost estimation algorithm as presented in the previous section is interfaced with a retrofitted milling system using the FANUC System 3M-Model A. The parts selected are chosen to demonstrate the capability and efficiency of the entire system.

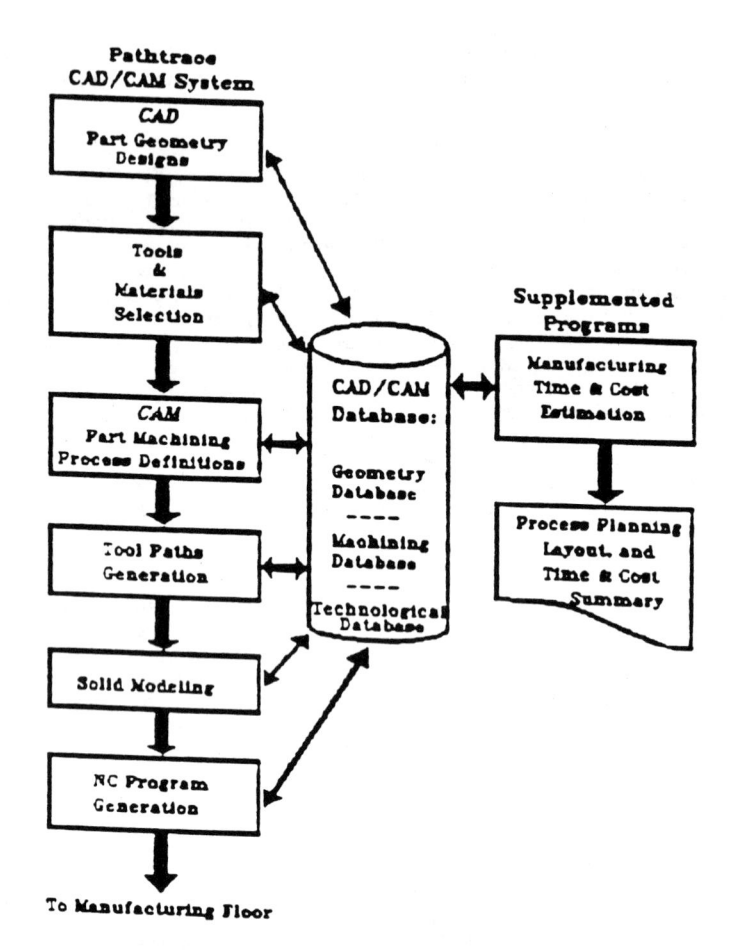

Figure 1 Structure of the PC-Based CAD/CAM System.

File Type	Prefix	Example
Geometry	None	TEST
Geometry Instructions	-G-	-G-TEST
Milling Instructions	-M-	-M-TEST
Turning Instructions	-T-	-T-TEST
Parametric Files	-X-	-X-TEST
CNC Program	$	$TEST

Table 1 Pathtrace Data File Structure.

Volume in drive C is GENESIS III
Directory of C:\CAMDATA

TEST	SEQ	TEST1	SEQ	TEST2	SEQ	TEST3	SEQ
TEST4	SEQ	TEST5	SEQ	TEST6	SEQ	TEST7	SEQ
TEST8	SEQ	TEST9	SEQ	TEST10	SEQ	TEST11	SEQ
TEST12	SEQ	TEST13	SEQ	TEST14	SEQ	TEST15	SEQ
TEST16	SEQ	TEST17	SEQ	TEST18	SEQ	TEST19	SEQ
TEST20	SEQ	TEST21	SEQ	TEST22	SEQ		

23 File(s)   13545472 bytes free

Figure 2A Illustration of the Pathtrace Data Files Under the MS-DOS CAMDATA Directory. (Here shows those DOS files named after the Pathtrace working directory called TEST.)

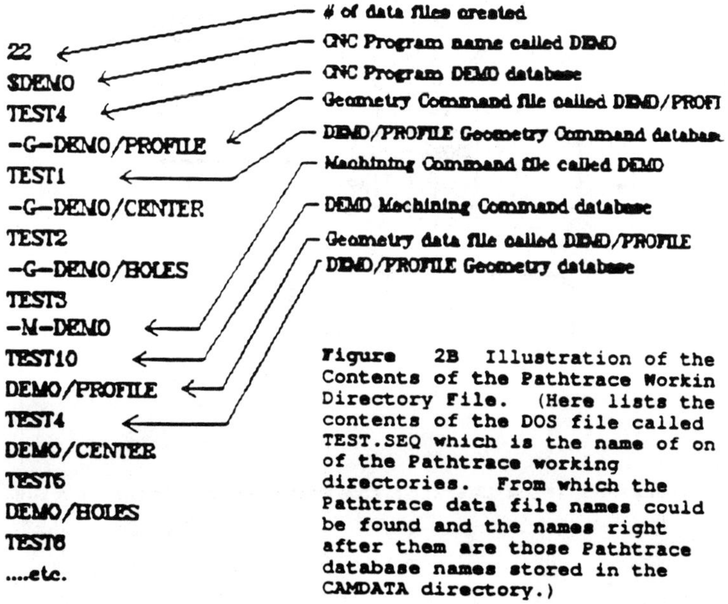

```
                          — # of data files created
                          — CNC Program name called DEMO
22  ←
$DEMO  ←                  — CNC Program DEMO database
TEST4  ←                  — Geometry Command file called DEMO/PROFI
-G-DEMO/PROFILE  ←        — DEMO/PROFILE Geometry Command database
TEST1                     — Machining Command file called DEMO
-G-DEMO/CENTER            — DEMO Machining Command database
TEST2
-G-DEMO/HOLES             — Geometry data file called DEMO/PROFILE
TEST3                     / DEMO/PROFILE Geometry database
-M-DEMO  ←
TEST10  ←
DEMO/PROFILE  ←
TEST4  ←
DEMO/CENTER
TEST5
DEMO/HOLES
TEST6
....etc.
```

Figure 2B Illustration of the Contents of the Pathtrace Working Directory File. (Here lists the contents of the DOS file called TEST.SEQ which is the name of one of the Pathtrace working directories. From which the Pathtrace data file names could be found and the names right after them are those Pathtrace database names stored in the CAMDATA directory.)

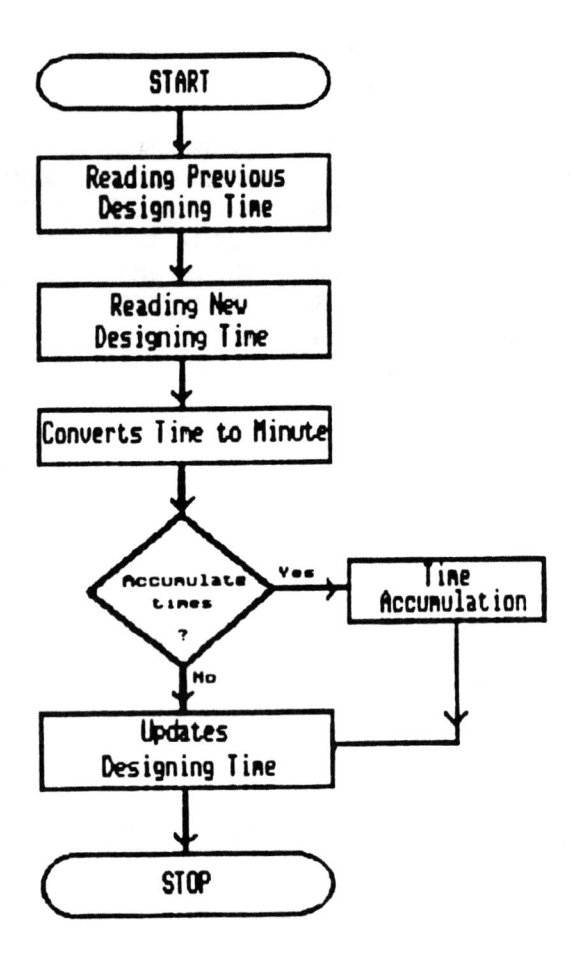

Figure 3 Flow Chart of the Part-Design-Timing Routine.

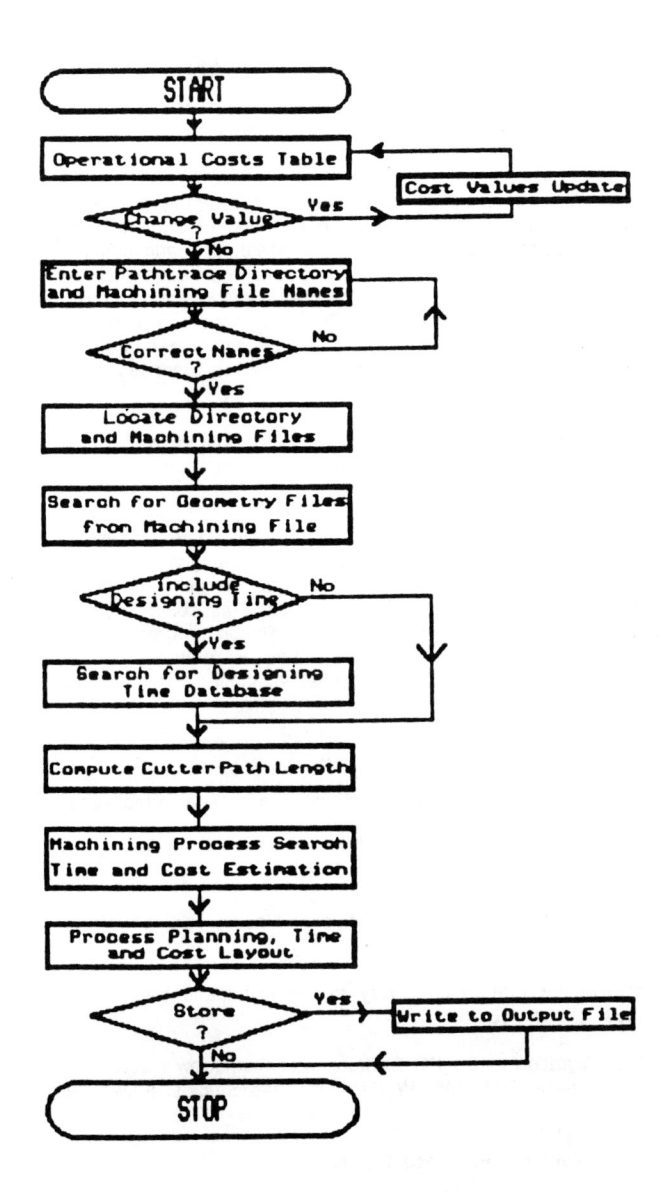

Figure 4 Flow Chart of the Time and Cost Estimation Routine.

After the parts has been designed and machining operations have been defined, some geometry databases and machining databases were created. By running the Pathtrace solid modeling routine visual verification of machine parts were made through 3D display, a CNC program was then generated and was downloaded to the CNC machine and executed.

Then the time and cost estimation program was invoked to compute the machining time (as well as the part designing time, if required), and the cost of manufacturing the part per piece. Table 2 shows typical operational costs for particular machining process, and is displayed on the computer screen once the time and cost estimation program is called upon. These cost values are used for computing the total manufacturing cost of a part. Finally, a summary output, which includes the process planning for manufacturing the typical part fluidity spiral pattern, time and cost for each machining operation, and total manufacturing time and cost, is generated. Table 3 shows the summary outputs of the fluidity spiral.

Table 4 lists the comparison among the machining time as computed by the time and cost estimation algorithm, the Pathtrace generated time, and the time recorded at real time manufacture in the CNC machine. The real machining time was recorded without taking the tool change time into account. On the contrary, the computed machining time in the Pathtrace CNC program included the tool change time. In fact, it is actually the time for a fully automated CNC system which has an automatic tool changes installed. The time and cost estimation program, on the other hand, provides the flexibility of inputing the actual tool change time; and it was set to zero in this case as comparing with the real machining time. Results show that the time and cost estimation program can give a very good approximation of the machining time. Its value is deviated only about ±5 percent,and this algorithm gives a good time and cost estimation for the shop floor and also the management references.

CONCLUSIONS

A databased time-cost estimation algorithm is developed. The effectiveness of the algorithm is demonstrated with examples.

REFERENCES

1. C.M. Foundyller, "CAD/CAM Systems in Transition" Machine Design, March 8 (1984) 199-203.

2. D.E. Hegland, "CAD/CAM Integration - Key to the Automatic Factory" Production Engineering, August (1981) 30-35.

3. R.T. Bannon, "CAD Migration to the PC Environment - Tomorrow's Low Cost Workstations" Advances in CAD/CAM Workstation Case Studies, Kluwer Acad. Publ. 1986 25-32.

4. Pathtrace Geometry Mannual, Pathtrace Limited, 1987.

5. Pathtrace Milling/Turning mannual, Pathtrace Limited, 1987.

Table 2 Operational Costs for Cost Estimation.

```
LIST OF OPERATIONAL COSTS ($ per min.)
****************************************
 1. FACE MILLING COST  : ..........  $   0.45
 2. PROFILING COST     : ..........  $   1.30
 3. AREA CLEARANCE COST : ........  $   1.50
 4. CROSS-HATCH AREA CLEAR. COST :  $   1.50
 5. SLOT PROFILING COST : ........  $   0.75
 6. DRILL CYCLE COST   : ..........  $   0.50
 7. CHIP BREAK CYCLE COST : ......  $   1.00
 8. REAM/BORE CYCLE COST : .......  $   1.00
 9. BORE CYCLE COST    : ..........  $   1.11
10. TAPPING CYCLE COST : ..........  $   0.12
11. TOOL COST (per piece) : ......  $  15.00
12. LABOR COST         : ..........  $  20.00
13. MATERIAL COST (per piece) : ..  $  15.00
14. DESIGN COST        : ..........  $  10.00
15. OTHERS             : ..........  $   9.20
```

Table 3 Summary Output of Fluidity Spiral Manufacturing.

```
**********************************************
** POST-PATHTRACE PROCESS LISTING & COST ESTIMATION **
**********************************************

** Working Directory is ....  TRUE
** Machining Filename is ...  -M-SPIRAL
** Group <A> is Geometry File SPIRAL/BILLET1
** Group <B> is Geometry File SPIRAL/BILLET2
** Group <C> is Geometry File SPIRAL/BILLET3
** Group <D> is Geometry File SPIRAL/BILLET4
** Group <E> is Geometry File SPIRAL/CENTER
** Group <F> is Geometry File SPIRAL/HOLE
** Group <G> is Geometry File SPIRAL1
** Group <H> is Geometry File SPIRAL2

** Dimensional Unit is ....  IMPERIAL
** Material chosen is .....  ALUMINUM

 GROUP                                    MACHINING
++TYPE++   ++PROCESS++      ++FEEDRATE++  ++TIME (min)++  ++COST($)++
-----------------------------------------------------------------
           TOOL CHANGE -- # 132 , DIAMETER .5                15.00
   A       AREA CLEARANCE CYCLE    16.5    1.538408          2.31
   B       AREA CLEARANCE CYCLE    16.5    1.43373           2.15
   C       AREA CLEARANCE CYCLE    16.5    1.468413          2.20
   D       AREA CLEARANCE CYCLE    16.5    2.197094          3.30
   E       AREA CLEARANCE CYCLE    16.5    .4012954          0.60
   F       AREA CLEARANCE CYCLE    16.5    5.757576E-02      0.09
   G       PROFILING CYCLE         16.5    3.094617          4.02
   H       PROFILING CYCLE         16.5    3.425666          4.45
   G       PROFILING CYCLE         16.5    3.094617          4.02
   H       PROFILING CYCLE         16.5    3.425666          4.45
           FEED MOVE CYCLE         16.5    2.540583          3.30
-----------------------------------------------------------------

TOTAL TIME USED ONLY IN MACHINING PART IS :    22.928 (min.)

TOTAL COST (excluding Designing Cost) IS :    $291.83   per piece.
```

Table 4 Machining Time Comparison of Parts Manufacture.

PART MANUFACTURE	MACHINING TIME AT PRE-DEFINED FEEDRATE (min.)		
	TIME & COST ESTIMATION PROGRAM	PATHTRACE CNC PROGRAM	REAL TIME CUTTING AT 100% FEEDRATE
Fluidity Spiral	22.9280	23.9712	24.1218
Rod Yoke	29.2168	30.8627	30.2023
Gear	1.8818	1.0291	1.1290

Improvement Curves in Manufacturing

R.C. CREESE and MADHU SUDHAN

Industrial Engineering Department
West Virginia University
Morgantown, WV

ABSTRACT

This study grew out of the importance of cost modeling to concurrent engineering (cooperative product development) and has resulted in the development of general forms of models for predicting the average unit cost, unit cost and cumulative cost of products. The cost predicting equations were derived from the unit curve model, which assumes that the unit time versus the production units generate a straight line on logarithmic paper. The unique contribution here is that the effect of multiple improvement curves on time saving were studied and compared with improvement curves with single improvement rate and curves with no improvement. This is applicable for both decreasing and increasing improvement rates as well as considering multiple improvement curves. Another important application of the improvement curve is to predict the loss of improvement during manufacturing interruption when long time periods of no production occur. These new cost prediction tools were introduced to help cost analysts practice concurrent engineering in a general manufacturing environment. A comparison of the differences between the unit cost model and average unit cost model will be made.

INTRODUCTION

Improvement curve theory is a tool to estimate cost, which allows accurate projections of future production costs based on relevant data. This technique, most popular in the aircraft industry, can also be used in industries where volumes are limited and a variety of products are made in a job-shop environment. Towards this end an improvement curve model is being developed to allow a user to accurately predict the time or eventual cost. Other terms for 'Improvement Curve' are experience curve, learning curve, manufacturing progress function, cost quality relationship, cost curve, efficiency curve, production acceleration curve, and performance curve.

The improvement curve concept was first developed by T. P. Wright in 1936. The observation Wright made was that as the number of units produced doubled, the average unit time to produce the units decreased at a specific rate. The rate of improvement that occurs is specific to the manufacturing process being considered [1]. In a 1944 investigation, Crawford, noticed in other cases as the number of units doubled, the unit time to produce additional units decreased at a specific rate.

Improvement curves can help in estimating future production costs and forecast labor time. This concept can be readily applied to industries were substantial manufacturing progress is expected [2,3], such as in the computer, heavy machinery, and shipbuilding industries. The improvement or learning effect has also been noted in the production planning models [4] and several industries have recently conducted experiments leading to varied and more specialized application of improvement curve theory [5].

The improvement curve is one of several strategic planning tools used by manufacturers to plan for the future. Many firms use improvement curves as a planning tool to control costs and predict prices. Some important factors that affect the rate of improvement in a manufacturing environment [7] other than personnel or human learning are as follows:
1. Improved or advanced product design.   2. Improved processes and tooling.
3. Improved manufacturing methods.   4. Incentives to employees.
5. Increased availability of raw materials and equipment.

UNIT TIME ANALYSIS (Crawford Approach):
Unit Time:

The improvement curve can be represented in the form of an equation as follows:

$$Y(u) = a N^b \tag{1}$$

where, $Y(u)$ = unit time for N units (hours)
   a = first unit time (hours) I = improvement rate as a percent
   N = specific production level of concern L = learning rate = 100-I
   b = exponent for the curve; $b = \dfrac{\log[(100-I)/100]}{\log(2)}$

The equations for multiple improvement rates can be derived in a similar fashion as follows, and is illustrated using three improvement rates.

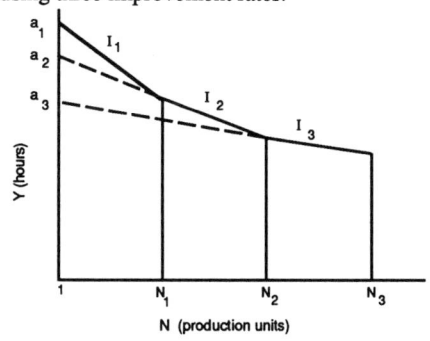

Figure 1. Improvement Curve with Three Learning Rates on Log-Log Scale.

Figure 1 illustrates an improvement curve with three learning rates I_1, I_2, and I_3, where $I_1 > I_2 > I_3$. The development of the expression for Y in terms of the slopes of the improvement curve and the ranges in which the particular improvement rates are applicable follows:

let, I_1 = improvement rate for the range 1 to N_1 $\quad$ a_1, a_2, a_3 = intercepts of the improvement curves
I_2 = improvement rate for the range N_1 to N_2 $\quad$ b_1, b_2, b_3 = slopes of the improvement curves
I_3 = improvement rate for the range N_2 to N_3

Considering the improvement curve with three improvement rates, one obtains,

$$Y(u) = a_1 N_1^{b_1-b_2} N_2^{b_2-b_3} N^{b_3} = a_3 N_3^{b_3} \qquad (2)$$

This allows the determination of the unit time Y for the Nth unit of production in terms of the first production unit time a_1, the slopes of the improvement curves b_1, b_2, and b_3, and the production limits for the improvement rates N_1 and N_2. These models have assumed non-decreasing learning rates as the production quantity increases and no step changes in the unit production times.

In a few instances, such as the aircraft industry, improvement may decrease as the production quantity increases. In this case, there is low improvement during the production of the first few items and greater improvement takes place in the second stage and then reduces as the production quantity increases in the third stage.

CUMULATIVE TIME:

The unit cost improvement curve can be used to calculate the cumulative unit cost which will be discussed in this section. To calculate the total cost or the cumulative unit cost of N units, it is simply the sum of all Y(u)'s from equation (1). The sum of all Y(u)'s can be represented as shown in Figure 2.

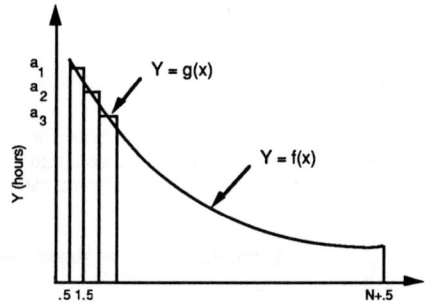

Figure 2. Cumulative Unit Cost Assumed as a Continuous Summation.

From Figure 2 cumulative unit cost can be assumed to be a continuous summation. Since $\int_{0.5}^{N+0.5} g(x)\,dx = a_x$, the partial sum $\Sigma\, a_x = \int g(x)\,dx$. We may now compare f(x) with the step function g(x). Therefore, the cumulative unit cost can be written in the general form as:

$$Y(c) = a \sum X^b = a \int_{0.5}^{N+0.5} X^b\, dN = a \left[\frac{(N+.5)^{b+1}}{b+1} - \frac{(0.5)^{b+1}}{b+1} \right] \qquad (3)$$

where,
    Y(c) = time for cumulative units   b = exponent for the curve
    a = first unit time (hours)   N = Specific production level of concern

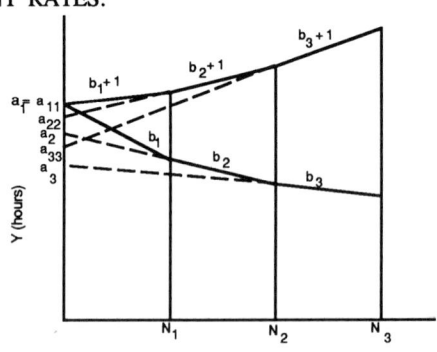

THREE IMPROVEMENT RATES:

Figure 3. Improvement Curve with Cumulative Time in Log Scale.

Let us consider the improvement curve in Figure 3 with three improvement rates I1, I2, I3, where I1 > I2 > I3. The development of the expression for Y in terms of the slopes of the curve and the ranges in which they are applicable are shown below.
let,
I_1 = improvement rate for the range 1 to N a_{11}, a_{22}, a_{33} = intercepts of the cumulative curves
I_2 = improvement rate for the range N_1 to N_2 b_1, b_2, b_3 = slopes of the improvement curves
I_3 = improvement rate for the range N_2 to N_3 a1,a2,a3 = intercept of the unit improvement curve

Considering the improvement curve with three improvement rates, we get,

$$Y(c) = \frac{a1}{b1+1}\left[(N_1+.5)^{b1+1} - (.5)^{b1+1}\right] + \frac{a2}{b2+1}\left[(N_2+.5)^{b2+1} - (N_1+.5)^{b2+1}\right]$$
$$+ \frac{a3}{b3+1}\left[(N_3+.5)^{b3+1} - (N_2+.5)^{b3+1}\right] \qquad (4)$$

where,
$$a2 = a_1 N_1^{b_1-b_2} \;;\; a3 = a_2 N_2^{b_2-b_3}$$

This allows the determination of the cumulative unit time for the Nth (unit of concern) in terms of the intercepts of both unit and cumulative improvement curves, the slopes of the improvement curves, and the production range for the improvement rates.

Average Unit Time:
Average unit time is calculated from cumulative time as shown below,

$$Y(a) = a\left[\frac{(N+.5)^{b+1}}{b+1} - \frac{(0.5)^{b+1}}{b+1}\right] / N \qquad (5)$$

where,
  Y(a) = average unit time   N = unit of concern
  a = first unit   b = slope of curve

An example in Table 1 illustrates the use and effect of improvement curves. The effect of individual improvement rate as well as the combined improvement rates is illustrated. It shows the effect of improvement at 25th unit for 0%, 5%, 10%, 20% and a combined rate of 20-10-5%, 10-20-5% in the cumulative, average and unit time basis.

Table 1 Improvement at 25th unit for the Unit Time Analysis.

Improvement rate Unit	Average Unit Time (hrs)	Unit Time (hrs)	Cumulative Time (hrs)	Average Unit Savings (hrs)	Unit (hrs) Savings	Cumulative Savings (hrs)
0%	100.00	100.00	2500	00.00	00.00	0
5%	84.40	78.80	2110	15.60	21.20	390
10%	70.88	61.31	1772	29.12	38.69	728
20%	49.36	35.48	1234	50.64	64.52	1266
20-10-5%	61.20	54.63	1530	38.80	45.37	970
10-20-5%	62.48	53.66	1562	37.52	46.34	938

AVERAGE UNIT TIME (Wright Approach):

If equation (1) represents the average unit time, instead of the unit time, then equation (7), (8), (9) as represented in Table 2 will result for the average unit time, cumulative unit time, and unit time.

TABLE 2.   SUMMARY OF TIME PREDICTION EQUATIONS
UNIT TIME ANALYSIS   Equation
Unit Time $Y(u) = a_1 N_1^{b1-b2} N_2^{b2-b3} N^{b3}$ (2)
Cumulative Unit Time $Y(c) = \frac{a1}{b1+1}[(N_1+.5)^{b1+1} - (.5)^{b1+1}]$
$+ \frac{a2}{b2+1}[(N_2+.5)^{b2+1} - (N_1+.5)^{b2+1}] + \frac{a3}{b3+1}[(N_3+.5)^{b3+1} - (N_2+.5)^{b3+1}]$ (4)
Average Unit Time $Y(a) = a\left[\frac{(N+.5)^{b+1}}{b+1} - \frac{(0.5)^{b+1}}{b+1}\right] / N$ (5)
AVERAGE UNIT TIME ANALYSIS
Average Unit Time $Y(a) = a_1 N_1^{b1-b2} N_2^{b2-b3} N^{b3}$ (7)
Cumulative Unit Time $Y(c) = a_1 N^{b_1+1}$ (8)
Unit Time $Y(u) = T_n - T_{n-1} = a[N^{b+1} - (N-1)^{b+1}]$ (9)

An example which illustrates the use and the effect of the Average Unit Curve is presented in Table 3. Here one assumes that average unit time versus production units generates a straight line on lograthimic scale. The effect of individual improvement rates as well as the combined improvement rate is illustrated The Multiple improvement of 20%, 10% and 5% occurs in the range 1-3, 3-10,10-30 production units respectively. Table 2 shows the effect of improvement at 25th unit for 0%, 5%, 10%, 20% and a combined rate of 20-10-5 %, 10-20-5% improvement.

Table 3 Improvement at 25th Unit for Average Unit Time Analysis.

Improvement rate Unit	Average Unit Time (hrs)	Unit Time (hrs)	Cumulative Time (hrs)	Average Unit Savings (hrs)	Unit (hrs) Savings	Cumulative Savings (hrs)
0%	100.00	100.00	2500	00.00	00.00	0
5%	78.80	73.10	1970	21.20	26.90	530
10%	61.31	52.15	1533	38.69	47.85	967
20%	35.48	24.22	887	64.52	75.78	1613
20-10-5%	54.63	50.67	1366	45.37	49.33	1134
10-20-5%	53.66	49.77	1342	46.34	50.23	1158

There are inconsistencies in the average unit time analysis improvement rates for multiple curves. These inconsistencies are apparent when calculating unit times from the cumulative average expressions. This can be illustrated using an example. Consider the unit times for 10th and 11th unit. From unit time equation for average unit time analysis, for a curve with initial unit time of 100 hours and multiple improvement rates of 20%, 10%, 5% occur at ranges 1-3, 3-10, 10-30 production units respectively, therefore we get,

$$Y_{10} = T_{10} - T_9 = a_2 [10^{b_2+1} - 9^{b_2+1}] = 50; \quad Y_{11} = a_3 [10^{b_3+1} - 9^{b_3+1}] = 54 \text{ hours}$$

where, $a_2 = 82.96$ $b_2 = -0.152$ $a_3 = 69.32$ $b_3 = -0.074$

The unit times are 50 hours and 54 hours for 10th and 11th unit respectively. This shows that the unit times actually increases even though the average unit time decreases and this case is noted only at the point were the improvement rate changes. This is a serious problem and leads to misinterpretation of future unit times when using multiple rates with cumulative average curves.

MANUFACTURING INTERRUPTIONS

Interruptions in improvement or learning generally occur [1] when new model changes are introduced, change in product design, or in case of intermittent production. These manufacturing interruptions due to long time periods of no production lead to loss of improvement acquired during previous production [6]. Loss of improvement will result in direct increase in production time and ultimately increased cost.

Let us consider the Figure 4 to illustrate the meaning of manufacturing interruption. It shows that the interruption took place at unit X, that is, production was stopped at unit X and took several months to resume production. The unit (X+1) is the 1st unit after interruption. The first unit before interruption was a. The amount of improvement lost due to interruption is proportional to the increase in time Z, one method used [6] to calculate the loss is discussed as follows.

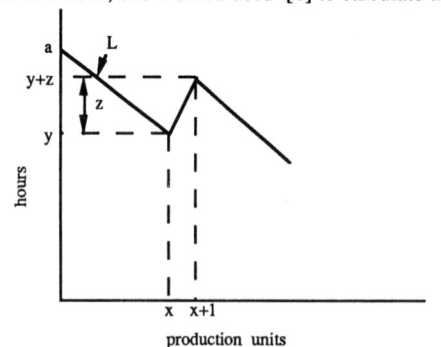

Figure 4. Improvement Curve with Manufacturing Interruption in Log Scale

That is, $Z = (a - y)*t /12$ (5)

where, Z = increase in time due to interruption a = first unit time
y = last unit time before interruption t = time period of interruption in months

This equation is a modified form of one presented by Smith [6]. The loss of improvement is therefore a function of the time period of interruption. This time period can be as short as a weekend break and still can have considerable effect on the improvement rate, this was clearly shown by Hancock [8].

The amount of improvement lost is equivalent to going backwards on the curve to a point 'L' as shown in the Figure 6. The improvement starts again from the point 'L' after interruption and proceeds as usual. Some important materials/data to be stored/recorded to prevent the loss of improvement other than personnel learning are: tooling used in the manufacturing process; the assembly methods used; training others to become familiar with the process; and handwritten instructions, key data, equations, etc.

CONCLUSION

The general forms of the cost prediction equations were derived for both the unit and average unit improvement curves. These models were tested and found to give good results. Using these equations one can solve or predict future costs or times to produce a part in a manufacturing environment based on improvement curves. These concepts can be applied profitably in a concurrent engineering environment to predict production costs for new products or when there is a product design change. New cost estimation functions are needed for the new processes, methods, and products. Improvement curves are a necessary item for predicting accurate time standards and the associated costs.

ACKNOWLEDGEMENTS

This work has been sponsored by Defense Advanced Research Project Agency (DARPA), under contract No. MDA972-88-C-0047 for DARPA Initiative in Concurrent Engineering (DICE).

REFERENCES

[1]   Louis E. Yelle, "The Learning Curve: Historical Review and Comprehensive Survey", Decision Sciences, Vol. 10, pp 302-324, 1979.
[2]   F. J. Andress, "The Learning Curve as a Production Tool", Harvard Business Review, Vol. 32, No.1, pp 87-97, 1954.
[3]   W. Z. Hirch, "Firm Progress Ratio", Econometrica, Vol. 24, No. 2, April 1956.
[4]   D. E. Kroll, K. R. Kumar, "The Incorporation of Learning in Production Planning Models", Annals of Operations Research, pp 291-304, 17 (1989).
[5]   Sal M. Kadri, "Learning Curve-Their Theory and Application", AACE Bulletin, Vol. 5, No.4, p-83, December 1963.
[6]   J. Smith, Learning Curve for Cost Control, IIE, Norcross, GA 30092, 1989.
[7]   H. Hall, "Experience with Experience Curve for Aircraft Design Change", N.A.A Bulletin, pp 59-66, December 1957.
[8]   W. M. Hancock, "The Prediction of Learning Rates for Manual Operation", The Journal of Industrial Engineering, January 1967.

Chapter XII

Materials: Composite

Introduction

Engineering applications using polymer matrix composites have increased not only in number but also in the range of applications because of their outstanding strength, stiffness, and light weight, and the ability to tailor material properties through the variation of fiber orientations.

The first paper in this chapter deals with the development of a model for interlaminar stress predictions in the regions of high stress gradients. The second paper develops a parametric model for the optimum design of a laminated composite flange focused on the quick evaluation and performance optimization of a composite flange with the optimum selection of material system, layup pattern and gore strip angle. The third paper describes a method of calculation of a stress-strain state in a multilayered rectangular plate manufactured from isotropic or orthotropic materials under the influence of a transversal dynamic load. The fourth paper discusses computer aided dynamic analysis of laminated composite plates. In the last paper, an architecture of developing engineering database system is presented and discussed with the rigid-plastic simulation engines.

The Payoffs of Concurrent Engineering in Advance Material Development

JACKY C. PRUCZ

West Virginia University
Concurrent Engineering Research Center (CERC)
2000 Hampton Center
Morgantown, WV 26506

APPLICATION DEVELOPMENT OF ADVANCED COMPOSITE MATERIALS

The technical community broadly recognizes today that widespread applications of advanced composite materials are inhibited by four major types of technical barriers - design, manufacturing, inspection and repair. [1].

The <u>Design</u> - related barriers stem, primarily, from the fact that most composite parts are being still designed today the same way as equivalent metal parts are. The current engineering design practice with composite materials employs effective properties, that define the relations between averages of field variables such as stress and strain when their spatial variation is statistically homogeneous. The heterogeneous microstructure of composites is, therefore, replaced by an equivalent continuum characterized by effective properties which are calculated from micromechanical models. [2]. Since the investment costs in advanced composite materials are usually higher than those of conventional materials, they cannot be competitive for a broad range of applications, from the cost-effectiveness viewpoint, unless all their potential advantages are fully exploited. One such advantage is the optimal tailoring of material properties to specific functional requirements, which cannot be leveraged as long as "equivalent homogeneous properties" are used in design. The performance and cost parameters of composite materials are strongly dependent on their microstructural characteristics associated with both material properties (e.g. internal phase geometry, or physical properties of the phases) and processing conditions (e.g. fiber coating or consolidation). Such information should be properly accounted for in the design process since it dictates the range of realizable properties and the ability to control them. This need has added a new wrinkle to the optimization problem of design and manufacturing of engineering components, namely that the material properties and processing conditions have now become variables in such problems. [3]. Close interactions between design, materials and manufacturing specialists are necessary throughout the product development process to materialize the potential opportunities of creating microstructural materials tailored to their applications.

The <u>Manufacturing</u> - related barriers are linked both to cost and performance. Low-volume production and low productivity have driven the manufacturing costs to almost prohibitive levels for many applications, such as the construction, heavy machinery or automotive industries. Labor costs play a heavy role in this regard, because of both a higher percentage of manual labor and higher skill requirements than in part manufacturing with traditional materials. The poor consistency of material properties is a major concern of all potential users of advanced composite materials. Unless the manufacturing processes are thoroughly understood, automated and controlled in such a way that the resulting material properties are reproducible, tailorable and verifiable, the designers cannot gain sufficient confidence in these materials to bring the "safety factors" down and make their utilization more effective. The emerging "Intelligent Processing of Materials" (IPM) technology is expected to facilitate broad-scale commercialization of advanced materials by improving their manufacturing processes, both from the cost and quality standpoints. It relies on an integrated framework of on-line quality control, process simulation, data reduction and analysis tools, whose components are not all yet fully developed for transitioning to routine applications.

The <u>Inspection</u> - related barriers can be traced, mainly, to the "lack of confidence" and "poor property repeatability" issues that were mentioned above. They not only trigger a larger frequency of inspections and more complex inspection procedures than may actually be needed, but also drive the acceptance criteria for inspection standards to levels that may be unnecessarily high from the tradeoff viewpoint between cost and performance. Selective use of evolving NDE (non-destructive evaluation) technology, along with modern IPM techniques may reduce, gradually the inspection requirements and, therefore, the manufacturing costs, of advanced composite materials. However, such advancements must occur in conjunction with the development of practical methods for "Damage Tolerant Design" and "Life-Prediction" of these materials.

In general, composites are considered to be more tolerant to damage than isotropic materials due to unique mechanisms of inhibiting damage growth and propagation. Effective utilization of this capability in practical applications requires, however, a rational methodology for characterization and control of frequent damage states. Such a methodology may include, for example, practical design and manufacturing guidelines to promote crack branching and deflection, favorable phase transformations and other features landing to resistance to crack-growth. [4]. Its implementation requires continuous collaboration between component designers, material scientists, manufacturing specialists and, obviously, the customer throughout the component development process, from concept to deployment.

The Repair - related barrier is raised, primarily, by the well-known difficulties of performing "secondary fabrication" operations on composite parts, like drilling, bending, joining and others. Besides the intensive research and development efforts that are currently directed towards improving the "repair" techniques and enhancing our understanding of how various repair approaches may affect the residual performance of the part, this barrier can be effectively addressed also through appropriate design and manufacturing methods. Early consideration of reliability and maintainability aspects at the design stage of a composite part will not only reduce the probability that the part may fail and need repair in service, but also simplify the repair procedures if and when they are needed.

THE CONCURRENT ENGINEERING ENVIRONMENT

The concept of "Concurrent Engineering" was introduced by the U.S. Department of Defense as a systematic approach towards improving the overall quality and reducing the life-cycle cost of new weapon systems through an integrated environment that enables, simultaneous consideration of all elements of the product life cycle, from concept through disposal. There are many ways to formally define the meaning of "Concurrent Engineering", but the most "popular" definition appears to be that proposed in the IDA Report [5] published in 1988. A wide variety of practicing forms of Concurrent Engineering have been documented and analyzed starting with this report and continuing with more recent publications, that either address global implementation issues [6] or describe successful case studies for specific applications [7]. Despite the multiple facets of Concurrent Engineering and the multiple approaches, methods and tools that it embodies, five key features have been identified as essential elements for its implementation:

1. The use of multi-disciplinary teams that cover all the various perspectives involved in the product life-cycle, including design, manufacturing and support.

2. Effective, continuous communication across the different disciplines within multi functional design teams, between different companies involved in the product development (including early communication between primes, subcontractors and suppliers) and between the product developer and its customer.

3. Application of quality engineering methods and a "continuous improvement" strategy to ensure the use of the "corporate history" available from the knowledge accumulated through past experience, as well as early detection of potential problems and efficient tradeoffs of multiple product and process alternatives.

4. Computer-based modeling of the product, its operation in simulated service environments and its simulated manufacturing and inspection processes in order to support the "rapid prototyping" approach and cooperative work based on information sharing between all the disciplines involved in the product development cycle.

5. An integrated environment that links appropriate CAD/CAE/CAM tools with the product/process simulation models to enable effective information management and decision support throughout the entire process of new product development and deployment.

The use of one or more of the above principles, under various combinations, has led to impressive benefits in cost savings (40-60%), cycle time reduction (about 50%) and quality improvement (30-60%) in a broad spectrum of applications, ranging from agricultural equipment to electronics and aircraft development. [5,6]. Recent advances in computational technology, primarily in the areas of object-oriented information management, trans-network communications, graphics and knowledge-based systems are expected to enhance even further the payoffs that integrated, computer-based concurrent engineering environments can bring to the product development enterprises of the U.S. industry. Such an environment is being developed at the Concurrent Engineering Research Center (CERC) of West Virginia University (WVU) under the sponsorship of the DICE (DARPA Initiative in Concurrent Engineering) program. It is aimed at facilitating cooperative product development through a heterogeneous and distributed system architecture based on a modular collection of information management services and decision support tools, whose main components are described in several technical papers included in Reference. [8].

THE PAYOFFS OF CONCURRENT ENGINEERING TO ADVANCED MATERIALS DEVELOPMENT

The development of advanced composite materials and products offers the opportunity to draw from the emerging concurrent engineering technology benefits that may be even more impressive than those demonstrated in other application domains. This assertion relies on two unique characteristics of the advanced materials enterprise:

1. All the major barriers to broad-scale application development of advanced materials, as listed in the first section of this paper, call for an open, integrated cooperative environment that spans over all the disciplines and all the stages involved in the development cycle of composite products, including material design, processing, inspection, and repair. Such an environment, as visualized in Fig. 1, can be provided by the concurrent engineering framework developed under the DICE program.

2. High competitiveness on international markets requires tight development schedules and continuous improvement of life-cycle performance, which often lead to the need to initiate the component design process while potential candidate materials are not yet fully developed and characterized. The only practical solution to such a need is a concurrent engineering environment which enables the implementation of a "rapid prototyping" strategy based on parametric models, off-line quality assurance, statistical methods, uncertainty and constraint management. [9].

A recent report by the Office of Technology Assessment (OTA) [10] raises the concern that "although the United States has achieved a strong position in advanced materials technologies, largely as a result of military programs, it is by no means certain that the United States will lead the world in commercialization of these materials". Besides identifying certain key policy objectives to accelerate the commercialization of advanced materials technologies in the United States, the report outlines research and development priorities to overcome technical and economical barriers associated with the development and application of advanced composite materials. They address each of the four major product commercialization phases discussed in the first section of this paper, namely design, manufacturing, inspection and repair, with emphasis on cost reductions and quality improvement through effective modelling, testing and processing techniques. The development of such techniques is the cornerstone of the emerging Intelligent Processing of Materials (IPM) technology [11] which is expected to play a major role in enhancing the market competitiveness of the US industry in the area of advanced materials. However, the potential benefits of IPM are not likely to be fully exploited without its integration within a broader concurrent engineering environment, as described in the second section of this paper. Besides addressing critical barrier to commercialization that may not be covered by IPM, like conceptual design or product support, a computer-based concurrent engineering framework may be regarded as an enabling approach to IPM, that provides the necessary infrastructure for its implementation through two main types of tools:

1. A computer framework that provides suitable utilities for knowledgebase representation and "real time" communications between various

specialized stations that form the development cycle of tailored materials-engineering specifications, macromechanics, processing, on-line inspection, testing, macromechanics, microstructural characterization of material constituents.

2. Processing and functional models that provide quantitative relationships between the engineering properties of the material on one hand and its microstructural, macrostructural and processing characteristics on the other hand.

The major payoffs expected from the implementation of Concurrent Engineering (CE) technology to the commercialization of advanced materials are synthesized concisely in Table 1, in the form of "CE Response" to the technical and economical needs for promoting such commercialization. Although this table outlines the contributions of CE to each of the four commercialization stages of advanced materials, design, manufacturing, inspection and repair, the key CE characteristic that provides the foundation to most of its benefits, is the integrated environment for effective coordination and flow of information across multiple disciplines and organizations throughout the product development cycle. This allows concurrent analysis of multiple interacting considerations, like material composition, production processes, secondary fabrication techniques, design guidelines, starting from the early stages of the product commercialization process and continuing through its completion.

SUMMARY

The multiple payoffs that the CE technology is expected to bring to the area of advanced materials development can be expressed, in summary, in the form of a more rapid and cost-effective transition of laboratory technology to production and deployment. This goal can be achieved through the cost reductions and quality improvements enabled by an integrated CE framework between materials science, processing, component design, testing and field maintenance, supported by effective tools and knowledgebases for information management and decision support. It is likely to have a strong positive impact on the market competitiveness of US manufacturing of advanced materials, by facilitating both the technical and business aspects of their commercialization.

References

1. Wilkins, D.J., "What's Wrong with Composites and What Can We Do About It", The Newsletter of the Center for Composite Materials, University of Delaware, July-Aug. 1990

2. Tsai, S.W., Composites Design, 4th Edition, Dayton, OH, Think Composites 1988.

3. Prucz, J.C., D'Acquisto, J and Smith, J., "Elastodynamic Tailoring of Motion Conversion Mechanisms by Using Fiber Reinforced Composites", Journal of Reinforced Plastics and Composites, Vol 8, pp. 398-409, July 1989.

4. Kanniaen, M.F. and Popelar, C.H., Advanced Fracture Mechanics, Oxford Univ. Press/Clarendon Press, Oxford, 1985

5. Winner, R.I., J.P. Pennell, H.E. Bertrand, and M.M.G. Slusarczuk, "The Role of Concurrent Engineering in Weapons System Acquisition". IDA Report R-338, Institute for Defense Analyses, Alexandria, VA, December 1988.

6. Meredith, J.W. and Blanchard, B.S., "Concurrent Engineering: Total Quality Management in Design", published by the Society of Naval Architects and Marine Engineers, April 1990

7. Schrage, D.P., McConville, J., Martin, C.L. and Craig, J.I., "An Example of Concurrent Engineering Principles Applied to the Preliminary Design of a Light Commercial Utility Helicopter", Proceedings of the First Annual Symposium on Mechanical System Design in A Concurrent Engineering Environment, Hang, E.J., editor, The University of Iowa, Iowa City, October 24 & 25, 1989.

8. "Emerging Prototypes for Concurrent Engineering", Proceedings of the Second National Symposium on Concurrent Engineering, Concurrent Engineering Research Center, West Virginia University, Morgantown, WV, February 7-9, 1990.

9. Karandikar, H. and Mistree, F., "A Method for Concurrent and Integrated Material Selection and Dimensional Synthesis", submitted to Transactions of ASME, Journal of Mechanical Design, May 1990.

10. Advanced Materials by Design, Congress of the United States, Office of Technology Assessment, Report OTA-E-351, June 1988.

11. O'Brien, D.W. and Payne, J.E., "Concurrent Design and Intelligent Control for Production of Advanced Materials", Proceedings of the Second National Symposium on Concurrent Engineering, Concurrent Engineering Research Center, West Virginia University, Morgantown, WV, February 7-9, 1990.

Table 1 - Summary of CE Payoffs to Advanced Materials Development

Commercialization Stage	Technical and Economical Needs	CE Response
A. Design	1. Modelling of correlations between material properties, microstructure and processing conditions 2. Optimum tailoring of materials to specifications 3. Cheap materials 4. Life prediction	a. Tailoring material selection to design requirements via integrated CE environment b. Dynamic distributed data-base with efficient browsing utilities c. Continuous interactive analysis of dynamic microstructure/processing/ property relationships via CE integrated framework d. Soft prototyping of service conditions
B. Manufacturing	1. Consistent properties 2. Statistical and Adaptive Process Control 3. Cheap processes	a. Tailoring processing conditions to design requirements via integrated CE environments b. On-line, adaptive process control c. Expedient comparisons between alternate processes through knowledge-based systems d. Flexibility to change e. Simulations of manufacturing processes

C. Inspection	1. Advanced NDE	a. Automated downloading and storage of test data in large distributed data bases with broad accessibility and efficient browsing
	2. Statistical Methods	b. Continuous refinement of testing methods through rapid prototyping and evaluation
	3. Model-based testing	c. Reliable inspection techniques based on damage tolerance and robust design criteria
	4. Damage-tolerance criteria	d. Dynamic comprehensive databases of material properties process parameters, and life-cycle performance
D. Repair	1. High reliability	a. Reduce the need for repairs by integrating reliability and life prediction considerations at all stages of the product development cycle
	2. Repair techniques	b. Facilitate repairs by empowering "Design for maintainability" tools and strategies
		c. Library of service data from previous similar applications

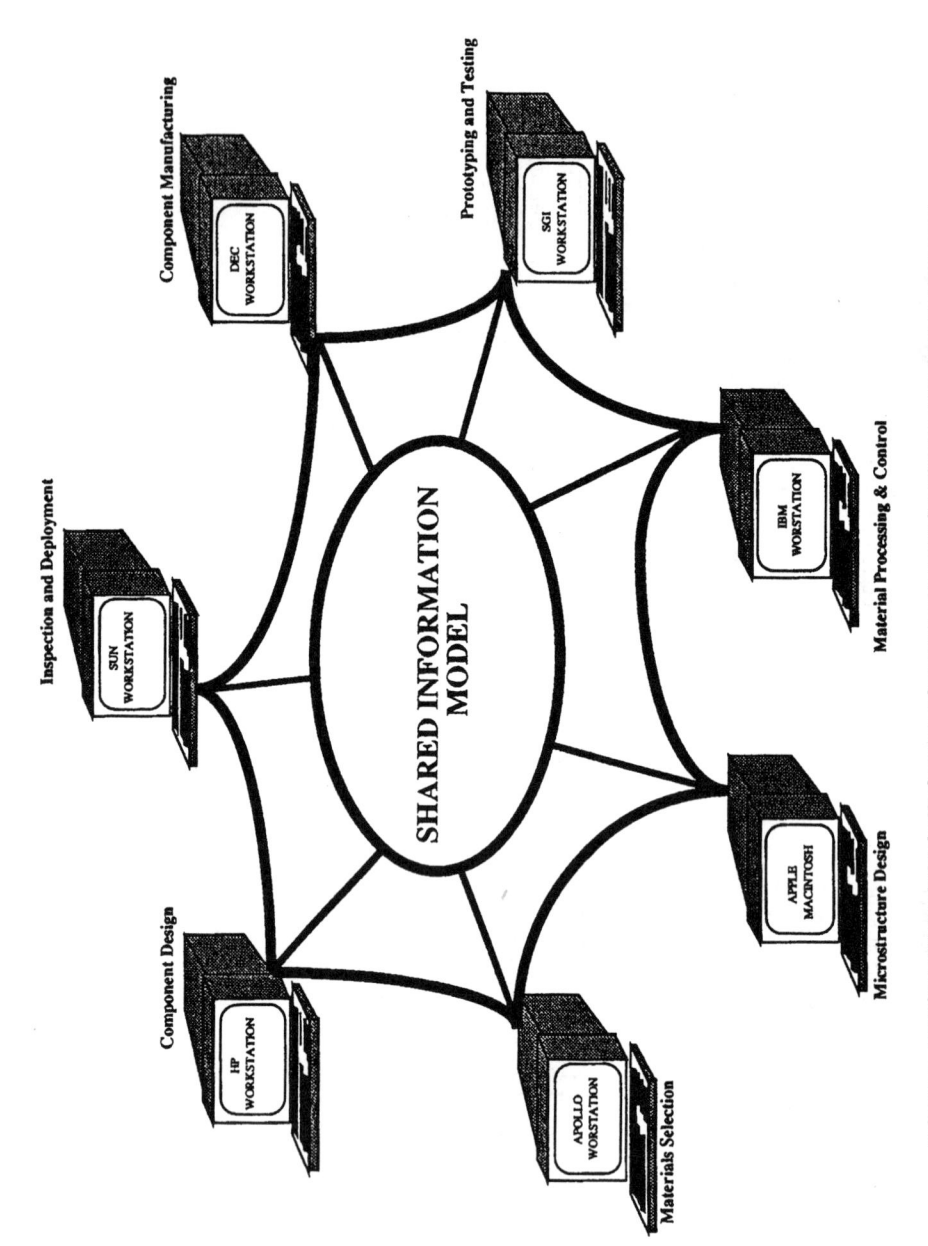

FIGURE 1: INTEGRATED FRAMEWORK FOR DEVELOPMENT AND APPLICATION OF ADVANCED MATERIALS

A Practical Engineering Approach for Predicting Interlaminar Stresses in Composites

JACKY PRUCZ and MARIOS LAMBI

Department of Mechanical and Aerospace Engineering
West Virginia University
Morgantown, WV

ABSTRACT

This paper deals with the development of a model for the interlaminar stress predictions in the regions of high stress gradients. The model is based on the modified Donnel approach where the stresses obtained form the classical engineering theory are improved by adding a series of corrections. These corrections are determined by satisfying the stress equilibrium and compatibility conditions of two-dimensional elasticity. Accuracy of the model is verified by comparing the results of some sample cases with those obtained from a finite element code.

INTRODUCTION

Laminated composite structures exhibit various failure modes such as fiber-matrix debonding within individual layers, delamination or separation of layers, cracks through one or more layers and fiber fracture. Most failure mechanims are associated with a complex state of stress with steep gradients, and are dominated by interlaminar stresses that exist in regions near free edges, ply terminations, cutouts, voids and holes. Interlaminar and intralaminar cracks often initiate at these sites due to high local and shear stresses. Expedient, but reliable predictions of the stress field in such regions are essential for developing practical design methods based on a clear understanding of the complex behaviour and failure mechanims of composite structures. Unfortunately, the existing design tools and practices do not provide such a predictive capability. On one hand, closed form solutions of the exact elasticity equations are limited to a few simple geometric and loading configurations. On the other hand, classical engineering theories, which may be simple enough for routine use in design, fail to account for such essential physical effects associated with composite material behaviour as interlaminar stresses. Three dimensional theories for laminated composites, which yield accurate predictions for such effects, become intractable in a practical design environment. Three dimensional numerical simulations are expensive, time consuming and often inappropriate for interlaminar stress analysis.

A laminated field model that is reliable and yet simple enough to suit a practical design environment is presented in this paper. The model follows the modified Donnell approach [1]. Similar approaches have been used in [2,3], but results are restricted only to a first degree refinement over the classical bending theory approximations. In this paper results are presented icorporating a second order refinements in the formulation to improve the stress prediction capabilities of the model, especially in the high stress gradient regions.

The results from the model are compared with the corresponding finite element predictions for some sample cases. Good agreement is obtained. The model is expected to serve as an important tool for the parametric design approach, where a large number of possible configurations are to be evaluated expediently with a reasonable level of accuracy.

ANALYSIS APPROACH

The analysis is restricted to orthotropic materials with principal material directions corresponding to the axes of the ply, subjected to a plane stress state. A plane strain situation can be analyzed by proper transformation of the elastic constants.

Consider a laminate made of N prefectly bonded plies, each ply having a plane of material symmetry parallel to the plane of the laminate. A particular ply that is singled out for study is shown in Figure 1 with the notation and sign convention used as shown.

The appropriate form of Hooke's law (constitutive equations) for plane stress is

$$\varepsilon_{xx} = S_{11}\sigma_{xx} + S_{13}\sigma_{zz}$$
$$\varepsilon_{zz} = S_{13}\sigma_{xx} + S_{33}\sigma_{zz} \qquad (1)$$
$$\gamma_{xz} = S_{55}\sigma_{xz}$$

where $S_{11} = \frac{1}{E_{11}}$, $S_{13} = -\frac{\nu_{13}}{E_{11}}$, $S_{33} = \frac{1}{E_{13}}$, $S_{55} = \frac{1}{G_{13}}$

Overall equations of equilibrium for the kth ply, in terms of force and moment resultants derived by integrating the 2-D equilibrium equations, are

$$N^k_{,x} + T^k_2 - T^k_1 = 0$$
$$Q^k_{,x} + P^k_2 - P^k_1 = 0 \qquad (2)$$
$$M^k_{,x} - Q^k + \frac{h^k}{2}(T^k_1 + T^k_1) = 0$$

where $_{,x}$ denotes differentiation with respect to x.

The force, moment and shear stress resultants for the kth ply are

$$(N^k, M^k, Q^k) = \int_{-\frac{h^k}{2}}^{\frac{h^k}{2}} (\sigma_{xx}, z\sigma_{xx}, \sigma_{xz})^k \, dz \qquad (3)$$

superscript 'k', which indentifies the ply, is dropped in the subsequent equations for convenience.

Following the modified Donnel's approach[1], based on an approach briefly outlined by Donnell in [4], the stresses in a generic ply (Figure 1) subjected to interfacial shear and peel stresses on its top and bottom surfaces, denoted as P and T respectively, are expressed as infinite series in which the first terms correspond to the classical engineering theory and later terms to increasingly minor refinements. These series satisfy equilibrium and compatibility conditions of two dimensional elasticity theory and the boundary conditions on the top and bottom surfaces.

Assuming that the generic ply is subjected to a transverse normal stress P_2 only, as shown in case 1 of Figure 2, the stress disribution for this case is assumed as

$$\sigma_{xx} = 12\frac{M}{h^3} z + f_1(z) P_2$$
$$\sigma_{xz} = \frac{3Q}{2h} \left(1 - \frac{4z^2}{h^2}\right) + f_2(z) P_2 \qquad (4)$$
$$\sigma_{zz} = 0 + f_3(z) P_2$$

The first terms in equations (4) represent the classical stress predictions for this particular loading case. Functions $f_1(z)$, $f_2(z)$ and $f_3(z)$ represent a first order improvement over the classical stress predictions. These functions are determined by satisfying the equilibrium and compatibility equations to a certain order derivatives of the applied interfacial stress. For this set of improvements, this order implies neglecting terms involving $P_{2,x}$. A second set of improvements are obtained later, based on satisfying the equilibrium and compatibility equations to a higher order, neglecting terms $P_{2,xxx}$. For case 1, following the procedure outlined in reference [1], the stress distributions, upto a first order refinements, finally appear as

$$\sigma_{xx} = 12\frac{M}{h^3} z + \frac{\alpha z}{h}(\frac{4z^2}{h^2} - \frac{3}{5}) P_2$$
$$\sigma_{xx} = \frac{3Q}{2h}(1 - \frac{4z^2}{h^2}) \tag{5}$$
$$\sigma_{zz} = (\frac{1}{2} + \frac{3z}{2h} - \frac{2z^3}{h^3}) P_2$$

where $\alpha = (S_{55} + 2S_{13})/(2S_{11})$; it is unity for an isotropic material.

The stress distribution for the general case (Figure 1), where the ply is subjected to both interfacial transverse normal and shear stresses is obtained by combining four different cases as shown in figure 2. Following a similar approach for all the cases and combining the results, it can be shown that the stress distribution for the general case is

$$\sigma_{xx} = \frac{N}{h} + 12\frac{M}{h^3} z + \alpha[\frac{h}{12}(1 - \frac{12z^2}{h^2}) N_{,xx} + \frac{3z}{5h}(1 - \frac{20z^2}{3h^2}) M_{,xx}]$$
$$\sigma_{xz} = \frac{Q}{h} - \frac{z}{h} N_{,x} + \frac{1}{2h}(1 - \frac{12z^2}{h^2}) M_{,x} \tag{6}$$
$$\sigma_{zz} = r - \frac{z}{h} Q_{,x} - \frac{h}{8}(1 - \frac{4z^2}{h^2}) N_{,xx} - \frac{z}{2h}(1 - \frac{4z^2}{h^2}) M_{,xx}$$

where $r = (P^1 + P^2)/2$

If a second set of improvements is carried out over the stresses given by equation (6), following exactly an identical preocedure as for the first order of improvements, the final stress distributions for a generic ply (Figure 1) come out as

$$\sigma_{xx} = \frac{N}{h} + 12\frac{M}{h^3}z + \alpha[\frac{h}{12}(1 - \frac{12z^2}{h^2}) N_{,xx} + \frac{3z}{5h}(1 - \frac{20z^2}{3h^2}) M_{,xx}]$$
$$+ h^2\alpha^2[(\frac{27z}{1400h} - \frac{z^3}{5h^3} + \frac{2z^5}{5h^5}) M_{,xxxx} + \frac{h}{6}(\frac{7}{240} - \frac{z^2}{2h^2} + \frac{z^4}{h^4}) N_{,xxxx}] +$$
$$h^2\frac{S_{33}}{S_{11}}[\frac{1}{24}(1 - \frac{12z^2}{h^2}) r_{,xx} - \frac{z}{h}(\frac{1}{40} - \frac{z^2}{6h^2}) Q_{,xxx} - (\frac{11z}{1120h} -$$
$$\frac{z^3}{12h^3} + \frac{z^5}{10h^5}) M_{,xxxx} - \frac{h}{8}(\frac{3}{80} - \frac{z^2}{2h^2} + \frac{z^4}{3h^4}) N_{,xxxx}]$$

$$\sigma_{xz} = \frac{Q}{h} - \frac{z}{h} N_{,x} + \frac{1}{2h}(1 - \frac{12z^2}{h^2}) M_{,x}$$
$$+ ah[(\frac{1}{80} - \frac{3z^2}{10h^2} + \frac{z^4}{h^4}) M_{,xxx} - \frac{z}{12}(1 - \frac{4z^2}{h^2}) N_{,xxx}]$$

$$\sigma_{zz} = r - \frac{z}{h} Q_{,x} - \frac{h}{8}(1 - \frac{4z^2}{h^2}) N_{,xx} - \frac{z}{2h}(1 - \frac{4z^2}{h^2}) M_{,xx}$$
$$- ah^2[(\frac{z}{80h} - \frac{z^3}{10h^3} + \frac{z^5}{5h^5}) M_{,xxxx} + \frac{h}{12}(\frac{1}{16} - \frac{z^2}{2h^2} + \frac{z^4}{h^4}) N_{,xxxx}] \qquad (7)$$

By satisfying the three-strain displacement and constitutive equations, the relationship for the axial force and moment resultants in terms of average axial and transverse displacements are obtained. These appear as

$$N = \frac{h^3}{S_{11}}[\bar{u}_{,x} - S_{13}r + \frac{h}{12}S_{13} N_{,xx} + \frac{\alpha h^3}{360}S_{13} N_{,xxxx}]$$

$$M = \frac{h^3}{12S_{11}}[-\bar{w}_{,xx} + \frac{1}{h}(S_{55} + S_{13}) Q_{,x} + \frac{2\alpha}{5h}S_{11} M_{,xx} -$$
$$\frac{S_{33}h}{60} Q_{,xxx} + \frac{h}{175}(\alpha^2 S_{11} - \frac{5}{6}S_{33}) M_{,xxxx}] \qquad (8)$$

where $\bar{u}$ and $\bar{w}$ are the average axial and transverse displacements defined as

$$\bar{u} = \int_{-\frac{h}{2}}^{\frac{h}{2}} \frac{u}{h} \, dz \quad \text{and} \quad \bar{w} = \int_{-\frac{h}{2}}^{\frac{h}{2}} \frac{w}{h} \, dz \quad (9)$$

The corresponding displacement equations derived by appropriate integration of the constitutive equations (1) for plane stress may be shown as

$$u = \bar{u} - z\, \bar{w}_{,x} + \frac{z}{h} S_{55} Q + \frac{1}{2}(S_{55} + S_{13})[\frac{h}{12}(1 - \frac{12z^2}{h^2}) N_{,x}$$
$$+ \frac{z}{h}(1 - \frac{4z^2}{h^2}) M_{,x}] + \frac{\alpha h^2}{12}(S_{55} + S_{13})[h(\frac{7}{240} - \frac{z^2}{2h^2} + \frac{z^4}{h^4}) N_{,xxx}$$
$$+ \frac{6z}{5h}(\frac{1}{8} - \frac{z^2}{h^2} + \frac{2z^4}{h^4}) M_{,xxx} + \frac{h^2 S_{33}}{48}[2(1 - \frac{12z^2}{h^2}) r_{,x} - 3h(\frac{3}{40}$$
$$- \frac{z^2}{h^2} + \frac{2z^4}{3h^4}) N_{,xxx} - \frac{2z}{h}(1 - \frac{4z^2}{h^2}) Q_{,xx} - \frac{z}{h}(\frac{7}{10} - \frac{4z^2}{h^2} + \frac{24z^4}{5h^4}) M_{,xxx}]$$

$$w = \bar{w} + \frac{z}{h} S_{13} N - \frac{S_{13}}{2h}(1 - \frac{12z^2}{h^2}) M + \alpha h S_{13}\{ -(\frac{1}{80} - \frac{3z^2}{10h^2} + \frac{z^4}{h^4}) M_{,xx}$$
$$+ \frac{h}{2}(\frac{z}{6h} - \frac{2z^3}{3h^3}) N_{,xx} + \frac{h^3 \alpha}{2}[\frac{7z}{3}(\frac{7z}{240h} - \frac{z^3}{6h^3} + \frac{z^5}{5h^5}) - \frac{S_{33}}{6S_{13}}(\frac{z}{16h} - \frac{z^3}{6h^3}$$
$$+ \frac{z^5}{5h^5}) - \frac{S_{33}}{4\alpha S_{11}}(\frac{3z}{80h} - \frac{z^3}{6h^3} + \frac{z^5}{15h^5})] N_{,xxxx}\} + hS_{33}[\frac{z}{h} r - \frac{z}{8}(1 - \frac{4z^2}{3h^2}) N_{,xx}$$
$$+ \frac{1}{24}(1 - \frac{12z^2}{h^2}) Q_{,x} + (\frac{7}{480} - \frac{z^2}{4h^2} + \frac{z^4}{2h^4}) M_{,xx}] \quad (10)$$

METHOD OF SOLUTION

To show the application of the model, the above formulation is utilized here for the development of the appropriate solution for a specimen with two elements or group of plies, (known as sublaminates). Continuity of displacements and stresses is implied by the perfect interface assumption between the two plies. The equilibrium equations (2), the equations for the resultant force and moment in terms of average variables (8), the displacement distributions (10), the continuity requirements and the appropriate surface conditions make-up the desired system of equations which are solved. This system of ten defferential equations is reduced further by expressing eight of the ten dependent variables with respect to the remaining two, which, namely, are M^1 and M^2. This reduction in the number of equations yields a 2x2 system of non-homogeneous ordinary differential equations in the from

$$\begin{bmatrix} A_{611} & A_{612} \\ A_{621} & A_{622} \end{bmatrix} \begin{Bmatrix} M^1 \\ M^2 \end{Bmatrix}_{,xxxxxx} + \begin{bmatrix} A_{411} & A_{412} \\ A_{421} & A_{422} \end{bmatrix} \begin{Bmatrix} M^1 \\ M^2 \end{Bmatrix}_{,xxxx} + \begin{bmatrix} A_{211} & A_{212} \\ A_{221} & A_{222} \end{bmatrix} \begin{Bmatrix} M^1 \\ M^2 \end{Bmatrix}_{,xx} + \begin{bmatrix} A_{11} & A_{12} \\ A_{21} & A_{22} \end{bmatrix} \begin{Bmatrix} M^1 \\ M^2 \end{Bmatrix} = \begin{Bmatrix} \alpha^1 \\ \alpha^2 \end{Bmatrix} x^2 + \begin{Bmatrix} \alpha^3 \\ \alpha^4 \end{Bmatrix} x + \begin{Bmatrix} \alpha^3 \\ \alpha^4 \end{Bmatrix}$$

(11)

where A_{ijk} and α^i are constant coefficients.

The general solution of each dependent variable of the system (for example,. M^1 and M^2, where superscripts 1 and 2 denote the upper and lower elements of the specimen), consists of the sum of two parts (i) a complementary solution form the homogeneous part and (ii) a particular solution from the non-homogeneous system. Since the equations are linear differential equations with constant coefficients, the complimentary solution, M_c, for each dependent variable consists of a series of terms in the general form

$$M_c = F\, e^{\lambda x} \qquad (12)$$

where F are constants derived from the system of differential equations.

The values of λ are found by setting the determinant of the coefficients (equation 11), of the homogeneous system, to zero. Algebraic expressions for the expansion of the determinant are not written for simplicity. In a condensed form the determinant (characteristic equation) of the system is

$$E_1 S^{10} + E_2 S^8 + E_3 S^6 + E_4 S^4 + E_5 S^2 + E_6 = 0 \qquad (13)$$

where E's are constant coefficients. The roots of the polynomial equation (13) furnish the values of λ and are calculated with a standard numerical mothod.

Part (ii) of the solution involves the particular solution. For this a polynomial of the second degree is assumed to be the desired particular solution that satisfies the non-homogeneous system of differential equations. This assumed solution is of the from

$$M_p = \begin{Bmatrix} \beta 1 \\ \beta 2 \end{Bmatrix} x^2 + \begin{Bmatrix} \beta 3 \\ \beta 4 \end{Bmatrix} x + \begin{Bmatrix} \beta 3 \\ \beta 4 \end{Bmatrix} \qquad (14)$$

where β are constant coefficients which are determined by substituting the above solution in the system of equations and then by balancing the terms of either side. Combining equations (12) and (14) the general solution becomes

$$M = M_c + M_p \qquad (15)$$

APPLICATION

The accuracy of the model for the stress predictions is demonstrated by applying the proposed formulation on a specimen with two elements or plies. The specimen is fixed at one end and is subjected to a uniform axial displacement at the other free end. The boundary conditions appropriate to this model are prescibed at the element ends. Plywise edge boundary conditions are satisfied in an overall sense. Each element is treated as a homogeneous orthotropic region under a plane stress condition.

The results are obtained with the improved model (second order refinements) and also with the equations that account for the first order refinements only, which are of the same type as those used in [2,3]. For the sample case chosen here, attention is focused near the fixed boundary which is the region of high stress gradients. The specimen in the present analysis has the following parameters

Lenght/Thickness Ratio = 5

Top Layer	**Bottom Layer**
E_{11} = 3.19 Msi	E_{11} = 19.0 Msi
E_{33} = 3.19 Msi	E_{33} = 1.54 Msi
v_{21} = 0.11	v_{21} = 0.36
G_{12} = 0.57 Msi	G_{12} = 0.86 Msi
Thickness = 0.1 inches	Thickness = 0.1 inches

Results presented in figures 3 and 4 clearly show the importance of the underlined terms in capturing the true responce in regions of high stress gradients. In order to verify the accuracy of the present formulation, the problem was solved using a finite element code developed in house. Good agreement was obtained as shown in figures 5 through 7.

Apart from the sample case discussed above, the formulation was tested on other cases such as a cantilever and a simply supported beam with uniformly distributed loads on the top surface. In all the cases investigated, in the absence of second order refinement terms, incorporated in this paper, solutions correlate very poorly near the region of high stress gradients (say, fixed end). Of course at distances sufficiently away from such regions of steep stress gradients, even with only first order refinement terms [1] reasonable accuracy is obtained. This happens because in the absence of second order improvement terms, the roots of the characteristic equation are not dependent on the transverse compliance term. This, in tern, affects the accuracy of the solution near the fixed boundary. This clearly indicates that for situations where the transverse effects are more severe the second order refinement terms must be included in the analysis.

CONCLUSIONS

An analytical model is developed that incorporates the important effects of tranverse shear and normal stresses. The model is simple to use and predicts the interlaminar stresses much more effectively in composite laminates. It is an important tool for evaluating many possible configurations quickly and efficiently. It is also well suited for preliminary design studies, with a large number of configurations. The method has been validated using a finite element analysis code.

ACKNOWLEDGEMENTS

This work has been sponsored by the Defense Advanced Research Projects Agency (DARPA), under contract No. MDA972-88-C0047 for DARPA Initiative in concurrent Engineering (DICE).

REFERENCES

1. Armanios, E. A., New Methods of Sublaminate Analysis for Composite Structures and Applications to Fracture Processes, Ph.D Dissertation, Georgia Tech, December 1984

2. Armanios, E. A., Rehfield, L. W. and Reddy, A. D., Design Analysis and Testing for Mixed-Mode and Mode II Interlaminar Fracture of Composites, Composite Materials: Testing and Design (Seventh Conference), ASTM STP 893, J. M. Whitney, Ed., Americal Society for Testing and Materials, Philadelphia, 1986, pp. 232-255.

3. Rehfield, L. W, Armanios, E. A., and Changli, Q., Analysis of Behaviour of Fibrous Composite Compression Specimen, Recent Advances in Composites in the United States and Japan, ASTM STP 864, J. R. Vinson and M. Taya, Eds., American Society for Testing and Materials, Philadelphia, 1985, pp. 236-252.

4. Donnell, H. L., Bending of Rectangular Beams, Journal of Applied Mechanics, Vol. 19 (1952), p. 123.

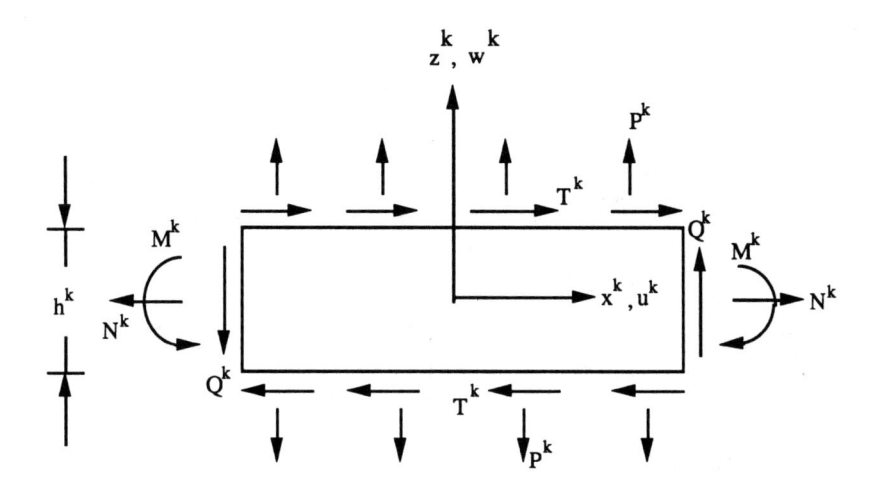

Figure 1 : Notation and Sign Convention for the k Ply

CASE I

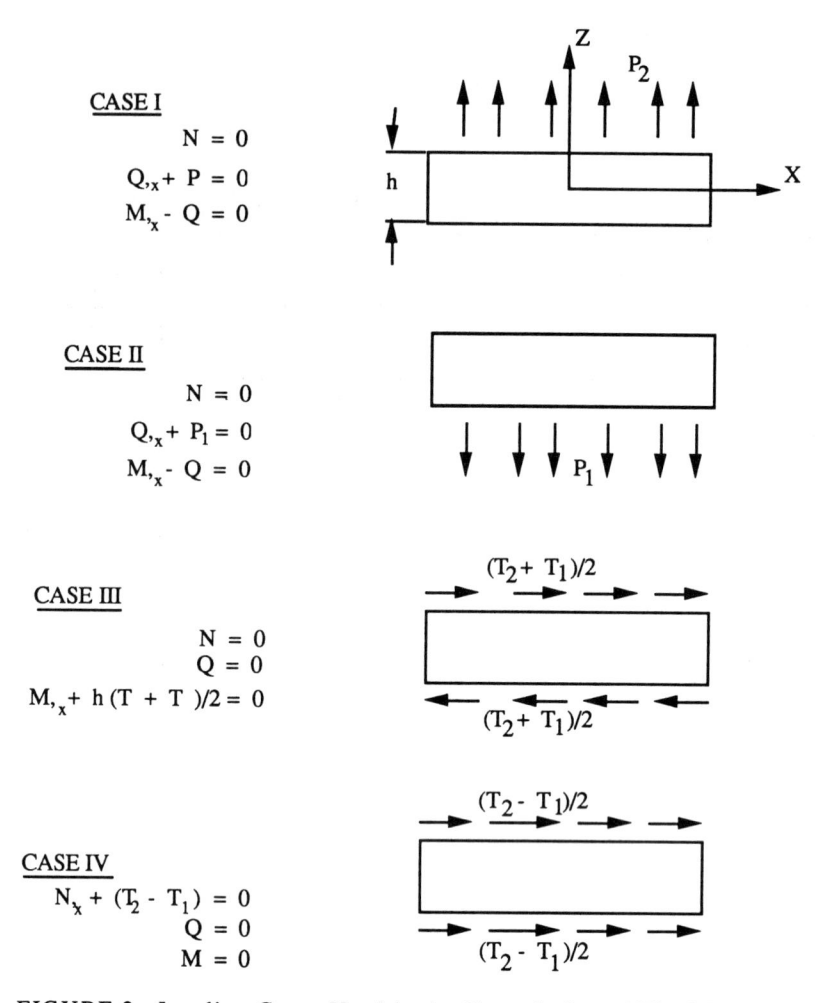

$N = 0$
$Q_{,x} + P = 0$
$M_{,x} - Q = 0$

CASE II

$N = 0$
$Q_{,x} + P_1 = 0$
$M_{,x} - Q = 0$

CASE III

$N = 0$
$Q = 0$
$M_{,x} + h(T + T)/2 = 0$

CASE IV

$N_x + (T_2 - T_1) = 0$
$Q = 0$
$M = 0$

FIGURE 2 : Loading Cases Used in the Formulation of Final Equations

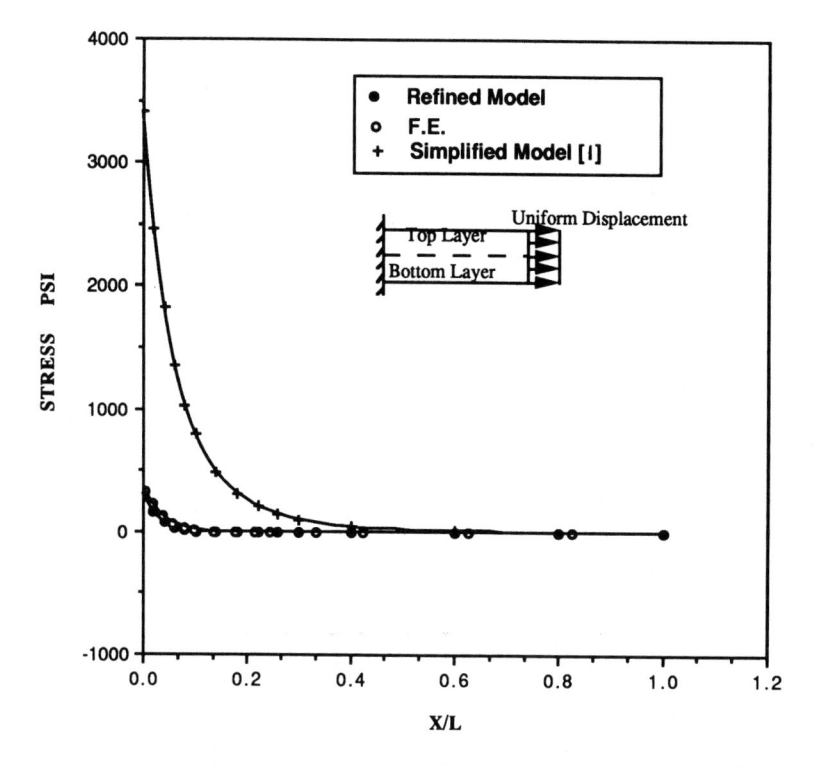

FIGURE 3 : Transverse Normal Stress at Midplane of Top Layer

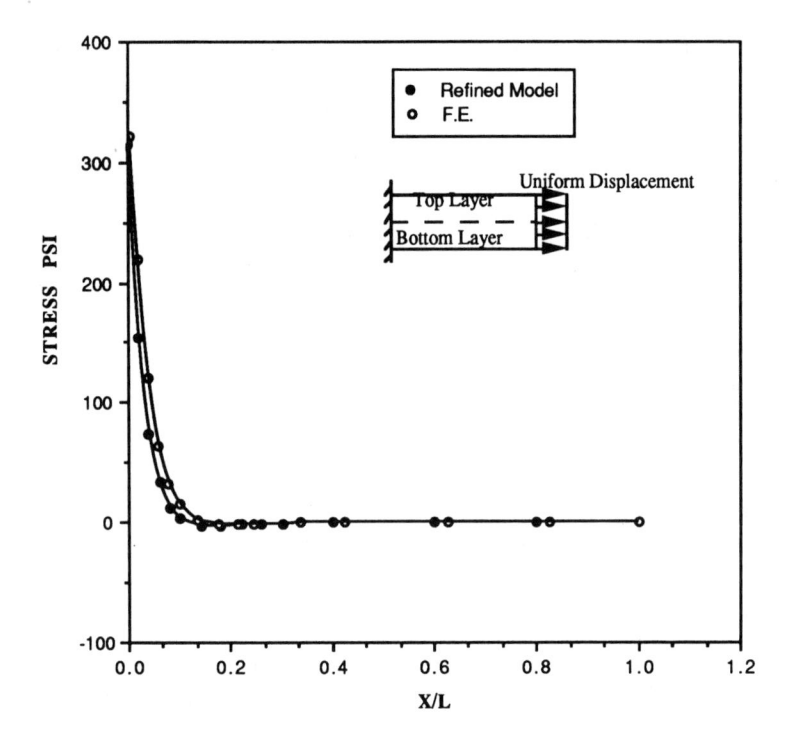

FIGURE 4 : Comparative Results for the Transverse Normal Stress as in Figure 3

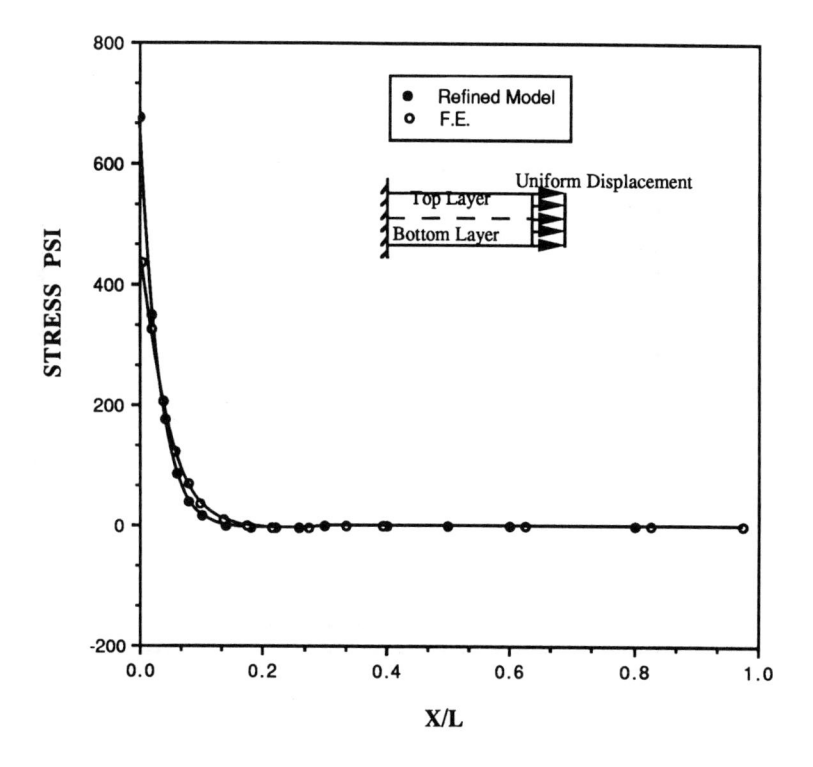

FIGURE 5 : Transverse Normal Stress at the Laminate Interface

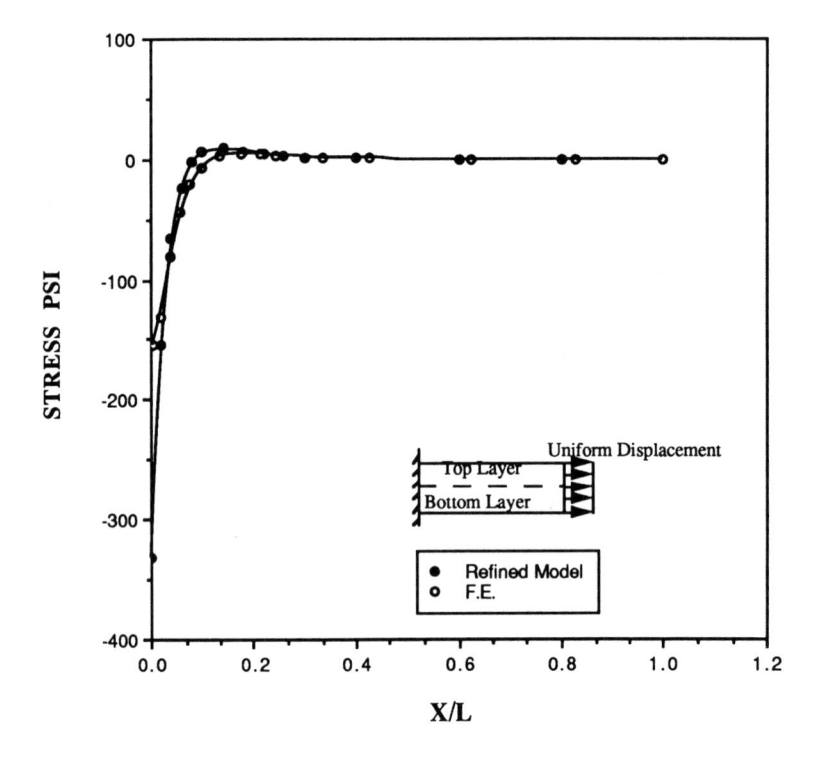

FIGURE 6 : Transverse Shear Stress at Laminate Interface

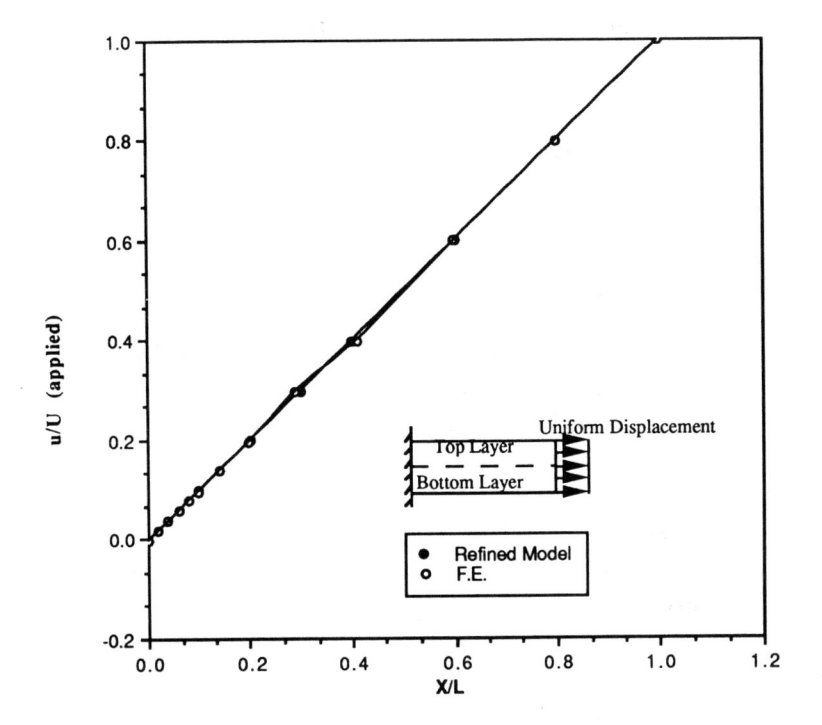

FIGURE 7 : Interface Axial Displacement

Interactive Optimum Parametric Design of Laminated Composite Flange

B.S.-J. KANG, JACKY PRUCZ, and F.K. HSIEH

Department of Mechanical and Aerospace Engineering
West Virginia University
Morgantown, WV 26506

ABSTRACT

A parametric model has been developed for the optimum design of laminated composite flange. The study focuses on the quick evaluation and performance optimization of composite flange with the optimum selection of material system, layup pattern and gore strip angle. The effect of butt-joint between the gore strips is also evaluated. Classical laminate field theory and maximum stress failure criteria are employed in the analysis. An interactive optimum flange design computer code has been developed for on-line design performance evaluation of the laminated composite flange.

1. INTRODUCTION

In recent years, engineering applications using polymer matrix composites, especially in aerospace structures, have increased not only in number, but also in the range of applications. The widespread use of fiber-reinforced composite laminates in structural applications is due to its outstanding strength, stiffness, lightweight and the ability to tailor material properties through the variation of fiber orientations [1]. For composite structural design, one of the fundamental optimization problems in fiber-reinforced composite structures is the design of laminates subjected to in-plane loading conditions. A few theories have been proposed in the past two decades to analyze the stress field in composite laminates [2,3,4,5]. The most popular of these is the classical laminate theory [6,7,8] which has been shown to yield reasonably accurate stress calculations, especially for thin laminate composite structures. In general, for practical engineering structural design using laminated composites, if the side to thickness ratio is greater than seven, L/H > 7, classical laminate theory, comparing with higher order theory, is adequate to provide quite accurate design analysis [1,2,4] and therefore remains widely used in pratical engineering composite design works [9,10].

One major obstacle in designing structural components using laminate composites is the lack of a suitable, easy-to-use engineering design program which can provide quick performance estimations as well as efficient and robust tailoring of design parameters to requirements. A typical example is the design and fabrication of a laminated composite flange which requires to assemble several identical pieces of gore strips to form a sublayer of the composite flange, as shown in Fig. 1. Typically, the gore strips are cut from a large size unidirectional angle lamina (ply) and therefore, the fibers are not continuous at the intersection (butt-joint) between two gore strips (Fig. 1). Moreover, the fiber orientation of each layer changes as a function of circumferential location, i.e. material properties will vary from position to position and thus the flange geometry dictates the reinforcement trace attainable with the composite material. Also, with the existence of a butt-joint in a laminate, complex stress states with a rapid change of stress gradients will occur at the butt-joint region due to geometric discontinuity. The abrupt change of material properties within the laminate increases the complexities in the composite laminated flange design [11]. To the best of our knowledge, there has been no similar study on laminated composites flange design with the existence of butt-joint in the flange. Although, papers concentrating on the geometrical discontinuity [12,13] and optimal laminated design [9,10,14] have been discussed before. The objective of this research is to develop an engineering module for optimized laminated composite flange design. The engineering module developed is (i) to provide quick design performance estimation and (ii) to perform optimization. Using classical laminate field theory and failure criterion, we have developed an interactive computer program to estimate laminated composite flange performance as a function of the choice of design variables. Performance optimization was also incorporated in the computer program by varying the layup patterns to achieve maximum strength consistent with given loads and flange dimensions.

2. GEOMETRIC MODELING

Design variables for a laminated composite flange (as shown in Fig. 1) include gore angle, start angle, ideal angle, location angle, butt-joint location, layup pattern, flange dimensions, applied loads and types of composite material. Two geometric modeling are involved in this analysis; (1)

single layer modeling and (2) laminate modeling.

SINGLE LAYER MODELING

For a single layer in a laminated composite flange, which is made of N pieces of identical gore strip cut from an unidirectional angle lamina (see Fig. 1), the fiber angle at the middle circumferential location of gore strip is the same as the unidirectional angle of lamina and will be called ideal angle (IDA) here in order not to be confused with the fiber angle which is varied along the circumferential location. The gore strips are layed up at any start angle (SA) to form a layer. The fiber angle at any circumferential location can be determined by the following equations.

1. At circumferential location AB, rotating β degree counter-clockwise from the middle of gore strip, the fiber angle is

$$FA = IDA + \beta \quad \text{(1a)}$$

2. At circumferential location CD, rotating α degree clockwise from the middle of gore strip, the fiber angle is

$$FA = IDA - \alpha \quad \text{(1b)}$$

Table 1 shows a typical example of fiber angle at different circumferential locations for a 30-degree gore strip with 0-degree ideal angle.

LAMINATE MODELING

The geometrical characteristic of the laminated flange is the combination of layers with different ideal angle and start angle as described in the previous section. Its typical in-plane configuration is also shown in Fig. 1. In combining plate theory with laminate theory, the assumption is that each composite lamina can be analyzed as a thin plate. For each lamina of the flange, the stress-strain relation is [15]

$$\begin{Bmatrix} \sigma_\theta \\ \sigma_r \\ \sigma_{r\theta} \end{Bmatrix}^k = [Q]^k \begin{Bmatrix} \varepsilon_\theta \\ \varepsilon_r \\ \varepsilon_{r\theta} \end{Bmatrix}^k \quad (2)$$

and the classical constitutive equations of a laminate is [15]

$$\begin{Bmatrix} N_\theta \\ N_r \\ N_{r\theta} \\ M_\theta \\ M_r \\ M_{r\theta} \end{Bmatrix} = \begin{bmatrix} A_{11} & A_{12} & A_{16} & B_{11} & B_{12} & B_{16} \\ & A_{22} & A_{26} & B_{21} & B_{22} & B_{26} \\ & & A_{66} & B_{61} & B_{62} & B_{66} \\ & & & D_{11} & D_{12} & D_{16} \\ & \text{symm.} & & & D_{22} & D_{26} \\ & & & & & D_{66} \end{bmatrix} \quad (3)$$

or

$$\begin{Bmatrix} N \\ \overline{M} \end{Bmatrix} = \begin{bmatrix} A & \vdots & B \\ \cdots & \vdots & \cdots \\ B & \vdots & D \end{bmatrix} \begin{Bmatrix} \varepsilon^o \\ \overline{K} \end{Bmatrix}$$

where N = Total number of layer
[A] = Extensional stiffness matrix
[B] = Coupling matrix
[D] = Flexural stiffness matrix

STRESS/MOMENT RESULTANTS

The flange is considered to be used to connect two ducts and is subjected to internal pressure, bending moment and torque transferred from the duct. The stress/moment resultants due to these loads are presented in the following.

(i) Stress resultant due to internal pressure

The stresses of a hollow cylinder subjected to internal pressure (P_i) are given by [16], and the stress resultants N_r and N_θ distributed circumferentially over the flange cross section are

$$N_r = \frac{R_i^2 P_i}{R_o^2 - R_i^2} \left(1 - \frac{R_o^2}{r^2} \right) \times t$$

$$N_\theta = \frac{R_i^2 P_i}{R_o^2 - R_i^2} \left(1 + \frac{R_o^2}{r^2} \right) \times t \quad (4)$$

$$N_{r\theta} = 0$$

where R_o and R_i are the outer and inner radii of the flange respectively.

(ii) Moment resultant due to bending moment and torque

The moment resultants M_T and M_r due to bending moment (BM) applied to the flange are

$$M_T = \frac{BM}{2 R_i} \times (\frac{R_o - R_i}{2}) \qquad (5)$$

$$M_r = \frac{BM \times (R_o - R_i)}{8 \pi R_i^2} \qquad (6)$$

and the moment resultant due to torque (T) applied to the flange is

$$M_{r\theta} = \frac{T}{2 \pi R_i} \qquad (7)$$

3. DESIGN PROCEDURE

An interactive computer program was developed to obtain the optimal flange design parameters. Fig. 2 shows the flow chart of the program. This interactive computer program can be run on VAX or IBM PC, and real-time on-line graphics is supported for the IBM PC version.

Initially, geometrical dimensions of flange, loading conditions, layup configuration and types of material are input into the program. The program then begins stress calculation and the calculated stresses coupled with failure criterion are used to determine number of layers required to resist the applied loads.

4. RESULTS AND DISCUSSIONS

In the following, for demonstration purpose, selected numerical results of different layup patterns under a given loading condition are presented and discussed. AS-4397 polymer fiber material and applied loads of internal pressure, P_1=72 psi, bending moment, BM=437,600 lb-in, and torque, T=10,000 lb-in were used in this numerical analysis. The material properties of AS-4397 and polymer (matrix) PMR15 are listed in Table 2.

Four layup patterns, 0/90, 30/-30, 45/-45, and 0/45/-45/0 were analyzed and four gore strip angles were selected for evaluation. Since fiber angle and stiffness matrix vary along the flange circumferential location, the number of layers required at each circumferential location will differ as well. If butt-joint effect is taken into consideration, material properties at the butt-joint location are set to equal matrix properties. Results of required layers for each layup pattern are shown in Figs. 3 and 4.

As shown in Fig. 3, for 0/90 layup pattern, 15-degree gore angle is the best choice whereas 60-degree gore angle is the worst one. Also, for 60-degree gore angle, the number of layer required increases from 64 to 124 if the effect of butt-joint is considered. For layup pattern of 30/-30, if butt-joint effect is taken into consideration, the total number of layers required is the same for the four gore angles considered, i.e. critical location is at the butt joint location. For layup pattern of 45/-45, the trend is the same as that for layup pattern of 30/-30, i.e. total layers required are the same for the four gore angles considered and butt joint is the critical location. For layup pattern of 0/45/-45/0, the total number of layers required increases with the increase of gore angles, with or without considering the butt-joint effect.

Fig. 4 shows pieces versus layer plots which were obtained by combining the figures similar to Fig. 3 for every layup pattern and gore angle. These plots show that the number of layer and the corresponding number of gore strip required. Generally, 15-degree gore strip requires minimum number of layer and is the best choice. For gore angle of 15, 30 and 45 degrees, the best layup is 0/90 while for gore angle of 60 degrees layup 30/-30 will be the better choice.

4 CONCLUSIONS

In this investigation, based on classical laminate theory and maximum stress failure criterion, a parametric model was developed for predicting the strength of laminated composite flange subjected to combined external loads.

An interactive computer program was developed as a design tool that can provide the following information and advantages:

(i) The program takes into account the loading conditions, variations of gore angles, start angles, location angles, layup patterns, butt-joint effect, flange dimensions and types of composite material for on-line optimum design of laminated composite flange.

(ii) The pieces versus layers plots (Fig. 4) can be used as a guideline for designer to determine and choose the best layup configuration for optimum performance and minimum fabrication cost.

(iii) The design code is capable of designing laminated composite flange in the presence of butt joint.

Although only limited layup configurations and one loading condition were presented in this analysis, the design program can be easily extended to more sophisticated configurations and other loading conditions.

ACKNOWLEDGEMENT

This work was sponsored by Defense Advanced Projects Agency (**DARPA**), under contract No. MDA972-88-C-0047 for DARPA Initiative in Concurrent Engineering (**DICE**).

REFERENCES

[1] J,N. Reddy, 'Energy and Variational Methods in Applied Mechanics', 1984.

[2] K. H. Lo, R. M. Christensen, E. M. Wu, stress solution determination for high order plate theory', Int. J. Solids Structures, Vol.14, 655-662, 1978.

[3] J. N. Reddy, 'A simple higher-order theory for laminated composite plates', J. of Applied Mechanics, Vol.52, 745-752, 1984.

[4] N.D. Phan, J.D. Reddy, 'Analysis of laminated composite plates using a higher-order shear deformation theory', Int. J. For Numerical Methods In Eng.,Vol.21, 2201-2219, 1985.

[5] H. Murakami, 'Laminated composite plate theory with improved in-plane responses', J. Applied Mechanics, Vol.53, 661-666, 1986.

[6] N. J. Pagano, 'Stress fields in composite laminates', Int. J. Solids Structures, Vol.14, 385-400, 1978.

[7] E. Reissner, Y. Stavsky, 'Bending and stretching of certain types of heterogeneous aeolotropic elastic plates',J. Appl. Mech., Vol.28, p.402, 1961.

[8] S. B. Dong, K. S. Pister, R. L. Taylor, 'On the theory of laminated anisotropic shells and plates', J. Aero Sci., Vol. 28, P.969, 1962.

[9] W. J. Park, 'An optimal design of simple symmetric laminates under the first ply failure criterion', J. Comp. Mat'l. Vol.16, 341-353, 1982.

[10] R. M. Christensen, E. M. Wu, 'Optimal design of anisotropic (fiber-reinforced) flywheels', J. Comp. Mat'l, Vol.11, P.395,1977.

[11] S. S. Wang, I. Choi, 'Boundary-layer effects in composite laminates', J. Applied Mechanics, Vol.49, 541-560, 1982.

[12] C. C. Lin, C. C. Ko, 'Stress and strength analysis of finite composite laminate with elliptical holes', J. Comp. Mat'l, Vol.22, P.373, 1988.

[13] J. H. Lee, S. Mall, 'Strength of composite laminate with reinforced hole', J. Comp. Mat'l, Vol.23, 1989.

[14] W. J. Park, 'Symmetric three-directional optimal laminate design', J. Comp. Mat'l, Vol.21, P.532,1987.

[15] R. M. Jones, 'Mechanics of composite materials', 1975.

[16] S. P. Timoshenko, J. N. Goodier, 'Theory of elasticity', 3rd, 1970.

Table 1   Fiber angle at different location and SA for IDA=0

FOR GORE ANGLE= 30 DEGREES
     IDEAL ANGLE= 0 DEGREE

START ANGLE	0	5	10	15	20	25	30
LOCATION ANGLE	\multicolumn{7}{c}{FIBER ANGLE}
0	-15/15	-10	-5	0	5	10	15
5	10	-15/15	-10	-5	0	5	10
10	5	10	-15/15	-10	-5	0	5
15	0	5	10	-15/15	-10	-5	-10
20	-5	0	5	10	-15/15	-10	-5
25	-10	-5	0	5	10	-15/15	-10
30	-15	-10	-5	0	5	10	-15/15

Table 2   Composite Material properties

property	AS-4397	IM7/CE9220	PMR 15	CE9220
E_x	18.27 Msi	20 Msi	0.47 Msi	1.2 Msi
E_y	1.392 Msi	1.2 Msi	0.47 Msi	1.2 Msi
G_{xy}	1.029 Msi	0.7 Msi	0.24 Msi	0.7 Msi
ν_{12}	0.3	0.35	0.36	0.37
X_t	203.15 Ksi	320 Ksi	8 Ksi	8 Ksi
X_c	205.9 Ksi	200 Ksi	16 Ksi	24 Ksi
Y_t	5.37 Ksi	8 Ksi	8 Ksi	8 Ksi
Y_c	29.87 Ksi	24 Ksi	16 Ksi	24 Ksi
S	13.49 Ksi	14 Ksi	8 Ksi	14 Ksi

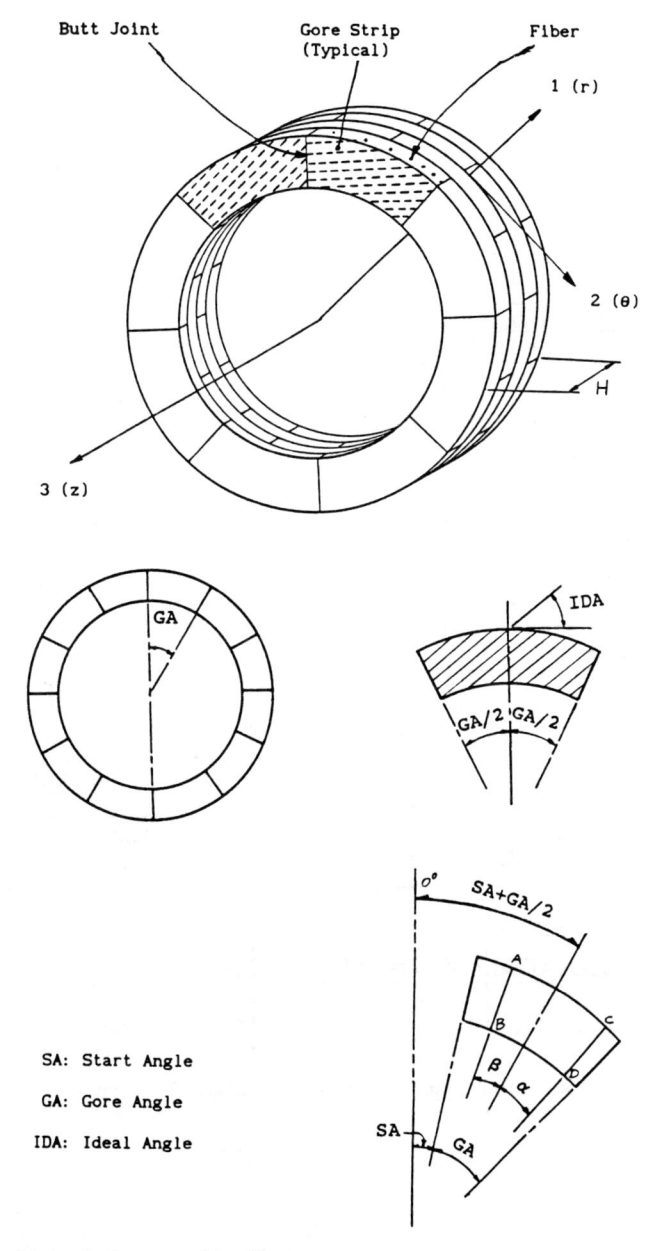

Fig. 1 A typical laminated composite flange

SA: Start Angle
GA: Gore Angle
IDA: Ideal Angle

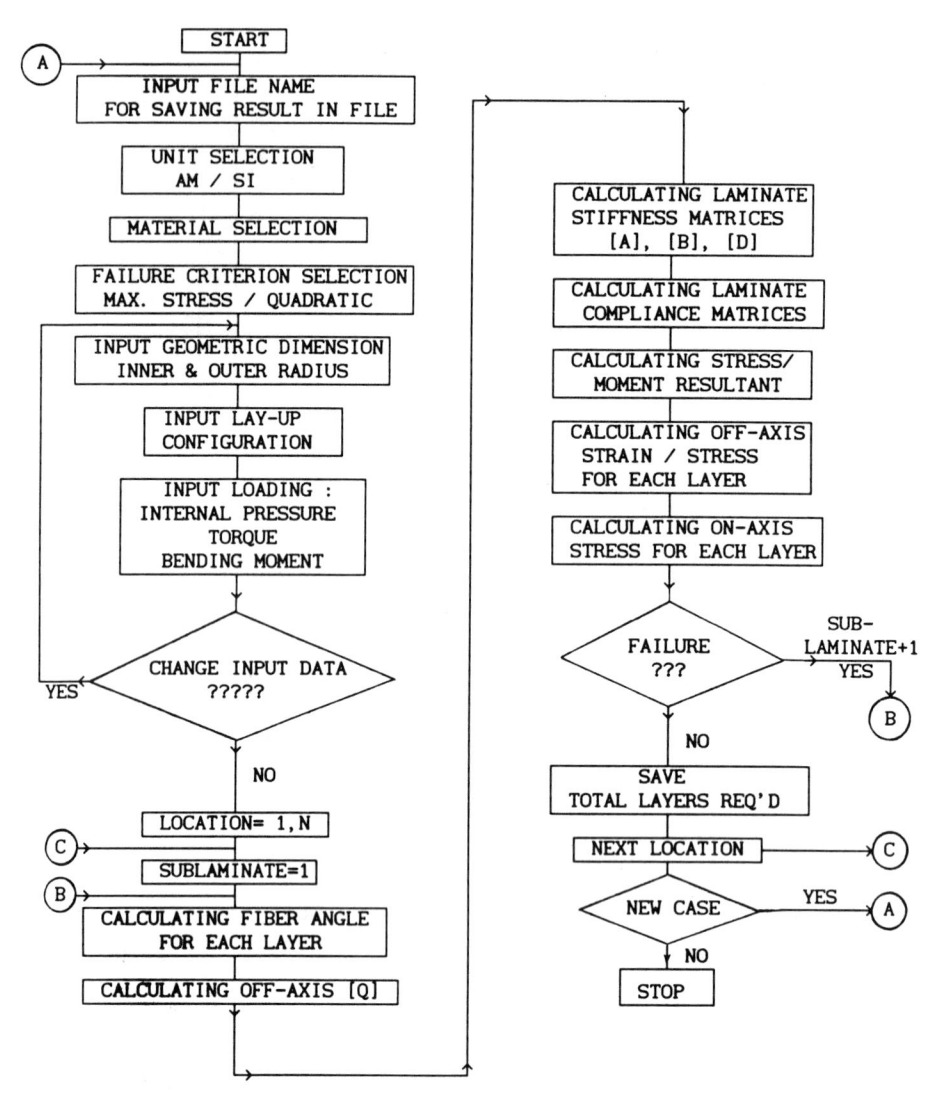

Fig. 2 Laminated composite flange design flow chart

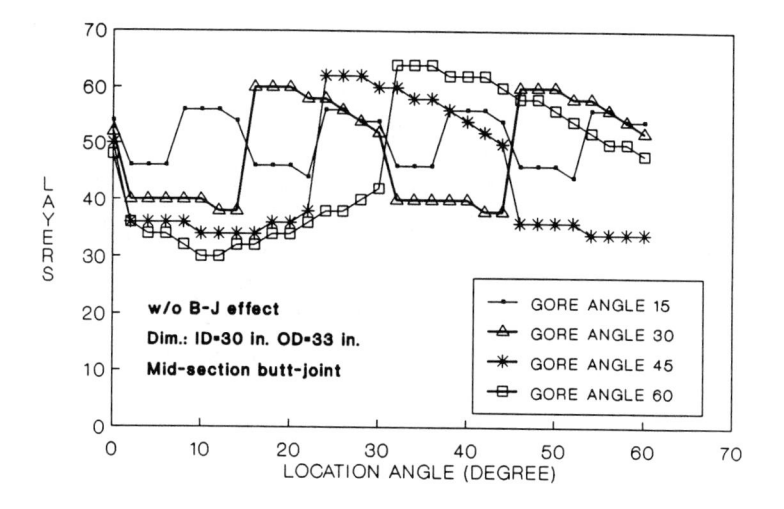

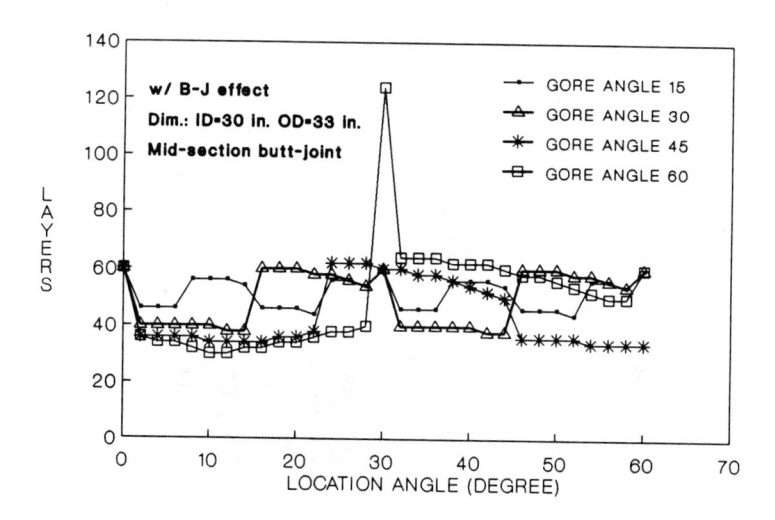

Fig. 3 Layers vs. location angle curves for layup pattern of 0/90

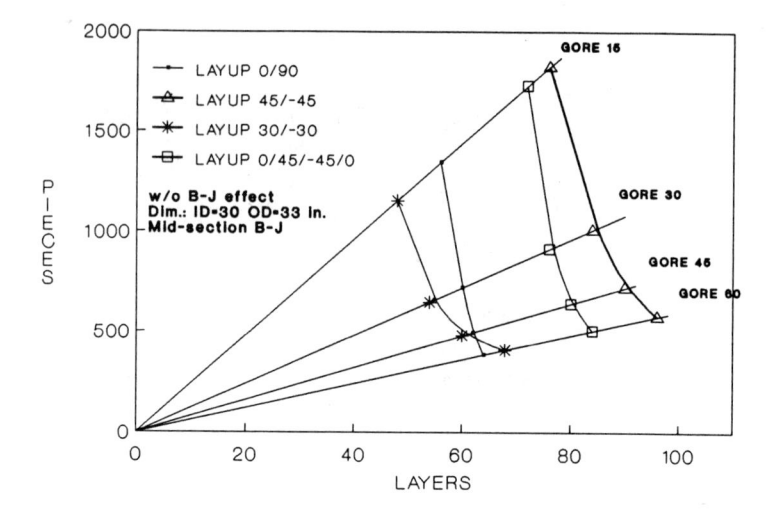

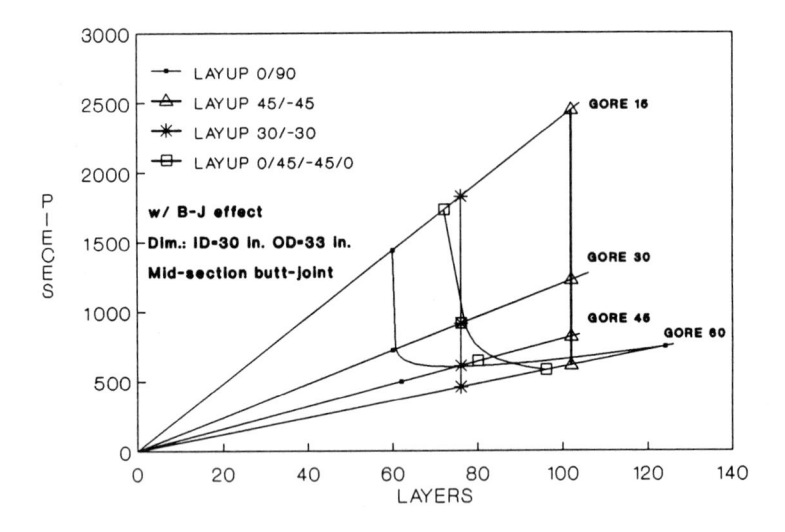

Fig. 4 Pieces vs. layers curves

Computer Aided Dynamic Analysis of Laminated Composite Plates

ALEXANDER E. BOGDANOVICH and ENDEL V. IARVE
Latvian Academy of Sciences
Engineering and Technology Center
Riga, Latvian SSR, USSR

SUREN N. DWIVEDI
Department of Mechanical and Aerospace Engineering
West Virginia University
Morgantown, WV

ABSTRACT

A method of calculation of a stress-strain state in a multilayered rectangular plate manufactured of isotropic or orthotropic materials, under the influence of transversal dynamic load, is proposed. The calculational procedure is based on specific approximations of displacements along all the coordinates, by use of specially elaborated spline functions. Computer aided numerical realization of the method is provided for the two - dimensional case (plane deformation of a plate) with approximation of displacements by special second - degree (with respect to transversal, z - coordinate) and third - degree (with respect to longitudinal, x - coordinate) spline functions. The analysis of transverse stresses arising under short - time impulse in graphite/epoxy and organic glass/polymeric adhesive laminated plates is performed. Numerical examples illustrate also the applicability of the method to the calculation of a stress - strain state in a plate together with characteristics of the impact contact interaction. The results illustrating dependencies of a contact force on time are presented.

INTRODUCTION

The problem of calculation of a stress - strain state in multilayer and reinforced structural elements is becoming ever more actual. The complexity of the problem is explicable by the principal structural specifics of such an elements, to be taken into account. Namely, the mechanical inhomogeneity of reinforced materials and the presence of layers having different mechanical characteristics in a laminate. These specific features require that special approaches to the analyses would be elaborated. Thus, for example, when solving the problem on the base of the theory of elasticity in terms of displacements, using finite element or finite difference methods, one is to add certain special procedures as to calculate interfacial stresses. The alternative approach, based on a hybrid finite element with independent displacement and stress approximations, leads to high - order systems of equations, that provides difficulties in practical application of the approach. Therefore, for the purpose to elaborate more efficient numerical analysis for various dynamic problems of laminated composite structural elements, some new ideas are required.

Considering the problem of dynamic bending of a multilayer plate due to the impact by a rigid body, a computer simulation procedure based on some theoretical model describing the process of contact interaction, has to be evolved. Such a model, in its simplest version, usually is based on the classic Hertz's formula, which constitutes the relation between a contact force and depth of penetration of an indentor into the target. The applicability of the approach is restricted by the condition that the interaction time is substantially greater than the time of one way run of a stress wave through the characteristic size of a targed body (in particular, its thickness).

The more general and, correspondingly, more complex approach for solving the problem of impact - contact interaction, is based on a formulation of the joint dynamic problem for the system "impactor - target". There are two versions of such an approach known from the literature. The first one is tightly connected with the particular numerical method to be applied: finite element or finite difference method [1]. The second is based on some variational principles [2] and seems to be more promising for complex dynamic contact problems. It is worth to note, that particular numerical results have been obtained by use of both these approaches only for rather simple three - dimensional quasi - static and two - dimensional dynamic problems.

In this paper, the method combining calculation of stress - strain state in a laminated plate on the basis of special spline - approximations of displacements, with the variational approach for the calculation of contact impact interaction between a plate and rigid impactor [3,4], is applied.

THE BASIC MATHEMATICAL PROCEDURE

The three - dimensional problem of the theory of elasticity for a laminated rectangular plate, consisting of isotropic or orthotropic layers, is under consideration. On the part of an upper surface (Fig. 1) acts some localized or distributed dynamic load. The bottom surface is free of loading. Along all four edge surfaces any necessary type of boundary conditions can be imposed. At the initial time instant all the displacements and their velocities in the whole volume of a plate, have to be prescribed.

Let us introduce Cartesian coordinates x,y in the plane of a plate, and z - coordinate through the thickness. The corresponding plate dimensions are: L, A and H (see, Fig. 1). It is important to underline, that no presupposed restrictions have to be imposed on the mutual relations of L, A and H values.

The total number of plies in a laminate is n, their thicknesses are h_k, k=1,...,n. The direction of the reinforcing fibers in a layer coinsides with the Ox or Oy axis ($0°$ and $90°$ plies, accordingly).

The first step of the solution is to divide each of the layers into certain number of sublayers l_k by the planes $z=z_i$, i=1,...,N-1; $0=z_0 < z_1 < ... <z_{N-1}<z_N=H$. Here, $N= \sum_{k=1}^{n} l_k$ is the total number of sublayers. The coordinates of interlaminar planes are z_{I_s}. Let us represent the dependencies of displacements versus z-coordinate in the form

$$u_x(x,y,z,t)=\sum_i U_i(x,y,t)\Phi_i(z); \quad u_y(x,y,z,t)=\sum_i V_i(x,y,t)\Phi_i(z); \quad u_z=\sum_i W_i(x,y,t)\Phi_i(z), \quad (1)$$

where $\Phi_i(z)$ is a set of linearly - independent spline functions, having local supporter, and providing following necessary properties to displacements and deformations:
(a) displacements u_x, u_y, u_z are continuous through the thickness of a package;
(b) transversal deformations $\varepsilon_{xz}, \varepsilon_{yz}, \varepsilon_{zz}$ have the first - order disruptions on the interlaminar planes.

The condition (b) is necessary, but not sufficient as to fulfil the requirement of continuity of the transverse stresses $\sigma_{xz}, \sigma_{yz}, \sigma_{zz}$ through the thickness of a package.

For the purpose to satisfy the above conditions on displacements and deformations, a special recurrent procedure for constructing linearly - independent polynomial spline functions had been elaborated. The spline functions, designated as $\Phi_{m,l}(z)$, are of m - th degree and have the defect $k_1 = 1$ in z_i -nodes (i ≠ I_s, s =1,..., n-1) and the defect k_2, m ⩾ k_2 ⩾ k_1 in z_{I_s} - nodes. If k_2=m, then in accordance with (1), displacements can be m-1 times continuously differentiated with respect to z - coordinate, on the intervals] $z_{I_{s-1}}$, z_{I_s} [, s=1,..., n, but their first derivatives can be discontinuous at z= z_{I_s}. For the case m ⩾ 2, the proposed approximation gives continuous field of displacements through the thickness of a package, as well as continuous stresses and strains through the thickness of each physical layer. On the interlaminar planes, there remain disruptions in the magnitudes of transverse stresses that can be reduced to the prescribed value, by use more and more dense subdivision into sublayers. The second degree (m=2) basic spline functions having the defect $k_2 = 2$ at z - values, corresponding to the interlaminar planes, are used in the following calculations.

The second step in the solution is to approximate $U_i(x,y,t)$, $V_i(x,y,t)$, $W_i(x,y,t)$ functions with respect to x and y - coordinates by polynomial spline functions. Let us introduce on Ox and Oy axes two sets of nodal points: $0 = x_0 < x_1 < ... < x_M = L$ and $0 = y_0 < y_1 < ... < y_P = A$. Due to presupposed homogeneity of all the mechanical characteristics of a monolayer along x and y coordinates, the $U_i(x,y,t)$, $V_i(x,y,t)$, $W_i(x,y,t)$ functions have to be continuously differentiated any times with respect to x and y coordinates.

Finally, the displacements can be introduced in the following form:

$$u_x(x,y,z,t) = \sum_{i=0}^{N_m} \sum_{k=0}^{P+m2-1} \sum_{j=0}^{M+m1-1} U_{ijk}(t) \Phi_{m,i}(z) \chi_{m1,j}(x) \chi_{m2,k}(y); \tag{2}$$

$$u_y(x,y,z,t) = \sum_i \sum_k \sum_j V_{ijk}(t) \Phi_{m,i}(z) \chi_{m1,j}(x) \chi_{m2,k}(y); \quad u_z(x,y,z,t) = \sum_i \sum_k \sum_j W_{ijk}(t) \Phi_{m,i}(z) \chi_{m1,j}(x) \chi_{m2,k}(y),$$

where $\{\chi_{m1,j}(x)\}$ and $\{\chi_{m2,k}(y)\}$ are two sets of linearly - independent spline functions of the m_1 and m_2 - degrees, accordingly. They are to have the $k_1=1$ order defect. The recurrent procedure of calculation of the basic spline functions is described in [1,2].

By use of (2), one can calculate three - dimensional field of displacements in a laminated rectangular plate at homogeneous or inhomogeneous boundary conditions, imposed on the six boundary surfaces. Particularly, for a laminate as a whole there can be formulated free support, clamping, free edge conditions, or their combination. It is possible also to impose an individual set of boundary conditions for each ply in a laminate, or for a certain group of plies.

After the basic approximation of displacements is specified, the problem can be treated as following. The kinetic K and potential Π energies of a plate have to be expressed through the coefficients $U_{ijk}(t)$, $V_{ijk}(t)$, $W_{ijk}(t)$. In the case of a prescribed transverse load, applied on the top surface z=H, we are to satisfy the following boundary conditions:

$$\sigma_{xz}(x,0,t) = \sigma_{yz}(x,0,t) = \sigma_{zz}(x,0,t) = 0 \; ; \; \sigma_{xz}(x,H,t) = \sigma_{yz}(x,H,t) = 0 \; ; \; \sigma_{zz}(x,H,t) = -q(x,t), \tag{3}$$

and also to require that

$$\delta \int_0^t [K(\frac{dU_{ijk}}{dt}, \frac{dV_{ijk}}{dt}, \frac{dW_{ijk}}{dt}) - \Pi(U_{ijk}, V_{ijk}, W_{ijk})] dt = 0 \tag{4}$$

By solving the set of ordinary differential equations obtained from (4), one can calculate the $U_{ijk}(t)$, $V_{ijk}(t)$, $W_{ijk}(t)$ functions. After that, displacements, strains and stresses have to be calculated in a direct way, in accordance with the general scheme of solution, shown in Fig. 2.

The method proposed can be used also for the problem of dynamic contact interaction between an elastic laminated plate and rigid indentor. The mathematical background and algorithm of calculations for this case are described in [3,4]. The problem is reduced finally to the unconditional minimization of the functional

$$\int_0^t [K(\frac{dU_{ikj}}{dt}, \frac{dV_{ikj}}{dt}, \frac{dW_{ikj}}{dt}) - \Pi(U_{ikj}, V_{ikj}, W_{ikj}) + \frac{1}{2}MV^2 + \lambda \int\int_\Omega \Lambda^2(x,y,t) dx dy] dt, \tag{5}$$

where M and V(t) are mass and velocity of impactor, λ - Lagrange's indeterminate multiplier,

$\Lambda(x,y,t)$ - some function depending on the displacement of impactor. Variation of (5) with respect to U_{ikj}, V_{ikj}, W_{ikj}, V, ε, λ leads to the interrelated system of ordinary differential equations. The system is solved by the numerical algorithm based on Wilson's θ - method, with special additional iterative procedure elaborated to satisfy the necessary interaction conditions at each step of numerical integration in time.

NUMERICAL EXAMPLES

Let us consider two - dimensional problem of the dynamic bending of a laminated plate, presupposing that the plate is infinitely long in y - direction (cylindrical bending). The load is prescribed by the formula

$q(x,t) = q_0 Q(t) \beta(x)$, where $Q(t) = t/t_0$ at $t \leq t_0$, $2-t/t_0$ at $t_0 < t \leq 2t_0$, 0 at $t > 2t_0$;

$\beta(x) = [(x-L/2)^2 - \varepsilon^2]^2 / \varepsilon^4$ at $|x-L/2| \leq \varepsilon$, and 0 at $|x-L/2| > \varepsilon$.

Two typical plates are considered. The first is manufactured of a cross - ply graphite/epoxy composite with unidirectionally reinforced monolayers. The material of a monolayer has following characteristics: $E_1 = 1.94.10^{11}$ N/m^2, $E_3 = 7.72.10^9$ N/m^2, $v_{13} = 0.3$, $G_{13} = 4.21.10^9$ N/m^2, $\rho = 1.63.10^3$ kg/m^3 (subscript 1 corresponds to the direction of reinforcement in a monolayer). The second plate is manufactured of isotropic organic glass layers with characteristics: $E=6.10^{10}$ N/m^2, $v=0.3$, $\rho=1.5.10^3$ kg/m^3, and adhesive polymeric layers with $E=2.8.10^9$ N/m^2, $v=0.33$, $\rho=10^3$ kg/m^3. The values L=1m, H=0.01m, ε =0.02L are used.

The magnitudes of σ_{zz}/q_0 for three-ply laminates under short-time impulse load are shown in Fig.3 and Fig. 4 for several consecutive time instants. The direction of motion of the input impulse is indicated by arrows, as well as the direction of motion of the impulses passing through and reflected by the interfaces, upper and bottom free surfaces. It is seen from Fig.3, that in the case of orthotropic three-ply laminate having [0/90/0°] layer layup, distorsion of the input impulse is negligible. There are no visible impulses reflected from the interfaces. This effect can be explained by identity of transverse elastic characteristics of a monolayer, presupposed in this calculation. From the results shown in Fig.4, it is obvious that in the case of a strong inhomogeneity of elastic characteristics through the thickness of a laminate, it is possible to govern stress wave propagation process and, therefore, to find the optimum structure of a laminate, which provides best resistance against the prescribed short-time pressure impulse.

The results of calculation of a contact interaction force between a laminated plate and rigid impactor are shown in Fig.5. The comparison of numerical results, presented in [5] and obtained by the method proposed, is carried out for the 11-ply graphite/epoxy composite. The impactor in this case is a spherical steel ball having radius R=6.35 mm and mass M=8.18 g. The initial impactor velocity V=35 m/sec. In [5], the deformation process was calculated by use of finite element method, on the basis of a theory of thin orthotropic plates. Such an approach doesn't allow to analyze stress wave propagation process through the thickness of a laminated plate. Interaction between plate and impactor is modelled in [5] by use of a special indentation low, which combines Hertz's-type formula and some additional data obtained from experiment on static indentation of a steel sphere into the composite under consideration.

The main conclusion from a comparison of the curves presented in Fig.5, is that the contact force is depending strongly on the wave propagation process through the thickness of a plate. This process is responsible for the first maximum on the dependency "contact force - time", which corresponds to the t=2t$_H$ time instant, where t$_H$ is the one-way wave trip through the thickness of a plate. This process is also responsible for the specific high-frequency oscillations on the F(t) curve. The example above shows the importance of accounting for a finite thickness of a plate in the analysis of its resistance to transverse impact loading.

By use of the method proposed, one can calculate all the components of displacement vector, strain and stress tensors in an arbitrary point inside the volume of a laminated plate, during the whole impact event. The calculated values can be compared at each time step to the corresponding ultimate values. As a result of such a comparison, the critical parameters of impact load can be determined. Several examples of failure analysis of laminated composite plates under the effect of a low-velocity impact by a rigid body were presented in [3,4]. Some recent numerical results show, that even at rather low impact energy, ranging from 2.0 to 2.5 J, it is possible to obtain all the experimentally observed modes of failure, namely fiber breakage, matrix cracking, delamination, spalling. The realization of a particular failure mode depends on the structural parameters of a plate material and a combination of mass and velocity of impactor.

Some other results obtained by the method proposed, show also that a reasonable increase in impact damage resistance of a graphite/epoxy composite plates can be achieved by such a promising technological improvements, as spatial (out-of-plane) reinforcement and incorporating soft polymeric interleaves between the plies of a basic composite.

REFERENCES

1. A. I. Gulidov and I. I. Shabalin, "Numerical Realization of Boundary Conditions in Dynamic Contact Problems", Institute of Theoretical and Applied Mechanics, Preprint (in Russian), Novosibirsk, 1987, 37 p.
2. A. S. Kravchuk, "Variational Method in Dynamic Contact Problems", Mechanics of Deformable Bodies and Structures, School - Seminar on the Theory of Elasticity and Visco-Elasticity (in Russian), Erevan, 1985, p.p. 235-241.
3. A. E. Bogdanovich and E. V. Yarve, "Numerical Analysis of Impact Deformation of Laminated Composite Plates", Mechanics of Composite Materials (in Russian), 1989, No 5, p.p. 804-820.
4. A. E. Bogdanovich and E. V. Yarve, "Numerical Analysis of Laminated Composite Plates Subjected to Impact Loading", Proceedings of the American Society for Composites, Fourth Technical Conference, 1989, p.p. 399-409.
5. B. V. Sankar and C. T. Sun, "Low Velocity Impact Response of Laminated Beams Subjected to Initial Stress", AIAA Journal, 1985, Vol. 23, p.p. 1962-1969.

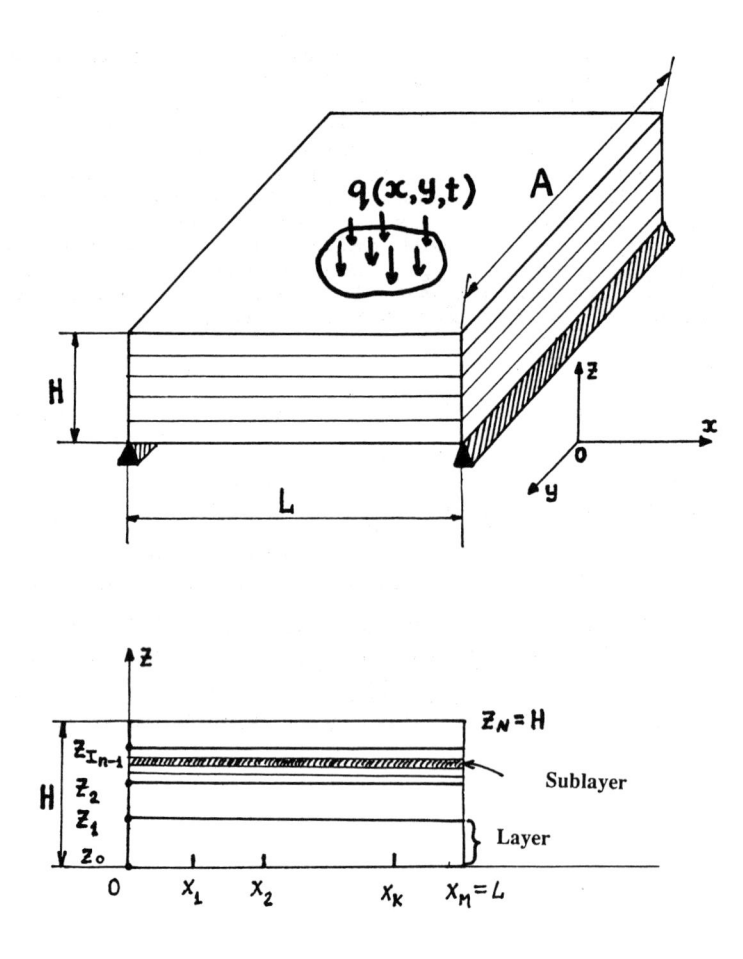

Fig. 1. Laminated Rectangular Plate Under Transverse Impact

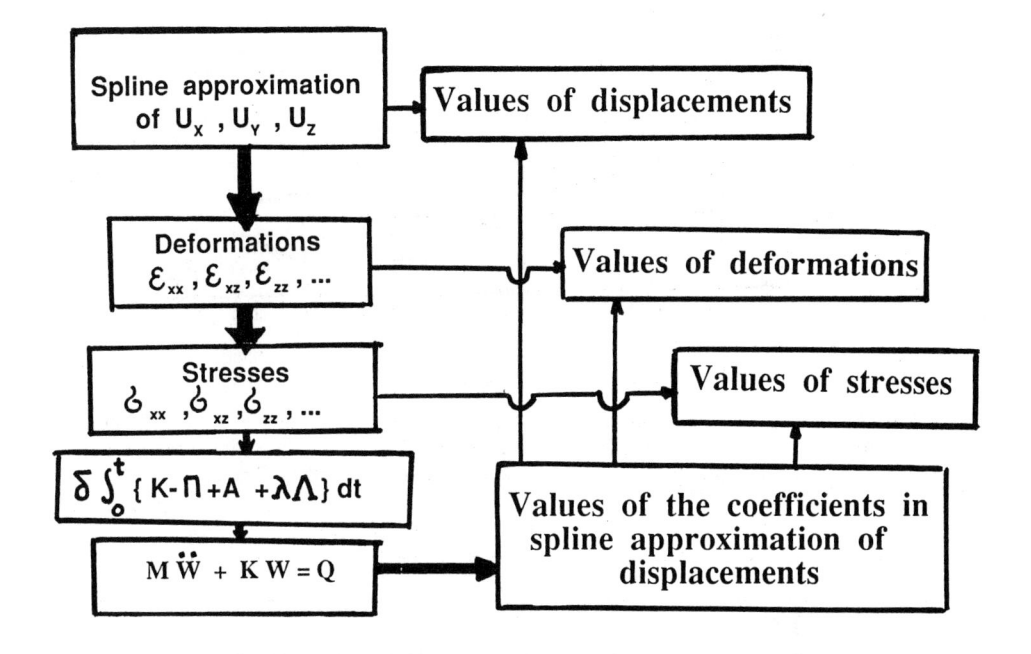

Fig. 2. The General Scheme of Computer Aided Analysis

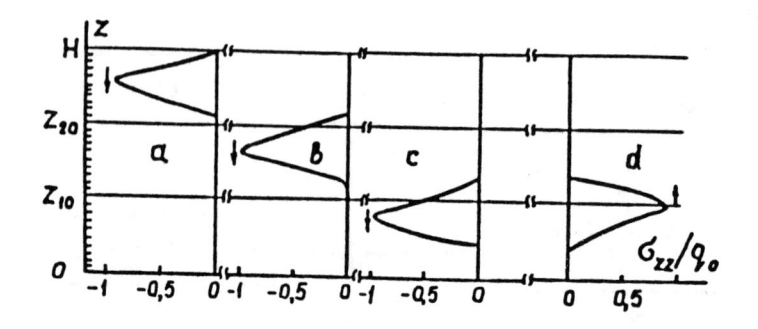

Fig. 3. Dependencies of Transverse Normal Stress on Z - Coordinate at the Middle Section of a Three - Ply [0/90/0°] Graphite/Epoxy Plate at Several Time Instants: $t = 2t_0$ (a), $4t_0$ (b), $6t_0$ (c), $10t_0$ (d); t_0 - the Loading Time

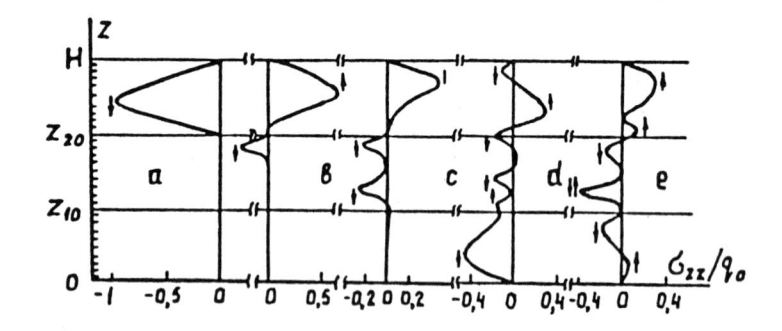

Fig. 4. Dependencies of Transverse Normal Stress on Z - Coordinate at the Middle Section of a Three - Ply Organic Glass/Adhesive/Organic Glass Plate at Several Time Instants: $t = 2t_0$ (a), $4t_0$ (b), $8t_0$ (c), $11t_0$ (d) $12t_0$ (e); t_0 - the Loading Time

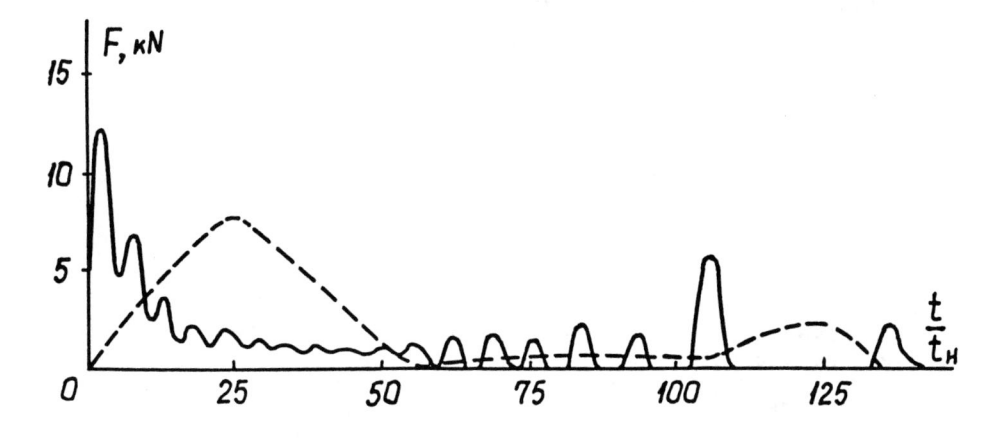

Fig. 5. Dependencies of a Contact Force on Time, Calculated in [5] (Dashed Line) and by Use of the Proposed Method (Solid Line)

Integration of Rigid-Plastic Simulation Engines into Engineering Database System for Advanced Forging

TATSUHIKO AIZAWA and JUNJI KIHARA

Department of Metallurgy
University of Tokyo
Tokyo, Japan

Summary
In the advanced forging, nondestructive forgeability evaluation as well as productive modeling are indispensable to predict the limit of working especially for new materials, to describe the change of geometries and dimensions in process and to make preform designs. For those purposes, engineering database system is preferable to process thus obtained data by both rigid-plastic simulations and experiments and to provide thus evaluated informations through retrievals and functional operations. Architecture of our developing engineering database system is presented and discussed with the rigid-plastic simulation engines. Through crack sensitivity evaluation, the limit of upsetting is estimated with comparison to experimental results.

Introduction
In the forging process design, the elasto-plastic or rigid-plastic simulations [1,2] become powerful tools to provide the mechanical behaviors of materials in forging even before the actual forging process; the rigid-plastic analysis, where elastic deforming medium is assumed to be rigid, has been widely used as the forging design tool [3]. Since relatively huge amount of simulation data are output by such rigid-plastic finite element analysis, accurate data management is required for the precise forging evaluation in the form of neutral data file. This is why 'Database' is required for forging process evaluation. In the database, 1) any free format data can be dealt with in the form of table or relation and 2) thus obtained data by simulations can be arranged into data structure by using the schema. Furthermore, through retrieving and archiving both the original and the processed data, forgeability evaluation of simulated results can be accommodated for reliability assurance of the existing forging process and improvement of processes with use of database functions [4,5]. Especially, various mechanical models are installed on thus created database to make synthetical forgeability evaluations with coupling effects, inhomogeneities of materials or cracking behaviors taken into account. In addition, we can afford to construct some technological evaluation axes for the present forging process to discuss over coming form of forging with aid of thus evaluated data relations which have been grown-up in the above database through the feasibility round-robin tests. In the present study, both the rigid-plastic simulation engines and the architecture of engineering database for forgeability evaluation will be briefly stated with some comments on the specific features of the present database system; finite element method is used for the rigid-plastic analysis and

RTI-RIM relational database for construction tools of database. In particular, fracture mechanics model will be discussed for evaluation of defect or crack sensitivity of materials in working on the basis of linear fracture mechanics; the predicted limit of upsetting is compared to experimental data.

Rigid-plastic simulation engine
Since large deformations and strains are often observed in the usual forging processes, rigid-plastic state can be assumed for materials in working: 1) elastic strains are neglected and elastic medium is rigid, 2) rate-from formulation is used, and 3) updated lagrangian form is employed with incompressibility condition taken into account. Our developing simulation engines have the following features: [F1] Both plane strain and axisymmetric mechanical states are modeled by our developing finite element library, [F2] Penalty function method is used to consider the incompressibility or constant volume condition of plastic strain rate, [F3] Arbitrary Lagrangian Eulerian method or ALE method is installed to deal with the singular points at the corner between rigid tools and deformable work of materials. As illustrated in Fig. 1, this simulation engine is working not only to provide the original mechanical data for evaluation but also to make reanalysis on the database. For further improvements, both element control function and three dimensional analysis are to be installed to the present system.

Architecture of engineering database system for forging
To complete precise description of forging behaviors and forgeability evaluation prior to actual production, 1) necessary mechanical informations are calculated with sufficient accuracy, 2) through powerful data management, forgeability evaluation items are processed on the basis of thus simulated results, 3) through the field tests, lots of actual forging examples are dealt with in order to create adaptive and valid frame of forgeability evaluation to true forging behaviors. As shown in Fig. 1, both the plastic flow informations of shape, velocity and displacement and the mechanical data like strains and stresses

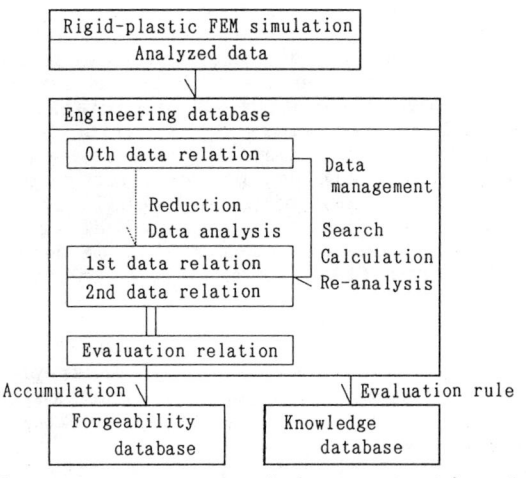

Fig. 1 Architecture of forgeability evaluation database

are straightforwardly obtained by the rigid-plastic engines, while the traction distribution applied to the tool surfaces or other elaborate mechanical parameters should be estimated by data processing on the database. That is, a hierarchical data structure must be created where 0-th data denote the simulated results or a little processed data, 1-st data the highly processed data through retrieval or arithmetic operations, and 2-nd data the newly predicted data by model evaluation. In the present system, 0-th data are stored into neutral files and represented in the visualized image on the screen. 1-st data are arranged and listed in the relational table for further use; the estimated tractions are to be used for elastic response analysis of dies and tools. Since the operating functions of database is not so powerful to construct the 2-nd data, programs and specific schemata should be designed to make reanalysis on the database. In this case, Both 0-th and 1-st data are transferred to reanalysis program from neutral file by several schemata, and thus obtained 2-nd data are stored again into database. For the present forgeability evaluation, construction of these 2-nd or higher classes in architecture leads to precise description of mechanical and metallurgical behaviors in actual forging processes.

Fracture mechanics model of forgeability evaluation

Main concern of issues appearing in the whole forging process including heat treatments is prediction of cracking or failure in the successive plastic deformation; reduction or pass schedule in forging or heat treatment conditions should be all reconsidered to be free from such failures. One approach to predict the ductile cracking is use of empirical relation of the accumulated strain and hydrostatic pressure with materials constant: due to Ref. [6], the ductile fracture criterion is held if the following inequality in the total equivalent strain $\bar{\varepsilon}$ and the pressure p is satisfied, or,

$$\int_0^{\varepsilon_1} <\bar{\varepsilon} + ap + b> d\bar{\varepsilon} \geq c \qquad (1)$$

where a, b, and c are materials constants to be determined by experiments. This approach is easy to be applied to actual situations if those materials properties are known; however, lots of experiments are necessary to determine those constants with sufficient accuracy. To be noted, since no relations are considered between microstructure of materials and cumulative damage or precracking behaviors, the effectiveness of thus obtained empirical relation is severely deteriorated in practical situation where inhomogeneities, microstructure or texture have strong influences on the mechanical and metallurgical properties of materials in working. Hence, another alternative methodology is indispensable to reconsider sensitivity of materials in work to geometric defect and flaws or materials inhomogeneities.

Authors [7,8] have been concerned with the application of a frame of fracture mechanics to this kind of workability evaluation. Through some fundamental studies, the first-phase frame of fracture mechanics models is proposed in what follows:
[A1] A flaw originating from an intergranular precrack or inclusion is assumed to be an implicit defect in materials,
[A2] Limit of toughness KIc even in plasticity or the surface energy release rate is obtained by fine-controlled uniaxial

tensile test,
[A3] Stress intensity factor K in application is calculated by reanalysis on the present database. Then, when K > KIc, brittle type fracture is assumed to take place posterior to significant amount of plastic deformation.

With respect to [A1], various precrack initiators exist in the real materials: participation of sulfur or sulfide, inclusions or micro porosities. These initiators are subject to both shear deformation and compressive pressure in forging process. Hence, at the critical condition, the intergranular cracking could occur from these deformed initiators. In general, it is difficult to determine the critical value for cracking behaviors in plastic deformation. Due to the precise uniaxial tensile testings at relatively high temperature, it is found that 1) the intergranular cracking takes place after 10 to 20 % plastic strains, and 2) crack initiation point corresponds to onset of branching in the uniaxial true stress vs true strain relation from the original stress-strain curve in ductile state where only dimple fracture surface is observed in failure. Then, KIc or critical energy release rate can be directly defined from the onset stress σ_c and the characteristic grain size c by the following equation:

$$K_{Ic} = \sigma_c \sqrt{\pi c} \quad \text{and} \quad \gamma_c = \frac{c}{2E}\sigma_c^2 \quad . \tag{2}$$

In evaluation on the database, the applied stress intensity factor K or energy release rate are calculated by the virtual crack extension method; as illustrated in Fig. 2, 1) the virtual crack is located perpendicular to the principal direction with the highest tensile stress, 2) the current element or set of elements are subdivided into smaller elements with mesh size equal to the grain size or the prescribed length, 3) with velocity field at the original element boundary kept constant, reanalysis is performed by one new element extension to calculate the total power of plastic work if a crack would initiate from the precrack, and 4) the energy release rate is determined by difference of powers between the cracked and the uncracked states due to the virtual crack extension.

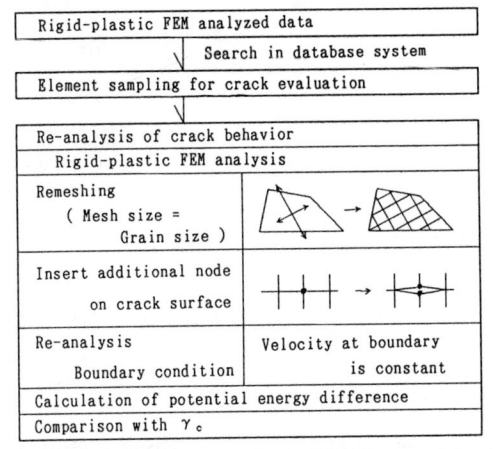

Rigid-plastic FEM analyzed data	
Search in database system	
Element sampling for crack evaluation	
Re-analysis of crack behavior	
Rigid-plastic FEM analysis	
Remeshing (Mesh size = Grain size)	
Insert additional node on crack surface	
Re-analysis Boundary condition	Velocity at boundary is constant
Calculation of potential energy difference	
Comparison with γ_c	

Fig. 2 Fracture mechanics model in the present database system

Table 1 Forgeability evaluation for the limit of upsetting

Stress criterion relation of forgeability evaluated in database

Reduction(%)	Element	Evaluation Point	σ (kgf/mm^2)
10	10	4	0.1084858D+02
20	10	4	0.2075828D+02
30	10	4	0.3450150D+02
40	10	4	0.5316217D+02
50	10	4	0.7631619D+02
60	10	4	0.9537270D+02

Experimental Data (Upsetting Test)

Limit Reduction	44.8 44.8 47.3 47.1 47.0
Average Limit Reduction	4 6. 2 %

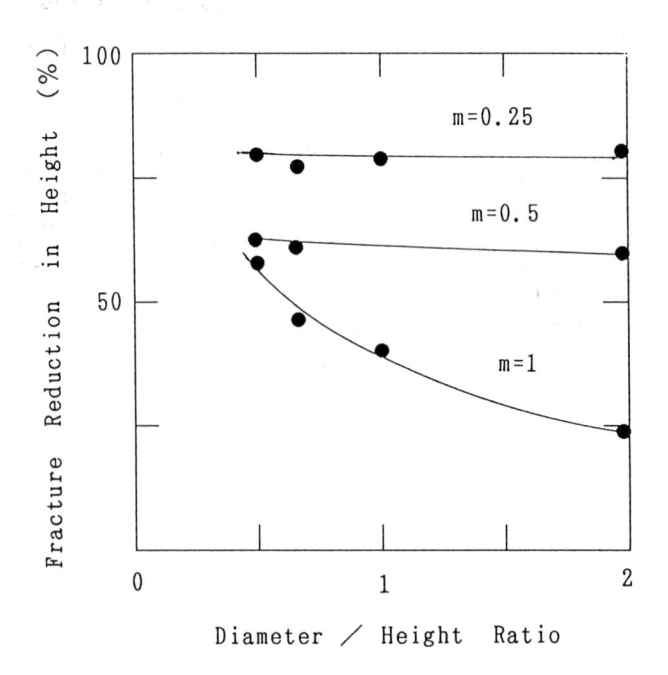

Fig. 3 Effect of aspect ratio to fracture reduction in height

Table 2 Forgeability evaluation of inner cracks

ε	4 %	1 2 %	
Crack Position [mm] (in radial direction)	0.981012	0.946805	
Crack Length [mm]	0.1962024	0.189361	× 3
γ_c (J)	2.18×10⁻⁵	2.03×10⁻⁵	6.09×10⁻⁵
Total Power [J/s] (Cracked Materials)	0.7863168D+01	0.162807D+02	0.1042628D+02
Total Power [J/s]	0.7883260D+01	0.1065565D+02	
Δ W (J/s)	-2.00×10⁻⁵	-2.76×10⁻⁵	-22.9×10⁻⁵

Let us evaluate the limit of upsetting with comparison to the experimental data; in experiment, YSM materials are employed where the ultimate strength is 67 Kg/mm2, the elongation 19%, the reduction in area 38% and the constitutive equation given by
σ = 29.9 (1+2339.7 ε)∧0.165   [Kg/mm2]. The cylindrical test piece is employed: 14 mm x 21 mm. Assuming the axisymmetric state in upsetting, crack extension could take place nearly uniaxially; in the present model, the fracture criterion leads to maximum tensile stress criterion. Table 1 shows that the estimated reduction when the maximum principle tensile stress reaches the ultimate strength is 40 to 50%, while the measured average limit of upsetting is 46.2 %. Fig. 3 depicts the effect of aspect ratio to such fracture reduction in height for various friction factors m. As observed in actual forging, critical reduction is further reduced with increase of m. Table 2 lists the calculated difference of power due to virtual extension in tensile testing when specimen fractures at 10% plastic strain.

Conclusion
Engineering database integrated with the rigid-plastic finite element analysis enables us not only to make precise, systematic forgeability evaluation for each forging process but also to accumulate and rearrange the previously simulated and evaluated data for further elaborate evaluation. Use of the fracture mechanics model on the present integrated database leads to quantitative discussions over embrittlement or loss of toughness during heat treatments or the materials process design with consideration of sensitivity to defects or inhomogeneities in shape and materials. Further improvement of the present system is being performed with use of object-oriented database with the rigid-plastic FEA which is adaptive to working on EWS and network circumstances.

References
[1] O. C. Zenkiewicz, et al., J. Strain Analysis, 10(3) (1975).
[2] R. Duggiraia et al., Near Net Shape (1988) 159-167.
[3] T. Aizawa et al., Proc. Symposium on Computational Methods in Structural Engineering and Related Fields 11 (1987) 151-156.
[4] T. Aizawa et al., Advanced Tech. Plast. 1 (1990) 59-64.
[5] T. Aizawa et al., Proc. Symposium on Computational Methods in Structural Engineering and Related Fields 13 (1989) 165-171.
[6] K. Osakada et al., Bull. JSME 21 (1978) 1236-1243.
[7] J. Kihara, Proc. 39th Joint Conf. Tech. Plast.(1988) 21-24.
[8] T. Nagasaki et al., Proc. 40th Joint Conf. for Tech. Plast. (1989) 615-618.

Chapter XIII

Implementation

Introduction

Although CE is a methodology with tremendous promise for the future, it can only be beneficial if the embodying philosophy of the enterprise is based on Total Quality Management (TQM). However, CE can be a superfluous investment and a counterproductive effort if the guiding philosophy of TQM is absent. Perhaps the fundamental requirement for the implementation of CE is the education of the individual because this instills "group" thinking rather than individualistic thinking. Such education must begin at the top of an organizational hierarchy and trickle down to the lower levels. Thus, the only change that will herald the newer design methodologies like concurrent engineering is a cultural one. In the absence of an accompanying cultural change, industrial productivity can never be enhanced, despite the design methodologies adopted. Much of this forms the subject of the first paper in this chapter.

The second paper deals with integration modeling of Flexible Manufacturing Systems (FMS) in Concurrent Engineering environments. The FMS integrated model is treated as a set of models with interfaces realized through special transformations of the models. The goal of the proposed technique is to provide flexibility of a FMS design process as well as a simultaneous access to the FMS project versions from both the designer's and manager's side.

If a company wants to be competitive in the global marketplace, it needs to consider incorporating the principle of concept to customer and customer-driven product development in the form of Quality Function Deployment. QFD is an effective planning tool that translates the customers' demands into appropriate company requirements. The last paper in this section discusses a specific investment in people and process improvement for the purpose of institutionalizing a company's commitment to customer-driven product development. The need for transformation from the concept of quality through inspection to customer-driven QFD and the goals to be achieved from this transformation are addressed. Moreover, an overview of the concepts of quality, evolution, implementation and applications of QFD at Ford Motor Company is given, with illustration of the working principles of QFD. Finally, the paper provides a close look at the application of QFD in simultaneous application of product and manufacturing process design practice of Concurrent Engineering.

Transitioning CE Technology to Industry

S.N. DWIVEDI, RAVI PRASAD and D.W. LYONS
Department of Mechanical and Aerospace Engineering
West Virginia University
Morgantown, WV

1 INTRODUCTION

The American Industrial productivity growth rate touched a new low recently when it fell to nearly 1%, the lowest among industrial nations. The situation is alarming, and demands revolutionary changes to enable US industries to retain their competitiveness. On the other hand, the Japanese have established an unbeatable hegemony over the world market. The wellspring of the impressive performance of Japanese industries is their management style and product development methodology. Using the Concurrent Engineering approach, the Japanese have succeeded in introducing high quality, inexpensive products faster and consistently in the market.

Concurrent Engineering (henceforth CE) is the harmonious blending of all the produce life-cycle disciplines with the objective of getting it 'right the first time'[42]. A comprehensive definition of Concurrent Engineering is given in the IDA report [10] on CE: Concurrent Engineering is the systematic approach to the integrated concurrent design of products and related processes including manufacture and support. This approach allows developers, from the outset, to consider all the elements of product life-cycle from conception through disposal including quality, cost, schedule and user requirement."

The approach has been proven to yield significant benefits[42]. It has resulted in better designed products which meet customer requirements and expectations. The lead time for product introduction has reduced significantly and so have costs, both of which have lead to a substantial increase in productivity of the organization.

In spite of all this, CE is not an absolute guarantee to success. Mere adoption of CE is not the stepping stone to success. It is the cultural milieu and the ethical values pervading in the organization that are the driving force behind the effectiveness of design methodologies like CE[4]. The Japanese have achieved this, for they have successfully interfaced advanced technology with their cultural and social traditions resulting in the emergence of Japan as the unchallenged economic superpower.

In the context of implementation of CE in American industry, it is very important to understand the Japanese work culture which is the primary reason for their astonishing productivity.

2 JAPANESE PRODUCTIVITY CULTURE

Japan's traditional culture is inseparably linked to its contemporary technology. Its unique strengths are derived from this cultural connection. This is the secret behind Japan's ability to produce outstanding industrial products very economically. Some of the key features of Japanese productivity culture are discussed below:

A) EMPLOYEE PARTICIPATION IN QUALITY CONTROL

The Japanese simply did not borrow quality control from West; rather, they recased it into a process of continuous quality improvement. Total participation of all the employees, with the aid of QC circles formed for this purpose, is the major characteristic of Japanese quality control.

'Jishu kanri' or the process of autonomous decision making sets the goals for the QC teams. The circles play a very important role in motivating the employees to contribute their best to the company. One of the main reasons for the overwhelming success of QC circles is the pursuit of creativity and the desire to learn more among the employees. Surprisingly, economic incentives happen to be weakest of all the motivations.

B) JAPANESE CONCEPT OF LABOR

Japanese do not consider work as a form of economic acitvity; rather, they perceive it as a form of religious devotion. 'Work' and 'Labor' has no distinction in Japanese culture. On contrary, in Western thinking while 'work' has a positive nuance, 'labor' has negative connotation. It is associated with drudgery and hardship. The western value system has a pronounced tendency to disparage labor, something completely alien to Japanese culture.

C) EQUALITY AND GROUP ACCOMPLISHMENT

In a Japanese enterprise, a companypresident is no different from other people in the company. The president is also an ordinary fellow worker; there are no class barriers in the company. This produces a sense of equality and unity among the employees, eliminating all possibilities of nepotism and discrimination.

Another unique feature is the sense of group accomplishment. In Japanese corporations, the stress is on defining group responsibilities rather than individual responsibilities. The satisfaction and sense of participation that the employees perceive augment their dedication for the company.

D) INSTINCT FOR SURVIVAL AND PATRIOTISM

More than 80% of Japan's terrain is inhospitable. The country is totally dependent on foreign exports for food-grains and raw materials. All this has induced a very strong feeling of anxiety and urge of protectionism among the Japanese. The society is glued together very strongly with the instinct for survival and national love. This has made a great contribution to the steadfast dedication of the people to economic prosperity and self-reliance.

E) REALIZATION OF ZERO-DEFECT PRODUCTION

The reason that there are so few defects in Japanese products and breakdowns are so rare is the strong Japanese dislike for even slight blemishes. Even small mistakes are regarded as a matter of shame reflecting on personal/company's honor. Another reason is fastidiousness in their character and strong aversion to criticism. The casual acceptance of defective merchandise or faulty products is out of the question.

F) EMPHASIS ON PRODUCTION

A great strength of Japanese technology is unification of design and production i.e., practical and successful implementation of concurrent engineering. Opinions and ideas of shop floor employees are clearly reflected in design. The product design proceeds with the objective of making manufacturing easy. The importance of manufacturing is manifested in the fact that over 40% of company presidents and top executives in Japan are people with shop floor experience.

G) TRADITION OF MINIATURIZATION

The great success of Japanese consumer electronics products lies in the skill of miniaturization. The Japanese have amply demonstrated the verity of 'small is beautiful'. This again can be traced back to the characteristics of their lineage. The trait can be clearly seen in 'Netsuke'- carving tiny ornamental toggles, 'Bonsai'- growing dwarf trees, 'Haiku'- the 17 syllable traditional form of poetry and many other things. This trend also has its roots in resource conservationism which is natural for the Japanese people, especially since Japan is a land of scarce resources.

H) CAPACITY FOR ABSORBING TECHNOLOGY AND DIVERSIFICATION

Very few post World War-II inventions are attributed to Japan. The Japanese have basically imported technology and commercially exploited it. Japan has shown an extraordinary quality for blending and fusing disparate elements, as minifested in the articulate combination of the oriental and occidental in its culture. 'Wayo Setchu' -a blending of Japan and the West clearly expresses this. Japan has succeeded in skillfully fusing available technology and exploiting it to maximum benefit as, for example, the growth of its steel industry.

3 BARRIERS TO CONCURRENT ENGINEERING

The success story of CE depends upon the pragmatic approach of having harmonious coordination and cooperation at every level, exploiting every opportunity and employing every resource to competitive advantage[10]. In Japan high productivity depends on this simple rule whereas the implementation of the same to American industry is full of obstacles owing to its work culture and other barriers which are discussed in detail.

a) EDUCATION AND IN-PLANT TRAINING

..."The US is what I call a messianic society. We seem to think someone is going to come out of the hills and solve our problems.... The answer is it is our responsibility to develop requirements and programs for education..."

-Fred Garry, Vice-President, Corporate Engineering and Manufacturing, General Electric Co., commenting on America's lack of education.

This seems very appropriate in the implementation of Concurrent Engineering. CE calls for an indepth understanding of Engineering and Manufacturing issues. American education leads to narrow specializations helping in realization and enhancement of individual goals and more financial rewards. Also, American universities produce the best talent for top management with leadership qualities, but with a myopic vision of future. The emphasis is on short-term gains rather than long-term planning [5,8,11].

The Japanese industrial set-up places emphasis on engineering background through both corporate in-house training as well as University education. The emphasis is on developing 'quality product design' as the natural way of thinking of management [5,8]. This encourages people with engineering and technical backgrounds to strive for top management positions. In contrast, the American educational system leads to a general lack of interest in shop floor activities with the 'elite talent' developing an aversion for manufacturing and vying for non shop floor responsibilities. This results in a thinking of 'solving problems in hand', producing immediate gains and claiming credit for the results. The goals become 'individual oriented' and employees exhibit reduced loyalty for the company.Consequently, this defeats the basic objective of CE, which is attaining greater cooperation across the depth and width of the organization [5,11]. In addition, the willingness of Japanese firms to invest in technological education of the workers, concomitant with basic education in Science and Mathematics, enables easier absorption and diffusion of new technology. It also facilitates modification and adaptation of technology to a diverse nature of products and processes. The American system lacks this and thus hampers the effective application of new technologies.

b) SELF INTESTED GOALS AND NO COLLECTIVE PERFORMANCE:

The basic objective of CE is the formation of multi-disciplinary teams working together for integrated, concurrent design of the product and processes. Attaining this requires team work and contribution rather than individual performance. Japanese organizations, as mentioned earlier, promote and recognize group work, which automatically leads to tasks being performed concurrently.

The situation in the U.S. is different, for work responsibilities are very rigidly defined and lead to virtual specializations and little interaction among the employees. This further reduces individual contributions to the project if the individual perceives that he/she will not get the expected recognition. Desire for personal gains induces rivalries, personality clashes and egotist tendencies.

c) LACK OF TECHNOLOGICAL ASSIMILATION AND WORLD SCANNING

Assimilating new technologies and efficiently implementing them for developing inexpensive, high quality products rapidly has a direct bearing on the implementation of CE technique. The Japanese are adept in absorbing new research and refining it to suit their application. They have been religiously scanning the world for new technologies, analyzing them for commercial viability and assimilating them accordingly. Their strategy is to carefully scrutinize foreign research activities and supplement them with their own R&D; this approach has yielded rich dividends.

In contrast, since the U.S. has traditionally been the technological leader, neither the industry nor the government has made an effort to locate, evaluate and translate foreign technical information. The U.S. leads the world in the quantity and quality of research, but this lead has not yielded commercially competitive products. This is attributed to the inability of American companies to transform discoveries into high quality products and into processes for designing, manufacturing, marketing and distributing. The absence of available market for these technologies led to their sale abroad at cheap prices. According to a rough estimate Technologies that cost 250-500 billion dollars to develop were sold to Japan for as little as $5-$10 billion. This serious drawback of American industry is leading to its obsolescence and continuous market-position erosion.

d) NON INVOLVEMENT OF GOVERNMENT TOWARDS GROWTH OF INDUSTRY:

In Japan, the Ministry of International Trade and Industry (MITI) maintains close coordination between Government agencies and industry. It supports foreign import of technology, arranging research funding for large scale projects, facilitating collaborations among vendors for new ventures, establishing research consortiums for carrying out new research and, above all, providing a sort of parental support to all industries. In the U.S., the relations between government and industry are characterized, more often, by discord than by harmony. In the U.S., most of the federal spending is devoted to military and space programs while commercial and industrial development have been left to private industry. Because of stiff competition, the industries can't spend huge amounts on long-term R&D projects and are more worried about short-term profits. To improve productivity, raise general level of commercial technology, and make it inexpensive, the government has to contribute to industrial R&D.

e) NON PARTICIPATION OF VENDORS IN INDUSTRIAL GROWTH

Supplies and subcontractors play a significant role in the productivity of the main contractor. By providing valuable insights about specific components, the vendors can help the chief vendor to achieve better designed, inexpensive products. In Japan, companies usually have single suppliers, thus these suppliers provide quality components right at the assembly line at the required time [1,4]. The main supplier need not maintain a large inventory of these parts thus saving space and resources. The concept in industrial parlance is known as JIT (Just In Time) i.e. the vendor will supply the required amount of components as and when the main contractor needs it.

The scene in the U.S. is of cherished practices of competitive bids, many competitors for even single component, distrust, lack of quality and non-contribution of component suppliers in product (which uses the component) improvement.

f) NON-CONSIDERATION OF LONG-TERM PERSPECTIVES

American managers are expected to show immediate results. Thus, the thinking is one of short-term strategies that result in short-term gains. The managers' horizon of planning is usually 1-2 years, so long-term planning is neglected. Japanese managers first think of capturing the market; immediate gains are not given any priority. The result of implementation of developmental policies which progressively hone the company's productivity finally results in market control and huge profits. The time span given to a Japanese manager is 4-5 years. The figure below shows the duties of Management in the successful implementation of CE.

g) NON-INVOLVEMENT OF MANUFACTURING AT THE CONCEPTUAL STAGE

The concept of CE is integrated and concurrent design of products and their related processes, including manufacturing and support groups. Without participation of the manufacturing team, the concept of CE is defeated. Sequential engineering, followed by U.S. industries, leads in transforming a generic solution of researchers into a specific design by design team. This finally leads to the product development stressing improved performance, increased product development cycles and reduced quality.[9]

The Japanese system of early involvement of manufacturing team leads in describing such design alternatives, production, specific information on material selection, tolerances and finish requirements, and tooling considerations.

The early involvement of manufacturing calls for proper in-house educational system, involvement of management in fostering teamwork and committed involvement of vendors and individuals. [9]

h) NEED FOR ACCEPTANCE OF CE BY ALL TEAMS

CE is a pragmatic approach, unifying different divisions and many conflicting requirements. To maintain proper balance of this delicate issue there should be coordination, co-operation. Devastating effects arises even from a modest failure from any of the participating multifunctional teams to meet the objective.[10,11] Acceptance of the CE technique requires

a) dedicated involvement of management in fostering teamwork;
b) emphasis on thorough market-driven strategies;
c) setting up of wise goals and maintaining balance of interests into all participating teams.

4. IMPLEMENTATION TOOLS FOR CE

Understanding the behavior of any process, product, or mechanisms, requires certain tools of implementing them. Correct usage of these tools have a tremendous aid in design, production, and engineering aimed at sharp reduction of life cycle costs, and short design cycles with improved quality. Implementation of these tools requires

1) Identifying and analyzing the problem
2) Choosing the right tool for implementation to create order and regularity

Implementation of any of these tools depends upon the discretion of the company foreseeing the above mentioned advantages. Few of these implementation tools for CE (shown in Fig. 2) are discussed in detail.

4.1) DESIGN FOR ASSEMBLY (DFA)

Design for Assembly is a methodology which optimizes the relation among various functions such as materials, technology, process and cost at the conceptual stage [40].
---A.SANDY MANRO

The product functions, cost and quality result from the

combined efforts of Engineering and Manufacturing. As a CE tool, Design for Assembly offers a structural approach for incorporating objectives into the design optimization process. Even though design accounts for only 5% of the total product cost, it determines at least 70% of the total cost of a product [41,40]. The implementation of DFA needs [31] commitment and support of top management, knowledge of state-of-the-art design principles and tools, knowledge of geometric dimensioning and tolerancing requirements.

DFA focuses on the definition of a part and basic design principles. Certain rules and guidelines are applied to the product and process design for significant improvements [28,31].

1) Design for minimum number of parts;
2) Design for part handling and presentation;
3) Evaluate assembly methods;
4) Design for ease of assembly;
5) Select fasteners for ease of assembly;
6) Design for vertical assembly;
7) Design for modular assembly;
8) Eliminate or simplify adjustments;
9) Minimize product variations;
10) Eliminate electrical cables;

Various design assessment tools can be used to quantify the extent to which a product meets DFA objectives. This helps to determine the ease of product assembly, material handling, and parts feeding and orientation [31,27]. They are:

-The Boothroyd and Dewhurst Analysis program;
-Assembly insights for Designs;
-The manufacturing rating system;

The various benefits derived from the implementation of DFA technique are [31,29]:

-Elimination of Non-functional parts and reduced inventory;
-Reduction of Engineering changes;
-Reduction of assembly time and costs;
-Shortening of product cycle time and better product produced;
-Improved quality;
-Increased customers satisfaction and value.

4.2) VALUE ENGINEERING (VE)

"Value Engineering is an organized effort directed at analyzing the functions of system, equipment facilities, procedures, and supplies for the purpose of achieving required function(s) at the lowest cost of effective ownership, consistent with requirements for performance, reliability, quality, maintainability and safety" [42] - Defense management course.

VE identifies areas of excessive or unneccessary costs and attempts to improve the value of the product. The successful application of VE requires creativity to innovate, alternate designs, system methods, or processes that will perform the necessary functions at the lowest possible cost [25,42]. The methodology of VE involves:

1) Evaluation of function through various means by preparing functional flow block diagrams and secondary functions.

2) Synthesis of various alternatives that will perform the basic function and determination of cost of alternatives.

3) Representation of the alternative with the lowest overall cost, and the worth of that basic function.

The implementation of the above steps requires a phase-wise study, wherein each phase contributes in establishing the following requirements [25,26,22]:

Functional Phase - Indepth understanding of the project and determination of functions;
Creative Phase - Conceiving new designs, development of alternatives and evaluation of alternatives;
Development Phase - Developing action plans and identification of potential problems;
Recommendation Phase - Information and implementation of product development improvements.

The major benefits through implementation of VE are
1) Reduces the risk of new product development;
2) Reduces the overall per unit cost of new product by reducing the sunk costs development and reduction of recurring cost of manufacturing the product;
3) Helps in fostering team approach for identifying functions and designed especially to improve intergroup performance;
4) Application of VE at the conceptual stage helps in achieving optimum engineering changes and reduction of overall costs.

4.3) SOLID MODELING

In determining a solution to a real engineering problem, the creative art of conceiving the physical means of achieving an objective is first and analyzing the possible solution is next. Solid Modeling plays an effective role in the conceptual design phase both in conceiving and analyzing the solution. The ability to form a visual image of geometrical and physical configurations has a tremendous advantage in creating a physical means of achieving a technological objective. Development of Solid Modeling has provided this 'thinking through' process of judgement, conception and reflection in searching for a conclusion [35,36].

While other tools of CE are processes which help in attainment of the objective, Solid Modeling gives the visual representation of attaining the objectives. Lack of ambiguity in the displayed images is the key benefit of Solid Modeling.

This facilitates the promotion of tighter integration of design and manufacturing functions as it is easier to master and manipulate. The development of various solid modelers with features, simulation packages and incorporation of AI techniques has allowed the systems to acquire knowledge to design the part, plan and manufacture process, make the part and inspect the finished product on its own. During the conceptual stage, this helps in thoroughly understanding various degrees of difficulties in its development and gives a clear direction in the evolution of the part. It allows engineers to become productive more quickly because the flexibility of their design options minimizes the weaknesses [36,37].

The usage of Solid Modeling techniques provide:
1) Complete, unambiguous picture that can be understood by all the team members easily and help in better coordination, communication and assimilation of the objectives;
2) Help in reduction of costs and lead times with 100% improvement in quality;
3) Help in providing the Design Engineers with flexibility to quickly create, modify and iterate the conceptual design and Manufacturing Engineer with precise geometry needed to drive automated equipment and processes;
4) Helps in the study of relationships governing the part geometry and also the interaction of the parts making the assembly, thereby also helping the vendors in developing precise parts;
5) Help in accelerating design and manufacturing processes, thereby increasing the product development cycle

4.4) TOTAL QUALITY MANAGEMENT (TQM)

Quality consists of those product features which meet the requirements of customers and thereby provide cash-inflow, profitability, freedom from repair and replacement of defective parts. Total Quality management is defined as the attitude that produces a comprehensive company-wide system for achieving the desired product characteristics [17,18] Various elements of TQM are

-Statistical Process Control (DEWING)
-Quality Improvement Techniques (JURAN)
-Company Wide Quality Control (ISHIKAWA)
-Quality Engineering (TAGUACHI)
-Quality Function Deployment

4.4.1) QUALITY ENGINEERING BY DESIGN (TAGUACHI METHOD)

A range of values of a process characteristic which are acceptable are said to be within specified limits. The performance of a product is better if the dimensions of the individual components were made to conform to certain 'optimal value' than within the tolerance limits. Dr. Taguachirecommends statistically designed experiments that help in setting up of parameters which will result in a product whose characteristics are consistently close to the ideal target [12,24]. He stresses three design steps:

1) System design - the best production equipment and tentative production processes.
2) Parameter design - the values of the parameters which optimizes the product loss.
3) Tolerance design - selection of tolerance that should be implemented in manufacturing to assure minimum loss of product manufactured and used by customer.

Various steps involved in the robust design are shown in Figure 3 where the implementation of Dr. Taguachi's quality engineering methods lead to:

1) Reduction of performance variations about a target value.
2) Fewer engineering changes after product is designed.
3) Reduction of scrap and rework and increased customer satisfaction.

4.4.2) QUALITY FUNCTION DEPLOYMENT (QFD)

QFD is a means or way to connect the voice of the customer to many activities that can be deployed through product planning, engineering, manufacturing, assembly and service[23]. The foundation of QFD is the belief that products should be designed to reflect customers' desires and tastes. This instigates marketing, design and manufacturing divisions to work closely together from the time a product is first conceived [20,23].

QFD uses team work for creative brainstorming and identification of customers' demands through market research data, dealers' input, sales department wants, special customer opinion surveys, qhich includes placing their products in public areas and encouraging potential customers to examine them.

Different parameters such as various functions, mechanisms, failure modes, parts and assemblies, critical manufacturing steps related to customerrequirements are converted into engineering characteristics by using matrices. These characteristics are evaluated by multifunctional teams based on their engineering experiences, customer responses and statistical studies. To show positiveness, these parameters are and in turn lead to the establishment of target values to be achieved [23,24].

Realization of QFD yields:

1) Product development is based on customers requirements, hence that objective is carried through all the stages.
2) various strategies do not become vague or lost throughout its implementation from marketing, planning and manufacturing.
3) Even the minute details, based on the customers

requirements, are not overlooked.

4) High efficiency is achieved as a number of changes required at each stage are reduced (totally eliminated) giving clear direction to the objectives.

6 CAVEAT

Undoubtedly, CE is a methodology with great promise for the future, but it can be beneficial if, and only if, the embodying philosophy of the enterprise is based on the concepts of 'Total Quality Management' (TQM). CE can prove to be a superflous investment, a counterproductive effort if the guiding philosophy of TQM is absent. The fundamental requirement for CE implementation is the education of the individual, for this will change the 'individualistic' thinking of the people to one of 'group' thinking. This education has to begin from the highest rungs of the hierarchical ladder in the organization and trickle down to lower levels. Thus, it is the cultural change that will herald the fruition of newer design methodologies like concurrent engineering. The implementation of these approaches in traditional management settings will definitely boomerang. Whatever design methodologies are adopted, in the absence of the accompanying cultural changes, American industrial productivity can never be enhanced.

REFERENCES

1) Chacko, K.G. "Robotics/Artificial Intelligence/Productivity: US-Japan Concomitant Cinditions"Petrocelli Books, Princeton, New Jersey.

2) Freeman, C. "Technology Policy and Economic Performance - Lessons from Japan", Science Policy Research Unit, University of Sussex, London.

3) Ninety Eighth Congress, Second Session, "Japan Technological Advances and Possible United States Responses Using Research Joint Ventures", US Government Printing Office, Washington: 1984.

4) Moritani, M., Getting the Best for the Least - Japanese Technology" The Simul press, Inc., Tokyo.

5) "The Rise and Fall of R&D", Mechanical Engineering, April 1988, p. 28-33.

6) "Coming up with New Products that Beat the Odds", Mechanical Engineering, April 1988, p. 38-41.

7) "Pulling - Not Passing - for Higher Productivity", Mechanical Engineering, April 1988, p.42-44.

8) "The Quiet Path to Technological Preeminence", Scientific America, October 1989, p.41-47

9) Smalley, S. "Can We Make It", Westinghouse MC&TC,

February 89.

10) IDA Report R-38 "The Role of Concurrent Engineering in Weapons System Acquisition", December 1988.

11) Barkan, P., "Some Impediments to Successful Implementation of Simultaneous Engineering".

12) Ryan, P.Thomas, "Statistical Methods for Quality Improvement", John Wiley & Sons.

13) Sullivan, P. Lawrence, "The Beginning, the End and the Problem In-between", A Collection of Presentations and QFD case studies, American Suppliers Institute, Inc.

14) Tribus, Myron. "Quality - Deming's Way", Mechanical Engineering, Jan. 1988.

15) Box, G. and Bisguard, S. "Statistical Tools for Improving Designs", Mechanical Engineering, Jan. 1988.

16) "Search for Quality: From Taguchi to Customer", Datapro Manufacturing Automation Series, Oct. 1988.

17) "Quality First", Datapro Manufacturing Automation Series, Oct. 1988.

18) "Quality: A Continuing Revolution" Datapro Manufacturing Automation Series, June 1989.

19) "The Quality Function", Datapro Manufacturing Automation Series, Dec. 1988.

20) "The Philosophy of Value Added Manufacturing", Datapro Manufacturing Automation Series, Dec. 1988.

21) "Establishing Total Quality Control", Datapro Manufacturing Automation Series, March 1987.

22) Hauser, R.J. & Clausing, D. "The House of Quality", Harvard Business Review, May-June 1988, p. 63-73.

23) Sullivan, L.P., "Quality Function Deployment", Quality Progress, June 1986, p. 39-50.

24) Ross, P.J. "The Role of Tagauchi Methods and Design of Experiments in QFD", Quality Progress, June 1988, p. 41-47.

25) "Potential Applications of Value Engineering in a Simultaneous Engineering Concept", Proceedings of 1987 SAVE Conference, p. 51-66.

26) Wixson, J.R. "Improving Product Development with Value Analysis/Value Engineering", Proceedings of 1987 SAVE Conference, p. 51-66.

27) Boothroyd, G. "Making It Simple Design for Assembly",

Mechanical Engineering, Feb. 1988.

28) Boothroyd, G & Dewhurst, P., "Design for Assembly: Selecting the Right Method."

29) Whitney, D.E., "Manufacturing by Design", Harvard Business Review, July-August 1988.

30) "Design for Automation in Assembly", Datapro Manufacturing Automation Series, Feb. 1988.

31) Dwivedi, S.N. and Kein, B.R., "Design for Manufacturability Makes Dollars and Sense", CIM Review, Spring 1986, p. 53-59.

32) Horn, D.S., "CIM Inches Ahead", Mechanical Engineering, Dec. 1988.

33) Sink, D.S., "Management of Quality and Productivity in the Organization of the Future", Proceedings of 1989 IIE Integrated Systems and Society for Integrated Manufacturing Conference, p. 21-26.

34) Fabrycky, W.J., "Engineering and System Design: Opportunities for ISE Professionals", Proceedings of 1989 IIE Integrated Systems and Society for Integrated Manufacturing Conference, p. 65-71.

35) Lexans, A.S., "Graphics-Analysis and Conceptual Design", Mechanical Engineering, Jan 1989.

36) Daniel, D., "The Power of Parameters", Mechanical Engineering, Jan. 1989.

37) Schraft, R.D. and Bassler, R., "Possibilities to Realize Assembly Oriented Product Design", 90th International Conference on Assembly Automation, Paris, France, May 1984, p. 243-261.

38) Daniel, D., "The Power of Parameters", Mechanical Engineering, April 1989.

39) Wallach, J.M., "Design for Manufacturing and Solid Modeling", Proceedings of Second International Conference on DFM, Nov. 1989.

40) Sandy, Manro. A., "Simultaneous Engineering for Improved Product Design and Manufacturing Interface", Proceedings of Seminar of Society of Manufacturing Engineers, Oct. 11-13, 1988.

41) "Principles and Applications of Value Engineering", US Army Management Training Activity Course Book.

42) Ravi, L., Garg, R., Dwivedi, S.N., "Concurrent Engineering - Why and What."

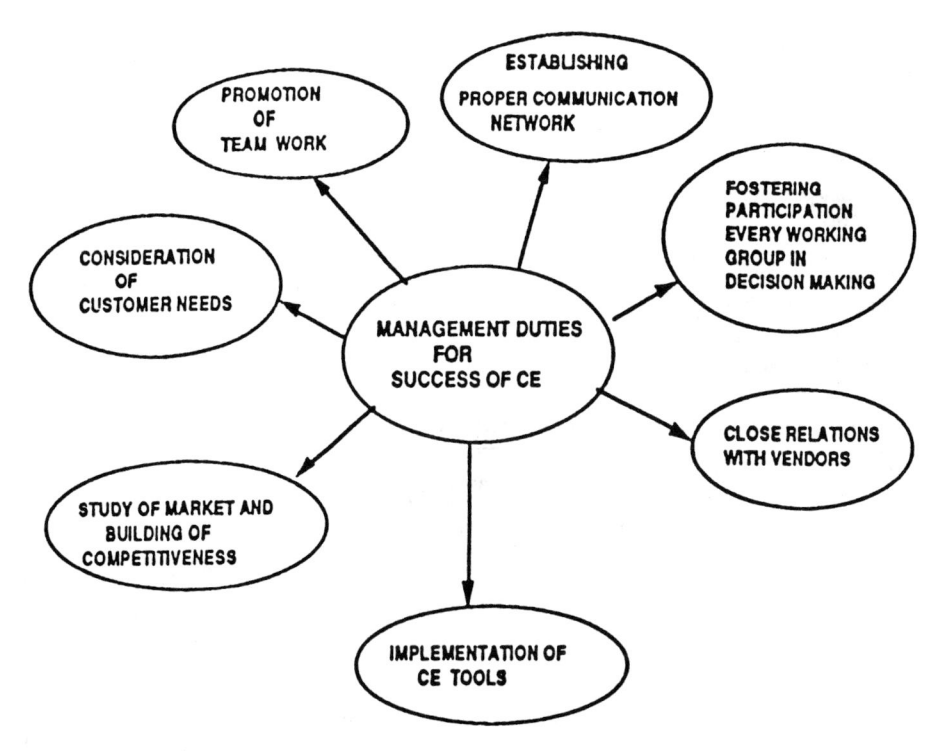

FIG 1. MANAGEMENT'S ROLE IN SUCCESS OF CE

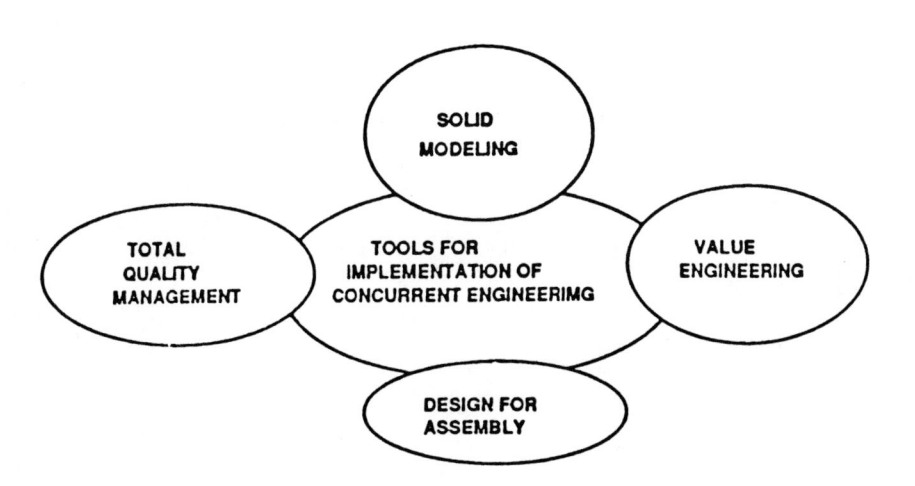

FIG 2. IMPLEMENTATION TOOLS OF CE

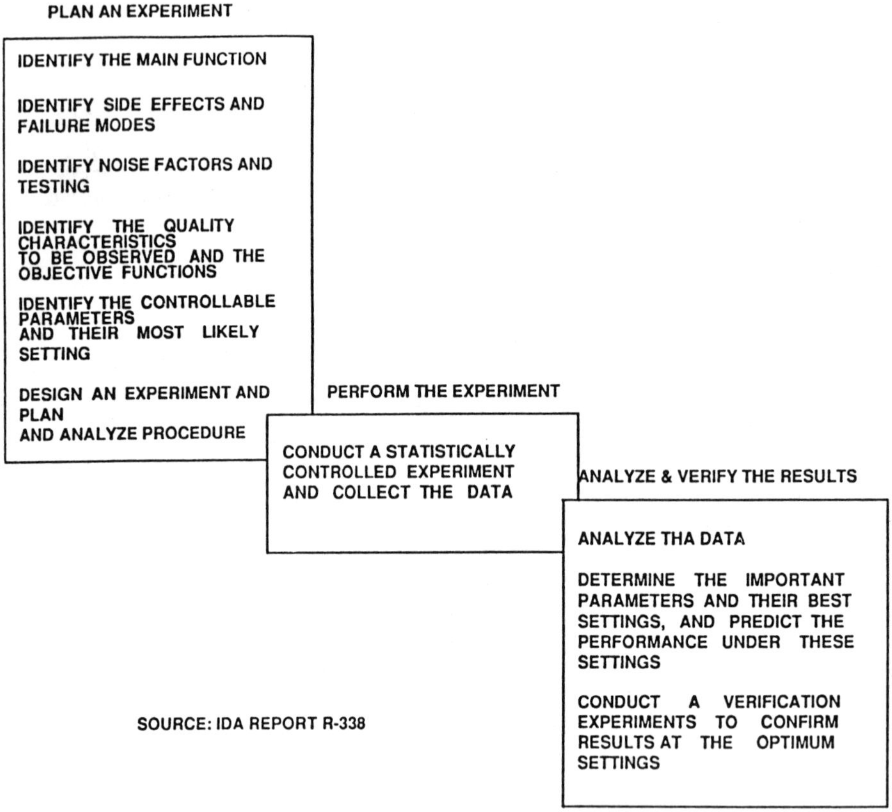

PLAN AN EXPERIMENT

IDENTIFY THE MAIN FUNCTION

IDENTIFY SIDE EFFECTS AND FAILURE MODES

IDENTIFY NOISE FACTORS AND TESTING

IDENTIFY THE QUALITY CHARACTERISTICS TO BE OBSERVED AND THE OBJECTIVE FUNCTIONS

IDENTIFY THE CONTROLLABLE PARAMETERS AND THEIR MOST LIKELY SETTING

DESIGN AN EXPERIMENT AND PLAN AND ANALYZE PROCEDURE

PERFORM THE EXPERIMENT

CONDUCT A STATISTICALLY CONTROLLED EXPERIMENT AND COLLECT THE DATA

ANALYZE & VERIFY THE RESULTS

ANALYZE THA DATA

DETERMINE THE IMPORTANT PARAMETERS AND THEIR BEST SETTINGS, AND PREDICT THE PERFORMANCE UNDER THESE SETTINGS

CONDUCT A VERIFICATION EXPERIMENTS TO CONFIRM RESULTS AT THE OPTIMUM SETTINGS

SOURCE: IDA REPORT R-338

FIG 3. VARIOUS STEPS INVOLVED IN THE ROBUST DESIGN

Integrated Models of FMS in Concurrent Engineering Environment

A.A. LESKIN

U.S.S.R. Academy of Sciences
Institute for Informatics and Automation
Leningrad 199178, U.S.S.R.

S.N. DWIVEDI

Department of Mechanical and Aerospace Engineering
West Virginia University
Morgantown, WV

SUMMARY

Concurrent Engineering as an approach for integrating all stages of a production life cycle involves the idea of implementing concurrent procedures in production process planning. In this framework, FMS design problems can be analyzed. The concepts of integrity and specific modularity of FMS representation are put forward when implementing this idea. An FMS integrated model is treated as a set of models with interfaces realized through special transformations of those models. The goal of the technique is to provide flexibility of FMS design process as well as a simultaneous access to the FMS project versions from designers' and managers' side.

Key words: Concurrent Engineering, FMS design, system models.

1. INTRODUCTION AND BACKGROUND

Computer science and computer facilities today are at the stage where efforts are being made to put forward a new methodology in engineering using a new way of { research/ design/ process planning/ quality control/ manufacturing } process organization. Production engineering is noted to have two benchmarks on the way of its development. The first one, which corresponds to the conveyor time, was introduced by G. Ford, with its principle of parts moving along tools. Then an opposite way became successful when tools in FMS started to move around the part, similar to the invention associated with the Japanese. We observe something of that kind in AI-supported engineering. The sequential way of tasks processing looks like a conveyor, with the product project moving through CAD/ CAPP/ CAM systems serving as an interlinking facility. The second way concerns fully computer integrated system where suitable tools come to the project improving both product performance and coordination due to continuously changing function allocations. This approach is naturally concerned with concurrent engineering (CE) [Dwivedi 1990].

The CE concept being applied to FMS design with an underlying integrated model (IM) is a supplement to the CE scheme. The concurrent engineering environment definitely involves challenges in carrying out process planning.That is why manufacturing system design procedures, closely associated with process planning, should reflect CE idea. Our goal is to build the proper conceptual basis for it .

Technological basis for an FMS design is considered in [Mitrophanov 1983], [Kusiak 1990], [Stecke 1985], etc. Specific features of the mentioned problem domain are considered in [Groover 1984], [Mitchell 1991], and analysis of the equipment selection problem is discussed in [Miller 1977, Hayes 1981]. As for computer science, a lot of results relevant to the problem are in use [Ohsuga 1989], [Milachich 1987] and many others. Partly scientific level of the results in this domain is characterized by DSS for FMS design [Harhalakis 1988, Issa 1987, Kusiak 1990, Kaltwasser 1989].

An advanced approach to knowledge representation in CAD FMS systems seems to be reasonable to develop on the basis of object-oriented programing ideas [Cox 1986, Goldberg 1983, Parsaye 1989]. Some results received in this direction [Combacau 1990], [Oshuga 1989], [Leskin, Ponomarev 1988, Bogdanov 1987] concerns a specific software structure underlying the technique suggested.

An important tool in FMS design is Matching procedure (MP) designed for finding candidate design variants corresponding to user requirements and process constraints. This procedure, without relation to the considered problem, was under study in [Ishii 1989]. A formal basis for creating MP was developed in [Leskin 1986, Bogdanov 1987]. The approach uses a special algebraical representation of technological objects and operations in a problem-oriented structure framework. Transformations of the structures are based on theory of categories and theory of graphs [Leskin 1986]. These results can be applied to the solution of the problem posed. A number of fundamental works related to this problem is available from theory of categories [Maclane 1971] including an A. Grotendick's schemata of diagram [Grotendick 1957], graph transformations methods [Roberts 1984, Rosenberg 1981, Tutte 1986] especially in F. Pham's interpretation [Pham 1967]. These works can serve as basis for a novel means of knowledge representation of sequenced jobs, and an example is considered in [Leskin 1986].

2. CONCURRENT ENGINEERING ENVIRONMENT FOR FMS DESIGN

Concurrent Engineering is an approach for integrating all stages of a production life cycle. Proper computer facilities constitute the CE environment (CEE). Both CE and CEE are being considered in terms of the whole production system. However, a bottom up view, based on CE ideas, forces the creation of a CE superstructure over a separate computer-aided system, despite what this system is designed for. Such a system should cooperate with groups of managers, designers and other industrial specialists supporting a variety of concrete jobs. Concerning FMS design or CAPP, such jobs are production process dispatching and scheduling, process planning, selection of production equipment, etc. All these jobs are definitely different appearances of the factory production process although they are often represented by different models. It is obvious that the product to be produced is an object which integrates parts of material and information flows. The models of these jobs are reasonable to represent in a computer taking into account that fact in order to provide flexible and efficient correspondence between them. This problem actually concerns developing an appropriate technique for knowledge representation. CEE is defined in this case by specific facilities which provide concurrent operations over mentioned models.

3. INTEGRATED MODEL APPEARANCE

FMS design process is carried out by a number of procedures, the base of which is specific models. It is necessary to operate with each model as well as with aggregates made up of them. According to parametrical technology, unified models interfaces should be defined in some way. However, one can suggest functional connections defined over fuzzy interfaces, some kind of non-evident interlinking, given by transformations of the entire models or their fragments. For example, if $G = (U, E)$ is a graph which represents production sequences, U is a set of nodes, E is a set of arcs, and T is a set of time segments; then a graph $G_t = (U \times T, E)$ is a new model describing a virtual timed production structure which can be visualized by a Gantt diagram. General features of the integrated model we introduce (IM) should be the following:

♦ keeping FMS intrinsic features, that is systems invariants;

♦ an ability to represent any production factor;

♦ IM must be open for including environmental factors.

When constructing IM, the following requirements must be provided:

♦ IM must serve as a constrictive tool for FMS research and design;

♦ IM must be designed as a macromodel, that is as a set interconnected models with functional and parametrical interfaces being defined;

♦ there must exist ways for creating IM for FMS design problem;

♦ IM models must be proper for specifying and prototyping FMS design process;

♦ IM must be open for including environmental factors;

♦ model interaction procedures should be defined constructively.

The essence of listed IM features is very much concerned with system invariants and an in-and-out way of FMS design process representation, the main demand for model transformations being a co-ordination between mapping model elements and mapping their connections. To provide this, a functoriality property from theory of categories is reasonable to use.

4. VIRTUAL AND PHYSICAL MODELS

Analysis of FMS design problems shows that a lot of design tasks can be solved by research of two interacted structures, a virtual structure, describing the goal object, and a physical structure which is an implementation of the virtual structure through available realizing objects. When representing elementary model units by semantic nets, the formal description of elements, fragments and structures, which together constitute IM, can be obtained by using a specific object, a supplemented graph, S-graph. Nodes of S-graph represent production operations (or equipment units types) and sets describing node characteristics. Arcs represent appropriate production sequences or they point to an operation (or equipment unit) attributes. An example of S-graph is shown in (Fig. 1), where M1 and M2 are machines, MX1 and MX2 are their attributes, B is a buffer, R is a robot, and BX4 and BX3 are their attributes, respectively.

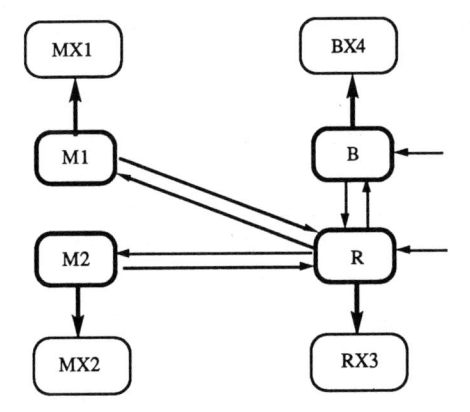

Fig. 1.

We consider a Virtual - Physical correspondence just as a tool realizing an idea of an assignment an appropriate facility to the task. In terms of S-graphs, this idea is realized by a covering operation. The Virtual Route Method (VRM) presented in [Leskin 1986] is designed to implement it in the direction of FMS design problems.

5. MODEL TRANSFORMATION OPERATIONS

Keeping intrinsic FMS features involves a demand of models functional similarity under model transformations. The Grotendick's concept of a diagram scheme [Grotendick 1957] in theory of categories can serve as a base to satisfy the demand. All operations are divided into two levels: elementary and meta-operations. Elementary transfer operations reflecting production content in FMS design process can be reduced to a few types operations such as "joining", "merging", etc., exactly relative to combining jobs within a virtual or physical machine-tool or FMC, taking into account providing performance indices. In Fig.2, mergin operation is shown with a sketch of the diagram describing transformation of attribute sets (for more details see [Leskin 1986].

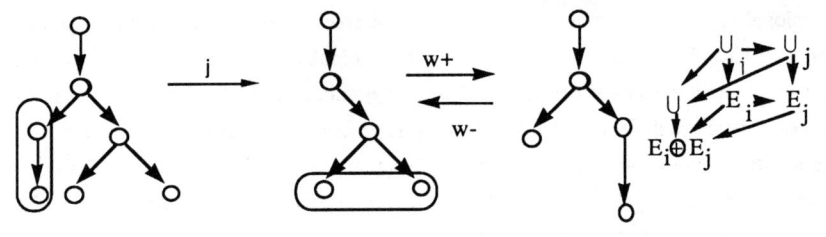

Fig. 2

Let us consider now an example of transformation of a model representation. Let S be a directed graph, S = (X, E, d) where X is a set of nodes, E is a set of arcs, d : E → X × X is a mapping. For an arc e, let d(e) be its direction with x as start and y as end points. The arc composition is defined in natural way for specifying the path on the graph. Let M be a set. The diagram in M with the scheme S is a function D which the relationship structure S transfers onto the set M. Graph S is an analog of Grotendick's scheme of diagram. Such a construction for a transition from one FMS representation to another support saving intrinsic system features. Specific meta-operations are introduced to realize a hierarchical way to handle with data as well as to provide expert groups to participate in the design process simultaneously. For example, computer specialists can be involved as early as at the stage where virtual production process graphs are dealt with. Suitable interfaces and cognitive aid should be organized on this level. Virtual and Physical sublevels are reasonable to introduce, with a scale of FMS to be designed being divided into three groups: System (S), Fragment (F), Element (E). Fig. 3 shows models interaction. Here EV_i is i-th virtual element, FV_m is a m-th virtual fragment, CFV_p is a p-th clastered virtual fragment, AFV_v is a v-th virtual fragment assigned onto a machine, SP_h is a physical FMS structure, J is ajoin operation, C is a claster operation, A is an operation of an assignment a production operation to a machine.

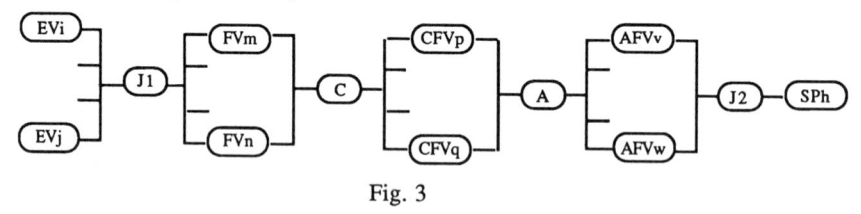

Fig. 3

6. EXTERNAL AND INTERNAL DYNAMIZING OF INTEGRATED MODEL

One of the most important factors in model integration is a transition to a dynamic FMS models. Two types of IM dynamizing are introduced. External IM dynamizing concerns interlinks between models as well as between FMS and environment. The 3-dimensional Gantt diagram accompanied by a suitable analytical tool is developed to support external IM dynamizing [Leskin 1989]. Internal IM dynamizing makes a time unfolding of separate models themselves. Dynamizing functor is used to implement such an operation. There are different ways to specify this functor. One way involves introducing some additional nodes with a new specific interpretation of a S-graph when each node becomes a dynamic system. All kinds of IM transformations are fulfilled under the functoriality constraint. All models above are accepted to be parts of IM, access to it in concurrent mode being created using two ways: through an FMS design data base and also simultaneously, directly to IM using specific interfaces. The latter way is for process planning specialists; it can either serve to handle with design process constraints in on-line mode or to participate in an operation result versions evaluation.

To combine separate models within a single FMS design process design, a specific algorithm is needed, and this was developed by the author earlier [Leskin 1988] with the use of commutative diagrams which constitute an FMS design network. An interaction both models and specialists with IM is organized on such a network.

We will show now how external dinamization can be carried out. Let S be a category of dynamical systems and G is a scheme of diagram [Grothendieck 1957] corresponding to an FMS structure, virtual and physical. The diagram in S with the scheme G is a function which relates pairs (U_i, U_j) connected by technological precedence relationship and appropriate dynamical systems (S_i, S_j). So an object in the category matches a node, and morphysm in the category S matches an arc. Transitions from one design stage to another as well as an interaction of models within IM are convenient to describe by hierarchical Petri nets. The net (shown in Fig.4) reflects the process of production equipment selection, where MHS is Material Handling System, VPS is Virtual Production System, FMC is Flexible Manufacturing Cell, PhPS0 is a basic and PhPS1 is a final variant of physical production structure.

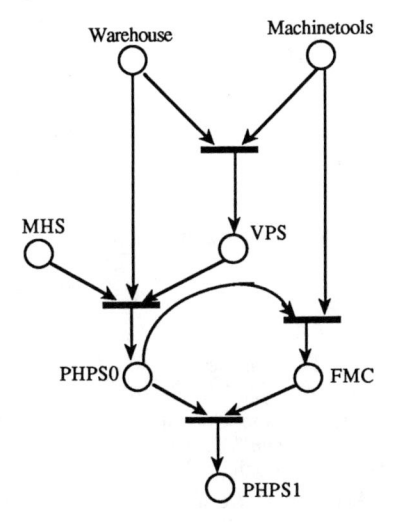

Fig. 4

Specific mode of modules interlinking requires defining design procedures as concurrent processes in Petri net framework. In particular, ports must be defined as well as appropriate data structures. A finite-state process for describing design network behavior according to external and internal model dynamizing ways can be introduced using an approach offered in [Smolka 1984].

7. SOFTWARE FRAMEWORK

An analytical manner of the whole approach helps to construct FMS design software. Well structured algorithms, where data and operations are described in constructive way, permit the exposure the CAD FMS system software architecture. An object oriented programming approach (OOPA) appears to be most promising for the implementation of the proposed idea, with classes of OOPA corresponding to the models. Functions supporting the models interaction are formed in accordance with proper demands. An algebra of commutative diagrams is useful for the development of software specifications [Leskin 1986].

CONCLUSIONS AND FUTURE WORK

A few FMS conceptual design system prototypes were developed according to the ideas considered, and this confirms the efficiency of the technique. Implementing IM/CE-based FMS design system permits: flexibility of FMS design process, better allocations of designer functions, high level formality of FMS variants descriptions, time efficiency of the design process, formal premises for creating an in-and-out concurrent engineering technique for FMS design and research. For the development of an intelligent CAD FMS system, it is necessary to apply coupling group decision making methods and the approach suggested with the use of a distributed system architecture.

REFERENCES

Bogdanov, K.I., Leskin,A.A., Ponomaryov, V.M. 1987. *Computer-aided AMS Conceptual Design System IDS 1.1*, Preprint , Leningrad, LIIAS, No. 45, 41 p.

Dwivedi, S.N., Lanka, R. 1990 "AI and Concurrent engineering in Factories of the Future," *International Conference on "Artificial Intelligence - Industrial Application"*: Abstracts of papers, 22-25 April, Leningrad

Findler, N.V. (ed.). 1979. *Associative Networks: The Representation and Use of Knowledge by Computers*. Academic Press, New York.

Groover, M.P., Zimmer, E.W. Jr. 1984. *CAD/CAM: Computer-Aided Design and Manufacturing,* Prentice Hall, Englewood Cliffs, N.J.

Grothendieck, A. 1957. Sur quelques points d'algebre homologique, *Tohoku Mathematical Journal,* - Second series, V.9, N 2,3.

Harhalakis, G., Lim, C.P., Mark, L, Cochrane, B.1988. "An Expert System Approach to Integrating Product Design and Manufacturing Execution Modules," *Preprints of the Seventh PROLAMAT Conference,* Dresden, GDR, V.1.

Hayes, G.M., Davis, R.P., Wysk, R.A. 1981. "A Dynamic Programming Approach to Machine Requirements Planning, " *AIIE Transactions,* Vol.13, 175-182.

Ishii, K., Hornberger, L., M. 1989. "Compatibility-based Design for Injection Molding," *Concurrent Product and Process Design, Chao,N.-H., Lu S.C.-Y.(Eds), The American Society of Mechanical Engineers, 57-63.*

Issa, T.N., Czajkiewicz, Z.J. 1987. "MRP 2 - Manufacturing Resource Planning: System Development, Implementation, and its Impact on Productivity at the Factory Level," A. Kusiak (Ed.), *Modern Production Management Systems,* Elsevier, New York, 711-727.

Kaltwasser, J., Leskin, A. A., Wirth, Z. 1989. Materialflusslogistik - wichtige Komponente des flexiblen automatisierten rechnerintegrierten Betriebes, *Fertigungstechnik und Betrieb.,* Berlin, N1., 7- 14.

Kusiak, A. 1990. *Intelligent manufacturing systems,* Prentice Hall, Englewood Cliffs, New Jersey, p.443

Leskin, A.A., Algebraic Models of Flexible Manufacturing Systems. - Leningrad: Nauka, 1986. - 150 p. (in Russian).

Leskin, A.A., Ponomarev, V.M. 1988. "Knowledge Representation in CAD FMS," *Preprints of the Seventh PROLAMAT Conference.* Dresden, GDR, V.2.

Maclane, S. 1971. "Categories for the Working Mathematician," *Springer Graduate Texts in Mathematics,* - 323 p.

Milachich, V. 1987. Teorja projektovanja tehnoloskih sistema. *Proizvodni sistemi ò.* - Beogradu: Univerzitet, - V.2.- 426 s.

Miller, D.M., Davis, R.P. 1977. "A Dynamic Resource Allocation Model for a Machine Requirements Problem," *IIE Transactions,* Vol.10, No.3, 237-243.

Mitchell, F. H. Jr. 1991. "CIM Systems: An Introduction to Computer-Integrate 4-D Manufacturing," Prentice-Hall, Inc., Englewood Cliffs, New Jersey, 542p.

Mitrofanov, S.P. 1983. *Scientific Organization of Batch Production.* Leningrad, Mashinostroenie, 2 v.786 p.

Ohsuga, S. 1989. *Toward Intelligent CAD Systems, Computer Aided Design,* 1989. - V.21, No. 5., P.315-337.

Pham, F. 1967. Singularities des processus de difussion multiple, *Annales de l'Institut Henri Poincare,* - Section A. Physique Theorique, v.VI, N 2.,89-204.

Roberts, F.S. 1984. *Applied Combinatorics.* Prentice-Hall, Inc., Englewood Cliffs, New Jersey, - 606 p.

Rosenberg, A.L. 1981. "Issues in the Study of Graph Embedding," *Lecture Notes in Computer Science,* Vol, 100, 150-176.

Smolka, S.1984. A Polinomial-Time Analysis for a Class of Communicating Processes, *LNCS,* 167,

Stecke, K.E., Solbert J.J.1985. "The Optimality of Unbalancing both Workloads and Machine Group Sizes in Closed Queuing Networks of Multiserver Queues," *Operations Research*,.882-910.

Tutte,W.T. 1986.*Graph Theory*, Menlo Park, California: Addison-Wesley Publishing Company, 424 p.

Implementing QFD at the Ford Motor Company

HAROLD F. SCHAAL

Ford Motor Company
Car Product Development
20000 Rotunda Drive
Dearborn, MI

WILLIAM R. SLABEY

Ivon Corporation
48720 Hanford
Canton, MI 48187

SUMMARY

During the 1980's the automotive industry began facing strong global competition while their labor and investment cost surpassed their foreign competition. Stringent government regulations, rising customer expectations, and productivity disappointments heightened the challenge of emphasizing quality and customer satisfaction. A close inspection of Japanese methods and management practices provides a significant insight. Japanese processes embody a philosophy of "Total Quality Excellence" focused on the voice of the customer. This made FORD rethink its philosophy and methods of doing business.

This paper deals with a specific investment in *people and process improvement* for the explicit purpose of institutionalizing FORD MOTOR COMPANY'S commitment to customer driven product development. It addresses the need for a transformation from a concept of quality based on inspection to customer driven QUALITY FUNCTION DEPLOYMENT.

PREFACE

In the late seventy's and early eighty's FORD MOTOR COMPANY had a significant emotional experience. Our market share was slipping, record losses were being recorded while our customers drove to our Japanese competition and traded in their Ford cars for imports. It's been said by some that FORD had a "near death experience." That experience alone was motivation to change!

FORD decided to invest in its future two ways:

    New Product Development -- Taurus/Sable.

    Invest in our people and processes.

FORD began to learn from the best of its Japanese competition. This paper deals with a specific investment in people and process improvement for the explicit purpose of

Copyright Ford Motor Company 1990 All rights Reserved

institutionalizing FORD MOTOR COMPANY'S commitment to customer driven product development.

FORD ACKNOWLEDGES IT HAS A PROBLEM

In the automotive industry, during the early 1980's the issues were clear. Companies were facing global competition, increasing labor and investment costs, shorter product life cycles, government regulation, rising consumer expectations and productivity disappointments. Headcount was restricted or reduced while companies were challenged by a renewed emphasis on quality and customer satisfaction. **Manufacturing and Design quality was improved using "find and fix" methods, usually a labor intensive and time consuming process. With restricted headcount, not enough time or resources were available to work upstream in the product development process to prevent these problems from happening again.**

NEED FOR TRANSFORMATION

During the 1980's Ford Motor Company experienced a transformation. The challenge in the automotive industry came from Japan. Their success caused American companies to look more seriously at the quality methods they used. Initially, it appeared very confusing. **It seemed like every time you turned around there was some new tool, method or process guaranteed to solve all your problems.** We heard about Statistical Process Control, Quality Circles, Just-In-Time, Total Preventive Maintenance, Taguchi Methods. The list goes on. This ever growing list prompted an in-depth investigation of our best competitors.

Several high performance companies were visited in Japan as members of the Dearborn, Michigan based *American Supplier Institute*. Yet each time we came away with the impression that we were looking at different pieces to a puzzle and not sure it was all the same puzzle. We (Ford) like many companies were trying to *lift* these quality techniques or tools out of an *unknown context*. **The context is a total quality control philosophy that exists inside high performance companies in both Japan and the United States.** This was a lesson learned. FORD began to rethink its mission, values and guiding principles.

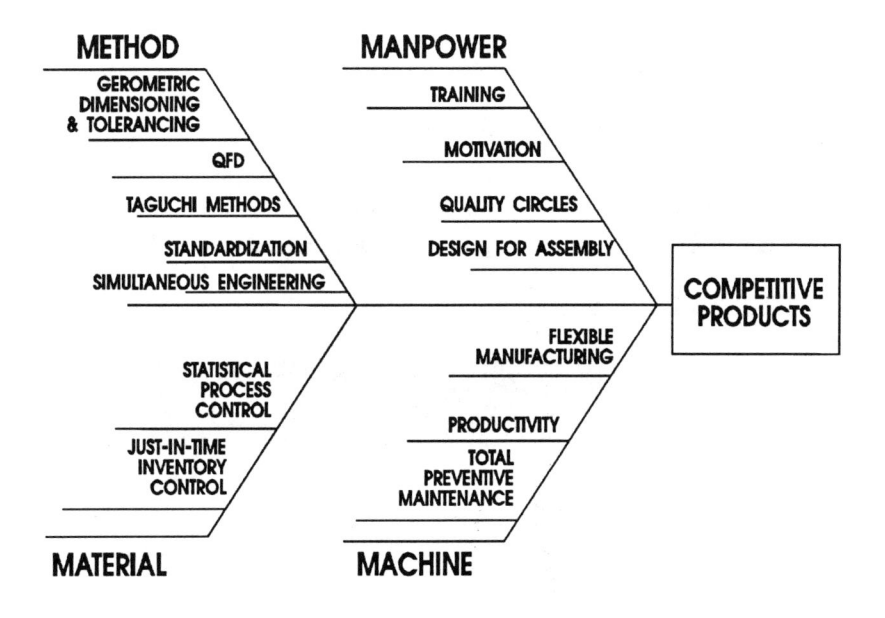

```
         METHOD              MANPOWER
         GEROMETRIC
         DIMENSIONING          TRAINING
         & TOLERANCING
              QFD              MOTIVATION
         TAGUCHI METHODS      QUALITY CIRCLES
            STANDARDIZATION   DESIGN FOR ASSEMBLY
         SIMULTANEOUS ENGINEERING                    ┌──────────────┐
                                                     │ COMPETITIVE  │
                                        FLEXIBLE     │   PRODUCTS   │
                                      MANUFACTURING  └──────────────┘
            STATISTICAL
              PROCESS          PRODUCTIVITY
              CONTROL
          JUST-IN-TIME           TOTAL
           INVENTORY           PREVENTIVE
            CONTROL            MAINTENANCE

         MATERIAL              MACHINE
```

Figure 1

THERE IS NO SHORTCUT

When the quality methods observed in Japan were organized on an Ishikawa Diagram, it started to make sense. Using an Ishikawa diagram helped identify the causes for "*Competitive Products at or below cost objectives, on time and good quality.*" The new methods observed in Japan can be organized into the major groupings of *man, machine, method, or material*. Each Japanese company was improving one or more aspects of the product development process. Improvement means to deliver competitive products on time, at or below cost objectives, and with good quality as defined by the customer.

Everyone at Ford knew we had to transform ourselves to be competitive in a global marketplace. But the task was harder than we thought. So, while Ford invested several billion dollars in its mainstream automotive business for the new Taurus/Sable, we also began working on the *giant jigsaw puzzle of quality improvement*.

Thus, began the transformation of Ford from a **quality through inspection company** to a company committed to **total quality excellence**.

Copyright Ford Motor Company 1990 All rights Reserved

WHERE DID FORD START?

Dr. Juran has said that *quality is like a circus tent with many entrances, but once you get in, the show is the same.* The entrance to the circus, i.e., *the quality method you use first* depends on your organization's strengths and weaknesses relative to competition, and where the strategic opportunities are for a competitive advantage. **Within this context, we will review the issues and opportunities that led FORD to implement a customer driven product development process supported by Quality Function Deployment.**

EXCESSIVE PRODUCT CHANGES

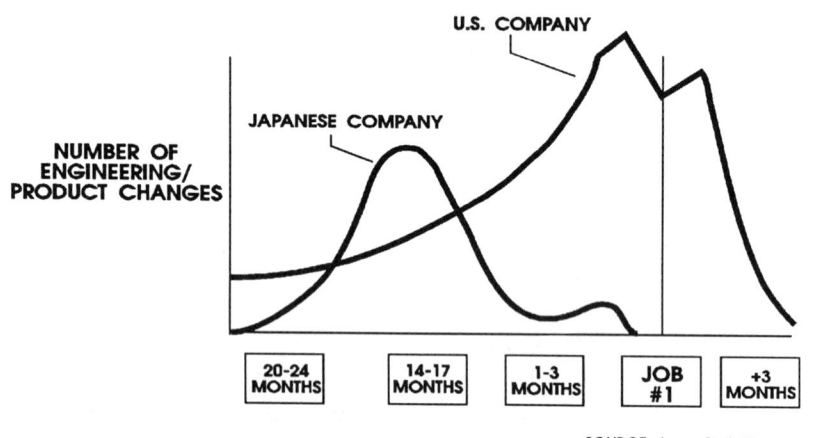

CHANGE COMPARISON

NUMBER OF ENGINEERING/PRODUCT CHANGES

JAPANESE COMPANY

U.S. COMPANY

| 20-24 MONTHS | 14-17 MONTHS | 1-3 MONTHS | JOB #1 | +3 MONTHS |

SOURCE: Larry P. Sullivan, American Supplier Institute, Dearborn, Michigan

Figure 2

An American car company was compared to its Japanese counterpart. The American car company made many more engineering changes during product development (see Figure 2). In addition, most of the changes in the Japanese company - 90% - occurred twelve to fourteen months before production start-up.

Copyright Ford Motor Company 1990 All rights Reserved

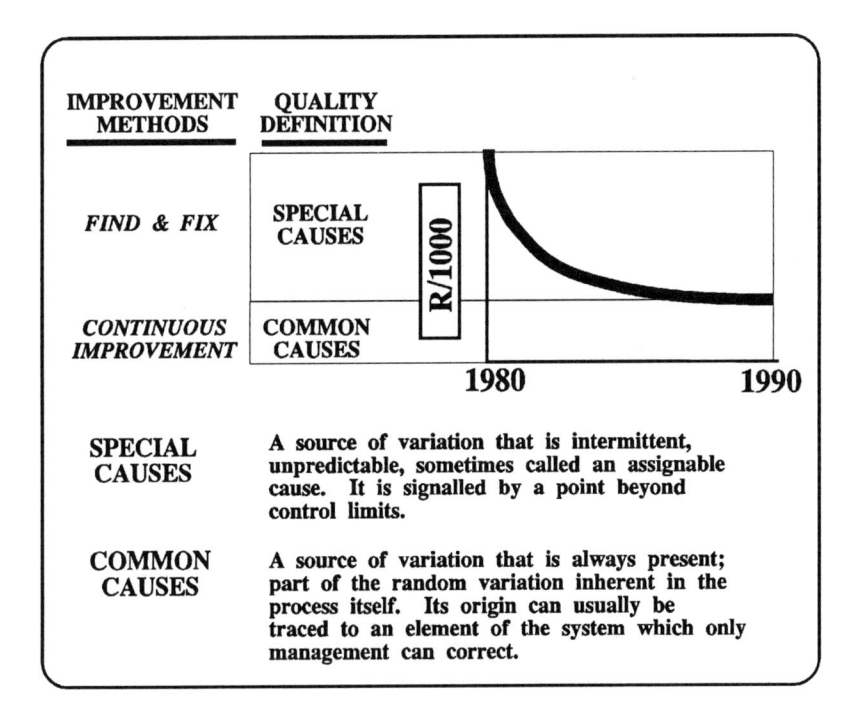

IMPROVEMENT METHODS	**QUALITY DEFINITION**
FIND & FIX	SPECIAL CAUSES
CONTINUOUS IMPROVEMENT	COMMON CAUSES

SPECIAL CAUSES	A source of variation that is intermittent, unpredictable, sometimes called an assignable cause. It is signalled by a point beyond control limits.
COMMON CAUSES	A source of variation that is always present; part of the random variation inherent in the process itself. Its origin can usually be traced to an element of the system which only management can correct.

Figure 3

PLATEAUING OF QUALITY EFFORTS

FORD began an aggressive quality improvement program in the early 1980's (see Figure 3). Initially, significant progress was achieved year after year. Problems were found and fixed. But, over time the rate of improvement decreased. Dr. W. E. Deming in his explanation of *common cause versus special cause* illustrates what was occurring. The situation is similar to the application of statistical process control to a manufacturing process. As special causes of variation are removed, the quality level being measured will stabilize, with some random variability occurring within the capability of the process. To get continued improvement, the process must be changed. *If you do what you did, you'll get what you have,* within the inherent variation of the process.

Copyright Ford Motor Company 1990 All rights Reserved

WHAT'S DIFFERENT ABOUT QFD?

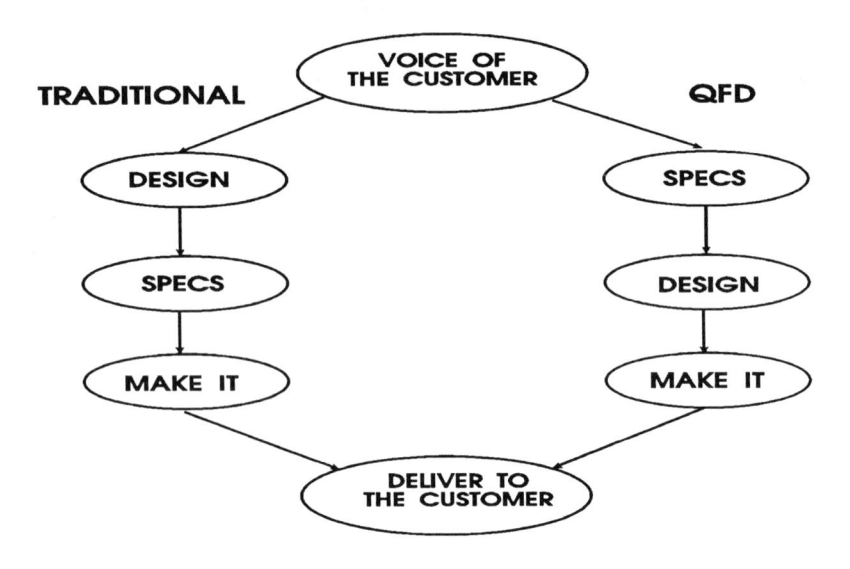

Figure 4

NEED TO REFOCUS THE DESIGN PROCESS

Design engineers are just like proud parents. Have you ever seen an ugly baby? The answer, of course, is yes. But never your baby. She is the most beautiful baby in the world. Why? SHE'S YOUR BABY! Product designs are like babies. Designers love their "babies." An examination of design processes by Dr. S. Pugh (see Figure 4) suggested a need to revise the sequence of events to insure that customer wants are clearly defined by the engineering specifications *before* proceeding with concept design. If the detail of a specific design is completed before establishing the engineering specifications, the capability of the design developed will be used to establish the specifications for product acceptance.

NEGATIVE VERSUS POSITIVE QUALITY

A little more subtle realization came to the forefront when a design engineer for clutch systems realized, through customer contact, that the focus of quality improvement was aimed primarily at "parts breaking," negative quality, than customers' desires for an "easy to operate," positive quality. Repairs per 1000 vehicles is a good quality indicator

Copyright Ford Motor Company 1990 All rights Reserved

for premature reliability failures or customer "dissatisfiers", but it does not capture the attributes of customer satisfaction. Customers want their manual shift clutch to be "easy to engage." Focusing on deviation from engineering specification of parts that were replaced by dealership mechanics did not create customer satisfaction, although it did eliminate customer dissatisfaction.

MAPPING THE VOICE OF THE CUSTOMER TO ENGINEERING SPECIFICATIONS

All too often the marketing people point their finger at engineers and say, "If you knew what you were doing, the customer would be satisfied." Conversely, engineers point their finger at marketing and say, "If marketing did their job, they would give us detailed product requirements before product development begins." Experienced QFD users understand the great philosopher POGO who said, *"We have met the enemy and the enemy is us."* If we continue with the clutch example started above, clutch design engineers knew the technical characteristics that effected customers' perception of a clutch that was "easy to engage," but had not taken the time to establish "target the best" values for those technical characteristics. The QFD approach pinpointed this bottleneck and the team created special vehicles that a sample of customers drove and rated for ease of engagement. The technical characteristics that related to customers' perception of "easy to engage" were varied from vehicle to vehicle. A designed experiment pinpointed the "target the best" value for those characteristics and established a "zone of indifference" to assist with setting engineering tolerances. Those "target the best" values judged any design alternative as to its capability to satisfy the customers' demand for an "easy to engage" manual clutch. This experience, and many others like it, cemented the notion that FORD marketing and engineering people had to work together to make a high quality car.

EXPANDED CONCEPT OF QUALITY

In 1987 FORD became aware of an expanded concept of quality, based on the work of Dr. Noriaki Kano. His model neatly compared customers' demanded quality characteristics with the company's achievement against same.

KANO MODEL OF QUALITY/FEATURES

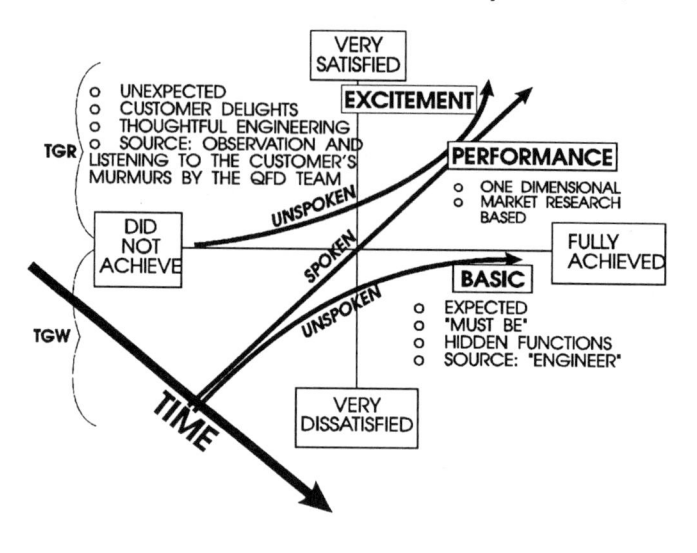

Figure 5

The *Kano Model of Quality/Features* (see Figure 5) defines three types of quality; Basic, Performance and Excitement. The model relates customer satisfaction to the degree of achievement.

Basic quality is the quality or function expected by the customer. It is unspoken unless violated. If basic quality is not achieved, the customer is very dissatisfied. Yet, when basic quality is fully achieved, the customer will not exhibit a high degree of satisfaction. The customer will tend to be neutral or ambivalent. Your car engine starting and not stalling are two examples of *basic quality*. Customers are ambivalent about these accomplishments because they are expected. If, however, the customer's car does not start or it stalls frequently the customer is very frustrated and dissatisfied with the car.

Performance quality measures how well the product meets or exceeds the customers' spoken wants. As the degree of achievement increases, customer satisfaction increases. Riding comfort and fuel economy are two examples of performance quality.

Excitement quality is best described as thoughtful engineering, providing unexpected pleasant surprises for the customer. If excitement quality is not provided by the product, the customer is not dissatisfied because the customer does not know to expect it. Yet, when excitement quality is provided, customer satisfaction will increase.

Copyright Ford Motor Company 1990 All rights Reserved

The Taurus/Sable vehicles contained many examples of excitement quality. Split sun visors and cargo nets in the trunk to hold down plastic grocery bags are two examples of excitement quality. Opportunities for excitement quality are discovered by observing how the customer uses your product and then developing options, features or functions to complement these latent customer wants.

It should be recognized that, over time, quality features move from **excitement quality** to **performance quality** to **basic quality**. An automotive example of this transition would be power steering. When initially introduced, power steering was **excitement quality**. Over time it became **performance quality** and subsequently, in many aspects, **basic quality**. New improvements in either function or performance such as variable assist power steering or four wheel steer can result in customer excitement quality.

The KANO Model of Quality/Features fit neatly into FORD's plan to improve its commitment to its customers. Quality Function Deployment and the KANO Model complemented one another.

MAJOR GOALS OF FORD'S TRANSFORMATION

Some major goals of FORD's transformation are:

1. The **number of engineering changes must be reduced** and problems uncovered in the product development process. Designs should be frozen much earlier in the development cycle. Upstream problem prevention is the goal.

2. A realization that the **systems and methods used during the product development process must be improved** to get significant improvement in quality, cost and timing.

3. The design process must be refocused to assure that the **voice of the customer is addressed early in the design process**. (The best way to capture the customer's voice it to develop *appropriate* product acceptance standards that can be used to guide design selection.) *Appropriate* implies benchmarking current and future standards to the customers' perception of both our product and our competitors. These improved design standards are then deployed into design concept selection and detailed design specifications.

4. The three dimensions of quality as outlined in the **Kano Model of Quality/Features must be consciously planned** and subsequently deployed into the product development process.

WHY DO WE NEED QUALITY FUNCTION DEPLOYMENT?

During the 1980's FORD initiated many changes in the way we conduct business. Emphasis on teamwork, employee involvement, participative management, significant cultural changes, and increased use of statistical and other analytical techniques resulted in a 65% improvement in quality from 1980 to 1988.

With QFD we realized both strategic and tactical benefits:

1. **Strategic** - QFD helped change the corporate culture to become more customer driven and team oriented.
2. **Tactical** - individual product designs and manufacturing methods were better as defined through the eyes of our customer.

The real value of QFD is its ability to direct the application of quality improvement tools and techniques. SPC, DOE/Taguchi methods and many more are extremely valuable to product improvement, but they are very time consuming and expensive. Without the *voice of the customer* as a key operating principle, the decision as to when and where to use these powerful analytical tools is left to the *voice of the engineer or executive*. The use of these tools should have the greatest impact on our ability to manufacture a product that **satisfies the demanded quality** characteristics of our customers, both external and internal.

Though QFD is an extremely flexible planning tool, there are some people who do not fully understand QFD and they characterize it as a control strategy, or a design approach, or an optimization for Quality Engineering. THEY ARE MISINFORMED! **Quality Function Deployment is <u>neither</u> a control strategy, or an approach to design, or an optimization for Quality Engineering.**

Quality Function Deployment is best characterized as a <u>planning tool</u> that identifies the significant few items on which to focus time, product improvement efforts and other resources. An important consideration in planning is that it requires *deployment* and follow up to make certain appropriate action is taken. **Existing methods and the people who have been using them are very slow to change.** Novice QFD users sometimes assume once the QFD has identified the significant few items on which to focus time and effort, the "QFD job is done." When in reality, the work has just begun! We have seen too many QFD projects identify major areas of improvement, but fail to act. Two years ago, one QFD project identified severe shortcomings in testing strategies needed to verify the design's capability of surviving its planned useful life. Two years have past and the tests have not changed. **Planning without deployment is a waste of time.**

The key word in this statement is **planning** (coupled with deployment). Planning selects the best course of action to achieve a goal. Planning requires *expert decisions* based on reliable information concerning the constraints that inhibit implementing the "optimum" approach. There are millions of decisions made to develop a new car. How many of those decisions require in-depth analysis and use of specialized analytical techniques such as Design of Experiments or SPC? All of them? None of them? Some of them? The answer is obvious -- some of them. But which ones? If the product development team has not systematically translated the voice of the customer, then the decision when and where to apply these analytical decisionmaking tools is left solely to the technical

Copyright Ford Motor Company 1990 All rights Reserved

specialist or executive. With QFD, the systematic translation of the customer's voice will identify when and where to apply these time consuming analytical techniques.

WHAT IS QFD?

At FORD the definition of QFD is:

> **Quality Function Deployment (QFD) is a planning tool for translating the customers' demanded quality characteristics (i.e., wants, needs, desires) into appropriate company requirements.** The intent of QFD is to incorporate the "voice of the customer" into all the phases of the product development cycle, through production and into the marketplace.

At FORD, quality is defined by the customer. Customers want products and service that, throughout their useful life, meet their needs and expectations at a cost that represents value. They want products that are good looking, modern, inexpensive, easy and comfortable to use, suitable for the environment, predictable and long life.

A very good synonym for QFD coined by Don Clausing, adjunct professor at M.I.T., is:

Customer Driven Product Development

The results of being customer driven are:

1. Total Quality Excellence

2. Greater Customer Satisfaction

3. Increased market share

WHAT ARE THE GOALS OF QFD?

There are two major goals of QFD:

1. **Product improvement.** Improving the product or service to increase customer satisfaction.

2. **Improve the complete product development process.** One key result of a company that uses QFD is that they will see a 1/3 to 1/2 reduction in the total product development cycle.

In order for FORD to maintain and increase its position in the marketplace, we must know the demanded quality characteristics of our customers. Yet, a thorough understanding of the customers' demands is a waste of time if these requirements cannot be incorporated into the product in a timely fashion. The time to bring an innovation to market is critical. The company that is first to the market usually get the largest market share.

Because QFD takes much more time understanding the demanded quality characteristics of the customer and defining the product in greater detail, QFD anticipates problems and avoids major "downstream" changes. If through team consensus, problems are prevented, then no one has to waste time solving the problems "downstream."

QFD ORIGINATED IN JAPAN

QFD was created in the late 1960's and one of the first industrial applications was in the early 1970's by Mitsubishi Heavy Industry at the **Kobe Shipyards in Japan.** Their products were oceangoing vessels built to military specification. The shipbuilding business requires a significant capital outlay to produce just one ship. This fact, combined with stringent government regulations, led the Shipyard's management to commit to some form of thorough upstream quality assurance.

To ensure that all government regulations, critical characteristics and customer requirements were addressed in their design, the Kobe engineers developed a QFD matrix that related these items to control factors of how the company would achieve them. The matrix also showed the relative importance of each entry, making it possible for the more important items to be identified and prioritized to achieve a greater share of available resources.

Other Japanese companies began using QFD in the mid 1970's. In the automotive industry, Toyota applied QFD to one specific problem, rust. Since that rust study, QFD usage has expanded and is now making inroads into many American businesses. It has become so popular because:

> 1. It is an aid in ensuring that products and processes are designed right the first time.
> 2. It is a logical next step for companies with a company wide approach to quality control.
> 3. It identifies the significant few decisions that require extraordinary effort and expertise to make the best decision based on the voice of the customer.

WHY QFD AT FORD MOTOR COMPANY

Ford Motor Company management concluded that the product development process must be improved by:

> 1. Changing from a "find and fix" method of product improvement to a "prevent" mode of operation.
>
> 2. Institutionalizing a customer focus in our product development process.
>
> 3. Continuing the commitment to continuous, never ending improvement.

EVOLUTION OF QFD AT FORD

Ford Motor Company initially became aware of QFD through visitations of management personnel to Japan. The initial pilot projects started in 1985. Mr. L. Sullivan, American Supplier Institute, and Dr. D. Clausing, adjunct professor at M.I.T., helped provide a broad management awareness of the benefits of QFD. In 1987, QFD was defined as a Concept to Customer (CTC) process tool and the QFD Subcommittee of the C-T-C Education and Training Committee was formed. During 1987, a review of QFD implementation findings at Ford resulted in the formation of a QFD Implementation Strategy Study Group in 1988. The study group membership represented all activities within North American Automotive Operations (NAAO).

IMPLEMENTATION OF QFD

Implementation of QFD at Ford to date can be viewed as a six phase implementation plan. The first five phases occurred from 1985 through the formation of the QFD Implementation Strategy Study Group in 1988. Phase VI, dealing with institutionalizing QFD in the organization, will be addressed later in this paper. Initial implementation focused on developing:

1. Awareness
2. Buy-In
3. Capability
4. Pilot Programs
5. Infrastructure
6. Consistent Ongoing Effort

1. AWARENESS

Awareness of QFD was achieved initially through three and four hour presentations by outside consultants. Organizations throughout NAAO were invited to send representatives. Participation by both management and general salary role personnel was encouraged. Subsequently, Ford personnel provided presentation and results of case studies.

2. BUY-IN

Initial buy-in to the pilot project was achieved by requesting volunteers. In many instances, senior management people would commit to conducting a pilot project within their activity and then through discussions with their staff, decide which of their activities

Copyright Ford Motor Company 1990 All rights Reserved

would take the lead role. Organizational buy-in is evolving over time as applications prove the effectiveness and value of the QFD methodology.

3. CAPABILITY

Training facilitation support and computer software were developed and pilot projects provided hands on capability.

4. PILOT PROGRAMS

Pilot projects were initiated in many organizations and at many levels of vehicle complexity, from part to total vehicle, to evaluate the process and gain a better understanding of how to do QFD.

5. INFRASTRUCTURE

The complexity of an organization as large as Ford required multi level networks of QFD users and support people to exchange information about pilot QFD projects. These networks were vital in accelerating momentum and instrumental to institutionalizing QFD within Ford.

PEOPLE TRAINED AND PROJECTS STARTED

Information on the evolution of training and QFD projects at Ford is provided in the following table:

Date	Number Trained	Number Projects
January, 1986	50	3
January, 1987	200	25
January, 1988	800	67
January, 1989	2,800	150
January, 1990	4,400	224
January, 1991	5,400*	400

* Projected

APPLICATIONS OF QFD

The initial applications of QFD at Ford were for *problem identification and resolution* of manageable size projects. These *problem identification and resolution* QFD projects created an experienced cadre of FORD people who successfully applied QFD in a simple, but not trivial area. Subsequent efforts addressing major *problem prevention* (upfront engineering) projects were much easier to complete. The latest applications of QFD at Ford are using QFD to support total vehicle planning. It is doubtful that total vehicle QFD could have been the first QFD projects because of the newness of QFD and the complexity of its application.

**APPLICATIONS OF QFD...
A SYSTEMS APPROACH**

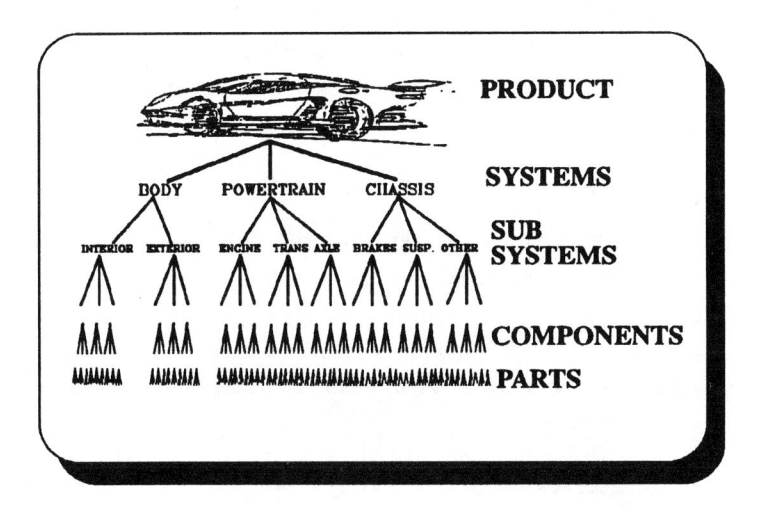

Figure 6

At FORD North American Automotive Operations we describe the type of QFD based on their relationship to the full vehicle (see Figure 6). Using a systems approach the hierarchy is:

Level 1. Full Vehicle

Level 2. Major System such as Body, Powertrain or Chassis

Level 3. Systems such as Engine

Level 4. Components such as the distributor

Level 5. Parts such as the distributor cap

Copyright Ford Motor Company 1990 All rights Reserved

Level 6. Raw materials such as the plastic in the distributor cap

Early QFD projects were at the component level. These were manageable in scope, but revealed to us the need to use a "top down" unified system approach to the use of QFD. Components could be *best in class*, but the system they operated in as components could be *worst in class*. Today, QFD is used to direct systematically a cross functional team to focus on the voice of the customer as the driving force behind "component level" engineering decisions. Without customer input on the quality of the system (eg., How easily the car starts.), engineers will consistently debate and fingerpoint about the root cause of any systems failure. For example, *without QFD* the engine distributor design engineer when confronted with customer complaints would routinely say, "I met all my engineering specifications, it passed the design verification tests, therefore it must be another component that is causing a problem."

QFD PROCESS IMPLEMENTATION STRATEGY

On May 12, 1989, the "Quality Function Deployment Implementation Strategy Committee" provided their recommendations to the NAAO and the Diversified Products Organization "Quality Strategy Committee." The report defined the actions required to *"implement Quality Function Deployment into the product development process at Ford Motor Company's North American Automotive Operations."*

6. CONSISTENT ONGOING EFFORT (INSTITUTIONALIZING QFD AT FORD)

It was recognized that the key to institutionalizing Quality Function Deployment at Ford was the integration of QFD into the **Concept-To-Customer** (C-T-C) process. Concept-To-Customer is a Ford Motor Company process developed to get:

1. Significant improvement in vehicle quality
2. Significant reduction in program execution costs
3. Significant reduction in program execution timing

The **C-T-C process** was developed concurrent with the introduction of QFD to FORD. Essentially, C-T-C resulted from a detailed review of FORD's total product development process to figure out the improvements needed to enhance product quality, time to market and cost effectiveness of the existing product development process. It should be noted that product development process is broadly defined as the cycle from idea to first produced article.

Copyright Ford Motor Company 1990 All rights Reserved

One result of the review was the realization that the existing product development process was essentially *inspection and correction of defects found* (i.e., find-and-fix). Each assembly, part, and component was designed and evaluated. Vehicles were then built from the assemblies and parts and the performance of the vehicle and systems were evaluated by tests. The evaluation criteria was defined by existing company objectives for function, performance and other quality attributes. Often these objectives were established in terms of maximum/minimum acceptable levels (i.e., *be no worse than specifications*) and not optimum target values related to customer wants. An example would be a parking brake. Foot pedal effort criteria of 125 foot pounds maximum for parking the vehicle on a hill. The design criteria should define the optimum target value, in terms of the customer want, for parking brake pedal effort. This suboptimization at the part and assembly level resulted in many engineering issues and concerns during the vehicle evaluation process. In addition, since the vehicle evaluations occur somewhat late in the program when resources and tooling were already committed, it became much harder and costly to correct these concerns and achieve optimum quality levels. Often trade-off decisions were required. Sometimes performance and excitement quality attributes were compromised to insure adequate functional and other performance wants of the customer were achieved.

Because of the review, the C-T-C process was developed. C-T-C operating principles included:

1. Customer Focus

2. Senior Management Involvement

3. Coordinated Planning

4. Effective Program Management

5. Solid Timing Disciplines

6. Necessary Skills and Expertise

7. Early Simultaneous Engineering Effort

Quality Function Deployment is seen as one of the tools to achieve the objectives of Concept-To-Customer at Ford Motor Company.

Copyright Ford Motor Company 1990 All rights Reserved

EXECUTIVE POLICY MANDATES QFD AS PREFERRED METHOD

The May 12, 1990 meeting with the Quality Strategy Committee resulted in Executive Management acceptance of the NAAO/DPO QFD Strategy. **NAAO/DPO management agreed that QFD is the best available tool for being customer driven.** The QFD implementation strategy was recognized as the preferred method of institutionalizing a customer-driven product development process. Use of QFD changes the focus of the total product development process to a prevent mode of operation.

QFD and C-T-C are both prevent disciplines. C-T-C is a comprehensive plan to improve the product quality and reduce program cost and time. QFD complements this effort by identifying, planning and prioritizing the deployment of critical items based on the voice of the customer. The C-T-C discipline requires new and existing technology to have achieved a specified level of maturity before incorporation into a program. The QFD effort complements this effort by the identification of current product weakness and/or breakthroughs needed to meet or exceed competition.

KEY CONCEPTS OF C-T-C & QFD

CTC	QFD
CUSTOMER FOCUS	DEPLOY VOICE OF CUSTOMER
SENIOR MANAGEMENT INVOLVEMENT	MANAGEMENT PARTICIPATES & USES QFD OUPUT FOR DECISIONS
COORDINATED PLANNING	TOTAL VEHICLE FOCUS
EFFECTIVE PROGRAM MANAGEMENT	EFFECTIVE COMMUNICATION OF OBJECTIVES/TARGETS
EARLY SIMULATANEOUS EMGINEERING	CROSS FUNCTIONAL TEAMS AT PHASE I
EARLY PROBLEM IDENTIFICATION AND RESOLUTION	IDENTIFIES BOTTLNECKS, PRIORITIZES RESOURCES
OBJECTIVE MEASURES AND COMPETITIVE BENCHMARKS	COMPETITIVE COMPARISION CUSTOMER AND ENGINEERING

Figure 7

The role of QFD in supporting the objectives of the C-T-C process can best be illustrated by comparison of the key concepts of each (see Figure 7).

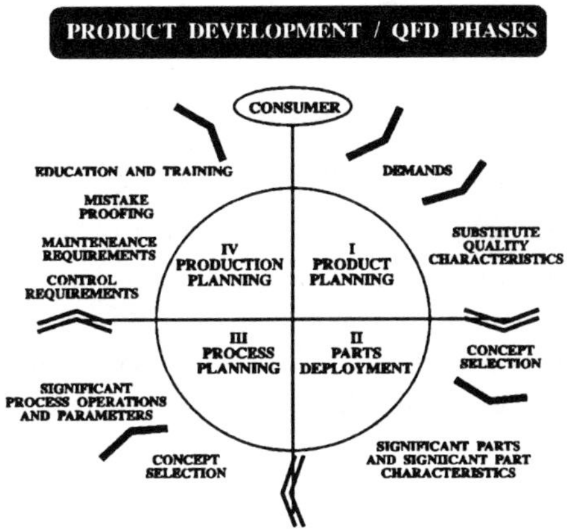

PRODUCT DEVELOPMENT / QFD PHASES

Figure 8

The relationship of the four phases of QFD to the product development process is best described by the accompanying illustration (see Figure 8).

During product planning (Phase I), customer demanded quality characteristics or wants are defined and related to substitute quality characteristics (SQC's). SQC's are the overall evaluative criteria used within the organization to define acceptability of the product. Targets for the SQC's are established based on competitive benchmarking and the customers' competitive assessment. Parts Deployment (Phase II) uses the SQC's defined in Phase I to evaluate and select a design from two or more alternatives and then identify the relationship between the SQC's and significant part characteristics.

Process Planning (Phase III) deals with the process concept selection and identification of critical process operations and parameters that cause the significant part characteristics identified in Phase II to be accomplished.

Production Planning (Phase IV) identifies control requirements, maintenance requirements, and also mistake proofing and operator training issues.

Phases I and II are "upstream" product design quality activities. New experiences for marketing and design people. Phases III and IV are a more traditional "downstream" quality control.

EXPECTED BENEFITS

The benefits of the QFD implementation strategy are:

Copyright Ford Motor Company 1990 All rights Reserved

1. **Common objectives** throughout the organization that are based on the voice of the customer.

2. An **integrated approach** to total systems and vehicle engineering.

3. **Improved effectiveness** will result from a consistent strategy.

4. The strategy will make the QFD effort more manageable by **focusing the effort** on new, important, and difficult market requirements.

The goal, over time, is to achieve continuous improvement in products and services. Improved quality should result. **The definitive judgement of our success shall be based primarily on the customer's assessment. Their perception is everything.** As we proceed on this path of continuous improvement we should become more efficient and reduce our costs.

SUCCESS CRITERIA

A question that is asked often is *"What is the criteria for successfully integrating QFD in everyday product development activities?"* The way to get this goal of continuous improvement is best illustrated by a Japanese short story read while in Japan.

> *There once was a rich and foolish man who envied the third floor of his neighbor's house. So, he hired a carpenter to build him one just like it. Upon returning home from a trip several weeks later, he found that the carpenter had completed the foundation of the first floor. "What are you doing"? He asked. "I wanted only the third floor!" The carpenter looked at him and shook his head while slowly replying, "You are indeed a rich, but foolish man. Do you not know that you need a first floor to support the second and a second floor to support the third floor? And most important is the foundation on which to build..."*

SUCCESS CRITERIA

1. CLEAR VISION
2. CONSISTENT PURPOSE
3. MOTIVATED WORKFORCE
4. CONTINUOUS NEVER ENDING IMPROVEMENT

Figure 10

Copyright Ford Motor Company 1990 All rights Reserved

The foundation of a house of quality is a management team with a **clear vision**. A vision of what **must** be done and a vision of what **can** be done.

The *first floor of a house of quality* is a **consistency of purpose**. Consistency of purpose that results in teamwork and timely communication.

The *second floor of a house of quality* is **motivation of the workforce**. Motivation at an individual level to internalize the goals and objectives of the organization.

Having established the *foundation and the first two floors of the house, the third floor of the house of quality*, **continuous never ending improvement**, can be established.

At FORD, we have defined *our mission*:

"*Our mission is to improve continually our products and services to meet customer needs...*"

We have established *our values*:

- o People
- o Products
- o Profits

And we are implementing our *guiding principles:*

- o Quality comes first
- o Customers are the focus of everything we do
- o Continuous improvement is essential to our success
- o Employee involvement is our way of life
- o Dealers and Suppliers are our partners
- o Integrity is never compromised

At Ford we are building a house of quality. QFD is a major part of that effort.

FORD MOTOR COMPANY'S mission is to "... *improve continually our products and services to meet customer needs...*" QFD provides a **planning and deployment tool** to accomplish that mission.

Copyright Ford Motor Company 1990 All rights Reserved

The question is, "Where is FORD going?" FORD is running with a strong conviction to be competitive and has the confidence and dedication to accomplish its mission.

THE CHALLENGE

Other companies must take the QFD concepts discussed in this paper and evaluate them against the systems and methods they presently use. The challenge is clear. **Beat these concepts, or, <u>you must use them</u>!**

BIBLIOGRAPHY

Akao, Yoji, 1987 ed., <u>Quality Deployment: A Series of Articles</u> (translated by Glen Mazur), GOAL/QPC, Methuen, Mass.

Clausing, Don, and John R. Hauser, "The House of Quality," <u>Harvard Business Review</u>, 1988, No. 6. (May-June 1988)

King, Bob, 1987, <u>Better Designs in Half the Time: Implementing QFD Quality Function Deployment in America</u>, GOAL/QPC, Methuen, Mass.

Sullivan, Lawrence P., "Quality Function Deployment." <u>Quality Progress</u>, Vol. XXI, No. 6 June 1988